FOOD MICROBIOLOGY AND BIOTECHNOLOGY

Safe and Sustainable Food Production

FOOD MICROBIOLOGY AND BIOTECHNOLOGY

Safe and Sustainable Food Production

Edited by

Guadalupe Virginia Nevárez-Moorillón
Arely Prado-Barragán
José Luis Martínez-Hernández
Cristóbal Noé Aguilar

APPLE
ACADEMIC
PRESS

Apple Academic Press Inc.
4164 Lakeshore Road
Burlington ON L7L 1A4
Canada

Apple Academic Press, Inc.
1265 Goldenrod Circle NE
Palm Bay, Florida 32905
USA

© 2020 by Apple Academic Press, Inc.

First issued in paperback 2021

Exclusive worldwide distribution by CRC Press, a member of Taylor & Francis Group

No claim to original U.S. Government works

ISBN 13: 978-1-77463-487-5 (pbk)
ISBN 13: 978-1-77188-838-7 (hbk)

Library and Archives Canada Cataloguing in Publication

Title: Food microbiology and biotechnology : safe and sustainable food production / edited by Guadalupe Virginia Nevárez-Moorillón, Arely Prado-Barragán, José Luis Martínez-Hernández, Cristóbal Noé Aguilar.
Names: Nevárez-Moorillón, Guadalupe Virginia, editor. | Prado-Barragán, Arely, editor. | Martínez-Hernández, José Luis, editor. | Aguilar, Cristóbal Noé, editor.
Description: Includes bibliographical references and index.
Identifiers: Canadiana (print) 20200197606 | Canadiana (ebook) 20200197630 | ISBN 9781771888387 (hardcover) | ISBN 9780429322341 (ebook)
Subjects: LCSH: Food—Microbiology. | LCSH: Food—Biotechnology.
Classification: LCC QR115 .F66 2020 | DDC 664.001/579—dc23

....

CIP data on file with US Library of Congress

....

Apple Academic Press also publishes its books in a variety of electronic formats. Some content that appears in print may not be available in electronic format. For information about Apple Academic Press products, visit our website at **www.appleacademicpress.com** and the CRC Press website at **www.crcpress.com**

About the Editors

Guadalupe Virginia Nevárez-Moorillón, PhD
Professor, School of Chemical Sciences,
Autonomous University of Chihuahua, Mexico

Guadalupe Virginia Nevárez-Moorillón, PhD, is a bacteriological chemist. She has been full-time Professor at the School of Chemical Sciences, Autonomous University of Chihuahua, Mexico, for the last 30 years. She is a member of the National Outstanding Researchers System of Mexico (Level II) and has been granted the National Award on Food Science and Technology. Her research interests include environmental microbiology as well as predictive microbiology applied to food systems, the role of lactic acid bacteria in traditional fermented foods, and antimicrobial properties of spices, such as clove and oregano. Dr. Nevarez-Moorillón is a member of the Institute of Food Technologists and the American Society for Microbiology and has served as Vice-President and President for the Mexican Association of Food Sciences and the Mexican Association for Food Protection. During the last 20 years, Dr. Nevárez-Moorillón has published numerous scientific articles and has reviewed manuscripts and research proposals for research publishers and organizations. She graduated as a bacteriological chemist from the Autonomous University of Chihuahua and took her PhD in Biology from the University of North Texas, doing research on bioremediation of contaminated soil.

Arely Prado-Barragán, PhD
Research Professor, Autonomous Metropolitan University,
Iztapalapa Campus, Mexico

Arely Prado-Barragán, PhD, is Research Professor at Autonomous Metropolitan University, Iztapalapa Campus (UAMI), Mexico, where she is a member of the research group in solid-state fermentation of the Department of Biotechnology. She has been a member of the National Outstanding Researchers System of Mexico since 1999 and of the Mexican Academy of Sciences since 2016. Dr. Prado-Barragán conducted two research stays—at the University of Victoria, British Columbia, Canada, and at the University of Provence in Marseille, France. She has published several articles in indexed journals, four books, several chapters, and two lab-teaching manuals. Her current research includes metabolites production by fermentative processes, valorization of agroindustry byproducts, and characterization and application of bioactive molecules in pharmaceutical and food industries. She has been leader of national and international projects such as the "Biotransformation of by-products from fruit and vegetables processing industry into valuable bioproducts," sponsored by the European Union. She is a member of the Institute of Food Technologists, Mexican Society of Biotechnology and Bioengineering, Mexican Association of Food Science, and the Mexican Academy of Research and Teaching in Chemical Engineering, she is the current President of the program the Science Behind Food. Dr. Prado-Barragán earned her BSc degree in Food Engineering at UAMI, and her PhM and PhD degrees from the Department of Applied Biochemistry and Food Science at the University of Nottingham, England.

José Luis Martínez-Hernández, PhD

Research Scientist and Full Professor, School of Chemistry,
Autonomous University of Coahuila (UAdeC), Mexico

José Luis Martínez-Hernández, PhD, is Research Scientist and Full Professor at the School of Chemistry, Autonomous University of Coahuila (UAdeC), Mexico. He is a research fellow of the nanobioscience group and food science and technology group. He has experience of more than 25 years and is recognized for developing different fermentative processes for obtaining fungal enzymes and metabolites applied in the food and agroindustry and for his leadership experience in the areas of biotechnology, nanobioscience processes, micro- and nano-immobilization, and enzyme technology. His experience focuses on the screening of fungal strains and the analysis, evaluation, and development of fermentative processes for the production of secondary metabolites, enzymes, and biomolecules for agricultural or industrial products. Dr. Martínez-Hernández earned his MSc in biotechnology and his PhD in technical sciences with a focus on bioprocesses and biotechnology from the Superior Polytechnical Institute Jose A. Echeverria in Havana, Cuba. Recently he was director of the office of technological innovation and training in the School of Chemistry at UAdeC, as well as head of the PhD program in food science and technology. He is the Head of the Graduate Studies Depatment, at UAdeC. He has participated actively in the Mexican Association of Food Science (AMECA) and Mexican Society of Biotechnology and Bioengineering (SMBB).

Cristóbal Noé Aguilar, PhD

*Dean of Research and Graduate Studies at the Autonomous
University of Coahuila, México*

Cristóbal Noé Aguilar, PhD, is Full Professor and Dean of the School of Chemistry at the Autonomous University of Coahuila, México. Dr. Aguilar has published more than 270 papers in indexed journals, more than 50 articles in Mexican journals, and 350 contributions to scientific meetings. He has also published many books and chapters. Professor Aguilar is a member of the National System of Researchers of Mexico (SNI) and has received several prizes and awards for his work, including including an Outstanding Researcher Award 2019 (International Bioprocessing Association, IBA-IFIBiop 2019); Prize of Science, Technology and Innovation Coahuila 2019; National Prize of Research 2010 of the Mexican Academy of Sciences; the Prize-2008 of the Mexican Society of Biotechnology and Bioengineering; National Prize AgroBio-2005; and the Mexican Prize in Food Science and Technology, CONACYT-Coca Cola México 2003. He is a member of the Mexican Academy of Science, the Mexican Association for Food Science & Biotechnology, the International Bioprocessing Association (IFIBiop), and several other scientific societies and associations. Dr. Aguilar has developed many research projects, including several international exchange projects. His research areas are the design of bioprocesses for bioproducts production (of importance in the horicultural/food/pharma/environmental sectors) and fermentation and & enzymatic technologies. He is associate editor of *Heliyon* and *Frontiers in Sustainable Food Systems,* and member of several editorial boards of international scientific journals.

Contents

Contributors .... *xi*

Abbreviations .... *xvii*

Preface .... *xxiii*

Introduction .... *xxvii*

Part I: Food Microbiology 1

1. **Hurdle Technologies for the Control of Microbial Growth in Food** 3
   Addí Rhode Navarro-Cruz, Raúl Ávila-Sosa, Obdulia Vera-López, and
   Carlos Enrique Ochoa-Velasco

2. **Production of Microbiologically Safe Fruits and Vegetables** 45
   Santos García and Norma Heredia

3. **Sustainable Production of Innocuous Seafood** 61
   Facundo Joaquín Márquez-Rocha, Carlos Alfonso Álvarez-González,
   Jenny Fabiola López-Hernández, and Guadalupe Virginia Nevárez-Moorillón

4. **Microbiological Quality and Food Safety Challenges in the
   Meat Industry** 85
   Daniela Sánchez-Aldana, Tomas Galicia-García, and Martha Yarely Leal-Ramos

5. **Fermentation as a Preservation Strategy in Foods** 103
   María Georgina Venegas-Ortega, Víctor Emmanuel Luján-Torres,
   Adriana Carolina Flores-Gallegos, José Luis Martinez-Hernández, and
   Guadalupe Virginia Nevárez-Moorillón

6. **Food Preservation Using Plant-Derived Compounds** 123
   Sofia Del Rosario Romero-Ramos, Diana B. Muñiz-Márquez,
   Pedro Aguilar-Zárate, and Jorge E. Wong-Paz

7. **Lactic Acid Bacteria in Preservation and Functional Foods** 137
   Ivan Salmerón, Samuel B. Pérez-Vega, Néstor Gutiérrez-Méndez, and
   Ildebrando Pérez-Reyes

8. **Fermented Milks: Quality Foods with Potential for Human Health** 163
   Blanca Estela García-Caballero, Olga Miriam Rutiaga-Quiñones,
   Silvia Marina González-Herrera, Cristóbal Noé Aguilar,
   Adriana Carolina Flores-Gallegos, and Raúl Rodríguez Herrera

9. **Training in Food Safety Practices and Food Manufacturing for
   Safe Food Production** 193
   José Rafael Linares-Morales, Arely Prado-Barragán, and
   Guadalupe Virginia Nevárez-Moorillón

Part II: Food Biotechnology..**207**

10. **Production, Recovery, and Application of Invertases and Lipases**........209
    Deicy Yaneth López Acuña, José D. García-García, Anna Iliná,
    Mónica L. Chávez González, Ayerim Hernández-Almanza,
    Rebeca Galindo Betancourt, Cristóbal Noé Aguilar, and
    José Luis Martínez-Hernández

11. **Recent Advances in the Bioconversion of 2-Phenylethanol Through
    Biotechnological Processes for Using as a Natural Food Additive**.........231
    Itza Nallely Cordero-Soto, Sergio Huerta-Ochoa, Marwen Moussa,
    Luz Araceli Ochoa-Martínez, Nicolás Oscar Soto-Cruz, and
    Olga Miriam Rutiaga-Quiñones

12. **Natural Antimicrobials from Vegetable By-Products:
    Extraction, Bioactivity, and Stability**..249
    Ricardo Gómez-García, D. A. Campos, A. Vilas-Boas, A. R. Madureira,
    and M. Pintado

13. **Fermentative Bioprocesses for Detoxification of Agri-Food Wastes
    for Production of Bioactive Compounds**..287
    Liliana Londoño-Hernandez, Mónica L. Chávez-González,
    Juan Alberto Ascacio-Valdés, Héctor A. Ruiz, Cristina Ramírez Toro,
    and Cristóbal Noé Aguilar

14. **Lipids as Components for Formulation of Functional Foods:
    Recent Trends**...319
    Leticia Xochitl Lopez-Martinez, José Juan Buenrostro-Figueroa,
    Edwin Rojo-Gutiérrez, Hugo Sergio García-Galindo,
    and Ramiro Baeza-Jiménez

15. **Milk-Clotting Enzymes: *S. elaeagnifolium* as an Alternative Source**........355
    Néstor Gutiérrez-Méndez, José Alberto López-Díaz, Dely Rubi Chávez-Garay,
    Martha Yarely Leal-Ramos, and Antonio García-Triana

16. **Pomegranate (*Punica granatum* L.) Nutritional and
    Functional Properties**...377
    Gerardo Manuel González-González, Jesus Antonio Morlett-Chávez,
    Adriana Carolina Flores-Gallegos, Juan Alberto Ascacio-Valdés,
    Sandra Cecilia Esparza-González, and Raúl Rodríguez-Herrera

17. **Biomass Fractionation to Bio-Based Products in
    Terms of Biorefinery Concept**...395
    Marcela Sofía Pino, Lorena Pedraza Segura, Rolando Acosta,
    María Evangelina Vallejos, Elisa Zanuso, Rosa M. Rodríguez-Jasso,
    Héctor Toribio Cuaya, Javier Larragoiti Kuri, María Cristina Area,
    Debora Nabarlatz, and Héctor A. Ruiz

Index..*429*

Contributors

Rolando Acosta
Chemical Engineering School, Industrial University of Santander, Bucaramanga, Colombia

Deicy Yaneth López Acuña
Nanobioscience Group, Food Research Department, School of Chemistry,
Autonomous University of Coahuila, Saltillo Campus, 25280 Coahuila, México

Cristóbal Noé Aguilar
Research Group of Bioprocesses and Bioproducts, Department of Food Research,
School of Chemistry, Autonomous University of Coahuila, 25280 Saltillo, Coahuila, México

Pedro Aguilar-Zárate
Master Program in Engineering, Technological Institute of Ciudad Valles,
National Technological Institute of Mexico 79010, Ciudad Valles, S.L.P., México

Carlos Alfonso Álvarez-González
Academic Division of Biological Sciences, Autonomous University Juarez of Tabasco, 86150
Villahermosa, Tabasco, Mexico

María Cristina Area
Institute of Materials of Misiones (IMAM), 3300, Posadas, Misiones, Argentina

Juan Alberto Ascacio-Valdez
Research Group of Bioprocesses and Bioproducts, Department of Food Research, School of Chemistry,
Autonomous University of Coahuila, 25280 Saltillo, Coahuila, México

Raúl Ávila-Sosa
Department of Biochemistry and Food, School of Chemistry, Meritorious Autonomous
University of Puebla, 72420 Puebla, Puebla, México

Ramiro Baeza-Jiménez
Food and Development Research Center, 33089, Delicias, Chihuahua, Mexico

Rebeca Galindo Betancourt
Reseach Center for Applied Chemistry, 25294 Saltillo, Coahuila, Mexico

José Juan Buenrostro-Figueroa
Food and Development Research Center, 33089, Delicias, Chihuahua, Mexico

D. A. Campos
Universidade Católica Portuguesa, CBQF - Centro de Biotecnologia e Química Fina – Laboratório
Associado, Escola Superior de Biotecnologia, Rua Arquiteto Lobão Vital 172, 4200-374 Porto, Portugal

Dely Rubi Chávez-Garay
School of Chemical Sciences, Autonomous University of Chihuahua, 31125 Chihuahua, Chih, Mexico

Mónica L. Chávez-González
Bioprocesses and Bioproducts Group, Food Research Department, School of Chemistry,
Universidad Autónoma de Coahuila, Saltillo, 25280, Coahuila, México

Itza Nallely Cordero-Soto
National Technological Institute of Mexico, Technological Institute of Durango Department of Chemical and Biochemical Engineering, 34080, Durango, Mexico

Héctor Toribio Cuaya
Ibero-American University, 01219 CDMX, Mexico

Sandra Cecilia Esparza-González
School of Medicine, Autonomous University of Coahuila, 25000 Saltillo, Coahuila, México

Adriana Carolina Flores-Gallegos
Research Group of Bioprocesses and Bioproducts, Department of Food Research, School of Chemistry, Autonomous University of Coahuila, 25280 Saltillo, Coahuila, México

Tomas Galicia-García
School of Chemical Sciences, Autonomous University of Chihuahua, 31125 Chihuahua, Chih, Mexico

Santos García
Department of Microbiology and Immunology, School of Biology, Autonomous University of Nuevo León, San Nicolás de los Garza, Nuevo León, 66455 Mexico

Blanca Estela García-Caballero
Research Group of Bioprocesses and Bioproducts, Department of Food Research, School of Chemistry, Autonomous University of Coahuila, 25280 Saltillo, Coahuila, México

Hugo Sergio García-Galindo
UNIDA, Technological Institute of Veracruz, National Technological Institute of Mexico 91897, Veracruz, Veracruz, Mexico

José D. García-García
Nanobioscience Group, Food Research Department, School of Chemistry, Autonomous University of Coahuila, Saltillo Campus, 25280 Coahuila, México

Antonio García-Triana
School of Chemical Sciences, Autonomous University of Chihuahua, 31125 Chihuahua, Chih, Mexico

Ricardo Gómez-García
Universidade Católica Portuguesa, CBQF - Centro de Biotecnologia e Química Fina – Laboratório Associado, Escola Superior de Biotecnologia, Rua Arquiteto Lobão Vital 172, 4200-374 Porto, Portugal

Gerardo Manuel González-González
Department of Food Research, School of Chemistry, Autonomous University of Coahuila, 25280 Saltillo, Coahuila, México

Silvia Marina González-Herrera
National Technological Institute of Mexico, Technological Institute of Durango Department of Chemical and Biochemical Engineering, 34080, Durango, México

Néstor Gutiérrez-Méndez
School of Chemical Sciences, Autonomous University of Chihuahua, 31125 Chihuahua, Chih, Mexico

Norma Heredia
Department of Microbiology and Immunology, School of Biology, Autonomous University of Nuevo León, San Nicolás de los Garza, Nuevo León, 66455 Mexico

Ayerim Hernández-Almanza
Bioprocesses and Bioproducts Group, Food Research Department, School of Chemistry, Autonomous University of Coahuila, Saltillo Campus, 25280 Coahuila, México

Sergio Huerta-Ochoa
Autonomous Metropolitan University, Iztapalapa Campus, 09340 CDMX, Mexico

Anna Iliná
Nanobioscience Group, Food Research Department, School of Chemistry, Autonomous University of Coahuila, Saltillo Campus, 25280 Coahuila, México

Javier Larragoiti Kuri
Ibero-American University, 01219 CDMX, Mexico

Martha Yareli Leal-Ramos
School of Chemical Sciences, Autonomous University of Chihuahua, 31125 Chihuahua, Chih, Mexico

José Rafael Linares-Morales
School of Chemical Sciences, Autonomous University of Chihuahua, 31125 Chihuahua, Chih, Mexico

Liliana Londoño-Hernández
Bioprocesses and Bioproducts Group, Food Research Department, School of Chemistry, Universidad Autónoma de Coahuila, Saltillo, 25280, Coahuila, México

José Alberto López-Díaz
Institute of Biomedical Sciences, Autonomous University of Ciudad Juarez, Ciudad Juárez, Chihuahua, México

Jenny Fabiola López-Hernández
Academic Division of Biological Sciences, Autonomous University Juarez of Tabasco, 86150 Villahermosa, Tabasco, Mexico

Leticia Xóchitl López-Martínez
CONACYT - Food and Development Research Center, 80110, Culiacán, Sinaloa, México

Víctor Emmanuel Lujan-Torres
School of Chemical Sciences, Autonomous University of Chihuahua, 31125 Chihuahua, Chih, Mexico

A. R. Madureira
Universidade Católica Portuguesa, CBQF - Centro de Biotecnologia e Química Fina – Laboratório Associado, Escola Superior de Biotecnologia, Rua Arquiteto Lobão Vital 172, 4200-374 Porto, Portugal

Facundo Joaquín Márquez-Rocha
Regional Center for Cleaner Production, National Polytechnic Institute, Tabasco Business Center, 86691 Cunduacan, Tabasco, Mexico

José Luis Martinez-Hernández
Nanobioscience Group, Food Research Department, School of Chemistry, Autonomous University of Coahuila, Saltillo Campus, 25280 Coahuila, México

Jesús Antonio Morlett-Chávez
Department of Food Research, School of Chemistry, Autonomous University of Coahuila, 25280 Saltillo, Coahuila, México

Marwen Moussa
UMR 782 Genetic and Microbiology of Food Processes (GMPA), AgroParisTech, INRA, Université Paris-Saclay, F-78850, Thiverval-Grignon, France

Diana B. Muñíz-Márquez
Master Program in Engineering, Technological Institute of Ciudad Valles, National Technological Institute of Mexico 79010, Ciudad Valles, S.L.P., México

Debora Nabarlatz
Chemical Engineering School, Industrial University of Santander, Bucaramanga, Colombia

Addi Rhode Navarro-Cruz
Department of Biochemistry and Food, School of Chemistry, Meritorious Autonomous
University of Puebla, 72420 Puebla, Puebla, México

Guadalupe Virginia Nevárez-Moorillón
School of Chemical Sciences, Autonomous University of Chihuahua, 31125 Chihuahua, Chih, Mexico

Luz Araceli Ochoa-Martínez
National Technological Institute of Mexico, Technological Institute of
Durango Department of Chemical and Biochemical Engineering, 34080, Durango, Mexico

Carlos Enrique Ochoa-Velasco
Department of Biochemistry and Food, School of Chemistry, Meritorious Autonomous
University of Puebla, 72420 Puebla, Puebla, México

Ildebrando Pérez-Reyes
School of Chemical Sciences, Autonomous University of Chihuahua, 31125 Chihuahua, Chih, Mexico

Samuel B. Pérez-Vega
School of Chemical Sciences, Autonomous University of Chihuahua, 31125 Chihuahua, Chih, Mexico

Marcela Sofía Pino
Biorefinery Group, Food Research Department, Faculty of Chemistry Sciences,
Autonomous University of Coahuila, 25280, Saltillo, Coahuila, México

M. Pintado
Universidade Católica Portuguesa, CBQF - Centro de Biotecnologia e Química Fina – Laboratório
Associado, Escola Superior de Biotecnologia, Rua Arquiteto Lobão Vital 172, 4200-374 Porto, Portugal

Arely Prado-Barragán
Autonomous Metropolitan University, Iztapalapa Campus, 09340 CDMX, Mexico

Raúl Rodríguez-Herrera
Department of Food Research, School of Chemistry, Autonomous University of Coahuila, 25280
Saltillo, Coahuila, México

Rosa M. Rodríguez-Jasso
Biorefinery Group, Food Research Department, Faculty of Chemistry Sciences,
Autonomous University of Coahuila, 25280, Saltillo, Coahuila, México, and Cluster of Bioalcoholes,
Mexican Centre for Innovation in Bioenergy (Cemie-Bio), México

Edwin Rojo-Gutiérrez
Food and Development Research Center, 33089, Delicias, Chihuahua, Mexico

Sofia del Rosario Romero-Ramos
Master Program in Engineering, Technological Institute of Ciudad Valles,
National Technological Institute of Mexico 79010, Ciudad Valles, S.L.P., México

Héctor A. Ruiz
Biorefinery Group, Food Research Department, Faculty of Chemistry Sciences,
Autonomous University of Coahuila, 25280, Saltillo, Coahuila, México, and Cluster of Bioalcoholes,
Mexican Centre for Innovation in Bioenergy (Cemie-Bio), México

Olga Miriam Rutiaga-Quiñones
National Technological Institute of Mexico, Technological Institute of Durango
Department of Chemical and Biochemical Engineering, 34080, Durango, México

Iván Salmerón
School of Chemical Sciences, Autonomous University of Chihuahua, 31125 Chihuahua, Chih, Mexico

Daniela Sánchez-Aldana
School of Chemical Sciences, Autonomous University of Chihuahua, 31125 Chihuahua, Chih, Mexico

Lorena Pedraza Segura
Ibero-American University, 01219 CDMX, Mexico

Nicolás Oscar Soto-Cruz
National Technological Institute of Mexico, Technological Institute of Durango
Department of Chemical and Biochemical Engineering, 34080, Durango, México

Cristina Ramírez Toro
School of Food Engineering, Universidad del Valle, Cali, 25360, Valle del Cauca, Colombia

María Evangelina Vallejos
Institute of Materials of Misiones (IMAM), 3300, Posadas, Misiones, Argentina

María Georgina Venegas-Ortega
Research Group of Bioprocesses and Bioproducts, Department of Food Research, School of Chemistry, Autonomous University of Coahuila, 25280 Saltillo, Coahuila, México

Obdulia Vera-López
Department of Biochemistry and Food, School of Chemistry, Meritorious Autonomous University of Puebla, 72420 Puebla, Puebla, México

A. Vilas-Boas
Universidade Católica Portuguesa, CBQF - Centro de Biotecnologia e Química Fina – Laboratório Associado, Escola Superior de Biotecnologia, Rua Arquiteto Lobão Vital 172, 4200-374 Porto, Portugal

Jorge E. Wong-Paz
Master Program in Engineering, Technological Institute of Ciudad Valles,
National Technological Institute of Mexico 79010, Ciudad Valles, S.L.P., México

Elisa Zanuso
CEB-Centre of Biological Engineering, University of Minho, Campus Gualtar, 4710-057 Braga, Portugal

Abbreviations

AA	arachidonic acid
ACE	angiotensin-converting enzyme
ALA	a-linolenic acid
APs	aspartic proteases
A_w	water activity
BBP	bio-based polymer
BCs	bioactive compounds
BHA	butylated hydroxyl anisole
BHT	butylated hydroxyl toluene
BioN6	bio-nylon 6
BioPCL	synthesize polymers such as poly-e-caprolactone
BioPE	bio-polyethylene
BioPET	bio-polyethylene terephthalate
BioPO	bio-polyolefins
BioPP	bio-polypropylene
BioPVC	bio-poly vinyl chloride
C	carbon
CB	cocoa butter
CBA	cocoa butter alternatives
CBE	CB equivalents
CBEX	CB extenders
CBI	CB improvers
CBR	CB replacers
CBS	CB substitutes
CCK	cholecystokinin
CCP	critical control points
CD	Crohn's disease
CDC	Center for Disease Control and Prevention
CDL	compounds derived from lignin
CEDA	Central de Abasto
ClO_2	chlorine dioxide
CM	chicken manure
CO_2	carbon dioxide
COFEPRIS	Comisión Federal Para la Protección Contra Riesgos Sanitarios

CP	control points
CPs	cysteine proteases
Da	Dalton
DAEC	diffuse-adhering *E. coli*
DAG	diacylglycerols
DHA	docosahexaenoic acid
DMD	Duchenne muscular dystrophy
DP	degrees of polymerization
DW	dry weight
EAE	enzymatic assisted extraction
EGEC	enteroaggregative *E. coli*
EHEC	enterohemorrhagic *E. coli*
EIEC	enteroinvasive *E. coli*
EN	enteral nutrition
EOLs	essential oils
EPA	Environment Protection Agency
EPEC	enteropathogenic *E. coli*
ETEC	enterotoxigenic *E. coli*
EU	European Union
FA	formaldehyde
FAO	Food and Agriculture Organization
FAs	fatty acids
FCM	fermented chicken manure
FDA	Food and Drug Administration
FFAs	free fatty acids
FO	fish oil
FOS	fructo-oligosaccharides
FOSHU	Food for Specified Health Uses
FSMA	Food Safety Modernization Act
FW	fresh weight
GAE	gallic acid equivalents
GAP	glyceraldehyde-3-phosphate
GFSI	global food safety initiatives
GHP	good hygienic practices
GIT	gastrointestinal tract
GMOs	genetically modified microorganisms
GMP	good manufacturing practices
GRAS	generally recognized as safe
H_2O_2	hydrogen peroxide
HACCP	hazard analysis and critical control point

HCN	hydrogen cyanide
HHP	high hydrostatic pressure
HMF	human milk fat
HMF	hydroxymethylfurfural
HPHT	high pressure, high temperature
HSPs	heat shock proteins
HTPEF	high temperature pulsed electric field
HUS	hemolytic uremic syndrome
IBDs	inflammatory bowel disease
ILCNC	International Lipid Classification and Nomenclature Committee
ISPR	*in situ* product removal techniques
IZSPLV	Istituto Zooprofilattico Sperimentale of Piemonte, Liguria, and Valle d'Aosta
KAB	knowledge-attitude-behavior
LA	linoleic acid
LaA	lauric acid
LAB	lactic acid bacteria
LC	liquid culture
LCFA	long-chain fatty acids
LC-PUFAs	long-chain unsaturated fatty acids
LCTs	long-chain triacylglycerols
LDH	lactate dehydrogenase
LE	lipid emulsions
LF	liquid fraction
MAE	microwave-assisted extraction
MAF	mango seed almond fat
MCA/PA	milk-clotting activity to proteolytic activity
MCFA	medium-chain fatty acids
MCTs	medium-chain triacylglycerols
MDF	medium density fiberboard
MFG	milk-fat globules
MFGM	milk-fat globules membrane
MNP	magnetic nanoparticles
MUFA	monounsaturated fatty acid
MW	microwaves
NAD	nicotinamide-adenine-dinucleotide
NEC	necrotizing enterocolitis
NLC	nanostructured lipid carriers
NOAA	National Oceanic and Atmospheric Administration

OA   oleic acid
OFMSW   organic fraction of municipal solid wastes
OO   olive oil
PA   palmitic acid
PDO   protected designation of origin
PEF   pulsed electric field
PHAs   polyhydroxyalkanoates
PKC   protein kinase C
PL   phospholipids
PMF   palm oil mid fraction
PN   parenteral nutrition
PO   palm oil
POP   palmitic, oleic, palmitic
POS   palmitic, oleic, stearic
PROFEPA   procuraduria federal de protección al ambiente
PS   polystyrene
PUFA   polyunsaturated fatty acids
QPS   qualified presumption of safety
RF   radiofrequency
RTE   ready to eat
SAF   surface adhesion fermentation
SCFA   short-chain fatty acids
SDS   sodium dodecylsulfate
SEMARNAT   Secretaría De Medio Ambiente y Recursos Naturales
SF   submerged fermentation
SFA   saturated fatty acid
SFE   supercritical fluid extraction
SL   structured lipids
SLN   solid lipid nanoparticles
SLP   solid lipid particle
SO   soybean oil
SOS   stearic, oleic, stearic
SPs   serine proteases
SSC   solid-state culture
SSF   solid-state fermentation
SSOP   sanitation standard operating procedures
STEC   Shiga toxin producer *E. coli*
STP   sodium tripolyphosphate
TAG   triacylglycerols

TPA	tetradecanephorbol 13-acetate
UAE	ultrasound-assisted extraction
UC	ulcerative colitis
UF	ultrafiltration
USDA-FSIS	United States of Agriculture's Food Safety and Inspection Service
VTEC	verotoxic *E. coli*
XOS	xylooligosaccharides

Preface

Food Microbiology and Biotechnology: Safe and Sustainable Food Production is a modern book published by Apple Academic Press, presenting original research contributions and scientific advances in the field of food microbiology and food biotechnology. Chapters cover broad research areas and offer original and novel highlights in microbiology and biotechnology and other related sciences. The book was born after several working meetings of the editors, where the needs of students, teachers, and researchers were analyzed in detail. An important characteristic of the editors is that they are all professors and scientists of several Mexican universities, and they have more than 20 years of experience in teaching and research, therefore, understanding the requirements and needs, both for updating and for the depth. The training is part of the context and the reality faced by the courses of microbiology and food biotechnology; so the book was born.

The book explores the most important advances in food microbiology and biotechnology, making special emphasis on the challenges that the industry faces in the era of sustainable development and food security problems. At least five of the seventeen sustainable development goals are related to safe food production, clean water, industrial innovation, and responsible consumption; they all are related to food production, innocuous food for all, reduction of food loss and waste, and reuse of food residues. These challenges are at the end, related to food microbiology and food biotechnology. A detailed and up-to-date revision is a useful tool in teaching and research related to sustainable agriculture and food production. This book demonstrates the potential and actual developments across the innovative advances in food microbiology and biotechnology.

Important advances are currently developed in food microbiology and biotechnology, which are solving challenges and opening up numerous opportunities for the food industry. The targeted use of microbiological and biotechnological methods can, among other strategies, support the increment of quantity and quality reducing the number of nutritional and technological problems associated to the food industry as well the use of inconvenient ingredients. Omic research studies and targeted breeding also greatly facilitate the progress in food microbiology and biotechnology contributing significantly to saving resources, optimizing harvest yields, and producing better foods. One of the most important challenges of the world population, which grows as never before, is the food production and supply in a sustainable and complex framework attending severe requirements of quality and nutritional demands. The necessity of new healthy food represents highly challenging issues for the food industry.

Current developments in food microbiology cover the relationship between the production, processing, service, and consumption of foods and beverages with the bacteriology, mycology, virology, parasitology, and immunology. The modern food scenario requires knowing the incidence and types of food and beverage microorganisms, microbial interactions, microbial ecology of foods, intrinsic and extrinsic factors affecting microbial survival and growth in foods, and food spoilage; the microorganisms involved in food and beverage fermentation (including probiotics and starter cultures); food safety, indices of the sanitary quality of foods, microbiological quality assurance, biocontrol, microbiological aspects of food preservation and novel preservation techniques, predictive microbiology, and microbial risk assessment; foodborne microorganisms of public health significance and microbiological aspects of foodborne diseases of microbial origin; methods for microbiological and immunological examinations of foods, as well as rapid, automated and molecular methods when validated in food systems; and the biochemistry, physiology, and molecular biology of microorganisms as they directly relate to food spoilage, foodborne disease, and food fermentations.

The food biotechnology field contributes enormously to the production of nutritious food and already has a long tradition in the food sector. People have used the catalytic properties of microorganisms and their enzymes in food production more or less consciously for thousands of years. The economic importance of food biotechnology becomes particularly evident when one looks at food products produced with the biotechnological method, permitting that biotechnology has become a major factor in the food processing industry. Food biotechnology employs state-of-the-art methods that make a significant contribution to making food safer, more tolerable and palatable.

This book is the sum of several works about the topics of food microbiology and biotechnology divided into two important areas: (i) food microbiology and (ii) food biotechnology. The book includes relevant information about food bioprocesses, fermentation, food microbiology, functional foods, nutraceuticals, and extraction of natural products, nano- and micro-technology, innovative processes/bioprocesses for utilization of by-products, alternative processes requiring less energy or water, among other topics. Due to its qualified and innovative content, there is no doubt it will be an important reference for food technology research, for undergraduate and graduate students, and also for professionals in the food industry.

We are sure that the topics covered by the book will be of great interest and support for readers, because they have been particularly focused to meet the demands and needs of students, teachers, and researchers of science and technology in the areas of biotechnology and food microbiology.

We deeply thank all the authors who responded enthusiastically to the call issued by contributing original and novel documents. Finally, we also appreciate the facilities granted by the institutions for which we work, the Autonomous University of Chihuahua, the Autonomous Metropolitan University, and the Autonomous University of Coahuila.

Guadalupe Virginia Nevárez-Moorillón, PhD
Arely Prado-Barragán, PhD
José Luis Martínez-Hernández, PhD
Cristóbal Noé Aguilar, PhD

Introduction

GUADALUPE VIRGINIA NEVÁREZ-MOORILLÓN,[1]
DANIELA SÁNCHEZ-ALDANA,[1] JOSÉ L. MARTÍNEZ-HERNÁNDEZ,[2]
ARELY PRADO-BARRAGÁN,[3] AND CRISTÓBAL NOÉ AGUILAR[1]

[1]Facultad de Ciencias Químicas, Universidad Autónoma de
Chihuahua, 31125, Chihuahua, México

[2]Food Research Department, School of Chemistry,
Universidad Autónoma de Coahuila, 25280, Saltillo, México

[3]Department of Biotechnology, Universidad Autónoma Metropolitana,
Iztapalapa, México City, 09340, México

Since ancient times, human societies have been closely related to the ability to acquire enough food, in such a way that not only the basic survival needs were met, but also the conservation of the same for its total disposition at any time and place and be able to devote time to the arts, crafts, and sciences. The development of one of the oldest activities of the human being, agriculture, is closely linked to the ability to preserve food, first by techniques developed over the centuries by observation type trial and error, and more recently by the application growing science and engineering.

The basis of these advances is our knowledge of food microbiology and biotechnology. Long before the description of the living animals for the first time, many of the conditions that controlled microbial spoilage had been identified empirically. However, it was the emergence of the science of microbiology and biotechnology that promoted the preservation of food from an art to a science, allowing food to be processed, distributed and marketed with a high degree of confidence concerning product quality and safety. Thus, the food microbiology and biotechnology have been an important part of the discipline since its early days.

The scopes of food microbiology and biotechnology are highly inclusive since it interacts with all subdisciplines among which we can find, the public health microbiology, microbial genetics, fermentation technologies, microbial physiology, and biochemistry. Also, food microbiologists/biotechnologists have been at the forefront of many microbiological concepts and advances.

The development of biofilms and the ability to detect low numbers of metabolically stressed microbes from highly complex matrices are two areas where food microbiologists are providing critical ideas in the behavior of microbiological systems. In addition, new research topics have been raised because of the unique challenge, in the face of food microbiologists, such as predictive microbiology, probiotics, microbial risk assessments, and natural antimicrobials.

This book has been prepared to offer updated and detailed scientific information on Food Microbiology. The book is organized into 10 main sections or chapters, five of which focus on the two fundamental aspects of the subject: food and microorganisms.

Each section consists of detailed information, from the generalities to the particular aspects of each topic of relevance, including basic microbiology, safety, food, pathogenic microorganisms, food preservation, sanitation and hygiene procedures, etc.

The microbial diversity found in foods is described from the classification by kingdoms and the main groups of microorganisms present in them. Although the main topic discussed is about microbial food pathogens, the book also covers another important aspect of food microbiology, such as food ecosystems and measures to prevent and control food, diseases transmitted by food consumption, etc.

Uncontrolled and unwanted microbial growth destroys large quantities of food, causing significant losses both economically and with respect to nutrient content. Besides, the consumption of food contaminated with particular microorganisms or microbial products can also cause serious illnesses, such as food-mediated infections and food poisoning. Every minute, there are around 50,000 cases of gastrointestinal diseases, and many individuals, especially children, die of these infections. The most important preventive measures are for the development and continued application of effective interventions to improve overall food safety.

Food biotechnology is undoubtedly one of the key technological areas in contemporary food industry development. The term food biotechnology is considered as the set of techniques that use living organisms or substances from them to develop or modify a food product, improve plants or animals, or to develop microorganisms for food production.

The ancient stage of food biotechnology can be considered empirical and is when biotechnology was born with the establishment of human societies and their need to develop organisms that would allow them to maintain food, industry and achieve territorial expansion. A second important stage referred to as the transition phase is presented with the intervention of science and technology in the development of biotechnological industries that contribute to the development of the great empires. And the third stage occurs with the birth of modern biotechnology occurs with the conjunction of two relevant situations: the first is the emergence of molecular biology, a discipline that helped on the discovery of the DNA structure. The genetic material of living cells and the genes that conform it, as well as of the mechanisms to translate the genetic information that is located in the DNA, in proteins.

This set of knowledge allows today, to have a precise subcellular image of the functioning of the living cell. The second situation of molecular biology is the awareness that science is transformed into a much more multi-disciplinary type of activity giving the convergence of various strategies, knowledge, and tools, glimpsing success to solve scientific and social problems. For thousands of years, humanity has been doing biotechnology in an empirical way, which only in modern times acquires a scientific basis.

The human population is growing dramatically, and it will probably reach 10 billion world inhabitants sooner than estimated. The challenges that will be faced soon might not even be considered now, but certainly many will be related to food production. The need for safe and sufficient food for all humans requires not only efficient food production and distribution system but also to assure its microbiological safety, nutritional value, and equal distribution worldwide.

Along with food production, several challenges are also related to assuring Food Security and are related to universal access to health care, clean water, clean energy, proper housing, and basic education. In recent years, the main challenges facing humanity have been described in the 17 Sustainable Development Goals, with a dateline of 2030 to achieve those goals.

In order to provide strategies to achieve the Sustainable Development Goals, the scientific community is proactively working on new food and nutrition solutions to ensure global food sustainability and nutrition security in the future. For this purpose, innovative solutions need to be considered throughout the whole food chain inclusive of food choices and dietary patterns in order to make any significant improvements in the food supply, nutritional, and health status. In the case of foods

This book has two complementary areas, Food Microbiology, and Food Biotechnology. Particular emphasis has been focused on the most important challenges in both areas analyzing them from the perspective of Sustainable Development. Food and microorganism's relationship can be of different nature: in one hand, food can be the means of transportation of pathogenic microorganisms, but also, microorganisms can be responsible for food spoilage. Foodborne pathogens, like many other human pathogens, are in constant evolution, so there are always new strains or groups that are identified as emerging pathogens. Evidently, there is a constant need to develop new identification methods for measuring microbial activity in foods. The production of safe food, from a sustainable point of view, is also included for some of the most important groups of foods. A review of the most advanced knowledge of food preservation is also included, as well as the training programs for safety in the food industry.

The up to date advances in food fermentation for food production or preparation of bioactive compounds is analyzed. Of particular importance is the production of food enzymes to be used in the preparation of food supplies. The mass production of food has led to the generation of large amounts of agroindustrial by-products. The use of those byproducts to obtain valuable compounds is one of the most important actual challenges. Therefore, the book will cover the biotechnological processing of agroindustrial residues to obtain phenolic compounds, organic acids or peptides. Biorefinery, as the process to obtain valuable products from residues, even for the production of energy, is also analyzed in this section.

In the first section dedicated to food Microbiology, we have included information describing the traditional and novel food processes used in hurdle technologies: an analysis of the most common food safety problems regarding produce production, the main pathogenic bacteria involved, and the most effective methods of pathogen detection and control. Also, a review of the sustainable production of innocuous seafood; a special section of the microbiological quality and food safety challenges in meat industry; a study about some of the most important benefits of preservative-fermentation, including economic aspects; an evaluation of the present use of plant-derived compounds in the preservation of food, and pay particular attention to the primary agents that cause spoilage and represents a risk for food products. It is also included, a review on the use of lactic acid bacteria in the preservation and functional foods; an analysis of the fermented milk as quality foods with potential for human health; and finally, the relevance of the training in good safety practices and food manufacturing for safe food production.

In the second section dedicated to food biotechnology, we have incorporated information on the production, recovery, and application of invertases and lipases; a description of the recent advances in the bioconversion of

2-phenyl ethanol through biotechnological processes for using as a natural food additive. Also, a review about the extraction, bioactivities, and stability of natural antimicrobials from vegetables by-products; a special report focused on the fermentative bioprocesses for detoxification of agri-wastes for production of bioactive compounds; an analysis of the recent trends of the use of lipids and fatty acids for obtaining functional foods; a contribution about the milk clotting enzymes; a description of the nutritional and functional properties of the pomegranate; and finally, a relevant analysis of the biomass fractionation to bio-based products in terms of biorefinery concept.

We are sure that the information provided in this book can be used as a reference for researchers, teachers, and under/post-graduate students, particularly as educational and training material. The effort made by both the editors and the authors of each contribution, allowed to elaborate this book of high scientific and technological quality in the areas of food microbiology and biotechnology. Finally, we want to thank the trust placed in this editorial project that we hope manages to pass the basic reaches of the scientific literature, becoming an educational tool in the formation of high-quality human resources.

PART I
Food Microbiology

CHAPTER 1

Hurdle Technologies for the Control of Microbial Growth in Food

ADDÍ RHODE NAVARRO-CRUZ, RAÚL ÁVILA-SOSA,
OBDULIA VERA-LÓPEZ, and CARLOS ENRIQUE OCHOA-VELASCO

Department of Biochemistry and Food, School of Chemistry,
Meritorious Autonomous University of Puebla, 72420 Puebla, Puebla,
México, Tel.: +52 222 295500, E-mail: carlosenriqueov@hotmail.com
(C. E. Ochoa-Velasco)

ABSTRACT

Microbial growth in food can cause deterioration and eventual spoilage of foods; food can also be the carrier of foodborne pathogens, causing a considerable public health problem. The control of microbial growth in food can be achieved by the modification of intrinsic and extrinsic factors, but processing technologies can have deleterious effects on food properties. The use of combined processing methods with mild antimicrobial capacities can lead to control of microbial growth, but without severe damages to food. The description of traditional and novel food processes used in hurdle technologies is revised in this chapter.

1.1 INTRODUCTION

The growth of microorganisms in food is one of the multiple causes of deterioration and is also related to foodborne diseases. The term food preservation refers to all the measures taken against any food spoilage microorganisms [1]. Unfortunately, in developing countries, more than 50% of fresh food is wasted due to the limited knowledge of food preservation [2]. Several intrinsic and extrinsic factors can affect microbial growth in food. Endogenous enzymes, substrates, sensitivity to light, oxygen content, pH, water activity (A_w), and oxide-reduction potential are considered intrinsic properties, while the relative humidity, temperature, presence, and activities of other microorganisms are extrinsic factors. The intrinsic and extrinsic factors can have either bactericidal/fungicidal or bacteriostatic/fungistatic effects. The combination of some or all of these factors and their control is one of the remaining challenges for food science [1, 3]. In recent years, it has been observed that pathogens associated with foodborne infections are more resistant and host-adapted to traditional food preservation methods. Therefore, different strategies such as the application of different factors simultaneously or sequentially called hurdles (generally known as combined methods, a combination of preservation factors, combined processes, barrier technology or combination techniques), have been developed to achieve a more gentle and effective microbial growth control. The combination of these factors can act synergistically in their antimicrobial capacity, using different mechanisms for inhibition or inactivation and assuring reduced costs of energy and maximizing processing output [4–7]. Therefore, this chapter gives an overview of the most important hurdle technologies (novel and conventional) to control foodborne and deteriorative microorganism's growth in food.

1.2 HURDLE TECHNOLOGY

Hurdle technology from the food safety point of view should be understood from two different aspects that may include conventional methods [8]:

1. **Common Methods of Food Processing and Preservation:** That implies a change in product nature that reduces microbial load or limits its growth.
2. **Conventional Methods of Extending Shelf Life:** It is based on changing the product storage and packaging conditions to inhibit microbial growth.

Both methods may include heat treatment, smoking, irradiation, freezing, canning, meat curing, as well as salting, drying, and dehydration, osmotic concentration, the addition of antimicrobials and other additives [9]. The exact method to be used will depend on the product, product safety and process facilities [10], since when hurdle technology is intelligently selected and then intentionally applied, causes a hostile environment to the microorganisms inhibiting their growth or shorter their survival or cause their death [5, 11, 12]. No single factor is responsible for making the product stable, but rather the stability results from the synergism among the combined factors [13]. The physiological response of microorganisms during food preservation (homeostasis, metabolic exhaustion, and stress reactions) is the basis for the application of hurdle technology [5, 13]. Homeostasis is the process which maintains the stability of the living cell's internal environment. Hurdle technology success fundamentally depends on guaranteeing the metabolic depletion of microorganisms; treatments to be used as barriers must be based on different principles in order to alter one or more of homeostasis mechanisms, and to prevent microorganisms from multiplying, whether they remain inactive or are eliminated [1, 5]. Therefore, each technique used as a barrier must be in proper combination and at optimum level to minimize the damage to the overall food quality [14]. Microorganisms stress reactions are active processes that involve the expenditure of energy mainly by stress proteins synthesis induced by heat, pH, aw, ethanol [15].

Metabolic combining inhibitory factors can result in a significant improvement in securing microbial safety and stability as well as the sensory, nutritional quality, and economic viability of foods [16, 17]. In recent years, food safety research has often focused on growth/no growth factors in almost all food products preserve by hurdle methods, and empirical experiments determine the number of hurdles and their types. However, it is still a challenging issue, the use of methods for assessing food stability [2, 18]. In developing countries, hurdle technology is proved to be useful for novel foods which despite minimally processing are ambient stable, as well as in modifications of traditional intermediate-moisture foods to make them ambient stable high-moisture foods [13]. The magnitude of microbial reduction achieved through the use of hurdles will vary depending on circumstances in primary production whether a substantial microbial burden (both pathogens and spoilage organisms) is assumed or not [17].

1.3   CONVENTIONAL HURDLE TECHNOLOGIES

Hurdle technology involves the use of several preservation techniques in combination, which reduces the intensive use of a conservation technique that produces a lesser impact on sensory quality. However, it should not be forgotten that a positive reaction to one or several treatments depends on the food that is used. It is necessary to carry out studies that allow identifying which is the sequence of treatments necessary to obtain a synergic effect and in this way produce a barrier effect that allows the extended shelf life of any food [19–23]. Each food process method has the goal to preserve or extend food useful life, such as the addition of preservatives as a commonly used technique. Fruits and vegetables are especially fragile and highly susceptible to rot, so there are many reports on their conservation by hurdle technology, jams, and jellies are the most classic example of hurdle technology applications, which combine the use of high temperatures, low pH, decreased A_w and anaerobic packaging to reduce the initial microbial load and inhibit the growth of survivors [24–27]. Those extreme modifications might not be necessary using hurdle technology; since obstacle technology is based on minimally processed methods, such as in mixtures of vegetables, combining disinfection, heat treatment, edible coating, modified atmosphere, and refrigeration [28–31]. In other examples, the combined methods include the addition of preservatives that are expected to have synergistic effects, along with A_w reduction, aseptic packing, with or without refrigeration [32–35]. This combination of obstacles has also been useful in bakery products [36], and reduced A_w is the fundamental obstacle for conservation. Recent applications of conventional hurdle technology are included in Table 1.1.

Heat processing is done to increase microbial food safety, is highly effective for killing microorganisms, and is frequently applied in combination with other preservation methods. However, some food products are very heat sensitive, and quality indicators such as color and texture are usually degraded during thermal treatment [1, 37]. Heat can reduce the population of fungal spores or weakened their ability to germinate and produce vegetative cells capable of overcoming. Moreover, it can bring a variety of morphological and structural changes in the outer membrane, including: the formation of vesicles and bullae on the cell surface, the release of lipopolysaccharides and proteins that alter the permeability of the barrier, causing an efflux on periplasmic enzymes, and sensitivity to surfactants and hydrophobic compounds. Mild heat sub-lethally damage microorganisms in a way that is unable to overcome other hurdles present in the food environment [3, 38, 39].

Juice pasteurization is a conventional method to preserve juices, but even when microbiological control is achieved, the quality characteristics can be adversely affected. As an alternative, thermal treatments applied in sub-lethal levels, in combination with antimicrobials were applied in juices. Sugarcane juice has low acidity, high A_w, and sugar content and therefore it deteriorates rapidly even when refrigerated. Kunitake et al. [40] combined heat treatments (85, 90, and 95°C, 30s) with passion fruit pulp in sugarcane juice, managing to reduce total coliforms, eliminate *Salmonella*, and mesophiles

TABLE 1.1 Conventional Hurdle Technology

Food	Hurdles	Microorganism	Effect	References
Acai pulp	Heat treatment, pH, and Aw reduction	Mesophiles, molds, yeasts, coliforms, and *Salmonella*	Microbiological stability for 5 months at 25°C	[26]
Pumpkin puree	Heat treatment, pH, Aw reduction, and preservatives	Mesophiles, molds, yeasts, and lactic acid bacteria	Shelf life extension for 3 weeks at 25°C	[24]
Anco pumpkin puree	Heath treatment and pH	Mesophiles, thermophiles, molds, and yeasts	Growth inhibition	[27]
Minimally processed vegetable mix	Heat treatment, edible coating, sodium hypochlorite, modified atmosphere and refrigeration	Mesophiles, molds, yeasts, coliforms, and Salmonella	Increase in shelf life of up to 7 days	[29]
Sugarcane juice	Heat treatments and the addition of passion fruit pulp	Coliforms and *Salmonella*	Reduction of coliforms and total inhibition of *Salmonella*	[40]
Green Olives	Mild heat treatment, salt, and preservatives	*Monascusruber*	The spoiling by the fungus was avoided	[34]
Orange and apple juice	Mild heat and extract combination (*Zingiber officinale* and *Xylopia aetiopica*)	*A. niger, A. flavus*, and *R. stolonifer*	Fungicide effect	[38]
Orange juice	Mild heat treatments (57–61°C), citral, and vanillin	*L. innocua*	Logaritmic reduction (5 log)	[41]
Orange juice	Mild heat treatments (54–60°C), limonene, and orange essential oil	*E. coli* O157:H7	Logaritmic reduction (5 log)	[42]

TABLE 1.1 *(Continued)*

Food	Hurdles	Microorganism	Effect	References
Sugarcane juice	Mild heat treatments with potassium metabisulfite, ginger, mint, and lemon extracts	Mesophiles, molds, and yeasts	Reduction of 90% of MO	[37]
Black truffles	Low temperatures and gallic acid	*Pseudomonas* spp.	Reduction of 90% of MO	[35]
Cauliflower	Low temperatures, salt, and preservatives	Molds and yeasts	Microbial safe till 180 days	[33]
Cauliflower	pH, salt, and preservatives	Molds, yeasts, and coliforms	Sensitive decrease in bacteria, fungi, and yeast	[30]
Analogue of fermented dough	pH, preservatives, and Aw reduction	*Eurothium* spp., *Aspergillus* spp., *Penicillium corylophilum*	Growth inhibition	[36]
Apple juice	pH, EDTA, sodium tripolyphosphate, enterocins, and mild heat	*E. coli* O157:H7	Logaritmic reduction (5 log)	[52]
Food substrates (carrot juice, chicken soup, pumpkin cream)	pH, tymol, and cymene	*B. cereus*	Inactivation	[51]

TABLE 1.1 *(Continued)*

Food	Hurdles	Microorganism	Effect	References
Cashew, soursop, peach, mango, passion fruit, orange, guava, and cupuassu	pH, nisin, and low temperatures	*S. aureus* and *L. monocytogenes*	Reduction of 90% of MO	[7]
Grated carrot	pH, Aw reduction, preservatives, and partial dehydration	Mesophiles, molds, yeasts, coliforms, and spores	Growth reduction	[32]
Cod and salmon juice	Modified atmosphere packing, chitosan coating, super chilling, lactic acid bacteria	*Shewanella baltica, Photobacterium phosphoreum, Brochothrix thermosphacta, Lactobacillus sakei, Hafnia alvei, Serratia proteamaculans, L. monocytogenes*	Inhibitory effect	[48]
Minimally processed potatoes	Modified atmosphere packaging, cold storage and chemical treatments	Mesophiles, coliforms, and spores	Shelf life extension to 2 weeks	[31]
Tomato juice	Chitosan combined with extracts of *Lithospermum erythrorhizon, Rheum palmatum, Thymus vulgaris, Lippiacitriodora, and* low temperatures	*B. cereus, E. coli, S. aureus, S. cerevisiae,*	Bacteriostatic effect	[58]

TABLE 1.1 *(Continued)*

Food	Hurdles	Microorganism	Effect	References
Orange and mango juice	*Panax ginseng* extracts in combination with malic acid and potassium sorbate	*E. coli* O157:H7 and *Salmonella enterica*	Inactivation	[57]
Watermelon juice	Blend of acidic juices (apple and lemon) with eugenol and trans-cynnamaldehyde	Mesophiles and coliforms	Reduction of 90% of MO	[43]

levels remained low; the addition of fruit pulp helps to low juice pH. On the other hand, Rajendran, and Bharathidasan [37] managed to increase the shelf life (up to 60 days) and the decrease of deteriorating microorganisms (<35 CFU/mL in BMA and <10 CFU/mL in fungi and yeasts) in sugarcane juice using the combination of mild temperatures (70°C), potassium metabisulfite (1,000 ppm) and ginger, mint or lemon extracts (0.6, 0.3 and 0.7% v/v). In orange juice, the combination of mild heat treatments (50–60°C) with natural antimicrobials (citral, vanillin, limonene, ginger, and *Xylopia aetiopica*) act synergistically in the inhibition of pathogens (*L. innocua* and *E. coli* 0157H:7) and deteriorative microorganisms (*A. niger*, *A. flavus*, and *R. stolonifer*) [38, 41, 42].

Moreover, Espina et al. [42] described that the presence of shoulders in thermal inactivation kinetics has been related to the need to surpass a certain threshold of damage for the cells to be inactivated, or to a higher cell ability to repair sub-lethal damages. Antimicrobials avoided the shoulder effect, so the cost-effectiveness of the process would be expected to rise, due to higher energy efficiency. Since the composition of fresh and commercial orange juice was different, the efficacy and antimicrobials ratio could be reduced by interaction with food components that could be active in fresh orange juice but inactive in thermally treated commercial orange juice. The magnitude of the synergistic effect does not depend on treatment temperature but increases with the concentration of antimicrobials [42].

In food science, a food product with a pH higher than 4.6, requires treatment to control bacterial growth, specifically spore-forming pathogenic growth [43, 44]. Acidic conditions may affect microorganisms' growth by interfering with the synthesis of cellular components and can be lethal because of the damage to the outer membrane, cell surface protein denaturation, cytoplasmic pH value, and the subsequent DNA and enzymes damage. The protective mechanisms of acid tolerance need further clarification, but it is known that acid shock induces pH homeostasis and protein repair systems. This inducible pH homeostasis appears to be the result of preventing net proton movement across the membrane of adapted cells rather than increased internal buffer capacity. At extremely acidic pH, proton leakage across the membrane proceeds faster than the bacterial pH homeostasis systems can remove them, leading to intracellular acidification and subsequent disruptions of biochemical processes [39, 44, 45].

Seafoods are especially delicate for conservation; their half-life from fishing time to consumption does not exceed one or two weeks, and even if some mild preservation form is applied, they can only last up to three to four weeks. Some of the techniques for preserving marine foods are with low pH, low A_w, and antimicrobials [46], biopreservation [47] or a combination of moderate conservation technologies such as biopreservation, modified atmosphere packaging and super chilling [48]. Meat products widely differ in their formulation, rheological, and sensory characteristics, so the conservation method used for one, might not be appropriate for another; hurdle technology could help reduce potentially harmful ingredients such as nitrites [2]. Among the combination of treatments used in meat products is the decrease in pH, a low A_w, [49], and vacuum packing [50].

Delgado et al. [51] reported that control of pH used in combination with thymol and cymene has a synergistic effect against *B. cereus* in food substrates similar to carrot juice, chicken soup and especially in pumpkin cream. The combination applied in watermelon juice was high acidity juices (apple and lemon) and mixture natural antimicrobials (eugenol and trans-cinnamaldehyde). This combination kept the microbial count stable below the norm values for 14 days. At low pH values, natural antimicrobials may become more hydrophobic, thus accentuating their antimicrobial activity as they can interact more thoroughly with bacterial membranes [43]. In apple juice, the combination of pH, EDTA, sodium tripolyphosphate (STP), and enterocin (AS-48) generates a permeabilization of outer membrane of *E. coli* 0157:H7 [52]; moreover, the use of nisin in different juices (with different pH values) such as cashew, soursop, peach, mango, passion fruit, orange, guava, and cupuassu, achieves 90% of *L. monocytogenes* and *S. aureus* reduction. The best reduction was obtained with mango and peach mixtures; pH may have a strong influence on bacteriocin activity, especially acidic pH. These hurdles could change peptide oligomerization as well as surface charge changes on target bacteria [7].

Preservatives are often added as an acid salt to increase their solubility in aqueous solution. Their effectiveness depends on food pH since the antimicrobial effect of the undissociated acid is much stronger than the dissociated acid. At low pH values, it is believed that the undissociated sorbic acid penetrates the cell membrane altering the internal pH of the microorganisms, causing denaturation and inactivation of the enzymes [36]. The maximum pH for the activity is around 6.0–6.5 for sorbate, 5.0–5.5 for propionate and 4.0–4.5 for benzoate [53]. However, there are consumer concerns about the possible health effects of these compounds; thus, research has focused on alternative compounds derived from plants [54]. Natural preservatives like organic acids and plant extracts (essential oils (EOs) and spices) have shown the ability to reduce foodborne pathogens and spoilage microorganisms in food. Organic acids and plant extract are good potential antimicrobials, and they are considered GRAS (generally recognized as safe) and can be obtained at low cost. Also, they are currently favored over synthetic ones because of consumer concerns over possible health effects of the latter [54]. Nevertheless, they are often less attractive commercially due to their ability to react with other food ingredients, and some may have low water solubility. They can also change sensorial food properties and have a narrow activity spectrum [55]. Organic acids have antimicrobial effects because of the ability to penetrate the cell membrane and to decrease intracellular pH [56, 57].

The combination of *Panax ginseng* extracts (2% v/v) with malic acid (0.5% v/v) and potassium sorbate (0.05% v/v) in orange juice and mango can inhibit the growth of *Salmonella enterica* and *E. coli* O157:H7. Natural components can intercalate into the plasma membrane. This leads to changes membrane fluidity, and thus affects membrane function producing a cellular response and directly interacts with specific membrane proteins. Pathogenic microorganisms have a low adaptation to an acidic environment [57], although a bacteriostatic effect was reported in tomato juice with the combination of different extracts (*Lithospermum erythrorhizon, Rheum palmatum, Thymus vulgaris, Lippiacitriodora, Rosmarinus officinalis, Salvia lavandulifolia,* and *Thymus mastichina*) with chitosan [58].

1.4 NOVEL HURDLE TECHNOLOGIES

In conventional thermal processing, a large amount of energy is transferred to food, which can trigger unwanted reactions, leading to undesirable sensorial changes and nutrient degradation. In contrast, non-thermal innovative processing technologies have emerged as promising preservation methods, showing as highlights safety assurance and shelf-life extension without compromising the quality of processed foods [7]. Some of these methods could include the application of high pressures, ultrasound, high voltage electric fields, electromagnetic waves in moderate doses and light pulses, causing least damage to quality parameters [59].

1.4.1 HIGH HYDROSTATIC PRESSURE (HHP)

High hydrostatic pressure (HHP) is a novel technology widely used to inactivate pathogens and spoilage microorganisms in a significant number of food products due to its economic and technological viability [60]. Appropriate HHP application may retain freshness, sensorial, and nutritive properties of food, which are characteristics currently demanded by consumers [61]; thus, HHP has been applied in the food industry since 1990 as a pasteurization method [62]. Nowadays, thanks to technology development, HHP equipment can reach pressures between 100–800 MPa [63]. Although pressure is the most important experimental parameters in HHP process, treatment time and temperature are also important [64].

As many other novel technologies, the use of HHP as hurdle technology to inactivate microorganisms in food have been studied in order to reach an adequate inactivation (> 5 log cycles), reducing the damage caused to food properties. It is well known that HHP microorganism inactivation is related to its effects on cell membrane components, proteins, enzymes, and ribosomes. Thus, the higher the size of microorganisms the higher the effect of HHP; therefore, molds, and yeasts are more affected than bacteria (Gram-positive are more resistant than Gram-negative bacteria). Moreover, spores and ascospores are resistant to application of HPP alone, the combination of treatment must be applied [65].

Table 1.2 includes reports on the effect of HHP with thermal treatment against native microorganisms, inoculated pathogenic microorganisms, spores, and ascospores. For example, HHP treatment and middle-high temperatures were evaluated against native microorganisms (aerobic bacteria, molds, and yeasts) in tomato puree [66], strawberry puree [67] and fruit puree or baby food [68]: In all the examples, an increase in HHP processing time, temperature, and/or pressure, increased microbial inactivation.

HHP and thermal treatment were also combined against inoculated microorganisms in juices and nectar. *Lactobacillus plantarum* was inoculated in mandarin juice processed with HHP at 0–400 MPa, and different temperatures (15, 30 and 45°C). Results indicated that at 15 and 30°C the microbial reduction was not affected until a pressure of 325 MPa was reached, achieving a maximum microbial reduction of 4 log cycles; however, application of high temperature (45°C) and 400 MPa of pressure reduced 6 log cycles of the inoculated microorganism [69]. In other juices such as apricot (pH = 3.80), orange (pH = 3.76), sour cherry (pH = 3.30) and apple juices (pH = 3.50) the effect of temperature (30 and 40°C) and HHP (350 MPa) were evaluated; no microbial growth was observed at the highest tested temperature (40°C) and HHP; moreover, at lower temperature, an inverse linear relation was observed between pH and microbial reduction of *S. aureus* 485 and *E. coli* O157:H7 [70].

To inactivate spores or ascospores, the application of high pressure and high temperatures (70–120°C) are needed; this combination is called high-pressure sterilization or HPHT (High pressure, high temperature) [71].

TABLE 1.2 High Hydrostatic Pressure Combined Process

Food	Hurdles	Microorganism	Effect	References
Tomato puree Strawberry puree Fruit puree	High hydrostatic pressure and middle-high temperatures	Molds, yeasts, and mesophiles	Growth inhibition	[66–68]
Mandarin juice	High hydrostatic pressure and mild temperatures	*Lactobacillus plantarum*	6 log reduction	[69]
Apricot, orange, sour cherry, and apple juices	High hydrostatic pressure, mild temperatures and pH	Molds, yeasts, mesophiles, *S. aureus*, *E. coli* O157:H7	Growth inhibition	[70]
Tomato juice	High pressure, high temperature	*B. coagulans* 185A	Reduction of 90% of MO	[74]
Pineapple juice and nectar	High hydrostatic pressure and high temperatures	*Byssochlamys nivea* ascospores	3–5 log reduction	[73]
Hami melon juice	High hydrostatic pressure and high temperatures	*B. subtilis* spores	5 log reduction	[75]
Apple juice	High hydrostatic pressure and high temperatures	*Neosartorya fischeri* ascospores	5.2 log reduction	[76]
Milk	High hydrostatic pressure and high temperatures	*C. sporogenes* 7955 spores	Spore reduction	[77]
Beff slurry	High hydrostatic pressure and high temperatures	*B. cereus*	5 log reduction	[78]
Guava and mango juice	High hydrostacic pressure and *Mentha piperita* L. EO	*E. coli* O157:H7	Inhibitory effect	[79]
Fruit juices	High hydrostacic pressure, EOs (*C. reticulate L., C. sinensis L., C. lemon L.) and limonene*	*E. coli* O157:H7 and *L. monocytogenes*	Growth inhibition	[80]
Milk	High hydrostatic pressure, nisin, and lysozyme	*E. coli*	4.6 log reduction	[81]

TABLE 1.2 (*Continued*)

Food	Hurdles	Microorganism	Effect	References
Milk and orange juice	High hydrostatic pressure, mild temperatures and bacteriocins	*S. aureus* and *L. monocytogenes*	8 log reduction	[65]
Turkey breast ham	High hydrostatic pressure and carvacrol	Lactic acid bacteria, psychrotrophic, and *L. innocua*	Significant extension of shelf life	[61]
Duck meat	High hydrostatic pressure and high temperatures	Mesophiles	Reduction below detection limit	[82]
Beef steak	High hydrostatic pressure and mild temperatures	*E. coli O157:H7*	5 log reduction	[83]
Ready to eat ham and turkey	High hydrostatic pressure, salt concentration and sodium nitrite	*L. monocytogenes*	4 log reduction	[84]
Cooked ham	High hydrostatic pressure, enterocin, and refrigeration	*L. monocytogenes* and *Samonella enteritidis*	Extended shelf life and microbial inhibition	[85]
Beef carpaccio	High hydrostatic pressure, lactoperoxidase, and lactoferrin	*E. coli O157:H7, L. monocytogenes* and *Salmonella enterica*	3 log reduction	[86]
Chorizo sausage	High hydrostatic pressure and a_w	*L. monocytogenes*	3 log reduction	[87]
Cooked ham	High hydrostatic pressure, reuterin, lactoperoxidase, and lactoferrin	*E. coli O157:H7, L. monocytogenes* and *Salmonella enterica*	5 log reduction	[88]
Meat sausage	Mild pressure, lactic acid bacteria and bacteriocin	*L. innocua*	3 log reduction	[89]
French cheese	High pressure and thyme	*L. monocytogenes*	5 log reduction	[91]
Filled chocolates	High hydrostatic pressure and refrigeration	Molds, yeasts, and mesophiles	Reduction of 90% of MO	[92]
Pumpkin	High pressure thermal and pulsed electric fields	Molds, yeasts, and mesophiles	Reduction below detection limit	[93]

HPHT is recommended due to several reports indicated that the application of HPP (600–900 MPa) alone for less than 15 min was not enough to inactivate different microorganisms ascospores; instead, an activation process was reported [72, 73]. HPHT have applied in juices and milk. For example, tomato juice inoculated with spores of *Bacillus coagulans* 185A was treated with HPHT under 600 MPa of pressure and different temperatures (75–105°C). As expected, increasing treatment temperature increases spore inactivation, reaching less than 10 CFU/mL with 10, 7, 4, 3 min and 40 s at 75, 85, 95, 100, 105°C, respectively at 600 MPa of constant pressure [74].

In pineapple juice and nectar, *Byssochlamys nivea* ascospores inactivation was evaluated using HHP (600 MPa continuous or in cycles) and heat (20–90°C). Results indicated that ascospores inoculated in juice were more sensible to HHP than those inoculated in nectar; however, at 600 MPa of pressure (15 min) and 90°C, no microbial count was detected in both products. Similar results were obtained with 600 MPa and 80°C for three cycles of 5 min or five cycles of 3 min each one [73]. Chen et al. [75] evaluated the effect of HHP at different temperatures (45–65°C) during holding times (10–20 min) in *B. subtilis* spores in Hami melon juice. The combination of 464 MPa, 54.6°C of heat for 12.8 min was enough to reach a five-log reduction of the tested spores. In apple juice, Evelyn [76] showed that combined treatment (600 MPa, 75°C for 40 min) reduced (5.2 log cycles) *Neosartorya fischeri* ascospores, while higher temperatures (85°C for 70 min and 95°C for 20 min) were needed if only thermal treatment was applied. In milk, the effect of HHP (700, 800 and 900 MPa) at different temperatures (80, 90 and 100°C) against *C. sporogenes* 7955 spores was evaluated, results indicated that increasing process pressure and temperature increase the spore reduction, reaching values of D = 38.2 min and D = 1.3 min with the lowest (700 MPa at 80°C) and highest (900 MPa at 100°C) factors combination [77]. In other liquid products such as beef slurry, HHP (600 MPa) plus heat (70°C) was compared with only heat at the same temperature against *B. cereus* ICMP 12442 spores; results clearly indicated that the use of HHP plus thermal treatment reached >5-log cycles of the spores compared to < 1-log cycle with only heat after 30 min treatment [78].

HHP effectiveness was also evaluated in combination with EOs obtained from different plants to inactivate microorganisms in juices. De Carvalho et al. [79] described that the combined effect of EOs and novel technologies depends on tested concentrations, treatment kind and intensity, and food matrix. For example, at 150 MPa (15 min) the use of 0.63 (μL/mL) of *Mentha piperita* L. EO showed a higher inactivation process (two-fold) of

E. coli O157:H7 in guava and mango juice than the inactivation obtained with 150 MPa alone [79]. In other studies, the inactivation of *E. coli* and *L. monocytogenes* was evaluated using HPP (300–550 MPa) and different EOs (*C. reticulate* L., *C. sinensis* L., *C. lemon* L.) and limonene. Results indicated that *L. monocytogenes* was more affected by the HHP than *E. coli*; while, EOs improved the inactivation of *E. coli* O157:H7, although the higher inactivation was reached with limonene (200 µL/mL) and 300 MPa or the application of 500 MPa alone, which indicated that the use of Eos might reduce the pressure and cost of HHP treatment [80].

Nisin (400 IU/mL), lysozyme (400 µg/mL) and its combination were also used with HHP and mild heat treatment to inactivate different *E. coli* strains (MG1655 and three pressure-resistant mutants isolated from MG1655). Results showed that HHP alone only reach a maximal reduction of 2.3 log cycles at 700 MPa in whole milk, while the combination of lysozyme and nisin affects the microbial reduction of all the *E. coli* strains; moreover, a synergistic effect was observed when peptides were used together. The combination of both peptides, HHP (550 MPa) and mild temperature (50°C) increased microbial reduction up to 4.5 log cycles of the most resistant *E. coli* strain [81]. A similar result was reported by Alpas and Bozoglu [65] that evaluated the combination of HHP (345 MPa per 5 min), heat (50°C) and bacteriocins (5000 AU/mL) to inactivate *S. aureus* and *L. monocytogenes* strains and reported a microbial reduction higher than 8 log cycles; also, microbial growth was not detected during 30 days of storage at 25°C. As observed, the effect of EOs or bacteriocins may be improved due to the effect of HHP on cell wall.

In processed meat, HHP has been applied as a post-packaging tool for salt reduction and improving safety aspects. The most important challenge is to control emerging pathogens such as *L. monocytogenes* and to extend shelf life against spoilage bacteria [6, 82, 83]. Myers et al. [84] applied HHP (600 MPa for 3 min) on ready-to-eat sliced ham and turkey breast formulated with sodium nitrite (0 or 200 ppm), sodium chloride (1.8% or 2.4%), achieving a log reduction of 3.85–4.35 cycles for *L. monocytogenes*. Moreover, when HHP (200 MPa–400 MPa for 10 min) is combined with enterocin LM-2 in sliced cooked ham inhibit the growth of bacteria and inactivates *L. monocytogenes* and *S. enteritidis* [85]. Other combinations of HHP are with natural antimicrobials, such as lactoferrin, reuterin, and lactoperoxidase which are able to inhibit (3 to 5 log reduction) *E. coli* O157:H7, *L. monocytogenes* and *S. enterica* in beef carpaccio [86], chorizo sausage [87], and cooked ham [88]. Castro et al. [89] applied a combination of pressure (300 MPa, 5 min, 10°C) with lactic acid bacteria (LAB) (*Pediococcus acidilactici*, HA-6111-2) and its bacteriocin (bacHA-6111-2) to treat traditional Portuguese ready-to-eat meat sausage, *L. innocua* cells were inactivated by the traditional hurdles (e.g., salt, spices), and the reduction was higher when HHP was applied. EOs, extracts, and their isolated compounds may represent feasible alternatives, providing additional assurance to the natural claims of HHP. When pressure (600 MPa/180 s at 25°C) is combined with carvacrol (200 ppm) on low-sodium sliced vacuum-packed turkey breast ham, primary growth models showed a significant extension of the shelf life due to reduced growth rates and maximized lag phases of *L. monocytogenes* and other major spoilage groups of lactic acid and psychrotrophic bacteria [61]. However, when olive leaf extract was combined with HHP, processing was evaluated on sliced Iberian dry-cured meat; olive leaf extract was not effective to preserve sliced dry-cured shoulder alone or combined with HHP [90]. Finally, other food solid products treated with HHP are applied in French cheese with thyme inhibiting *L. monocytogenes* [91], with refrigeration in filled chocolates [92], and combined with pulsed electric fields (PEF) in pumpkin to reduce microbial load [93].

1.4.2   PULSED ELECTRIC FIELD (PEF)

PEF is a novel technology widely used for food processing products as a non-thermal treatment. PEF has been extensively investigated for food pasteurization due to its effective inactivation of pathogen and spoilage microorganisms, and without the adverse effects of thermal treatment [94–96]. As with other novel technologies, PEF is continuously evolving and innovating to become a more widely used technology in food industry. PEF microbial inactivation is related to electroporation (or electrical break-down) process, where an external electric field increases the membrane cell potential causing a local breakdown, this membrane structural changes lead to an increase in the electrical conductivity and permeability (pores), this effect is permanent when the electric field in membrane goes beyond critical strength of the same [97–99]. PEF process is affected by several factors such as energy input, field strength (> 20 kV/cm), treatment time (less than a second), product composition and temperature [100], and can be applied as bipolar, an exponentially decaying, oscillatory pulse, and square wave [101, 102]. Molds and yeasts are more sensitive to PEF than bacteria (Gram-positive bacteria are more resistant to Gram-negative bacteria), and is a technology used alone or in combination with thermal treatment, natural antimicrobials, and other novel technologies (Table 1.3) [103].

TABLE 1.3 Pulsed Electric Fields Combined Process

Food	Hurdles	Microorganism	Effect	References
Blended orange and carrot juice	Pulsed electric fields and heat	Molds, yeasts, and mesophiles	2.5–3.5 log reduction	[104]
Apple cider	Pulsed electric fields and heat	Molds, yeasts, and mesophiles	Growth inhibition	[105]
Apple juice	Pulsed electric fields and mild heat	S. Panama and S. cerevisiae	3–6 log reduction	[106]
Red apple juice	Pulsed electric fields and mild heat	S. enteriditis, E. coli, and S. cerevisiae	3–6 log reduction	[100]
Tomato juice	Pulsed electric fields, mild heat, and nisin	Mesophiles	4 log reduction	[107]
Chinese rice wine	Pulsed electric fields and low temperatures	S. cerevisiae	Growth inhibition	[108]
Skimmed milk	High temperature pulsed electric field	B. subtilis endospores	Growth control	[109]
Beer	Pulsed electric fields and heat	S. cerevisiae	3.5 log reduction	[110]
Guava and mango juice	Pulsed electric fields and Mentha piperita L. EO	E. coli O157:H7	Growth inhibition	[112]
Cranberry juice	Pulsed electric fields, potassium sorbate and sodium benzoate	Molds, yeasts, and mesophiles	Growth inhibition	[113]
Apple juice	Pulsed electric fields and high intensity light pulse	E. coli K12	More than 6 log reduction	[6]

PEF was typically applied in combination with middle or high heat treatment in order to reduce their treatment effects. In this regards, Rivas et al. [104] applied two treatments to blended orange and carrot juice using PEF and heat (T1 = 25 kV/cm, 280 μs, 68°C and T2 = 25 kV/cm, 330 μs, 70°C) and compared with a high thermal treatment (98°C, 21 s). The authors reported a microbial reduction of 2.46 log cycles, 3.26 log cycles and no-microbial growth in total plate counts, respectively, while a reduction of 2.67 log cycles, 2.85 log cycles and no-microbial growth in molds and yeasts. Although a microbial reduction was higher with thermal treatment; sensorial, quality was higher in PEF treated juice. A similar result in apple cider was reported by Evrendilek et al. [105] who informed a decrease in bacteria, molds, and yeasts; moreover, no microbial growth was detected during 68 days of cider storage at 4°C.

In apple juice inoculated with *S.* Panama and *S. cerevisiae* treated with mild temperatures at the beginning of the treatment (36°C) in combination with PEF (80 kJ/kg) were able to reduce at least 6 log cycles, while using a temperature of 20°C with PEF (80 kJ/kg) only reduced 3 log cycles [106]. Similar results were obtained when red apple juice was treated with PEF (25, 30, and 35 kV/cm) and mild temperatures (30, 40, and 50°C) to inactivate *S. enteriditis, E. coli,* and *S. cerevisiae* [100]. In tomato juice, different treatments such as thermal treatment (50°C), thermal treatment (50°C) plus PEF (80 kV/cm, 200 pulses), and thermal treatment (50°C), PEF (80 kV/cm, 200 pulses) and nisin (0.4 g/100 mL) were applied in order to evaluate their effect on total plate counts. Results indicated that the combination of the three hurdle factors reached a 4.4 log reduction, while the thermal treatment alone or in combination with PEF only reached 0.74 and 1.40 log reduction, respectively [107]. In Chinese rice wine, PEF was combined with low temperatures (25, 30, and 35°C) to reduce the microbial load of *S. cerevisiae*; results indicated that a slight increase of temperature might reduce time treatment and increase microbial reduction [108]. The higher temperature used of the combined treatment, the higher reduction reached, this is due to temperature affects the cell wall transition phase, conducting a change of phospholipids (PL) molecules which change from gel to liquid-crystalline phase, favoring the irreversible breakdown of the cell membrane. Moreover, nisin is a peptide that causes ion-permeable pore formation in the cytoplasmic membrane of cells improving cell death.

To inactivate spores and endospores, a combination of PEF (> 50 kV/cm) and higher temperature is needed; this process is called HTPEF (High temperature PEF). This hurdle technology combination was tested against *B. subtilis* endospores in reconstituted skimmed milk powder; results indicated that the application of PEF observed neither additive nor synergistic effect, thus the main factor was the high temperature of the process [109]. *S. cerevisiae* spores inoculated in beer were treated with PEF (45 kV) and heat (50°C) at different alcohol level (0, 4 and 7% v/v). Results indicated that the combination of PEF and heat, microbial inactivation was lower (0.6 log cycles); however, as alcohol content increases, microbial reduction increases reaching 3.5 of log reduction with PEF, alcohol (7%), and heat [110]. These results were also confirmed in food model where *B. subtili*s spores were higher inactivated with the increasing of temperature or PEF energy [111].

PEF effectiveness was also evaluated in combination with EOs obtained from different plants to inactivate microorganisms in juices. For example, at 150 µs (20 kV/cm) the use of 0.63 (µL/mL) of *Mentha, piperita* L. EO showed a higher inactivation process (almost two-fold) of *E. coli* O157:H7 in guava and mango juice than the inactivation obtained with PEF alone [79], this may be attributed to phenolic compounds presents in EOs [112]. Moreover, PEF was also combined with antimicrobials (900 mg of potassium sorbate and 1500 mg of sodium benzoate) added in the internal surface of the bottle for storage, and results indicated that during storage no microbial growth was observed in cranberry juice [113]. PEF was also combined with other technology such as high-intensity light pulse (HIPL) in apple juice, results indicated that the combination of PEF and HIPL in this order was more effective (> 6 log cycles of reduction) than combination alone or HIPL and PEF against *E. coli* K12 inoculated in apple juice [6].

1.4.3   *ULTRASONICATION AND THERMOSONICATION*

Ultrasonication is a novel technology applied to food products to improve their quality. It is widely applied as a pretreatment of different food processing such as freezing, frying, cutting, brining, drying, degassing, and homogenization among others [63, 114]. One of the most important applications of ultrasonication is inactivation of microorganisms in food; when sound waves application is associated with heat; it increases the effectiveness to inactivate different microorganisms and enzymes on fruit and vegetable products [115]. Ultrasound is generated by a transducer with a frequency of 20 kHz or more. The transducer produces waves through food which causes cavitation, which is the formation, growth, and collapse of micro-bubbles in the sonicated liquid food products [116]. Microbial inactivation is the result of mechanical effects (shear forces and microstreaming) or by the production of sonochemical products such as hydrogen peroxide which may oxidize macromolecules presented in cells, including DNA, lipids, and proteins [117–120]. Contrary to other new technologies, the use of ultrasound process for inactivating microorganisms does not depend on size, Gram type, or bacterial hydrophobicity. Instead, the biopolymer capsule that surrounds bacterial cells can avoid mechanical damage caused by the cavitation process. Thus, the use of ultrasound and mild heat is called thermosonication, while the use of high pressure, heat, and ultrasound is known as mano-thermosonication [121].

As with other hurdle technologies, the most combined barrier with ultrasound is middle thermal treatment. Ultrasound application against indicative microorganisms has been reported by several researchers (Table 1.4). For example, ultrasonication, and mild temperatures reported similar results (no microbial load) against total plate counts, molds, and yeasts in pear juice [122], in purple cactus pear juices [123, 124], orange juice [125], carrot juice [127], and apple juice [127]. On the other hand, the thermosonication process has been applied against microorganisms inoculated in different food products such as juices, nectars, milk, and its derivatives products. For example, the use of ultrasound (60–120 μm) and heat (20–60°C) for 3, 6, and 9 min were applied against *A. acidoterrestris*, *A. ochraceus*, *P. expansum*, *Rhodotorula* sp., *S. cerevisiae* in apple, cranberry, and blueberry juice and nectars. Results indicated that *A. acidoterrestris* was resistant to all treatment combination applied, reaching a maximum reduction of 0.155 log. Other microorganisms were reduced in 3.6–5.9 log cycles with the application of ultrasound (amplitude did not affect the microbial reduction), temperature (60°C), and treatment times of 3, 6, 9 min [128]. In green and purple cactus pear juice treated by ultrasound at different amplitudes (60–90%) and times (1, 3 and 5 min) with an increase of the temperature by the ultrasound process, a microbial reduction higher than 5 log cycles was reported for *E. coli* in all the combination tested. As control, the thermal treatment was used to evaluate the microbial reduction caused by the heat, and results indicated that an increase in 2.4–2.5 log cycles of reduction was achieved by the ultrasonication process [129]. Gabriel [130] reported a reduction lower than 5 log in *E. coli* O157:H7, *Salmonella* spp. and *L. monocytogenes* inoculated in orange juice of the combined effect of ultrasound and a maximum heat of 51.03°C after 40 min.

In milk and dairy products, thermosonication has also demonstrated satisfactory results by similar or higher reduction in microbial load. For example, a five log reduction of *L. innocua* inoculated in raw whole milk was reached after 10 min of combined treatment of US (400 W, 24 kHz) and heat of (63C), while a similar reduction was observed after 30 min of heat treatment at the

TABLE 1.4 Ultrasound Combined Process

Food	Hurdles	Microorganism	Effect	References
Purple cactus pear juice Orange juice Carrot juice Apple Juice	Ultrasonication and mild heat	Molds, yeasts, and mesophiles	Growth inhibition	[123–127]
Apple, cranberry, and blueberry juice	Ultrasonication and mild heat	*A. acidoterrestris, A. ochraceus, P. expansum, Rhodotorula* sp., *S. cerevisiae*	Reduction of 3.6–5.9 log cycles	[128]
Green and purple cactus pear juice	Ultrasonication and mild heat	*E. coli*	5 log reduction	[129]
Orange juice	Ultrasonication and mild heat	*E. coli* O157:H7, *Salmonella spp.* and *L. monocytogenes*	5 log reduction	[130]
Raw whole milk	Thermosonication	*L. innocua*	5 log reduction	[131]
Raw whole milk	Sonication and pasteurization	*A. flavithermus* and *B. Coagulans*	Growth inhibition	[132]
Fermented milk (ayran)	Thermosonication	*S. thermophilus* and *L. delbueckii*	7 log reduction	[133]
Strawberry juice	Ultrasonication and vanillin	Mesophiles, *Enterobacteriaceae*, total coliforms, lactic acid bacteria, molds, yeasts, *E. coli* O157:H7 and *L. innocua*	Growth inhibition	[134]
Milk and orange juice	Low power ultrasonication and cecropin P1	*E. coli*	3 log reduction in milk, 4 log reduction in orange juice	[135]
Lettuce	Ultrasound and hypochlorite	*S. typhimurium*	Reduction of initial levels (1 log)	[137]
Apples	Ultrasound and ClO$_2$	*E. coli* O157:H7 and *Salmonella*	Reductions between 2.5–3.9 log	[138]
Cherry tomato	Ultrasound and per acetic acid	*S. typhimurium*	4 log reduction	[139]

same temperature [131]. The application of a sonication process (5000 W, 20 kHz) for 10 min following by pasteurization (63.5°C and 30 min) reached a complete inactivation of *A. flavithermus* and *B. coagulans*; while heat treatment and ultrasound applied alone of showed a lower microbial reduction [132]. In ayran (fermented milk) samples inoculated with *S. thermophilus* and *L. delbueckii*, a similar reduction (~ 7 log cycles) was observed between the thermosonication process (80°C for 1, 3 and 5 min) compared to thermal treatment (90°C for 1 min) for *S. thermophilus*, while *L. delbrueckii* was completely inactivated using thermosonication (80C for 5 min) and thermal (90°C for 1 min) treatment [133].

  Ultrasound has also evaluated in combination with natural antimicrobials such as vanillin and cinnamaldehyde. For example, native flora (mesophilic bacteria, *Enterobacteriaceae*, and total coliforms, LAB, molds, and yeasts) or inoculated microorganisms (*E. coli* O157:H7 and *L. innocua*) were evaluated in strawberry juice enriched with prebiotic fiber that was processed with ultrasound (180 W, 40 kHz for 7.5 min) and vanillin (1.25 mg/mL). Results of combined treatment were scarcely (to non-inactivation for molds and yeasts to 0.48 log reduction for *Enterobacteriaceae* and total coliforms); however, in storage, microbial count was reduced, besides *E. coli* O157:H7 and *L. innocua* were not detected after 7 days of storage at refrigeration [134]. On the other hand, the use of low power ultrasonication (40 and 160 W) and antimicrobial peptide Cecropin P1 (20 µg/mL) was evaluated against *E. coli* in milk and orange juice. Results presented more than four log reduction of *E. coli* in orange juice and around three log cycles in milk. In orange juice, a synergistic effect was observed when the barriers were used in combination, as compared to each treatment alone. As mentioned above, ultrasound power produces pores in the cell wall which may be increased by the antimicrobials used, who also affects the cell wall of the microorganisms; thus combined effect facilitates the leakage of intracellular material [135]. There are only few studies done of ultrasound application in solid foods, and in most cases is necessary its combination with antimicrobials [136]. In a study carried out on lettuce, additional reduction of one log cycle of *S. typhimurium* was identified by combining ultrasound (32–40 kHz/10 min) with sodium hypochlorite at 50 ppm, compared to the exclusive use of ultrasound or chlorine [137]. Another report was done on apples with *E. coli* O157: H7 and *Salmonella* treated with ultrasound 170 kHz with variable amounts of ClO_2, obtained reductions of 1 to 4 log cycles. The highest reduction was achieved with a frequency of 170 kHz combined with 20 mg of chlorine dioxide [138]. Another successful combination was in cherry tomatoes with 45 kHz and 40 ppm of per acetic acid, when it was tested on the surface of cherry tomatoes reducing *S. typhimurium* (ATCC 14028) in approximate four logs [139].

1.4.4   UV-C AND PULSED LIGHTS

Ultraviolet-C light is a novel technology widely used for disinfecting water, air, and surfaces [140]; although, currently it has been investigated to inactivate microorganisms in food products such as meat, milk, honey, fruits, and vegetables, as well as in juices and nectars [141–145]. The main effect of the UV-C light is associated with the photons emitted at 254 nm by the UV-C lamps. The UV-C photons may penetrate through the cell membranes of microorganisms and affect the DNA, causing photoproducts (pyrimidine dimmers) which might block the DNA transcription and replication, conducting to cell death [146]. The application of UV-C light in food products is through the direct and continuous contact of the UV-C light with the food (liquid or solid) at different treatment times or doses. On the other hand, high-intensity light pulses or commonly known as light pulses is another novel technology which uses light by pulses of short duration (100–400 μs) in a range from ultraviolet to infrared wavelength (200–1100 nm) [147]. Pulsed light is used as a non-thermal treatment for inactivating different microorganisms in foods [148]. Approximately, 54% of the pulsed light falls into the ultraviolet wavelength that is why the main effect of it against microorganisms is associated to the photochemical action of the UV-C light (200–280 nm) [149]. For the application of pulsed light, the energy has to store in a high power capacitor and then liberated by high energy flashes in a short period of time [150].

As the inactivation mechanism is similar in both UV-C light and pulsed light treatments, as the number of membranes that the photons need to cross to reach microbial DNA is larger, the inactivation capacity of the UV-C light decreases; thus Gram-positive are more sensible than Gram-negative bacteria, yeasts, and molds [151]. This effect is also the main disadvantages of the application of light in food products because they are constituted of different compounds which may limited the pass of the light reducing its effect [152]. Recent studies evaluated the microbial reactivation process conducted mainly by the enzyme photolyase who may repair the damage caused by the UV-C light on the DNA [54, 153]; therefore, it is necessary the application of combined treatment to inhibit microbial reactivation. In this aspect, the use of different technologies in combination with UV-C light or pulsed light has been investigated. Some examples are listed in Table 1.5.

TABLE 1.5 UV-C and Pulsed Light Combined Process

Food	Hurdles	Microorganism	Effect	References
Carrot orange juice blend	UV-C and mild heat	E. coli, S. cerevisiae, P. fluorescens	4–5 log reduction	[154, 179]
Orange, apple juice	UV-C and mild heat	E. coli, S. cerevisiae	More than 6 log reduction	[156, 158, 180]
Coconut water	UV-C and mild heat	E. coli O157:H7, S. enterica and L. monocytogenes	5–6 log reduction	[159]
Coconut water	UV-C, mild heat, ph, and Aw	S. enterica	Lowest D_{uv-c} value (22.5 mJ/cm^2)	[160]
Coconut water	UV-C, cinnamaldehyde, and low temperatures	S. typhimurium	Maintained microbial load 30 days	[54]
Grapefruit juice	UV-C, cinnamaldehyde, and low temperatures	S. typhimurium	Bactericide treatment and bacteriostatic during storage	[153]
Apple juice	Pulsed light and ultrasound	A. acidoterrestris, S. Entenritidis, E. coli, and S. cerevisiae	6 log reduction	[161, 162]
Red pepper powder	UV-C radiation and mild heat treatment	E. coli and Salmonella typhimurium	Reduction of 90% of MO	[163]

The effectiveness of UV-C light treatment (0–10.6 kJ/m2), mild heat (40, 45, and 50°C for 15 min), and their combination was evaluated against *E. coli*, *S. cerevisiae*, and *P. fluorescens* inoculated in a carrot-orange juice blend. Results indicated that increasing the temperature of the thermal treatment increased the microbial reduction; however, an additive effect was observed when UV-C light was used in combination with heat (50°C); under this condition, for both bacteria *E. coli* and *P. fluorescens* at least five log cycles was reached; although, less than 4 log cycles was obtained for S. cerevisiae [154, 155]. A similar result was reported by Gayán et al. [156] in orange juice inoculated with *E. coli* and then treated with UV-C (27.5 J/ mL) light and mild temperature (55°C). This combination was sufficient to reduce more than six log cycles. Similar results were obtained by Gouma et al. [157] and Gayán et al. [158] in apple juice and orange and apple juice for *S. cerevisiae* and *E. coli*, respectively.

One interesting product is the coconut liquid endosperm because due to its intrinsic characteristics, has been widely treated with UV-C light alone or in combination with other technologies. For example, Gabriel et al. [159] evaluated the effect of UV-C light (0.42 mW/cm^2) and mild heat treatment (55–63°C) against *E. coli* O157:H7, *S. enterica* and *L. monocytogenes* inoculated in coconut water, results indicated that the combination of treatments increased the microbial reduction reaching 5.94, 5.62 and 6.20 log reduction at 55C. On the contrary, only 1.82, 0.16, and 2.02 log cycles reduction for *E. coli* O157:H7, *S. enterica,* and *L. monocytogenes* were associated to the UV-C light treatment alone. The combination of mild heat (40°C), acid (pH = 4.5), A$_w$ (0.96) and UV-C light showed the lowest D$_{uv-c}$ value (22.5 mJ/cm^2) on *S. enterica* inoculated in coconut water that all of these treatments applied alone [160]. Beristaín-Bauza et al. [54] indicated that coconut water treated with UV-C light (27.7 J/cm^2) and cinnamaldehyde (100 ppm) maintained the microbial load of *S. typhimurium* for 30 days of storage under refrigeration. A similar result was also verified by Ochoa-Velasco et al. [153] in grapefruit juice treated with UV-C light and cinnamaldehyde. Then the combination of UV-C light as bactericide treatment and antimicrobial may act as bacteriostatic during storage.

Pulsed light has also been used in combination with other novel technologies; the use of pulsed light and ultrasound are the combination most used. For example, Ferrario and Guerrero [161] evaluated the effect of pulsed light, ultrasound, and their combination against *S. cerevisiae* and *A. acidoterrestris* spores inoculated in commercial and natural squeezed apple juice. The use of pulsed light alone significantly decrease the microbial load of *S. cerevisiae* (2.0 and 3.7 log cycles) and *A. acidoterrestris* spores (3.0 log cycles); however, the use of US did not affect the microbial load of the spores. The use of combined methodology applied pulsed light prior to ultrasound treatment significantly increased the reduction in both juices reached up to six log cycles. Unfortunately, the combination of treatment was not enough to reached further inactivation in *A. acidoterrestris* spores that caused by the pulsed light treatment alone.

In another report, Ferrairo, and Guerrero [162] described that the combination of ultrasound before pulsed treatment increased microbial inactivation. The use of only ultrasound reduced 2.0, 1.4 and 1.1 log cycles for S. *entenritidis*, *E. coli,* and *S. cerevisiae* in apple juice, respectively, but the treatment using both technologies reached 6.3, 5.9, and 3.7 log reductions for the same microorganisms. The same author also validated the information against *S. cerevisiae* in apple juice. They concluded that the combination of both treatment, produce a severe damage in *S. cerevisiae*, which includes more rounded cell shape and swollen walls, complete coagulated lumen, dispersed intracellular material without visible plasma membrane and cell wall, commonly denominated "ghost cells," vacuolated inner content, broken cell wall with or without efflux of intracellular material [161]. Finally, the combination of UV-C (3.40 mW/cm^2) and mild heating (25–65°C) was useful to inactivate *E. coli* O157:H7 and *S. typhimurium* on powdered red pepper [163].

1.4.5   OTHER NOVEL TECHNOLOGIES

Another novel method that has been studied is the application of ozone due to its characteristics as a strong oxidant, in addition to its capacity as a disinfecting agent. It has been used for example (Table 1.6), on dried fruits and in the case of fresh fruits and vegetables for the destruction of the microflora [164]. Crowe-White et al. [165] treated wild blueberries (*Vaccinium angustifolium*), with rapid individualized freezing (IQF), 100 ppm of chlorine or 1 ppm of aqueous ozone spray with 60 seconds of contact time, it was observed that ozone treatments produced the most significant reductions in mesophils, yeasts, and molds. In Nile tilapia fish either in whole pieces or fillets, combining ozone at different concentrations with low temperatures (11°C) inhibited nearly 80% of mesophils [166].

Irradiation is an unconventional treatment that can be used for some food products, since this technique guarantee complete safety, while being

TABLE 1.6 Other Novel Hurdles Technologies Combinations

Food	Hurdles	Microorganism	Effect	References
Blueberries	Ozone and low temperatures	Molds, yeasts, and mesophiles	Reduction below detection limit	[165]
Nile tilapia	Ozone and low temperatures	Mesophiles	Reduction of 90% of MO	[166]
Kofte	Irradiation and low temperatures	*E. coli* O157:H7	5 log reduction	[169]
Lettuce	Irradiation and modified atmospheres	Molds, yeasts, and mesophiles	2 log reduction	[170]
Lettuce	Irradiation and hypochlorite	*E. coli* O157: H7	5 log reduction	[171]
Lettuce	Irradiation and low temperatures	Mesophiles and total coliforms	2.3 log reduction	[172]
Pear	Microwaves and antagonist mold	*Penicillium expansum*	Reduction of 90% of MO	[174]
Broccoli powder	Radiofrequency and mild heat	Mesophiles	4 log reduction	[177]
Red pepper powder	Radiofrequency thermal and indirect dielectric barrier discharge plasma	*E. coli* O157:H7 and *S. aureus*	Reduction below detection limit	[178]

lethal to vegetative form of pathogens and parasites, and has been designated with the term of cold pasteurization. FAO/WHO Commission and Codex Alimentarius recognize irradiation as safe technology. Currently, close to 40 countries that allow the commercialization of irradiated food products [167]. Irradiation sources that can be used in food industry are high-voltage electron beams of up to 10 MeV, X-rays of up to 5 MeV and gamma rays produced from radioisotopes 60 Co and 137 Cs. Irradiation has an excellent antimicrobial effect on microorganisms, especially on pathogens. Gram-negative bacteria are susceptible to irradiation, while spores are more resistant to ionizing irradiation compared to vegetative cells. In molds, irradiation also has an inhibitory effect; however, yeasts are more resistant so they could become the dominant flora in irradiated foods [168]. Gezgin and Gunes [169] tested inoculating samples of kofte (raw ground meat pie) with *E. coli* O157:H7 irradiated at different doses with low-temperature storage (4°C), and a decreased of five log cycles was observed. In lettuce cut and packed in a modified atmosphere with radiation (0.15 and 0.35 kGy), mesophiles, and yeasts were reduced by 1.5 and 1.0 log cycles, respectively [170]. The combination of radiation (0.1 to 0.6 kGy) and chlorination (200 ppm) in lettuce reduced up to 2.5 log cycles of *E. coli* O157:H7 [171]. This result is similar to those obtained by Likui et al. [172] who treated lettuce with irradiation (1 kGy) and cold storage (4°C) with a reduction of 2.35 and near 5 log cycles for total bacteria and total coliforms, respectively.

Electromagnetic waves such as microwaves (MW) and radiofrequency (RF) are currently applied in fresh fruits, and are useful in food treatment, since waves can penetrate to a greater depth, with the consequent volumetric heating. When microorganisms are attached to fruits pericarp, the application of organic solutions and disinfectants can be used for food safety; however, some insects and viruses are not eliminated as they grow from the embryo or maybe in a dormant state. Microwave treatment has been developed to control postharvest diseases of fruits and vegetables [173]. Zhang et al. [174] applied microwave power (0.45 kW), with the combination of antagonistic yeast (*Cryptococcus laurentii*) to control *Penicillium expansum* in pears, observing a reduction of *P. expansum* 100% to 65.5%. In RF, non-uniform heating of fruits due to its structure is the main obstacle for its use as disinfection methods [175], however, as an emerging technology; it has the advantage of larger depth of penetration, heat distribution, and low energy consumption. Despite the disadvantage described above, this technology has been positioned as a method to sterilize, pasteurize, and disinfect food and agricultural products such as fruits and nuts, also for contribution to the demands of consumers to use fewer conservatives in food [176]. Zhao [177], pasteurized broccoli powder by RF (6 kW, 27.12 MHz for 5 minutes), with polypropylene packing during RF heating, with a considerable microbial inactivation of 4.2 log cycles. Finally, a combination of RF and dielectric barrier discharge plasma treatment was applied on red pepper powder (*Capsicum annuum* L.) contaminated with *E. coli* O157:H7 and *S. aureus*. Microbiological analysis showed a microbial reduction below to the detection limits [178].

1.5 CONCLUSIONS

In a society increasingly concerned with consumer health, hurdle technology, by which two or more obstacles are involved in an appropriate combination, is a viable and innovative alternative in food preservation, with the advantage of product minimal sensorial and nutritional changes. Although more than a hundred hurdles could be combined, it should be borne in mind that hurdle technology success depends on guaranteeing microbial metabolic depletion, since there is evidence that microorganisms can acquire varying levels of resistance or tolerance to environmentally stress factors. Hurdle technology can provide different results, with a possibility that bacteria can become more resistant or even more virulent; therefore, the technologies used in microbial growth control, need to be carefully analyzed and optimized. It is necessary to identify the treatment sequences that are necessary to obtain a synergic effect and in this way produce a more effective barrier effect against microbial deterioration of foods.

KEYWORDS

- antimicrobials
- cinnamaldehyde
- food preservation
- hurdle technologies
- irradiation
- microwaves

REFERENCES

1. Lee, S. Y. Microbial safety of pickled fruits and vegetables and hurdle technology. *Internet J. Food Saf.,* **2004**, *4*, 21–32.
2. Rostami, Z., Ahmad, M. A., Khan, M. U., Mishra, A. P., Rashidzadeh, S., & Shariati, M. A. Food preservation by hurdle technology: A review of different hurdle and interaction with focus on foodstuffs. *J. Pure Appl. Microbiol.,* **2016**, *10*(4), 2633–2639.
3. Pundhir, A., & Murtaza, N. Hurdle technology-an approach towards food preservation. *Int. J. Curr. Microbiol. Appl. Sci.,* **2015**, *4*(7), 802–809.
4. Leistner, L. Further developments in the utilization of hurdle technology for food preservation. *J. Food Eng.,* **1994**, *22*(1–4), 421–432.
5. Leistner, L. Basic Aspects of food preservation by hurdle technology. *Int. J. Food Microbiol.,* **2000**, *55*(1–3), 181–186.
6. Caminiti, I. M., Palgan, I., Noci, F., Muñoz, A., Whyte, P., Cronin, D. A., Morgan, D. J., & Lyng, J. G. The effect of pulsed electric fields (pef) in combination with high intensity light pulses (HILP) on *Escherichia coli* inactivation and quality attributes in apple juice. *Innov. Food Sci. Emerg. Technol.,* **2011**, *12*(2), 118–123.
7. Oliveira, T. L. C., De Ramos, A. L. S., Ramos, E. M., Piccoli, R. H., & Cristianini, M. Natural antimicrobials as additional hurdles to preservation of foods by high pressure processing. *Trends Food Sci. Technol.,* **2015**, *45*(1), 60–85.
8. Rahman, M. S. Hurdle technology in food preservation. In: Siddiqui, M. W., & Rahman, M. S., (eds.), *Minimally Processed Foods* (pp. 17–34). Springer, New York, **2015**.
9. Leistner, L., & Gorris, L. G. M. Food preservation by hurdle technology. *Trends Food Sci. Technol.,* **1995**, *6*(2), 41–46.
10. Rahman, M. S. Hurdle technology in food preservation. In: Siddiqui, M. W., & Rahman, M. S., (eds.), *Minimally Processed Foods* (pp. 17–34). Springer: New York, **2015**.
11. Khan, I., Tango, C. N., Miskeen, S., Lee, B. H., & Oh, D. H. Hurdle technology: A novel approach for enhanced food quality and safety – A review. *Food Control,* **2017**, *73*, 1426–1444.
12. Luo, K., Kim, S. Y., Wang, J., & Oh, D. H. A combined hurdle approach of slightly acidic electrolyzed water simultaneous with ultrasound to inactivate *Bacillus cereus* on potato. *LWT – Food Sci. Technol.,* **2016**, *73*, 615–621.
13. Singh, S., & Shalini, R. Effect of hurdle technology in food preservation: A review. *Crit. Rev. Food Sci. Nutr.,* **2016**, *56*(4), 641–649.
14. Barwal, J. S. Hurdle technology for shelf stable food products. *Indian Food Ind.,* **1994**, *1*, 40–43.
15. Sharma, S., & Garmima, C. Use of hurdle technology in food preservation. *Curr. Res. Inf. Biotechnol.,* **2016**, *1*(1), 28–35.
16. Nura, A., Chukwuma, A. C., & Oneh, A. J. *Critical Review on Principles and Applications of Hurdle,* **2016**, 485–491.
17. Mogren, L., Windstam, S., Boqvist, S., Vågsholm, I., Söderqvist, K., Rosberg, A. K., Lindén, J., Mulaosmanovic, E., Karlsson, M., & Uhlig, E. The hurdle approach—a holistic concept for controlling food safety risks associated with pathogenic bacterial contamination of leafy green vegetables: A review. *Front. Microbiol.,* **2018**, *9*, 1–20.
18. Esser, D. S., Leveau, J. H. J., & Meyer, K. M. Modeling microbial growth and dynamics. *Appl. Microbiol. Biotechnol.,* **2015**, *99*(21), 8831–8846.
19. Leistner, L., & Gould, G. W. *Hurdle Technologies: Combination Treatments for Food Stability, Safety and Quality.* Springer: New York, **2002**.

20. Artés, F., & Allende, A. Minimal fresh processing of vegetables, fruits and juices. In: Da, W. S., (ed.), *Emerging Technologies for Food Processing* (pp. 677–716). Elsevier: New York, **2005**.

21. Hun, J. H. Edible films and coatings: A review. In: Han, J. H., (ed.), *Innovations in Food Packaging* (pp. 240–262). Academic Press: New York, **2005**.

22. Bico, S. L., Raposo, M. F., Morais, R. M. S., & Morais, M. M. Combined effects of chemical dip and/or carrageenan coating and/or controlled atmosphere on quality of fresh-cut banana. *Food Control,* **2009**, *20*(5), 508–514.

23. Denoya, G., & Ardanaz, M. Effect of the application of combined treatments of additives on the inhibition of enzymatic browning in apples cv. Granny Smith minimally processed (Aplicación de tratamientos combinados de aditivos sobre la inhibición del pardeamiento enzimático en manzanas cv. granny smith mínimamente procesadas). *Rev. Investig. Agropecu.,* **2012**, *38*(3), 263–267.

24. Gliemmo, M. F. F., Latorre, M. E. E., Gerschenson, L. N. N., & Campos, C. A. A. Color Stability of pumpkin (*Cucurbita moschata*, Duchesne Ex Poiret) puree during storage at room temperature: Effect of pH, potassium sorbate, ascorbic acid, and packaging material. *LWT – Food Sci. Technol.,* **2009**, *42*(1), 196–201.

25. Solvia-Fortuny, R. C., Alòs-Saiz, N., Espachs-Barroso, A., & Martin-Belloso, O. Influence of maturity at processing on quality attributes of fresh-cut conference pears. *J. Food Sci.,* **2004**, *69*(7), 290–294.

26. Alexandre, D., Cunha, R. L., & Hubinger, M. D. Acai conservation through obstacle technology (Conservação Do Açaí Pela Tecnologia de Obstáculos). *Ciência e Tecnol. Aliment,* **2004**, *24*(1), 114–119.

27. Sluka, E. F. Barrier technologies applied to the conservation of zapallo anco (*Cucurbita moschata* D.) purée (Tecnologías de Barreras Aplicadas a La Conservación de Puré de Zapallo Anco (*Cucurbita moschata* D.)). *Rev. Agronómica del Noroeste Argentino,* **2016**, *36*(2), 39–43.

28. Alzamora, S. M. *Combined Use of Ultrasound and Natural Antimicrobials to Inactivate Listeria monocytogenes in Orange Juice,* **2016**, *70*, 1850–1856.

29. Escobar-Hernández, A., Márquez-Cardozo, C. J., Restrepo-Flores, C. E., & Pérez-Cordoba, L. J. Application of barrier technology for conservation minimally processed vegetables mixtures (Aplicación de Tecnología de Barreras Para La Conservación de de Mezclas de Vegetales Mínimamente Procesados). *Rev. la Fac. Nac. Agron. Medellin,* **2014**, *67*(1280), 7237–7245.

30. Barwal, V. S., Sharma, R., & Singh, R. Preservation of cauliflower by hurdle technology. *Int. J. Food Sci. Technol.,* **2005**, *42*(1), 26–31.

31. İncedayi, B., Tamer, C. E., Suna, S., & Çopur, Ö. U. Hurdle technology for shelf stable minimally processed potato Cv. Agria. *J. Agric. Fac. Uludag Un.,* **2014**, *42*, 29–42.

32. Vibhakara, H. S. J., Das Gupta, D. K., Jayaraman, K. S., & Mohan, M. S. Development of a high-moisture shelf-stable grated carrot product using hurdle technology. *J. Food Process. Preserv.,* **2006**, *30*(2), 134–144.

33. Sinha, J., Gupta, E., & Chandra, R. Low cost preservation of cauliflower for 180 days through hurdle technology. *Int. J. Food Nutr. Sci.,* **2012**, *2*(1), 27–32.

34. Cappato, L. P., Martins, A. M. D., Ferreira, E. H. R., & Rosenthal, A. Effects of hurdle technology on monascus ruber growth in green table olives: A response surface methodology approach. *Brazilian J. Microbiol.,* **2018**, *49*(1), 112–119.

35. Sorrentino, E., Succi, M., Tipaldi, L., Pannella, G., Maiuro, L., Sturchio, M., Coppola, R., & Tremonte, P. Antimicrobial activity of gallic acid against food-related *Pseudomonas*

strains and its use as biocontrol tool to improve the shelf life of fresh black truffles. *Int. J. Food Microbiol.,* **2018**, *266*, 183–189.

36. Guynot, M. E., Ramos, A. J., Sanchis, V., & Marin, S. Study of benzoate, propionate, and sorbate salts as mould spoilage inhibitors on intermediate moisture bakery products of low pH (4.5–5.5). *Int. J. Food Microbiol.,* **2005**, *101*(2), 161–168.
37. Rajendran, P., & Bharathidasan, R. Standardization and preservation of sugarcane juice by hurdle technology. *Int. J. Adv. Agric. Sci. Technol.,* **2018**, *5*(2), 77–87.
38. Akpomedaye, D. E., & Ejechi, B. O. The hurdle effect of mild heat and two tropical spice extracts on the growth of three fungi in fruit juices. *Food Res. Int.,* **1998**, *31*(5), 339–341.
39. Galvão, M. F., Prudêncio, C. V., & Vanetti, M. C. D. Stress enhances the sensitivity of *Salmonella Enterica* Serovar *Typhimurium* to bacteriocins. *J. Appl. Microbiol.,* **2015**, *118*(5), 1137–1143.
40. Kunitake, M., Ditchfield, C., Silva, C., & Petrus, R. Effect of pasteurization temperature on stability of acidified sugarcane juice beverage. *Ciência e Agrotecnologia,* **2014**, *38*(6), 554–561.
41. Char, C. D., Guerrero, S. N., & Alzamora, S. M. Mild thermal process combined with vanillin plus citral to help shorten the inactivation time for *Listeria innocua* in orange juice. *Food Bioprocess Technol.,* **2010**, *3*(5), 752–761.
42. Espina, L., Condón, S., Pagán, R., & García-Gonzalo, D. Synergistic effect of orange essential oil or (+)-limonene with heat treatments to inactivate *Escherichia coli* O157:H7 in orange juice at lower intensities while maintaining hedonic acceptability. *Food Bioprocess Technol.,* **2014**, *7*(2), 471–481.
43. Yen, P. P. L., Kitts, D. D., & Pratap, S. A. Natural acidification with low-pH fruits and incorporation of essential oil constituents for organic preservation of unpasteurized juices. *J. Food Sci.,* **2018**, *83*(8), 2039–2046.
44. Nyhan, L., Begley, M., Mutel, A., Qu, Y., Johnson, N., & Callanan, M. Predicting the combinatorial effects of water activity, pH and organic acids on *Listeria* growth in media and complex food matrices. *Food Microbiol.,* **2018**, *74*, 75–85.
45. Zhu, J., Li, J., & Chen, J. Survival of *Salmonella* in home-style mayonnaise and acid solutions as affected by acidulant type and preservatives. *J. Food Prot.,* **2012**, *75*(3), 465–471.
46. Ghaly, A. E., Dave, D., Budge, S., & Brooks, M. S. Fish spoilage mechanisms and preservation techniques: Review. *Am. J. Appl. Sci.,* **2010**, *7*(7), 859–877.
47. Ghanbari, M., Jami, M., Domig, K. J., & Kneifel, W. Seafood biopreservation by lactic acid bacteria – A review. *LWT – Food Sci. Technol.,* **2013**, *54*, 315–324.
48. Wiernasz, N., Cornet, J., Cardinal, M., Pilet, M. F., Passerini, D., & Leroi, F. Lactic acid bacteria selection for biopreservation as a part of hurdle technology approach applied on seafood. *Front. Mar. Sci.,* **2017**, *4*, 1–15.
49. Leistner, L. Combined methods for food preservation, **2007**, *25*, 867–894.
50. Chawla, S. P., Chander, R., & Sharma, A. Safe and shelf-stable natural casing using hurdle technology. *Food Control,* **2006**, *17*(2), 127–131.
51. Delgado, B., Palop, A., Fernández, P. S., & Periago, P. M. combined effect of thymol and cymene to control the growth of *Bacillus cereus* vegetative cells. *Eur. Food Res. Technol.,* **2004**, *218*(2), 188–193.
52. Ananou, S., Gálvez, A., Martínez-Bueno, M., Maqueda, M., & Valdivia, E. Synergistic effect of enterocin AS-48 in combination with outer membrane permeabilizing treatments against *Escherichia coli* O157:H7. *J. Appl. Microbiol.,* **2005**, *99*(6), 1364–1372.

53. Saranraj, P., & Geetha, M. Microbial spoilage of bakery products and its control by preservatives. *Int. J. Pharm. Biol. Arch.,* **2012,** *3*(1), 38–48.

54. Beristaín-Bauza, S., Martínez-Niño, A., Ramírez-González, A. P., Ávila-Sosa, R., Ruíz-Espinosa, H., Ruiz-López, I. I., & Ochoa-Velasco, C. E. Inhibition of *Salmonella typhimurium* growth in coconut (*Cocos nucifera* L.) water by hurdle technology. *Food Control,* **2018,** *92.*

55. Zhou, G. H., Xu, X. L., & Liu, Y. Preservation technologies for fresh meat – a review. *Meat Sci.,* **2010,** *86*(1), 119–128.

56. Massey, L. M., Hettiarachchy, N. S., Martin, E. M., & Ricke, S. C. Electrostatic spray of food-grade organic acids and plant extract to reduce *Escherichia coli* O157:H7 on fresh-cut cantaloupe cubes. *J. Food Saf.,* **2013,** *33*(1), 71–78.

57. Raybaudi-Massilia, R., Zambrano-Durán, A., Mosqueda-Melgar, J., & Calderón-Gabaldón, M. I. Improving the safety and shelf-life of orange and mango juices using *Panax ginseng,* malic acid and potassium sorbate. *J. Fur Verbraucherschutz und Leb.,* **2012,** *7*(4), 273–282.

58. Giner, M. J., Vegara, S., Funes, L., Martí, N., Saura, D., Micol, V., & Valero, M. Antimicrobial activity of food-compatible plant extracts and chitosan against naturally occurring micro-organisms in tomato juice. *J. Sci. Food Agric.,* **2012,** *92*(9), 1917–1923.

59. Mohapatra, D., Mishra, S., Giri, S., & Kar, A. Application of hurdles for extending the shelf life of fresh fruits. *Trends Post Harvest Techology,* **2013,** *1*(1), 37–54.

60. Rendueles, E., Omer, M. K., Alvseike, O., Alonso-Calleja, C., Capita, R., & Prieto, M. Microbiological food safety assessment of high hydrostatic pressure processing: A review. *LWT – Food Sci. Technol.,* **2011,** *44,* 1251–1260.

61. De Oliveira, T. L. C., De Castro Leite, B. R., Ramos, A. L. S., Ramos, E. M., Piccoli, R. H., & Cristianini, M. Phenolic carvacrol as a natural additive to improve the preservative effects of high pressure processing of low-sodium sliced vacuum-packed turkey breast ham. *LWT – Food Sci. Technol.,* **2015,** *64*(2), 1297–1308.

62. Thakur, B. R., & Nelson, P. E. High-pressure processing and preservation of food. *Food Rev. Int.,* **1998,** *14*(4), 427–447.

63. Roobab, U., Aadil, R. M., Madni, G. M., & Bekhit, L. D. The impact of nonthermal technologies on the microbiological quality of juices: A review. *Compr. Rev. Food Sci. Food Saf.,* **2018,** 1–21.

64. Huang, H. W., Lung, H. M., Yang, B. B., & Wang, C. Y. Responses of microorganisms to high hydrostatic pressure processing. *Food Control,* **2014,** *40*(1), 250–259.

65. Alpas, H., & Bozoglu, F. The combined effect of high hydrostatic pressure, heat, and bacteriocins on inactivation of foodborne pathogens in milk and orange juice. *World J. Microbiol. Biotechnol.,* **2000,** *16*(4), 387–392.

66. Krebbers, B., Matser, A. M., Hoogerwerf, S. W., Moezelaar, R., Tomassen, M. M. M., & Van Den Berg, R. W. Combined high pressure and thermal treatments for processing of tomato puree: Evaluation of microbial inactivation and quality parameters. *Innov. Food Sci. Emerg. Technol.,* **2003,** *4,* 377–385.

67. Marszałek, K., Mitek, M., & Skąpska, S. The effect of thermal pasteurization and high pressure processing at cold and mild temperatures on the chemical composition, microbial and enzyme activity in strawberry purée. *Innov. Food Sci. Emerg. Technol.,* **2015,** *27,* 48–56.

68. Kultur, G., Misra, N. N., Barba, F. J., Koubaa, M., Gökmen, V., & Alpas, H. Effect of high hydrostatic pressure on background microflora and furan formation in fruit purée based baby foods. *J. Food Sci. Technol.,* **2018,** *55*(3), 985–991.

69. Carreño, J. M., Gurrea, M. C., Sampedro, F., & Carbonell, J. V. Effect of high hydrostatic pressure and high-pressure homogenization on *Lactobacillus plantarum* inactivation kinetics and quality parameters of mandarin juice. *Eur. Food Res. Technol.,* **2011**, *232*(2), 265–274.

70. Bayindirli, A., Alpas, H., Bozoglu, F., & Hizal, M. Efficiency of high pressure treatment on inactivation of pathogenic microorganisms and enzymes in apple, orange, apricot and sour cherry juices. *Food Control,* **2006**, *17*(1), 52–58.

71. Daher, D. Effect of High pressure processing on the microbial inactivation in fruit preparations and other vegetable based beverages. *Agriculture,* **2017**, *7*(9), 72.

72. Chapman, B., Winley, E., Fong, A. S. W., Hockin, A. D., Stewart, C. M., & Buckle, K. Ascospore inactivation and germination by high pressure processing is affected by ascospore age. *Innov. Food Sci. Emerg. Technol.,* **2007**, *8*(4), 531–534.

73. Ferreira, E. H., Da, R., Rosenthal, A., Calado, V., Saraiva, J., & Mendo, S. Byssochlamys Nivea inactivation in pineapple juice and nectar using high pressure cycles. *J. Food Eng.,* **2009**, *95*(4), 664–669.

74. Daryaei, H., & Balasubramaniam, V. M. Kinetics of *Bacillus coagulans* spore inactivation in tomato juice by combined pressure-heat treatment. *Food Control,* **2013**, *30*(1), 168–175.

75. Chen, J., Zheng, X., Dong, J., Chen, Y., & Tian, J. Optimization of effective high hydrostatic pressure treatment of *Bacillus subtilis* in Hami melon juice. *LWT – Food Sci. Technol.,* **2015**, *60*(2), 1168–1173.

76. Evelyn, K. H. J., & Silva, F. V. M. Modeling the inactivation of *Neosartorya fischeri* ascospores in apple juice by high pressure, power ultrasound, and thermal processing. *Food Control,* **2016**, *59*, 530–537.

77. Shao, Y., & Ramaswamy, H. S. *Clostridium sporogenes*-ATCC 7955 spore destruction kinetics in milk under high pressure and elevated temperature treatment conditions. *Food Bioprocess Technol.,* **2011**, *4*(3), 458–468.

78. Evelyn, & Silva, F. V. M. Modeling the inactivation of psychrotrophic *Bacillus cereus* spores in beef slurry by 600 MPa HPP combined with 38–70°C: Comparing with thermal processing and estimating the energy requirements. *Food Bioprod. Process,* **2016**, *99*, 179–187.

79. De Carvalho, R. J., De Souza, G. T., Pagán, E., García-Gonzalo, D., Magnani, M., & Pagán, R. Nanoemulsions of *Mentha piperita* L. essential oil in combination with mild heat, pulsed electric fields (PEF) and high hydrostatic pressure (HHP) as an alternative to inactivate *Escherichia coli* O157: H7 in fruit juices. *Innov. Food Sci. Emerg. Technol.,* **2018**, *48*(7), 219–227.

80. Espina, L., García-Gonzalo, D., Laglaoui, A., Mackey, B. M., & Pagán, R. Synergistic combinations of high hydrostatic pressure and essential oils or their constituents and their use in preservation of fruit juices. *Int. J. Food Microbiol.,* **2013**, *161*(1), 23–30.

81. Garcia-Graells, C., Masschalck, B., & Michiels, C. W. Inactivation of *Escherichia coli* in milk by high-hydrostatic pressure treatment in combination with antimicrobial peptides. *J. Food Prot.,* **1999**, *62*(11), 1248–1254.

82. Khan, M. A., Ali, S., Abid, M., Ahmad, H., Zhang, L., Tume, R. K., & Zhou, G. Enhanced texture, yield and safety of a ready-to-eat salted duck meat product using a high pressure-heat process. *Innov. Food Sci. Emerg. Technol.,* **2014**, *21*, 50–57.

83. Sun, S., Sullivan, G., Stratton, J., Bower, C., & Cavender, G. Effect of HPP treatment on the safety and quality of beef steak intended for sous vide cooking. *LWT – Food Sci. Technol.,* **2017**, *86*, 185–192.

84. Myers, K., Montoya, D., Cannon, J., Dickson, J., & Sebranek, J. The effect of high hydrostatic pressure, sodium nitrite, and salt concentration on the growth of *Listeria monocytogenes* on RTE ham and turkey. *Meat Sci.,* **2013**, *93*(2), 263–268.

85. Liu, G., Wang, Y., Gui, M., Zheng, H., Dai, R., & Li, P. Combined effect of high hydrostatic pressure and enterocin LM-2 on the refrigerated shelf life of ready-to-eat sliced vacuum-packed cooked ham. *Food Control,* **2012**, *24* (1/2), 64–71.

86. Bravo, D., De Alba, M., & Medina, M. Combined Treatments of high-pressure with the lactoperoxidase system or lactoferrin on the inactivation of *Listeria monocytogenes, Salmonella enteritidis* and *Escherichia coli* O157: H7 in beef carpaccio. *Food Microbiol.,* **2014**, *41*, 27–32.

87. Rubio, B., Possas, A., Rincón, F., García-Gímeno, R. M., & Martínez, B. Model for *Listeria monocytogenes* inactivation by high hydrostatic pressure processing in Spanish chorizo sausage. *Food Microbiol.,* **2018**, *69*, 18–24.

88. Montiel, R., Martín-Cabrejas, I., & Medina, M. Reuterin, lactoperoxidase, lactoferrin and high hydrostatic pressure on the inactivation of food-borne pathogens in cooked ham. *Food Control,* **2015**, *51*, 122–128.

89. Castro, S. M., Kolomeytseva, M., Casquete, R., Silva, J., Teixeira, P., Castro, S. M., Queirós, R., & Saraiva, J. A. Biopreservation strategies in combination with mild high pressure treatments in traditional Portuguese ready-to-eat meat sausage. *Food Biosci.,* **2017**, *19*, 65–72.

90. Amaro-Blanco, G., Delgado-Adámez, J., Martín, M. J., & Ramírez, R. Active packaging using an olive leaf extract and high pressure processing for the preservation of sliced dry-cured shoulders from Iberian pigs. *Innov. Food Sci. Emerg. Technol.,* **2018**, *45*(9), 1–9.

91. Bleoancă, I., Saje, K., Mihalcea, L., Oniciuc, E. A., Smole-Mozina, S., Nicolau, A. I., & Borda, D. Contribution of high pressure and thyme extract to control *Listeria monocytogenes* in fresh cheese – A hurdle approach. *Innov. Food Sci. Emerg. Technol.,* **2016**, *38*, 7–14.

92. Dias, J., Coelho, P., Alvarenga, N. B., Duarte, R. V., & Saraiva, J. A. Evaluation of the impact of high pressure on the storage of filled traditional chocolates. *Innov. Food Sci. Emerg. Technol.,* **2018**, *45*, 36–41.

93. García-Parra, J., González-Cebrino, F., Delgado-Adámez, J., Cava, R., Martín-Belloso, O., Elez-Martínez, P., & Ramírez, R. Application of innovative technologies, moderate-intensity pulsed electric fields and high-pressure thermal treatment, to preserve and/or improve the bioactive compounds content of pumpkin. *Innov. Food Sci. Emerg. Technol.,* **2018**, *45*(1), 53–61.

94. Barba, F. J., Parniakov, O., Pereira, S. A., Wiktor, A., Grimi, N., Boussetta, N., Saraiva, J. A., Raso, J., Martin-Belloso, O., & Witrowa-Rajchert, D. Current applications and new opportunities for the use of pulsed electric fields in food science and industry. *Food Res. Int.,* **2015**, *77*, 773–798.

95. Sitzmann, W., Vorobiev, E., & Lebovka, N. Applications of electricity and specifically pulsed electric fields in food processing: Historical backgrounds. *Innov. Food Sci. Emerg. Technol.,* **2016**, *37*, 302–311.

96. Wang, M. S., Wang, L. H., Bekhit, A. E. D. A., Yang, J., Hou, Z. P., Wang, Y. Z., Dai, Q. Z., & Zeng, X. A. A review of sublethal effects of pulsed electric field on cells in food processing. *J. Food Eng.,* **2018**, *223*, 32–41.

97. Zimmermann, U., Pilwat, G., Beckers, F., & Rieman, F. Effects of external electrical fields on cell membranes. *Bioelectrochem. Bioenerg,* **1976**, *3*, 58–83.

98. Ravishankar, S., Zhang, H., & Kempkes, M. L. Pulsed electric fields. *Food Sci. Technol. Int.*, **2008**, *14*(5), 429–432.

99. Saulis, G. Electroporation of cell membranes: The fundamental effects of pulsed electric fields in food processing. *Food Eng. Rev.*, **2010**, *2*(2), 52–73.

100. Katiyo, W., Yang, R., & Zhao, W. Effects of combined pulsed electric fields and mild temperature pasteurization on microbial inactivation and physicochemical properties of cloudy red apple juice (*Malus Pumila* Niedzwetzkyana (Dieck)). *J. Food Saf.*, **2017**, *37*(4), 1–9.

101. Hizal, M., Bayindirli, A., Damar, S., & Bozog, F. Inactivation and injury of *Escherichia coli* O157:H7 and *Staphylococcus aureus* by pulsed electric fields. *World J. Microbiol. Biotechnol.*, **2002**, *18*, 1–6.

102. Olatunde, O. O., & Benjakul, S. Nonthermal processes for shelf-life extension of sea foods: A revisit. *Compr. Rev. Food Sci. Food Saf.*, **2018**, *17*(4), 892–904.

103. Van Impe, J., Smet, C., Tiwari, B., Greiner, R., Ojha, S., Stulić, V., Vukušić, T., & Režek, J. A. state of the art of nonthermal and thermal processing for inactivation of micro-organisms. *J. Appl. Microbiol.*, **2018**, *125*(1), 16–35.

104. Rivas, A., Rodrigo, D., Martínez, A., Barbosa-Cánovas, G. V., & Rodrigo, M. Effect of PEF and heat pasteurization on the physical-chemical characteristics of blended orange and carrot juice. *LWT – Food Sci. Technol.*, **2006**, *39*(10), 1163–1170.

105. Evrendilek, G. A., Jin, Z. T., Ruhlman, K. T., Qiu, X., Zhang, Q. H., & Richter, E. R. Microbial safety and shelf-life of apple juice and cider processed by bench and pilot scale PEF systems. *Innov. Food Sci. Emerg. Technol.*, **2000**, *1*(1), 77–86.

106. Timmermans, R. A. H., Nierop, G. M. N., Nederhoff, A. L., Van Boekel, M. A. J. S., Matser, A. M., & Mastwijk, H. Pulsed electric field processing of different fruit juices: Impact of pH and temperature on inactivation of spoilage and pathogenic micro-organisms. *Int. J. Food Microbiol.*, **2014**, *173*, 105–111.

107. Nguyen, P., & Mittal, G. S. Inactivation of naturally occurring microorganisms in tomato juice using pulsed electric field (PEF) with and without antimicrobials. *Chem. Eng. Process. Process Intensif.*, **2007**, *46*(4), 360–365.

108. Huang, K., Yu, L., Liu, D., Gai, L., & Wang, J. Modeling of yeast inactivation of PEF-treated Chinese rice wine: Effects of electric field intensity, treatment time, and initial temperature. *Food Res. Int.*, **2013**, *54*(1), 456–467.

109. Cregenzán-Alberti, O., Arroyo, C., Dorozko, A., Whyte, P., & Lyng, J. G. Thermal characterization of *Bacillus subtilis* endospores and a comparative study of their resistance to high temperature pulsed electric fields (HTPEF) and thermal-only treatments. *Food Control*, **2017**, *73*, 1490–1498.

110. Milani, E. A., Alkhafaji, S., & Silva, F. V. M. Pulsed electric field continuous pasteurization of different types of beers. *Food Control*, **2015**, *50*, 223–229.

111. Siemer, C., Toepfl, S., & Heinz, V. Inactivation of *Bacillus subtilis* spores by pulsed electric fields (PEF) in combination with thermal energy – I. influence of process- and product parameters. *Food Control*, **2014**, *39*(1), 163–171.

112. Sanz-Puig, M., Santos-Carvalho, L., Cunha, L. M., Pina-Pérez, M. C., Martínez, A., & Rodrigo, D. Effect of pulsed electric fields (PEF) combined with natural antimicrobial by-products against *S. typhimurium*. *Innov. Food Sci. Emerg. Technol.*, **2016**, *37*, 322–328.

113. Jin, T. Z., Guo, M., & Yang, R. Combination of pulsed electric field processing and antimicrobial bottle for extending microbiological shelf-life of pomegranate juice. *Innov. Food Sci. Emerg. Technol.*, **2014**, *26*, 153–158.

114. Huang, G., Chen, S., Dai, C., Sun, L., Sun, W., Tang, Y., Xiong, F., He, R., & Ma, H. Effects of ultrasound on microbial growth and enzyme activity. *Ultrason. Sonochem,* **2017**, *37*, 144–149.

115. Anaya-Esparza, L. M., Velázquez-Estrada, R. M., Roig, A. X., García-Galindo, H. S., Sayago-Ayerdi, S. G., & Montalvo-González, E. Thermosonication: An alternative processing for fruit and vegetable juices. *Trends Food Sci. Technol.,* **2017**, *61*, 26–37.

116. Abdullah, N., & Chin, N. L. Application of thermosonication treatment in processing and production of high quality and safe-to-drink fruit juices. *Agric. Agric. Sci. Procedia.,* **2014**, *2*, 320–327.

117. Koda, S., Miyamoto, M., Toma, M., Matsuoka, T., & Maebayashi, M. Inactivation of *Escherichia coli* and *Streptococcus mutans* by ultrasound at 500 KHz. *Ultrason. Sonochem.,* **2009**, *16*(5), 655–659.

118. Gao, S., Lewis, G. D., Ashokkumar, M., & Hemar, Y. Inactivation of microorganisms by low-frequency high-power ultrasound: 1) Effect of growth phase and capsule properties of the bacteria. *Ultrason. Sonochem.,* **2014**, *21*(1), 446–453.

119. Gao, S., Lewis, G. D., Ashokkumar, M., & Hemar, Y. Inactivation of microorganisms by low-frequency high-power ultrasound: 2) A simple model for the inactivation mechanism. *Ultrason. Sonochem.,* **2014**, *21*(1), 454–460.

120. Gao, R., Jing, P., Ruan, S., Zhang, Y., Zhao, S., Cai, Z., & Qian, B. Removal of off-flavors from radish (*Raphanus sativus* L.) anthocyanin-rich pigments using chitosan and its mechanism(s). *Food Chem.,* **2014**, *146*, 423–428.

121. Kahraman, O., Lee, H., Zhang, W., & Feng, H. Manothermosonication (MTS) treatment of apple-carrot juice blend for inactivation of *Escherichia coli* 0157:H7. *Ultrason. Sonochem.,* **2017**, *38*, 820–828.

122. Saeeduddin, M., Abid, M., Jabbar, S., Wu, T., Hashim, M. M., Awad, F. N., Hu, B., Lei, S., & Zeng, X. Quality assessment of pear juice under ultrasound and commercial pasteurization processing conditions. *LWT – Food Sci. Technol.,* **2015**, *64*(1), 452–458.

123. Zafra-Rojas, Q. Y., Cruz-Cansino, N., Ramírez-Moreno, E., Delgado-Olivares, L., Villanueva-Sánchez, J., & Alanís-García, E. Effects of ultrasound treatment in purple cactus pear (*Opuntia ficus*-Indica) juice. *Ultrason. Sonochem.,* **2013**, *20*(5), 1283–1288.

124. Del Socorro, Cruz-Cansino, N., Ramírez-Moreno, E., León-Rivera, J. E., Delgado-Olivares, L., Alanís-García, E., Ariza-Ortega, J. A., De Jesús Manríquez-Torres, J., & Jaramillo-Bustos, D. P. Shelf life, physicochemical, microbiological and antioxidant properties of purple cactus pear (*Opuntia ficus* Indica) juice after thermo-ultrasound treatment. *Ultrason. Sonochem.,* **2015**, *27*(1), 277–286.

125. Valero, M., Recrosio, N., Saura, D., Muñoz, N., Martí, N., & Lizama, V. Effects of ultrasonic treatments in orange juice processing. *J. Food Eng.,* **2007**, *80*(2), 509–516.

126. Martínez-Flores, H. E., Garnica-Romo, M. G., Bermúdez-Aguirre, D., Pokhrel, P. R., & Barbosa-Cánovas, G. V. Physico-chemical parameters, bioactive compounds, and microbial quality of thermo-sonicated carrot juice during storage. *Food Chem.,* **2015**, *172*, 650–656.

127. Abid, M., Jabbar, S., Hu, B., Hashim, M. M., Wu, T., Lei, S., Khan, M. A., & Zeng, X. Thermosonication as a potential quality enhancement technique of apple juice. *Ultrason. Sonochem.,* **2014**, *21*(3), 984–990.

128. Režek, J. A., Šimunek, M., Evačić, S., Markov, K., Smoljanić, G., & Frece, J. Influence of high power ultrasound on selected moulds, yeasts and *Alicyclobacillus acidoterrestris* in apple, cranberry and blueberry juice and nectar. *Ultrasonics,* **2018**, *83*, 3–17.

129. Cruz-Cansino, N., Del, S., Reyes-Hernández, I., Delgado-Olivares, L., Jaramillo-Bustos, D. P., Ariza-Ortega, J. A., & Ramírez-Moreno, E. Effect of ultrasound on survival and growth of *Escherichia coli* in cactus pear juice during storage. *Brazilian J. Microbiol.*, **2016**, *47*(2), 431–437.

130. Gabriel, A. A. Inactivation behaviors of foodborne microorganisms in multi-frequency power ultrasound-treated orange juice. *Food Control,* **2014**, *46*, 189–196.

131. Bermúdez-Aguirre, D., Corradini, M. G., Mawson, R., & Barbosa-Cánovas, G. V. Modeling the inactivation of *Listeria innocua* in raw whole milk treated under thermo-sonication. *Innov. Food Sci. Emerg. Technol.,* **2009**, *10*(2), 172–178.

132. Khanal, S. N., Anand, S., Muthukumarappan, K., & Huegli, M. Inactivation of thermoduric aerobic sporeformers in milk by ultrasonication. *Food Control,* **2014**, *37*(1), 232–239.

133. Erkaya, T., Başlar, M., Şengül, M., & Ertugay, M. F. Effect of thermosonication on physicochemical, microbiological and sensorial characteristics of ayran during storage. *Ultrason. Sonochem.,* **2015**, *23*, 406–412.

134. Cassani, L., Tomadoni, B., Ponce, A., Agüero, M. V., & Moreira, M. R. Combined use of ultrasound and vanillin to improve quality parameters and safety of strawberry juice enriched with prebiotic fibers. *Food Bioprocess Technol.,* **2017**, *10*(8), 1454–1465.

135. Fitriyanti, M., & Narsimhan, G. synergistic effect of low power ultrasonication on antimicrobial activity of cecropin P1 against *E. coli* in food systems. *LWT – Food Sci. Technol.,* **2018**, *96*(4), 175–181.

136. De São José, J. F. B., De Andrade, N. J., Mota, R. A., Dantas, M. C., Stringheta, P. C., & Paes, J. B. Decontamination by ultrasound application in fresh fruits and vegetables. *Food Control,* **2014**, *45*(1), 36–50.

137. Seymour, I. J., Burfoot, D., Smith, R. L., Cox, L. A., & Lockwood, A. Ultrasound decontamination of minimally processed fruits and vegetables. *Int. J. Food Sci. Technol.,* **2002**, *37*(1), 547–557.

138. Huang, T. S., Xu, C., Walker, K., West, P., Zhang, S., & Weese, J. Decontamination efficacy of combined chlorine dioxide with ultrasonication on apples and lettuce. *J. Food Sci.,* **2006**, *74*(1), 134–139.

139. São José, J. F. B., & Vanetti, M. C. D. Effect of ultrasound and commercial sanitizers on natural microbiota and *Salmonella enterica typhimurium* on cherry tomatoes. *Food Control,* **2012**, *24*(1), 95–99.

140. Guerrero-Beltrán, J. A., & Barbosa-Cánovas, G. V. Review: Advantages and limitations on processing food by UV light. *Food Sci. Tech. Int.,* **2004**, *10*, 137–147.

141. Ochoa-Velasco, C. E., Cruz-Gonzalez, M., & Guerrero-Beltrán, J. A. Ultraviolet-C light inactivation of *Escherichia coli* and *Salmonella typhimurium* in coconut (*Cocos Nucifera* L.) milk. *Innov. Food Sci. Emerg. Technol.,* **2014**, *26*(2), 199–204.

142. Degala, H. L., Mahapatra, A. K., Demirci, A., & Kannan, G. Evaluation of non-thermal hurdle technology for ultraviolet-light to inactivate *Escherichia coli* K12 on goat meat surfaces. *Food Control,* **2018**, *90*(3), 113–120.

143. Roig-Sagués, A. X., Gervilla, R., Pixner, S., Terán-Peñafiel, T., Hernández-Herrero, M. M. Bactericidal effect of ultraviolet-C treatments applied to honey. *LWT – Food Sci. Technol.,* **2018**, *89*(11), 566–571.

144. Koca, N., Urgu, M., & Saatli, T. E. Ultraviolet light applications in dairy processing. *Technol. Approaches Nov. Appl. Dairy Process,* **2018**, 3–22.

145. Butot, S., Cantergiani, F., Moser, M., Jean, J., Lima, A., Michot, L., Putallaz, T., Stroheker, T., & Zuber, S. UV-C Inactivation of foodborne bacterial and viral pathogens and surrogates on fresh and frozen berries. *Int. J. Food Microbiol.,* **2018**, *275*, 8–16.

146. Ochoa-Velasco, C. E., & Guerrero-Beltrán, J. A. Short-wave ultraviolet-C light effect on pitaya (*Stenocereus griseus*) juice inoculated with *Zygosaccharomyces bailii*. *J. Food Eng.*, **2013**, *117*(1), 34–41.

147. Oms-Oliu, G., Odriozola-Serrano, I., Soliva-Fortuny, R., Elez-Martínez, P., & Martín-Belloso, O. Stability of health-related compounds in plant foods through the application of non thermal processes. *Trends Food Sci. Technol.*, **2012**, *23*(2), 111–123.

148. Palgan, I., Caminiti, I. M., Muñoz, A., Noci, F., Whyte, P., Morgan, D. J., Cronin, D. A., & Lyng, J. G. Combined effect of selected non-thermal technologies on *Escherichia coli* and *Pichia fermentans* inactivation in an apple and cranberry juice blend and on product shelf life. *Int. J. Food Microbiol.*, **2011**, *151*(1), 1–6.

149. Muñoz, A., Caminiti, I. M., Palgan, I., Pataro, G., Noci, F., Morgan, D. J., Cronin, D. A., Whyte, P., Ferrari, G., & Lyng, J. G. Effects on *Escherichia coli* inactivation and quality attributes in apple juice treated by combinations of pulsed light and thermosonication. *Food Res. Int.*, **2012**, *45*(1), 299–305.

150. Gómez-López, V. M., Ragaert, P., Debevere, J., & Devlieghere, F. Pulsed light for food decontamination: A review. *Trends Food Sci. Technol.*, **2007**, *18*, 464–473.

151. Gayán, E., Serrano, M. J., Raso, J., Álvarez, I., & Condón, S. Inactivation of *Salmonella enterica* by UV-C light alone and in combination with mild temperatures. *Appl. Environmental Microbiol.*, **2012**, *78*(23), 8353–8361.

152. Hernández-Carranza, P., Ruiz-López, I. I., Pacheco-Aguirre, F. M., Guerrero-Beltrán, J. Á., Ávila-Sosa, R., & Ochoa-Velasco, C. E. Ultraviolet-C light effect on physicochemical, bioactive, microbiological, and sensorial characteristics of carrot (*Daucus carota*) beverages. *Food Sci. Technol. Int.*, **2016**, *22*(6).

153. Ochoa-Velasco, C. E., Díaz-Lima, M. C., Ávila-Sosa, R., Ruiz-López, I. I., Corona-Jiménez, E., Hernández-Carranza, P., López-Malo, A., & Guerrero-Beltrán, J. A. Effect of UV-C light on *Lactobacillus rhamnosus, Salmonella typhimurium,* and *Saccharomyces cerevisiae* kinetics in inoculated coconut water: Survival and residual effect. *J. Food Eng.*, **2018**, *223*.

154. García, C. M., Ferrario, M., Guerrero, S., & García, C. M. Study of the inactivation of some microorganisms in turbid carrot-orange juice blend processed by ultraviolet light assisted by mild heat treatment. *J. Food Eng.*, **2017**, *212*, 213–225.

155. García, C. M., Ferrario, M., & Guerrero, S. Effectiveness of UV-C light assisted by mild heat on *Saccharomyces cerevisiae* KE 162 inactivation in carrot-orange juice blend studied by flow cytometry and transmission electron microscopy. *Food Microbiol.*, **2018**, *73*, 1–10.

156. Gayán, E., Serrano, M. J., Monfort, S., Álvarez, I., & Condón, S. Combining ultraviolet light and mild temperatures for the inactivation of *Escherichia coli* in orange juice. *J. Food Eng.*, **2012**, *113* (4), 598–605.

157. Gouma, M., Gayán, E., Raso, J., Condón, S., & Álvarez, I. Inactivation of spoilage yeasts in apple juice by UV-C light and in combination with mild heat. *Innov. Food Sci. Emerg. Technol.*, **2015**, *32*, 146–155.

158. Gayán, E., Serrano, M. J., Álvarez, I., & Condón, S. Modeling optimal process conditions for UV-heat inactivation of foodborne pathogens in liquid foods. *Food Microbiol.*, **2016**, *60*, 13–20.

159. Gabriel, A. A., Ostonal, J. M., Cristobal, J. O., Pagal, G. A., & Armada, J. V. E. Individual and combined efficacies of mild heat and Ultraviolet-c radiation against *Escherichia coli* O157:H7, *Salmonella enterica,* and *Listeria monocytogenes* in coconut liquid endosperm. *Int. J. Food Microbiol.*, **2018**, *277*, 64–73.

160. Estilo, E. E. C., & Gabriel, A. A. Previous stress exposures influence subsequent UV-C resistance of *Salmonella enterica* in coconut liquid endosperm. *LWT – Food Sci. Technol.,* **2017**, *86*, 139–147.
161. Ferrario, M., & Guerrero, S. Study of the inactivation of deteriorating microorganisms in apple and melon juices treated by pulsed light and ultrasound (Estudio de la inactivación de microorganismos deteriorativos en jugos de manzana y melón tratados por luz pulsada y ultrasonido). *Rev. del Lab. Tecnológico del Uruguay,* **2016**, *11*(11), 9–17.
162. Ferrario, M., & Guerrero, S. Impact of a combined processing technology involving ultrasound and pulsed light on structural and physiological changes of *Saccharomyces cerevisiae* KE 162 in apple juice. *Food Microbiol.,* **2017**, *65*, 83–94.
163. Cheon, H. L., Shin, J. Y., Park, K. H., Chung, M. S., & Kang, D. H. Inactivation of foodborne pathogens in powdered red pepper (*Capsicum annuum* L.) using combined UV-C irradiation and mild heat treatment. *Food Control,* **2015**, *50*, 441–445.
164. Habibi, M., & Haddad, M. H. Efficacy of ozone to reduce microbial populations in date fruits. *Food Control,* **2009**, *20*(1), 27–30.
165. Crowe-White, K. M., Bushway, A., & Davis-Dentici, K. Impact of postharvest treatments, chlorine and ozone, coupled with low-temperature frozen storage on the antimicrobial quality of low bush blueberries (*Vaccinium angustifolium*). *LWT – Food Sci. Technol.,* **2012**, *47*(1), 213–215.
166. De Mendonça, A., & Gonçalves, A. A. Effect of aqueous ozone on microbial and physicochemical quality of Nile tilapia processing. *J. Food Process. Preserv.,* **2017**, *41*(6), 128–135.
167. Molins, R. A., Motarjemi, Y., & Käferstein, F. Irradiation: A critical control point in ensuring the microbiological safety of raw foods. *Food Control,* **2001**, *12*, 347–356.
168. Atunescu, G., & Tofana, M. Effects of ionizing radiation on microbiological contaminants of foods. *Bull. UASVM Agric.,* **2010**, *18*, 1843–1853.
169. Gezgin, Z., & Gunes, G. Influence of gamma irradiation on growth and survival of *Escherichia coli* O157:H7 and quality of cig kofte, a traditional raw meat product. *Int. J. Food Sci. Technol.,* **2007**, *42*(9), 1067–1072.
170. Prakash, A., Guner, A. R., Caporaso, F., & Foley, D. M. Effects of low-dose gamma irradiation on the shelf life and quality characteristics of cut romaine lettuce packaged under modified atmosphere. *J. Food Sci.,* **2008**, *65*(3), 549–553.
171. Foley, D. M., Dufour, A., Rodriguez, L., Caporaso, F., & Prakash, A. Reduction of *Escherichia coli* O157:H7 in shredded iceberg lettuce by chlorination and gamma irradiation. *Radiat. Phys. Chem.,* **2002**, *83*(3), 391–396.
172. Likui, Z., Zhaoxin, L., Fengxia, L., & Xiaomei, B. Effect of γ-irradiation on quality-maintaining of fresh-cut lettuce. *Food Control,* **2006**, *17*, 225–228.
173. Karabulut, O. A., & Baykal, N. Evaluation of the use of microwave power for the control of postharvest disease of peaches. *Postharvest Biol. Technol.,* **2002**, *26*, 237–240.
174. Zhang, H., Zheng, X., & Su, D. Postharvest control of blue mold rot of pear by microwave treatment and *Cryptococcus laurentii. J. Food Eng.,* **2005**, *77*, 539–544.
175. Birla, S. L., Wang, S., Tang, J., & Hallman, G. Improving heating uniformity of fresh fruit in radio frequency treatments for pest control. *Postharvest Biol. Technol.,* **2004**, *33*, 205–217.
176. Mahendran, R. Radio frequency heating and its application in food processing: A review. *Int. J. Curr. Agric. Res.,* **2013**, *1*, 42–46.
177. Zhao, C. Microbial inactivation in foods by ultrasound. *J. Food Microbiol. Saf. Hyg.,* **2017**, *2*(1), 102–104.

178. Choi, E. J., Yang, H. S., Park, H. W., & Chun, H. H. Inactivation of *Escherichia coli* O157:H7 and *Staphylococcus aureus* in red pepper powder using a combination of radio frequency thermal and indirect dielectric barrier discharge plasma non-thermal treatments. *LWT-Food Sci. Technol.,* **2018**, *93,* 477–484.

179. García, C. M., Ferrario, M., & Guerrero, S. Effectiveness of UV-C light assisted by mild heat on *Saccharomyces cerevisiae* KE 162 inactivation in carrot-orange juice blend studied by flow cytometry and transmission electron microscopy. *Food Microbiol.,* **2018**, *73,* 1–10.

180. Gouma, M., Gayán, E., Raso, J., Condón, S., & Álvarez, I. Inactivation of spoilage yeasts in apple juice by UV-C light and in combination with mild heat. *Innov. Food Sci. Emerg. Technol.,* **2015**, *32,* 146–155.

CHAPTER 2

Production of Microbiologically Safe Fruits and Vegetables

SANTOS GARCÍA and NORMA HEREDIA

Department of Microbiology and Immunology, School of Biology, Autonomous University of Nuevo León, San Nicolás de los Garza, Nuevo León, 66455 Mexico, Tel.: +52 (81) 8376-3044, E-mail: norma@microbiosymas.com (N. Heredia)

ABSTRACT

The global production of fruits and vegetables grew 94% from 1980 to 2004 with a yearly average of 4.5%. Only in the United States, consumption per capita of fresh fruits and vegetables increased approximately 19 and 57%, respectively, between 1976 and 2007. This consumption trend has been adopted worldwide, provoking a concurrent growth of this economic activity. For example, in Mexico, a major producer of fruits and vegetables, approximately 8.9 million of their 118 million inhabitants are involved in food production and about 80% of those employed in food production work in agriculture. Mexican Agrifood systems use 11% of the country's surface to grow more than 500 vegetable species. Unfortunately, the high produce consumption has occurred concomitant with a rise in the number of produce-associated foodborne outbreaks worldwide. In the United States, between 1996 and 2010, approximately 23% of total foodborne illness outbreaks were produce-related. In Europe, 10% of the outbreaks, 35% of hospitalizations, and 46% of the deaths were linked to produce in the period 2007 to 2011, whereas in Australia, fresh produce was linked to 4% of foodborne disease outbreaks informed from 2001 to 2005. In this chapter, the most common food safety problems regarding produce production, the main pathogenic bacteria involved, and the most effective methods of pathogen detection and control will be discussed.

2.1 GLOBAL SITUATION OF PRODUCE PRODUCTION

The demand for fruits and vegetables among countries and the liberalization of agricultural markets has increased the exports of high-value fruit and vegetables, reaching 71.6 billion dollars in 2001. China is currently the world´s largest producer of fruits and vegetables in the world (34%) followed by Latin America and the Caribbean (11%), India (10%), and Africa together with the European Union (EU) (9%) (Figure 2.1) [1, 2]. Mexico, in particular, is the leading exporter of avocado in the world. It is also the second-leading exporter of tomato, watermelon, cantaloupe, papaya, and lime, and third-leading exporter of green pepper, cucumber, and onion, along with other vegetables. Moreover, Mexico is the most important supplier of produce to the USA [3]. Produce imported to the United States from Mexico generally has a comparable microbial quality to domestic samples produced in the United States [4]. Nevertheless, several outbreaks of foodborne illnesses in the United States have been associated with produce imported from Mexico and other parts of the world [5].

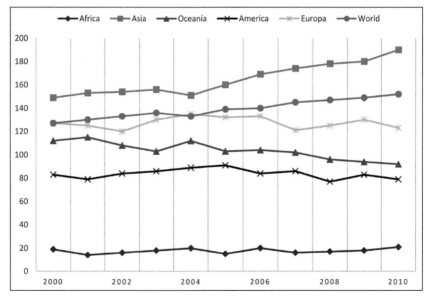

FIGURE 2.1   Production of fruits and vegetables by region.
Source: Modified from FAO [2].

Outbreaks of foodborne diseases in the developed world have been attributed with increasing frequency to fresh and fresh-cut fruits and vegetables.

The proportion of outbreaks of foodborne illness attributed to fresh produce also increased from 0.7% in the 1970s to 6% in the 1990s in the U.S.A. [6]. Outbreaks attributed to produce had a median of 40 illnesses, causing 20 million illnesses, and costing 38.6 $ billion every year [7]. During the period, 1990 to 2004 produce items were linked to 19.3% of foodborne outbreaks and 33.8% of cases documented in the USA [4]. Similarly, gastro-intestinal diseases due to the consumption of contaminated foods are among the top five causes of disease in Mexico. However, recent reports indicate that the rate of gastrointestinal diseases in children less than five years old and mortality rates in the general population is decreasing [8] which could be important for the production of safer foods.

2.2 MICROORGANISMS PRESENT IN PRODUCE

Multiple factors affect the survival or growth of the microbiota present on produce; these include the type and stage of produce, environmental conditions, and handling practices. However, changes in handling or processing methods are likely to change the microbiota as well. The types of microbes on fresh fruits and vegetables vary with commodity and processing. Analyses of vegetables indicate that bacteria are the predominant microorganisms with almost equal proportions of Gram-positive and Gram-negative organisms, and lesser number of molds and yeasts are present [9]. Most bacteria will inhabit the surface of fruits and vegetables, although internal tissue can also harbor microorganisms. Tissue components can help to neutralize compounds such as chloride, which is frequently used to control microorganisms [4].

The microbiota of fruits is somewhat different from those of vegetables; this is mainly due to the low pH that restricts the growth to acid-tolerant microorganisms, such as fungi and lactic acid bacteria (LAB) [9]. In general, *Pseudomonas flourescens, Erwinia herbicola,* and *Enterobacter agglomerans* are the major constituents of the microbiota of many vegetables; however, *Leuconostoc* spp., *Lactobacillus* spp., molds, and yeasts are also often present [4]. The interactions between saprophytic and pathogenic microbiota are poorly understood, and it can be considered different scenarios: may inhibit, not affected, or even promote the growth of pathogenic microorganisms [4].

One widespread practice is to offer for sale, mixtures of a variety of fruits and vegetables in the prepackaged salads format. In this situation, anticipating what types of microorganisms are present could be very difficult. Addition of other foods, such as meats, eggs, and dressing might contribute with their microbiota, and even more, the labor handling involved in making such products could provide an extra source of microbes [9].

2.3 PRINCIPAL PATHOGENIC BACTERIA AFFECTING PRODUCE

Pathogen contamination of fresh produce may originate before or after harvest, but once that produce is contaminated, it is difficult to be sanitized [7]. There are several agents such as chemicals, pathogens, and parasites, which may adulterate food at different points in the food production and preparation processes. *Salmonella, E. coli* O157:H7 and *Listeria monocytogenes* are among the most critical pathogenic bacteria of concern to produce food safety (Table 2.1). Indeed, they have been recovered from a wide variety of fruits and vegetables [6]. In the U.S.A. alone, the total economic costs of foodborne illnesses due to *L. monocytogenes*, non-typhoidal *Salmonella*, and *E. coli* O157:H7 have been estimated to be of $2.0 billion, $4.4 billion, and $607 million, respectively [6].

Also, these three pathogens provoke substantial costs to the food industry and the government such as those originated from recalls and destroyed products, sampling, and testing, and politics to reduce contamination [6]. Product recalls and affected consumer confidence are economically damaging for the food industry, which makes a high investment in microbiological surveillance. The recalls of fresh fruits and vegetables in the USA has increased over time, from 6 (2.2%) in 2003 to 57 (15.6%) in 2013. The most common reasons of these were the presence of *L. monocytogenes* or *Salmonella* serovars in foods [10]. The processing steps could increase the risk of foodborne illness by the consumption of minimally processed fruits and vegetables due to hygienic deficiencies during handling or by modifying the microenvironment that could provoke an increase or decrease of pathogens and their competitor microorganisms. An additional risk factor is when the growth of one organism produce better conditions that allow the growth of others (phenomena known as metabiotic effect) [9].

2.3.1 SALMONELLA

Salmonella is a member of the *Enterobacteriaceae* family, and are Gram-negative, non-spore forming, facultative anaerobe bacillus that grows between 35 and 37°C. These bacteria are classified by serotyping, according to differences in their somatic (O) and flagellar (H) antigens.

There are over 2600 different serotypes identified, all capable of cause infection in humans [11].

There are two dominant species of *Salmonella*: *S. enterica* and *S. bongori*. *Salmonella enterica* is divided into 6 subspecies: *enterica, salamae, arizonae, diarizonae, houtenae,* and *indica*. The subspecies *enterica* is responsible for more than 99% of human salmonellosis, and it includes more than 1,500 serotypes among which are *Salmonella typhimurium* and *Salmonella enteritidis*. *S.* Typhimurium is the most dominant serovar around the world and is associated with a high number of foodborne outbreaks [12]. The gastroenteritis is the predominant form of salmonellosis; appearing 12 to 72 h following ingestion and is characterized by nausea and vomiting, diarrhea, abdominal pain, and fever. The illness is often self-limiting with complete resolution in 4–7 days [11].

Constant reports evidenced the presence of *Salmonella* in fresh produce, especially during pre- and postharvest practices for the processing of ready-to-eat products (Table 2.1). These bacteria are the most commonly identified and associated with fresh produce-related infection. It was isolated in 48% of cases between 1973 and 1997 in the USA and 41% of cases during 1992–2000 in the UK [13]. Most of the *Salmonella* outbreaks (77%) were associated with produce, including fruits ($n = 14$), seeded vegetables ($n = 10$), sprouts ($n = 6$), nuts, and seeds ($n = 5$), spices ($n = 4$), and herbs ($n = 1$) [14].

A range of fresh fruit and vegetable products have been implicated in *Salmonella* infection, most commonly lettuce, sprouted seeds, melon, and tomatoes. However, *Salmonella* spp. are often isolated from produce sampled in routine surveys, including lettuce, cauliflower, sprouts, mustard cress, endive, spinach, and mushrooms. Elimination of *Salmonella typhimurium* from contaminated produce could be tough, since it has been demonstrated that bacteria are capable of penetrating the epidermis through open stomata in a process that involves flagellar motility and chemotaxis [15]. Therefore, prevention of contamination is the key measure to avoid problems with this bacterium.

2.3.2   E. COLI O157:H7

E. coli is a member of the *Enterobacteriaceae* family, is a Gram-negative, aerobic, facultatively anaerobic, non-spore-forming rod. They are regular inhabitants of the digestive tract of animals; however, some *E. coli* groups have acquired virulence factors and can cause diseases in humans. These virulent strains are classified in pathotypes: enteropathogenic (EPEC), enterotoxigenic

TABLE 2.1 Produce Related Foodborne Outbreaks from 1997 to 2017 where *Salmonella, E. coli* O157:H7, and *L. monocytogenes* Were the Responsible

Food Vehicle	Pathogen	Country	Number of Cases	Number of Hospitalizations	Number of Deaths
Sprouts (alfalfa, raw clover)	*Salmonella* spp.	U.S.A.	506	65	0
Pistachios	*Salmonella* spp.	USA	11	2	0
Cucumbers	*Salmonella* spp.	USA	991	221	6
Bean sprouts	*Salmonella* spp.	USA	115	ND	0
Mangoes	*Salmonella* spp.	U.S.A.	127	33	0
Pine nuts	*Salmonella* spp.	USA	43	2	0
Papayas	*Salmonella* spp.	U.S.A.	106	10	0
Maradol papaya	*Salmonella* Urbana	USA	7	4	0
Maradol papaya	*Salmonella* Newport and Infantis	USA	4	2	0
Maradol papaya	*Salmonella* Anatum	USA	14	5	1
Maradol papaya	*Salmonella*	USA	1210	67	1
Jalapeño and Serrano peppers	*Salmonella* spp.	USA	1442	286	2
Cantaloupe	*Salmonella* spp.	USA	332	113	3
Tomatoes	*Salmonella* spp.	USA	111	22	0
Tomatoes/peppers	*Salmonella* spp.	USA and Canadá	1442	ND	ND
Salads	*Salmonella* Singapore	UK	4	ND	ND
Sprouts (Alfalfa)	*Salmonella*	Australia	100	ND	ND
Rock melon	*Salmonella* Hvittingfoss	Australia	97	ND	ND
Lettuce (pre-packaged)	*Salmonella* Anatum	Australia	144	ND	ND

TABLE 2.1 *(Continued)*

Food Vehicle	Pathogen	Country	Number of Cases	Number of Hospitalizations	Number of Deaths
Sprouts (alfalfa)	*Salmonella* spp.	Canada	ND	ND	ND
Fruit Salad	*Salmonella* spp.	Canada	ND	ND	ND
Cantaloupes	*Salmonella* spp.	Canada	ND	ND	ND
Sprouts (onion)	*Salmonella* Cubana	Canada	20	ND	ND
Sprouts (alfalfa)	*Salmonella* Meleagridis	Canada	78	ND	ND
Cantaloupes	*Salmonella* Poona	Canada	2	ND	ND
Cucumbers	*Salmonella* Brandenberg	Canada	12	ND	ND
Roma tomatoes (suspected)	*Salmonella* Javiana	Canada	7	ND	ND
Sprouts (suspected mung beans)	*Salmonella enteritidis*	Canada	84	ND	ND
Sprouts (mung beans)	*Salmonella enteritidis*	Canada	8	ND	ND
Baby spinach	*Salmonella* Java	Sweden	177	46	0
Sprouts (alfalfa)	*Salmonella* Stanley	Sweden	51	0	0
Cantaloupes	*Salmonella*	Mexico	ND	ND	ND
Basil	*Salmonella*	UK	32	0	0
Sprouts (radish)	*E. coli* O157:H7	Japan	126 and 945	ND	13
Spinach	*E. coli* O157:H7	Canada	1	ND	ND
Lettuce romaine	*E. coli* O157:H7	Canada	29	ND	ND
Spanish onions	*E. coli* O157:H7	Canada	235	ND	ND
Watercress	*E. coli* O157	U.K.	ND	ND	ND
Imported salad	*E. coli* O157:H7	UK	161	60	2
Packaged salad	*L. monocytogenes*	USA	19	19	1

TABLE 2.1　*(Continued)*

Food Vehicle	Pathogen	Country	Number of Cases	Number of Hospitalizations	Number of Deaths
Apples	*L. monocytogenes*	USA	35	34	7
Sprouts (bean)	*L. monocytogenes*	USA	5	5	2
Cantaloupe	*L. monocytogenes*	USA	147	143	33
Diced celery	*L. monocytogenes*	USA	10	ND	5
Caramel Apples (pre-packaged)	*L. monocytogenes*	USA	35	31	7
Corn salad	*L. monocytogenes*	Italy	1566	292	0

ND: Not determined.

Source: Modified from Ref. [30].

(ETEC), enterohemorrhagic (EHEC), enteroinvasive (EIEC), enteroaggregative (EGEC), and diffusely adherent (DAEC). EHEC is a subset of Shiga toxin producer *E. coli* (STEC) and comprises pathogens of which the serotype O157:H7 is the most often implicated in infection worldwide [17].

E. coli O157:H7 was identified in 1975; however, it was first recognized as a pathogen in 1982 during an outbreak investigation of hemorrhagic colitis. In 1994 became a U.S.A. notifiable infection, and by 2000, reporting of cases was mandatory in 48 states of that country [18]. Infections with *E. coli* O157:H7 can cause severe illness and death. The infection can include diarrhea that is often bloody, severe stomach cramps, and vomiting; infection can progress to hemolytic uremic syndrome (HUS) and end in fatality [19].

The largest *E. coli* O157:H7 infection outbreak occurred in 1996 in Japan, when approximately 7,892 school children and 74 school staff and teachers became infected with bacteria from white radish sprouts [6]. Now is known that produce-associated outbreaks peaked in summer and fall (74%), and most commonly occurred in restaurants (39%). The most common products implicated in outbreaks include lettuce (34%), apple cider or apple juice (18%), salads (16%), coleslaw (11%), melons (11%), sprouts (8%), and grapes (3%) [18].

Knowledge of transmission routes and vehicles of this bacterium allows consumers to be educated on reducing risky behaviors that can decrease their probability for infection. Studies on the interactions between *E. coli* O157:H7 and vegetables such as lettuce leaves demonstrated efficient attachment of bacteria to the surface, trichomes, stomata, and cut edges. *E. coli* O157:H7 was also shown to colonize the inner tissues and stomata of cotyledons of radish sprouts, developed from seeds experimentally contaminated with the bacterium [16].

2.3.3 LISTERIA MONOCYTOGENES

L. monocytogenes is an infrequent but pernicious pathogen linked to disproportionally high levels of morbidity and mortality in susceptible individuals and remains a significant cause of foodborne illness around the world (Table 2.1). *L. monocytogenes* is small, Gram-positive, facultative anaerobic, flagellated, rod-shaped bacterium that occurs widely in agricultural, aquacultural, and food processing environments. The organisms can exist in an intracellular state within monocytes and neutrophils. Eight species comprises the genera *Listeria: L. monocytogenes, L. innocua, L. seeligeri,*

L. welshimeri, L. ivanovii, L. grayi, L. rocourtiae and *L. marthii*. Of them, only *L. monocytogenes* is generally regarded as capable of causing illness in humans, although sporadic cases have been attributed to *L. ivanovii, L. seeligeri, L. innocua,* and *L. welshimeri* [20].

L. monocytogenes was first recognized as a human foodborne pathogen in 1981 by an outbreak linked to contaminated coleslaw [21]. In adults, the disease listeriosis can provoke invasive and non-invasive syndromes. The non-invasive form could result in febrile gastroenteritis, whereas the invasive form is characterized by the onset of severe symptoms (meningitis, septicemia, primary bacteremia, endocarditis, non-meningitic central nervous system infection, conjunctivitis, and flu-like illness). Gastrointestinal symptoms are observed in around 33% of patients [20].

In general, non-invasive gastrointestinal illness is observed in immunocompetent individuals, while invasive listeriosis infections typically occur in immunocompromised individuals or pregnant women. There is high morbidity and mortality (20–40%) associated with invasive listeriosis infections [21].

The EU has seen an increase in notifications (of both listeriosis outbreaks and sporadic cases) through time; only in 2013, 1763 confirmed human cases of listeriosis was reported in 27 states member [22]. Five Member States of the EU reported one food-borne outbreak where a salad was the vehicle. A similar trend has been observed since 2010 in the U.S. where, the country has experienced a number of listeriosis outbreaks attributed to foods considered to be "moderate risk" or "low risk" by the existing risk assessments, including fruits and vegetables such as celery, lettuce, cantaloupe, sprouts, stone fruit and caramel apples [23].

Isolates of *L. monocytogenes* have unique characteristics that allow their persistence in unusual environments such as growth in a wide range of temperature (–1.5 to 50°C), and pH (4.3 to 9.6). It can survive freezing, and it is relatively resistant to heat. All these characteristics make it very difficult to control this organism [20].

2.4 CONTROL OF BACTERIAL PATHOGENS IN PRODUCE

Microbial contamination can occur during any of the steps in the farm-to-consumer continuum, including production, harvesting, processing, and transportation, retailing, and handling at home. This contamination could arise from environmental, animal, or human sources [7]. Bacterial pathogens in feces are transmitted directly or indirectly through water, soil, and hands

[24]. Improper handling during the production chain has been regarded as a critical step for microbial contamination of produce.

The main food safety problems in produce production in many countries are related to poor education, language differences, communication skills, unhealthy habits of handlers, poor water quality, use of manure or sewage (in a small number of regions), and lack of sanitary facilities (in some farms). To improve the microbiological condition of produce, the Mexican government has implemented a national program, called the System for Reduction of Contamination Risks. This program is in accordance with the U.S. Food and Drug Administration (FDA) Food Safety Modernization Act (FSMA) and outlines a series of rules and activities that should be followed by farmers, particularly those who want to export their products. The Mexican government has released good practices manuals, guides, and formats for farmers, and has encouraged the participation of third-party accredited consultants. In several areas, government audits are conducted on the registration and administrative processes, productive facilities, hygiene, handling of domestic fauna, farmworker training, internal evaluation, procedure validation, traceability, productive history, water management, fertilization procedures, the use and handling of agrochemicals, harvesting processes, packing, transportation, and organic production (if applicable), among others. If an audit reveals that certain standards are met, then the government would provide certification to the production facility. This program is expected to maintain high levels of microbiological quality of the produce grown in Mexico.

Since on many occasions fresh produce receives minimal processing and is eaten raw, pathogen contamination can represent a serious risk. Developed and developing countries face challenges in developing and implementing measures that will improve the microbial safety of vegetables. Although, growth of bacterial pathogens will be limited or prevented by cooling the products to 4°C or less, persistence of *Salmonella* and *E. coli* O157:H7 on cilantro, basil, chive, and other fresh herbs has been reported for at least 24 days of storage [25].

Different treatments have been proved to assure elimination of both surface and internal contamination of produce by pathogens, such as chlorine dioxide, electrolyzed water, UV light, cold atmospheric plasma, hydrogen peroxide, organic acids and acidified sodium chlorite [7]. Recent evidence suggests that fresh produce decontamination may be enhanced by technologies that overcome the barrier of diffusion, such as the application of gaseous agents, dispersion of chemical agents through some methods that involve ultrasound, non-thermal technologies (cold atmospheric plasma, or

UV light), or electrolyzed water [10]. Another technology that shows efficacy for its use in fruits and vegetables is low dose irradiation. The use of 1 kGy (100 had) or fewer dosages of ionizing radiation are useful in extending the shelf-life of some fruits and vegetables due to their capability to affect the survival of various classes of microorganisms, especially Gram-negative bacteria and molds [9].

In recent years, consumer trends have shifted toward using fewer synthetic food additives, so that products can be consumed in a more natural or all-natural state. A report demonstrated that extracts from oregano leaves and the peel and pulp of limes were as effective as chlorine or Citrol (a natural and commercially available disinfectant) in reducing by > 2 logs. The population of *Salmonella,* and *E. coli* O157:H7 on leafy greens and therefore, maybe a natural and edible alternative to chemicals to reduce the risk of contamination on leafy vegetables [26]. In another study, extracts of olive, apple, and hibiscus reduced 1 to 3 logs CFU/g *Salmonella* Newport on leafy greens by day 3 [27]. Also, the cocktail named BEC8/TC (composed by eight lytic *E. coli* O157:H7-specific phage strains: 38, 39, 41, CEV2, AR1, 42, ECA1, and ECB7 plus Trans-cinnamaldehyde) was highly effective against EHEC on baby romaine lettuce and baby spinach leaves [28].

It is important to consider that pathogens can adhere strongly to produce surfaces and form resistant biofilms or become internalized within pores and at cut surfaces, reducing the efficacy of cleaning and sanitizing treatments [29]. However, various compounds such as citrol and extract of *Haematoxylon brassiletto* affect the biofilm formation on *E. coli* pathotypes EHEC and EAEC [30].

2.5 METHODS TO DETERMINE PRODUCE QUALITY

An important aspect to consider for the methods available for pathogens analyses is that the dose required to cause illness in many cases is very low; this indicates that in some cases the microorganism only needs to contaminate the food and survive without necessarily reproduction [16].

The European Regulation 2073/2005 introduced two types of microbiological criteria, such as the food safety criteria, to evaluate the foodstuff microbiological quality, and the process hygiene criteria, to assess if production processes are correctly working. The quantitative detection of foodborne pathogens is generally carried out by conventional microbiological methods, which involve three steps: cultural enrichment, selective plating, and confirmation [4]. These are simple to apply, but with critical flaws, such as being time-consuming, depending on the microorganism ability to grow on synthetic media, and the possible false-negative results because of failed detection of cells in the viable but non-cultivable state, nonetheless able to cause disease [31]. Due to these limitations, rapid sampling and first-generation detection methods have been developed to improve the traditional methods, among these, a concentration and filtration method, and nucleic acid hybridization and immunoassays, decreasing the time of sampling and detection respectively. Rapid second-generation detection methods such as PCR, concentration by filtration and PCR [32], magnetic capture hybridization followed by multiplex real-time PCR [30] should replace the three steps by just one step, avoiding the enrichment and the selective culture. However, in practice, it is still necessary to growth bacteria on selective media [4].

An important fact is that most methods developed that investigate fecal contamination of vegetables do not address the origin of contamination. This aspect is critical to establish more adequate measures of prevention or control. Bacteria of the order *Bacteroidales* are common inhabitants in the intestine of human and animals. Isolates from specific animal species exhibit particular genome sequences conserved in their genomes. This characteristic is regarded as indicators for tracking the sources of fecal contamination of produce [33]. *Bacteroidales* are successfully detected in produce samples when 10 mg of feces contaminate the surfaces of strawberries and tomatoes [24]. The members of the order *Bacteroidales* have characteristics that make them ideals for source tracking: limited to warm-blooded animals, key components of their intestinal microbiota, do not proliferate in the environment, contain host-associated 16S rRNA gene sequences, and are commensal bacteria found at high levels (10^{10} cells/g of human feces) in the small intestines of humans and other higher mammals representing 30 to 40% of total fecal bacteria [24]. Thus, *Bacteroidales* markers are promising tools to identify sources of fecal contamination; however, more research in this area is required to determine their potential use to reduce the risks of contamination of produce. More sophisticated tools such as whole-genome sequencing can also help to generate hypothetical transmission networks and in some instances facilitate traceback of foods to their origin [14].

KEYWORDS

- *E. coli* 0157:H7
- enteropathogens
- European Union
- *Listeria monocytogenes*
- produce safety
- *Salmonella*

REFERENCES

1. Diop, N., & Jaffee, S. M. Fruits and vegetable: Global trade competition in fresh and processed products markets. In: Ataman, A. M., & Beghim, J. C. *Global Agricultural Trade and Developing Countries* (pp. 237–257). The World Bank, Washington, DC, **2005**.

2. FAO. *Statistical Yearbook of the Food and Agricultural Organization of the United Nations*, **2013**. Available at: http://www.fao.org/docrep/018/i3107e/i3107e03.pdf (accessed on 6 January 2020).

3. SIAP. Atlas Agroalimentario 2012-2018. Servicio de información agroalimentaria y pesquera. *Sagarpa*, **2018**. https://nube.siap.gob.mx/gobmx_publicaciones_siap/pag/2018/Atlas-Agroalimentario- (accessed on 6 January 2020).

4. León, S. J., Jaykus, L. A., & Moe, C. L. Food safety issues and the microbiology of fruits and vegetables. In: Heredia, N., Wesley, I., & García, S., (eds.), *Microbiologically Safe Foods* (pp. 255–290). John Wiley & Sons, Inc., Hoboken, New Jersey, United States, **2009**.

5. Scallan, E., Hoekstra, R. M., Angulo, F. J., Tauxe, R. V., Widdowson, M. A., Roy, S. L., Jones, J. L., & Griffin, P. M. Foodborne illness acquired in the United States—Major pathogens. *Emerg. Infect. Dis.*, **2011**, *17*, 7–15.

6. Park, S., Szonyi, B., Gautam, R., Nightingale, K., Anciso, J., & Ivanek, R. Risk factors for microbial contamination in fruits and vegetables at the preharvest level: A systematic review. *J. Food Prot.*, **2012**, *75*, 2055–2081.

7. Olaimat, A. N., & Holley, R. A. Factors influencing the microbial safety of fresh produce: A review. *Food Microbiol.*, **2012**, *32*, 1–9.

8. Health Secretary. Epidemiological bulletin. https://www.gob.mx/cms/uploads/attachment/file/415172/sem46.pdf (accessed on 6 January 2020).

9. Brackett, R. E. Microbiological consequences of minimally processed fruits and vegetables. *J. Food Qual.*, **1987**, *10*, 195–206.

10. Paramithiotis, S., Drosinos, E. H., & Skandamis, P. N. Food recalls and warnings due to the presence of foodborne pathogens—a focus on fresh fruits, vegetables, dairy and eggs. *Curr. Op. Food Sci.*, **2017**, *18*, 71–75.

11. Ricke, S. C., Koo, O. K., Foley, S., & Nayak, R. *Salmonella*. In: Labbé, R. G., & García, S., (eds.), *Guide to Foodborne Pathogens* (pp. 112–137). Wiley-Blackwell. Hoboken, NJ, **2013**.

12. Heredia, N., & García, S. Animals as sources of foodborne pathogens: A review. *Anim. Nut.*, **2018**, *4*, 250–255.
13. Heaton, J. C., & Jones, K. Microbial contamination of fruit and vegetables and the behavior of enteropathogens in the phyllosphere: A review. *J. Appl. Microbiol.*, **2008**, *104*, 613–626.
14. Gould, L. H., Kline, J., Monahan, C., & Vierk, J. Outbreaks of disease associated with food imported into the United States, 1996–2014. *Emerg. Infect. Dis.*, **2017**, *23*, 525–528.
15. Golberg, D., Kroupitski, Y., Belausov, E., Pinto, R., & Sela, S. *Salmonella typhimurium* internalization is variable in leafy vegetables and fresh herbs. *Int. J. Food Microbiol.*, **2011**, *145*, 250–257.
16. Alegbeleye, O. O., Singleton, I., & Sant'Ana, A. S. Sources and contamination routes of microbial pathogens to fresh produce during field cultivation: A review. *Food Microbiol.*, **2018**, *73*, 177–208.
17. Feng, P. *Escherichia coli*. In: Labbé, R. G., & García, S., (eds.), *Guide to Foodborne Pathogens* (2nd edn., pp. 222–240). Wiley Blackwell Hoboken, NJ, **2013**.
18. Rangel, J. M., Sparling, P. H., Crowe, C., Griffin, P. M., & Swerdlow, D. L. Epidemiology of *Escherichia coli* O157:H7 outbreaks, United States, 1982–2002. *Emerg. Infect. Dis.*, **2005**, *11*, 603–609.
19. Heiman, K. E., Mody, R. K., Johnson, S. D., Griffin, P. M., & Gould, L. H. *Escherichia coli* O157 outbreaks in the United States, 2003–2012. *Emerg. Infect. Dis.*, **2015**, *21*, 1293–1301.
20. Donnelly, C. W., & Diez-Gonzalez, F. *Listeria monocytogenes*. In: Labbé, R. G., & García, S., (eds.), *Guide to Foodborne Pathogens* (2nd edn., pp. 45–74). Wiley Blackwell Hoboken, NJ, **2013**.
21. Allen, K. J., Wałecka-Zacharska, E., Chen, J. C., Katarzyna, K. P., Devlieghere, F., Meervennec, E. V., King, J. O., & Bania, W. J. *Listeria monocytogenes* – An examination of food chain factors potentially contributing to antimicrobial resistance. *Food Microbiol.*, **2016**, *54*, 178–189.
22. European Food Safety Authority. *Listeria*. https://www.efsa.europa.eu/en/topics/topic/listeria (accessed on 6 January 2020).
23. Buchanan, R. L., Gorris, L. G. M., Hayman, M. M., Jackson, T. C., & Whiting, R. C. A review of *Listeria monocytogenes*: An update on outbreaks, virulence, dose-response, ecology, and risk assessments. *Food Control*, **2016**, *75*, 1–13.
24. Merino-Mascorro, J. A., Hernández-Rangel, L., Heredia, N., & García, S. *Bacteroidales* as indicators and source trackers of fecal contamination in tomatoes and strawberries. *Food Prot.*, **2018**, *81*, 1439–1444.
25. Jung, Y., Jang, H., & Matthews, K. R. Effect of the food production chain from farm practices to vegetable processing on outbreak incidence. *Microb. Biotechnol.*, **2014**, *7*, 517–527.
26. Orue, N., García, S., Feng, P., & Heredia, N. Decontamination of *Salmonella, Shigella,* and *Escherichia coli* O157:H7 from leafy green vegetables using edible plant extracts. *J. Food Sci.*, **2013**, *78*, M290–M296.
27. Moore, K. L., Patel, J., Jaroni, D., Friedman, M., & Ravishankar, S. Antimicrobial activity of apple, hibiscus, olive, and hydrogen peroxide formulations against *Salmonella enterica* on organic leafy greens. *J. Food Prot.*, **2011**, *74*, 1676–1683.
28. Viazis, S., Akhtar, M., Feirtag, J., & Diez-Gonzalez, F. Reduction of *Escherichia coli* O157:H7 viability on leafy green vegetables by treatment with a bacteriophage mixture and trans-cinnamaldehyde. *Food Microbiol.*, **2011**, *28*, 149–157.

29. Sapers, G. M., & Doyle, M. P. Scope of the produce contamination problem. In: Matthews, K. R., Sapers, G. M., & Gerba, C. P., (eds.), *The Produce Contamination Problem* (pp. 3–20). Academic Press, **2014**.

30. Garcia-Heredia, A., Garcia, S., Merino-Mascorro, J. A., Feng, P., & Heredia, N. Natural plant products inhibits growth and alters the swarming motility, biofilm formation, and expression of virulence genes in enteroaggregative and enterohemorrhagic *Escherichia coli*. *Food Microbiol.*, **2016**, *59,* 124–132.

31. Carloni, E., Rotundo, L., Brandi, G., & Amaglian, G. Rapid and simultaneous detection of *Salmonella* spp., *Escherichia coli* O157, and *Listeria monocytogenes* by magnetic capture hybridization and multiplex real-time PCR. *Folia Microbiol.,* **2018**, *63,* 735–742.

32. Gonzalez, N. J. A. Validación de un método molecular acoplado a filtración por membrana para la detección de patógenos en chile jalapeño y melón cantaloupe (Validation of a molecular methodcoupled to membranefiltrationforthedetection of pathogens in jalapeño pepper and cantaloupe). *Tesis de licenciatura (Bachelor Thesis)*. Facultad de Ciencias Biológicas, U.A.N.L., **2018**.

33. Ravaliya, K., Gentry-Shields, J., Garcia, S., Heredia, N., Fabiszewski, D. A. A., Bartz, F. E., Leon, J. S., & Jaykus, L. A. Use of *Bacteroidales* microbial source tracking to monitor fecal contamination in fresh produce production. *Appl. Environ. Microbiol.*, **2014**, *80,* 612–617.

CHAPTER 3

Sustainable Production of Innocuous Seafood

FACUNDO JOAQUÍN MÁRQUEZ-ROCHA,[1]
CARLOS ALFONSO ÁLVAREZ-GONZÁLEZ,[2]
JENNY FABIOLA LÓPEZ-HERNÁNDEZ,[2] and
GUADALUPE VIRGINIA NEVÁREZ-MOORILLÓN[3]

[1]*Regional Center for Cleaner Production, National Polytechnical Institute, Tabasco Business Center, 86691 Cunduacan, Tabasco, Mexico*

[2]*Academic Division of Biological Sciences, Autonomous University Juarez of Tabasco, 86150 Villahermosa, Tabasco, Mexico*

[3]*School of Chemical Sciences, Autonomous University of Chihuahua, 31125 Chihuahua, Chih, Mexico*

ABSTRACT

Seafood products are one of the most commonly associated foods associated with public health risks, because of the foodborne pathogens that they can harbor. In order to diminish those risks, the main problems that need to be controlled are the sanitary safety of the environment where the products are captured, as well as the inspection of seafood during storage and distribution. This chapter describes the different pathogens related to seafood, as well as the challenges associated with the sustainable production of safe seafood.

3.1 INTRODUCTION

The nature of risks associated with seafood in public health may be summarized as follow: (i) health risks at the point of capture in contaminated water bodies, and (ii) at the processing and distribution stages. The epidemiological data associated with foodborne diseases related to seafood can help on the determination of microbiological and chemical risks; therefore, national control programs that can supervise seafood-related diseases are mandatory. Seafood imported supply between nations is from one-fifth to half of the total seafood consumption; consequently, it is fundamental to ensure measures of similar safety control globally, with regulations to protect consumers. Most of the federal policies contain effective state-based programs, including assistance and specialized facilities.

The nature of seafood hazards in the foodservice sectors needs a vigorous campaign for information and education, especially for products such as raw fish and shellfish. An international surveillance system can provide information on the incidence of seafood-borne disease. In this regard, global, national agencies should give guidelines on microbial and natural toxin contamination, and also elaborate and update relevant information related to the origin of the product, including fishing industry. In the USA, there are agencies such as FDA (Food and Drug Administration), EPA (Environment Protection Agency) and NOAA (the National Oceanic and Atmospheric Administration), Meanwhile, in México there are agencies such as COFEPRIS (Comisión Federal para la Protección Contra Riesgos Sanitarios), PROFEPA (Procuraduria Federal de Protección al Ambiente), and SEMARNAT (Secretaría de Medio Ambiente y Recursos Naturales). Health agencies have better recognition in the occurrence of toxicity on fish and shellfish, red tides and the existence of vulnerable groups in specific locations, which include seafood advisories warnings for species-associated health risks, emitted by government authorities.

3.2　FOODBORNE ILLNESS ASSOCIATED WITH THE SEAFOOD CONSUMPTION

Seafood is well-known for its nutritional value, rich in protein and high in polyunsaturated fats [1]. The fish and shellfish habitat is highly exposed to domestic and hazards pollution, many fish shoals live or feed close to industrial or domestic discharges; therefore, fish and shellfish capture for human consumption are vulnerable to microbiological and organic pollution. The safety status of seafood is a concern to consumers and authorities, although the fishery industry and government agencies with its regulations have intensified efforts to control spreading pollution. Rapid and accurate identification of microbial and organic food contaminants is a focal point to trace bacterial pathogens within the food supply [2]. Environmentally endanger seafood material has many forms to express its impact on human health, from infections, food decomposition, organic, and enterotoxin toxicity, traceability, unsafe processes, health future challenges. In each topic, it is necessary for the generation of productive dialogue and collaboration between health authorities, food safety experts, multi-pathogen expertise, and the public in general [3].

The incidence of foodborne diseases associated with shellfish and the contamination with sewage at collecting points has been related for many years. Shellfish has been associated with pathogenic bacteria such as *Salmonella* spp., *Shigella* spp., *Campylobacter* spp., *Vibrio* spp., and virus such as Hepatitis A and norovirus [3]. Even when the presence of pathogenic parasites such as *Giardia* or *Cryptosporidium* has been associated with shellfish, no outbreaks associated with consumption of this seafood has been reported up to now [4].

Many of the human-related cases of Salmonellosis are associated with the consumption of seafood. Enteric *Salmonella* causes gastroenteritis, recognized by abdominal pain, nausea, vomiting, diarrhea, and headache [3]. The increase in case number of gastroenteritis worldwide has been attributed to factors such as a change in husbandry, feeding practice, agronomic processes, and food technology, as well as an increase of global trade and lifestyle. As a consequence, the number of susceptible populations to foodborne diseases has increased; seafood is reservoirs for *Salmonella* and other enteric bacteria. Fish species used for human consumption are widely distributed in the environment, and its mobility is also another factor to consider in the vulnerability to be a reservoir of *Salmonella* species. Human infections are acquired through consumption of undercooked food or contaminated water and food origin [5]. Bacterial species are widely distributed in the marine environment and the intestinal tracts of aquatic organisms; fish species are the biggest group used for the animal-based foods and are commercially used for food production [6].

The presence of pathogenic microorganisms in seafood is not always related to an infectious process; pathogenic bacterial strains can usually have virulent and virulent subspecies. As an example, *Vibrio* infections are associated with the strain pathogenicity, the harvesting conditions, as well as with the state of the shellfish tissue [7], host susceptibility [8], and shellfish preparation, among other factors [9–12]. The pathogenic species from the *Vibrio* genera are *Vibrio parahaemolyticus* [13], *V. vulnificus* [8], and *V. cholerae* [14], and the symptoms associated with the infection are diarrhea and gastroenteritis, although *V. vulnificus* can cause septicemia [8]. *V. parahaemolyticus* is related to foodborne outbreaks in Japan and China [13]. Gastroenteritis caused by *Vibrio* spp. is still an important foodborne disease in the USA [15], with an estimated 2800 cases per year by raw oyster's consumption [9]. *V. cholerae* still infects millions annually worldwide, though the incidence of the shellfish-borne disease is relatively low [16]. Due to climate changes, the patterns of geographical distribution of pathogenic *Vibrio* spp. can vary, and the microorganisms might be found in places not previously reported. Not all *Vibrio* strains are pathogenic, and both are usually present in the same marine environment [17]; virulence genes can help on the discrimination of non-pathogenic strains [18]. Still, research on the heterogeneity of estuarine populations is limited [19]; in the case of estuarine *V. parahaemolyticus* populations, it is considered that only a small fraction is pathogenic [13]. *V. vulnificus, V. parahaemolyticus*, and *V. cholerae* can co-exist within shellfish; the growth of *V. cholera* [20] on chitin oligomers promotes a state of natural competence allowing them to uptake foreign DNA. Other studies have shown evidence of horizontal gene acquisition between pathogenic *Vibrio* species [21].

An effective strategy to limit microbially contaminated shellfish is harvesting in areas with good water quality, where, microbial contamination is at minimal levels. Regulations protecting public health are essential tools to manage microbial contamination. Depuration of shellfish pathogenic viruses by monitoring bacteria does not accurately reflect shellfish safety [22, 23]. A more aggressive surveillance of food-associated virus is needed; actually, the emphasis has been placed on Norovirus and SARS [3], coliphage as a potentially reliable viral indicator [23].

The abundance of *Vibrio* strains in natural habitats, including bivalve shellfish [24] and seaweeds [25] is intimately tied to abioticmicrobial contamination [26]. Among the environmental factors associated with the presence of *Vibrio* strains, temperature, nutrients [27, 28], suspended solids, and dissolved organic carbons are mentioned [29]. Salinity is strongly correlated with Vibrio growth and survival and can correlate with the presence of the microorganism in harvested oysters [9, 10]. The incidence of fecal contamination has often been related to seasonal and meteorological events, although, domestic, and industrial discharges play an essential role, and a variety of factors affect the incidence, persistence, and severity of microbial contamination.

On the other hand, the presence of multiresistant bacterial strains is an enormous problem, not only associated with human health, but to the presence of these microorganisms in the environment. There are reports on the widespread distribution of antimicrobial-resistant *Salmonella* strains through the food chain [30]. Plasmids with resistance to common antimicrobials and resistance to quinolones and fluoroquinolones have reduced treatment options for human *Salmonella* septicemia [31]. Raufu et al. [1] reported 11.5% of the incidence of *Salmonella*, serovar Hadar comprising 87%, serovar 8%, and 5% *Salmonella eko*. Hadar isolates present antibiotic resistance to strep-tomycin (43.5%), sulfamethoxazole (34.8%), and trimethoprim (21.7%). These findings confirm the contamination of fish by *Salmonella*, could be due to microbiological pollution of the lake from sources that could include agriculture sediments, wildlife contamination or urban pollution. Also, the processing steps, from harvest to the market, could influence the quality of fish for human consumption. Prevalence of bacterial contaminants depends on several factors such as season, dietary habits, and individual immune status after exposure to the pathogen [32].

There is a need for better animal and human health surveillance programs in developing countries. It is necessary to identify sources, serovar distribu-tion, and prevalence of Salmonella and antibiotic-resistant bacterial strains that can help on the construction of control programs. In a report from Nigeria, the prevalence of *Salmonella* serovars, as well as the antimicro-bial susceptibility patterns was obtained from *Salmonella enterica* isolated from fish. Although antimicrobial resistance was low, some isolates were considered as serious health risks for the human population [1]. In another study, one strain of *Shewanella putrefaciens* identified by biochemical test, API20NE system, and 16S rRNA and intraperitoneally injection, showed 50% mortality in *Oreochromis niloticus*, and important aquaculture-grown fish. The study concluded that *S. putrefaciens* caused fish infection, a disease with septicemia and mortality in *O. niloticus* [33]. *S. putrefaciens* has been isolated from marine and brackish waters [34], sea bass fish (*Dicentrarchus labrax*) [35] and Goldfish (*Carassius auratus auratus*) [36]. In this sense, *S. putrefaciens* is most frequently isolated from marine fish [37] than freshwater fish [38, 39] and in tilapia's farm tank water [40]. There are some reports on fish diseased with unusual behavior, and fish displayed stagnancy, abnormal swimming, abdomen swelling, lack of appetite, internal illness such as pale liver, ascites, and spleen enlargement [33, 41].

Another possible case of fish outbreak bacterial infection was reported in a fish farm located on the southwest shore of Manzala Lake near El-Serw navigation canal, and a pathogenic *Aeromonas hydrophila* was isolated from Nilo tilapia and African catfish [42, 43]. This zone seems to be impacted with agrochemicals and other environmental factors [44]; the Enterobacteriaceae genera *Serratia* and *Citrobacter* have been isolated from *O. niloticus*, using molecular tools, identification by four universal primers 16S rRNA [45]. Histopathological alterations were found in the liver (vacuolar degenera-tion), kidney (pyknotic nuclei), and gills (epithelium interstitial edema) [45].

On the other hand, *Citrobacter freundii* is considered an opportunistic pathogen for humans and animals [46] and was isolated in Egypt, from tilapia and mullet [47]. On the other hand, *Citrobacter braakii* was first reported from fish was in Bursa [41], while *S. marcescens* and *Citrobacter* sp. were isolated from infected tilapia [48]. Tilapia infected by *Citrobacter freundii* showed symptoms such as tail necrosis and hemorrhage on the skin [49]. Histological differences in fish infected by *C. freundii* were found in rainbow trout fry, cyprinids, and catfish [50].

3.3 SEAFOOD AQUACULTURE

Shrimp aquaculture represents an industry in tropical countries, because of the high-density culture conditions and the number of aquaculture farms. However, because of this increased production of shrimp, the prevalence of diseases has also increased. Invertebrates have developed defense systems that respond against antigens of potential pathogens. The defense mechanisms depend entirely on the innate immune system that is activated by recognition of cell surface host proteins, such as lectins or peptides [51]. Innovation in fish nutrition drives toward sustainable development of the rapid aquaculture industry. A supplement such as probiotics can improve the health and nutrition of livestock [52].

Crustacean growth is an important aquaculture industry in developing countries; due to the high-density culture practices, the increase of infectious diseases is of particular concern. Crustaceans are a large invertebrate animal group with a complex and efficient innate immune response against a variety of microorganisms; however, they lack an immune system, since there is no evidence of acquired immunity. In this sense, invertebrate innate immune mechanisms include cellular and humoral responses. Nutrients, oxygen, hormones, as well as antimicrobial, antioxidant, and anti-inflammatory peptides are distributed through the hemolymph; several antimicrobial peptides have been characterized, including antimicrobial peptides, in crustaceans [53, 54], the proteins penaeidins in *L. vannamei* and *L. setiferus* [55], *L. stylirostris* [56], *Fenneropenaeus chinensis* [57], *P. monodon* [58], and *L. setiferus* [59]. Crustacean proteins participate in immune defense by specific recognition of carbohydrate-containing molecules, especially those found in the external surface to Gram-negative, Gram-positive bacteria, viruses, or fungi. The protective processes induced by these carbohydrate-driven recognition patterns, like agglutination, encapsulation, phagocytosis, clottable proteins, and bactericidal activity, have been studied [51, 60].

Worldwide fish is an important source of animal protein, where aquaculture feed production is not the exception [61]; in countries like Egypt, it is a critical component of the livestock sector and a significant source of animal protein [62]. Aquaculture helps on increasing fish production, which was achieved through higher fish stocking density and the application of artificial feeding. Food for fish is close to 50% of the associated costs in intensive culture systems; its composition can be modified by adding vegetable oil, or plant-based nutrients, that will cause a reduction in the level of expensive fishmeal [63, 64]. One of the problems associated with plant-based nutrients is fungal contamination during crop production; during processing, the feed can also be contaminated with fungal spores, mainly when grains are ground and the feed pelleted [64]. Also during storage, temperature conditions, and environmental humidity can increase fungal growth in the fish feed, with the consequent production of mycotoxins [65]. Mycotoxigenic fungi can cause severe damage to fish, including liver damage, compromised immune response, reduced growth rate, and increase mortality, which is a severe challenge to aquaculture development [66]. Contamination of food supply by mycotoxins is not exclusive of the aquaculture industry and is a difficult task to accomplish [67, 68].

In East Africa, the most common farmed fish are Nile tilapia (*Oreochromis niloticus*) and African catfish (*Clarias gariepinus*) [69]. It is important to consider that most farmers in Africa use locally-made commercial fish feed that due to climate conditions in the continent can be easily contaminated with mycotoxin-producing fungi [68]. Therefore, it is important to standardize and control the manufacturing and distribution of fish feed, since there are only a few reports on Sub-Saharan Africa on mycotoxin contamination in fish feeds [68, 70].

Since feed is one of the high costs associated with aquaculture, strategies such as the addition of animal manure have been used as a source of nutrients. Manure is applied directly, with the consequent risks associated with the presence of pathogens. Also, the high nutrient content of manure can promote algal or plant growth, and as such, it has been used as fertilizer. In a report by Elsaidy et al. [71], the authors used chicken manure (CM) and fermented chicken manure (FCM) for growth of Nile tilapia (*Oreochromis niloticus*) in aquaculture ponds. Bacterial counts of mesophilic and total coliforms were high in the CM amendment ponds, and *E. coli* and *Salmonella* were isolated from the CM ponds, but not from the FCM-treated ponds; therefore, it is recommended the fermentation treatment as a bacteriologically safe fishpond fertilizer [71]. It has also been reported that under thermophilic fermentation of CM, pathogens are destroyed [72].

As a food source, seafood is considered very important; they provide high-quality protein, polyunsaturated fatty acids (PUFA), and other nutrients.

The worldwide seafood demand can be met with a combination of fisheries and aquaculture industry [73]. Innovations in seafood nutrition provide a sustainable development with functional diets directed to specific targets to improve physiological health and immune systems optimizing growth, feed conversion, and reduce time-consuming. Dietary supplements may be obtained from natural resources such as microalgae, or from agroindustrial residues from oil or starch production [74, 75].

3.4   FISH CONTAMINATION WITH ORGANIC COMPOUNDS

Because of the rich protein content of seafood products, many proteolytic bacteria can contaminate them, and as a result, many toxic compounds can be produced. Also, seafood can accumulate toxic organic compounds that can be found in marine environments in low concentrations, but that can be bioaccumulated through the food chain.

One chemical compound that can be found naturally in many food products is formaldehyde (FA), which has been identified as a Group 1 carcinogen [76, 77]. The compound is formed and accumulated in fish due to the reduction of amines in post-mortem conditions, even during freezing. Natural concentrations of FA in fish are between 6.5–293 mg kg^{-1} and tolerable levels of FA for human are 100 mg kg^{-1} [77]. In a report by Bhowmik et al. [78], FA concentration of different fish markets from Dhaka was analyzed. Results demonstrated different concentrations of FA in the fish samples, with a range of 5.1–12.26 mg kg^{-1} and 10.8–39.68 mg kg^{-1} in freshwater and marine finfish samples respectively. The study revealed that the level of FA in the marketed fish species was within the expected level, which is below the tolerable levels for humans [78].

Histamine has an essential role as a neurotransmitter in humans, can be produced by the organism or can be ingested in the diet; histamine intoxication develops symptoms rapidly causing a rash, urticaria, edema, and inflammation. Bacteria produce histamine in fish from the group of Enterobacteriaceae as an inappropriate practice procedure, for this reason, it is necessary to establish safety practices to limit the histamine production [79].

3.5   ASSESSMENT OF SEAFOOD SAFETY AND QUALITY

Extensive knowledge of decrease in foodborne infectious is limited to the industrialized world, where there is evidence of infectious disease caused by contaminated food and water. Industrial developments in microbiologically safe food in developed countries remain limited. In this regards multiple factors along the food chain influence food safety. Sustainable food safety standards depend on monitoring, regulation, and vigilance by government agencies. Considering food as an excellent vehicle for pathogens (bacteria, virus, parasites, fungus, chemicals) moving to other food new host is highly risky. Food manage practice is full of opportunities, the well-recognized food-borne pathogens, such as *Salmonella, Campylobacter, E. coli,* and *Listeria monocytogenes* seems to generate new public health challenges, for example, the antimicrobial resistance issue and its environmental concern. On the other hand, viral foodborne pathogens, such as Norovirus, Hepatitis A, rotavirus, and SARS, are of importance in surveillance efforts for food products. Even more, there are many parasitic pathogens associated with the contaminated food, but there are only a few studies on their prevalence in food products, livestock, and wildlife. Among the challenges on food safety, the collaboration of the different areas related to the food production chain can be considered one of the most important, since is fundamental to understand the interaction of pathogens in their environment, so that new trends in epidemiology or the presence of emerging foodborne pathogens can be detected on time [3].

International agencies such as FAO [80] have compiled management resources providing details concerning risk mitigation and implementation practices of proper hygienic and manufacturing procedures (GHP/GMP) of hazard analysis and critical control point (HACCP) system to control biotoxins, pathogenic bacteria, viruses, and chemical pollutants. The minimum number of viable bacterial count to cause disease (MID) is high for *Vibrio* spp. ($>10^5$–10^6 CFU/g) but low for *Salmonella typhi* and *Shigella* spp. The concentration of bacteria that are pathogenic for human beings is usually low, but high concentrations are found in mollusks and the intestine of mollusk predators.

Contamination of fish products is frequently related to inadequate hygienic practices, including poor personal hygiene, inadequate water quality, and unsanitary processing conditions. To diminish the presence of pathogenic microorganisms, it is suggested the elimination of them in the food product. Cumulative time of the presence of bacterial pathogens growth on fish and fishery products depends on the internal temperature of exposure. The genus *Vibrio* pathogen species follow a similar pattern of cumulative time, safe for bacterial pathogens have a concern in seafood processing and handling. Increasing exposure temperature, increase exponentially bacterial growth [9].

3.5.1 *PATHOGENIC BACTERIA IN THE AQUATIC ENVIRONMENT AND NATURALLY PRESENT IN SEAFOOD*

There are several pathogenic bacteria related to seafood products, that have been associated with foodborne outbreaks, and that continue to be public health risks worldwide. Table 3.1 describes some outbreaks reported by national and international authorities.

Clostridium botulinum is ubiquitously present in diverse environments, soils, dust, marine, and freshwater sediments of wetlands, rivers, and lakes. Spores are transported by surface water in heavy rain or dust by the wind. There are at least five serotypes reported (A, B, C, D, E), and each have different geographical distribution. The mechanisms triggering a botulism outbreak remain poorly understood, but factor such as low water levels and high summer surface temperature has been correlated with outbreaks. In general, conditions such as pH between 7.5 and 9, negative redox potential and temperatures above 20°C botulism has a much higher prevalence [81]. The principal food conditions that can control the growth of *C. botulinum* are pH, water activity (A_w), temperature, and the presence of preservative substances. The presence of an associate (spoilage) microbiota may add to the risk of botulism since the microorganisms can use the oxygen present, and facilitate the growth and toxin production by *C. botulinum* [82].

Vibrio species can be naturally found in marine environments; environmental factors such as temperature, salinity, and the presence of algae, can be associated with their prevalence. Some species can cause human disease, including vibriosis, cholera, septicemia, and gastroenteritis [87]. Among the pathogenic *Vibrio* species are *Vibrio parahaemolyticus* and *V. cholerae* that causes gastrointestinal infections, while *V. vulnificus* can cause septisemia [13].

V. parahaemolyticus can be isolated from seafood products, in particular from bivalve mollusks. Higher prevalence of the pathogen is associated with warmer temperatures, as well as with high salinity content [87]. The virulence mechanisms of the microorganism are associated with hemolytic components, especially a thermostable direct hemolysin. Serotypes O3:K6, O4:K68, O1:R25 have recognized as the predominant groups responsible for most outbreaks since 1996 and considered more virulent than other serotypes [13]. Considering that is a mesophilic, halotolerant bacteria, *V. parahaemolyticus* will grow well in seafood stored at ambient temperature. The generation time at high temperatures (e.g., 12–18 minutes at 30°C)

TABLE 3.1 Pathogenic Bacteria Associated with Foodborne Outbreaks with Seafood as Source of Contamination

Microorganism	Associated Seafood	Location	Year	Comments	References
Vibrio cholerae	Multiple	Worldwide, with reports from 71 countries, particularly Yemen, Democratic Republic of Congo and Somalia	2017	Seventh reported pandemia	[83]
Salmonella	Fish and fish products	Europe	2017	Four outbreaks reported by European Union Members	[84]
Salmonella	Fish and fish products	Latvia, Poland, and Spain	2015	Six outbreaks reported	[85]
Salmonella	Frozen raw yellow fin tuna product	United States	2012	425 cases in 28 states	[86]
Vibrio parahaemolyticus	Fresh Crab Meat Imported from Venezuela	United States	2018	26 cases	[86]
E. coli pathogenic	Fish and fish products	Luxemburg	2015	43 cases	[85]
Staphylococcal enterotoxin	Fish and fish products	Belgium	2015	75 cases	[85]

allows the organism to proliferate rapidly, and has been demonstrated that can proliferate in live oysters [88]. In general, low-temperature storage will decrease bacterial numbers, but it will depend on the food matrix, salinity, and other factors [89].

Vibrio vulnificus can cause bacteremia and septicemia and is usually related to the consumption of raw seafood, especially bivalve mollusks (oysters). Outbreaks are reported from the United States [86], and the infection can be lethal in patients with conditions such as chronic cirrhosis, hepatitis, or alcohol abuse. The microorganism has several virulence factors that are responsible for the diverse symptoms, including a capsular polysaccharide, motility, cytotoxicity, and proteins involved in attachment and adhesion [90]

Regarding *Vibrio cholerae,* the microorganism is dispersed by water contaminated with fecal matter from infected patients. The bacteria has been associated to severe pandemic cases, with a large number of cases reported in many countries; in the present years, the seventh pandemia has been associated to the El Tor serotype and continues to present cases worldwide since 1970. In 2017, the countries with higher reported cases were Yemen (1,032,481 cases), Somalia (75,414 cases), and the Democratic Republic of Congo (56,190 cases) [83]. The microorganism can survive in the environment, and because of its ability to grow in high salinity, they are found in marine and estuarine environments and have been associated to zooplankton [91]; as such, there have been able to transport through the oceanic water bodies, dispersing the pathogenic strain to all continents. Seafood products are associated with *V. cholerae* transmission. It has been demonstrated that the microorganism can survive in estuarine water and rivers [92], and has been isolated from crab [93]. The microorganism is sensitive to heat and acid and those conditions are suggested for seafood treatments in order to control their prevalence [92].

Listeria monocytogenes is a Gram-positive microorganism that is widely distributed in the environment and has been recently related to foodborne outbreaks, that is especially dangerous to immunosuppressed patients. The microorganism can stimulate growth at refrigerating temperatures (psychrotolerant), and as such, it has been associated with food products that are maintained and transported in refrigeration, including fish and other seafood products [94]. According to Denny and McLauchlin, [95], the cases of listeriosis have diminished thanks to the surveillance mechanisms established in the European Union (EU) for foodborne pathogens. The presence of *L. monocytogenes* in seafood products is more related to the storage and processing of foods, more than to the presence of the pathogen in marine or estuarine environments. Therefore, control of *L. monocytogenes* in seafood products can be achieved by good hygienic practices (GHP) and the establishment of the HACCP (Hazzard Analysis and Critical Control Program) in the food processing facilities [96].

Salmonella causes two syndromes, enteric (typhoid) fever and gastroenteritis, *S. enterica* serovar Typhi and *S. enterica* serovar Paratyphi are responsible for a systematic infection with high fever, abdominal cramps in the first week, followed for watery diarrhea, and persistent abdominal pain. Incubation period around 8–28 days. Non-typhoidal Salmonella causes gastroenteritis with an incubation period of 8–72 h; symptoms are non-bloody diarrhea, abdominal pain, and muscle aches and fever [15]. Foodborne infections associated with Salmonella usually involve beef and chicken products, as well as non-cooked eggs; still, there are cases of outbreaks associated with fish and fish products, and the number of those cases has increased in recent years [84–86]. Salmonella can be widely distributed in the environment; therefore, it can be found in water contaminated with human or animal excreta, where shellfish can be found; this can be the cause of salmonellosis outbreaks associated with shellfish [4]. As with *Listeria*, bad handling, and processing procedures can contaminate seafood products [1]; also, the use of manure in aquaculture, can cause fish product contamination [71]. Persistence and dissemination of Salmonella are similar in saltwater and freshwater fish, and it can survive for long periods in soil and environments [97]. Critical temperature has been identified in countries with the tropical environment with deficient post-harvest management of the products, inadequate infrastructure to preserve at low temperature, low electricity availability, and no appropriate market facilities [98].

As with *Salmonella*, seafood products can also be contaminated with *Shigella*, a Gram-negative enterobacterium that cause gastrointestinal infections; *S. dysenteriae* is the etiological agent of dysentery. The microorganism is transmitted by the fecal-oral route, and seafood can be contaminated in water bodies contaminated with wastewater. Strategies for prevention of contamination with *Shigella* include the control of water quality and GHP in the processing plants and restaurants [99].

Other enterobacteria that have been associated with contaminated seafood products are *Escherichia coli*. There are several pathogenic strains of *E. coli*, that depending on their virulence factors and clinical symptoms, are identified as Enteropathogenic *E. coli* (EPEC), enterotoxigenic *E. coli* (ETEC), enteroinvasive *E. coli* (EIEC), diffuse-adhering *E. coli*

(DAEC), enteroaggregative *E. coli* (EAggEC) and enterohemorrhagic *E. coli* (EHEC) or verotoxic *E. coli* (VTEC) [100]. There are few reports on EPEC outbreaks associated with seafood [85]. EC legislation [101], implemented in 2005, determines the criteria for the production and harvesting of bivalve mollusks.

Another microorganism that has been associated with foodborne outbreaks by bacterial pathogens is *Staphylococcus aureus* enterotoxin producer, which causes a usually self-limited intoxication. The enterotoxin is heat stable, and is accumulated in the food product; the microorganisms could remain viable at the point of food consumption, but is the enterotoxin the responsible for the intoxication symptoms. Food contamination by *S. aureus* is associated to poor hygienic practices since the food handlers usually contaminate the food products by manipulation, but the microorganism has also been isolated from raw fish or fish products [102]. There are recent reports of intoxication by *staphylococcal enterotoxin* associated with fish and fish products [85].

3.6 NATURAL ANTIMICROBIAL FROM SEA PRODUCTS

Even when sea products are primarily considered as an excellent source of nutrients by direct consumption, the immense biodiversity of marine organisms can also be studied as a source of many other compounds. Algae is included in the diet in some Asiatic countries, but can also be used for the preparation of other useful materials.

Gracilaria verrucosa is an alga commonly found in seawater. It has been found in *G. verrucosa* bioactive compounds (BCs) against pathogenic fish bacteria to support fish farming. In this regard, fractions obtained from LC-MS chromatography with an ethanolic eluent contained antibacterial compounds of the polyphenol type. A fraction containing Quercetin-7-methyl-ether has a moderate antibacterial activity against *Aeromonas hydrophila*, *Pseudomonas aeruginosa*, *Pseudomonas putida*, and a weak antibacterial effect against *Vibrio harvey* and *V. algynoliticus* [103].

The mangrove crab from genus *Scylla* has shown the formation of antimicrobial peptides against Gram-positive and Gram-negative bacteria. These activities have found in muscle tissue, hemolymph, and shell. Attention for mud crabs aquaculture may result in a potential for therapeutic applications [104].

3.7 CONCLUDING REMARKS

The network FoodNet in the United States generate information that provides essential data for monitoring and guiding safety efforts, by identifying changes in the incidence of illnesses from a specific source. FoodNet is an effort that combines information from several state health departments: Centers for Disease Control and Prevention (CDC), United States FDA, United States of Agriculture's Food Safety and Inspection Service (USDA-FSIS) [105]. A compilation on 2015 by FoodNet confirmed incidence of pathogen infections, *Salmonella*, *Campylobacter*, *Shigella*, *Cryptosporidium*, *Vibrio*, *Yersinia*, *Listeria*, and *Cyclospora*, incidence rate were highest in children <5 years, Listeria had highest rates of deaths and hospitalized infection incidents (13 deaths per 100 infections and 96% hospitalization) [106]. These findings have a limitation because do not distinguish transmission source, some infection transmitted by food are not routinely identified, the surveillance area population might reflect specific locations. Another important consideration to be considered is the presence of antibiotic-resistant microbial strains, especially considering the actual crisis of multiresistant strains [107] (Table 3.2).

Several methodologies and procedures can help to visualize the entire scene around pathogenic seafood-borne infections, not only traditional biochemical, and microbiological pathogen analyses but analytical and molecular approaches as well (Table 3.2). Analytical tools may contribute to monitoring food safety and detect contaminants, which may cause health risks. Analytical techniques are accurate and advance procedures, but equipment and maintenance of it are quite expensive, usually used at the end of the pipe. Electrochemical biosensors provide high efficiency and accuracy measurements in the food industry. These biosensors challenge quantitative detection and screening of food contaminants and time-effective [112, 113]. In the case of visualizing the impacted area of anthropogenic organic compounds and waste, PCR technology may be used to detect the presence of enteropathogenic bacteria, pesticide, and hydrogen degrading bacteria. The sequencing of 16S rRNA is an important tool for the identification of unknown bacterial species of fish pathogen [45], pesticide-degrading bacteria [114], and hydrocarbon-degrading bacteria [110, 111, 115, 116], the specific increment on bacteria abundance is promoted by pollutants such as domestic waste, pesticide pollution, and oil pollution, respectively. Some research has been published pesticide degradation [108, 109].

TABLE 3.2 Primers Used to Identified Specific Bacteria Groups in Environment, as Good Detector of Pathogens or Pollutants

Bacteria	Primer Used	Type of Bacteria/Metabolic Pathway	References
Vibrio parahaemolyticus serotypes	wl66654(F)(5'-AGGAGCCCACAATGAACTG-3') wl66655(R)(5'-AAATGATACTTACGCACAAAC-3')	Pathogenic bacteria	[13]
Lysinibacillus fusiformis sp.	27F(5'-AGAGTTTGATCCTTGGCTCAG-3') 1492R(5'-TACGGCTACCTTGTTACGACTT-3')	Pesticide-degradation	[108]
Bacillus pumilus, *Stenutrophomonus* sp.	27F(5'-AGAGTTTGATCMTGGCTCAG-3') 1942R(5'-TACGG(C/T)TACCTTACGACTT-3')	Pesticide-degradation	[109]
Pseudomonas and *Bacillus*	F(5'-AGAGTTTGATCMTGGCTAG-3') R(5'-GGMTACCTTGTTACGAYTTC-3')	Hydrocarbon-degradation	[110]
Mycobacterium sp	27F(5'-AGAGTTTGATCCTTGGCTCAG-3') 1492R(5'-TACGGCTACCTTGTTACGACTT-3')	Hydrocarbon-degradation	[111]

KEYWORDS

- chicken manure
- Food and Drug Administration
- foodborne pathogens
- formaldehyde
- good hygienic practices
- seafood

REFERENCES

1. Raufu, I. A., Lawan, F. A., Bello, H. S., Musa, A. S., Ameh, J. A., & Ambai, A. G. Occurrence and antimicrobial susceptibility profiles of *Salmonella* serovars from fish in Maiduguri, sub-Saharah, Nigeria. *Egyptian J Aquatic Res.*, **2014**, *40*, 59–63.
2. Germini, A., Masola, A., Carnevali, P., & Marchelli, R. Simultaneous detection of *Escherichia coli* O175:H7, *Salmonella* spp., and *Listeria monocytogenes* by multiplex PCR. *Food Control*, **2009**, *8*, 733–738.
3. Newell, D. G., Koopmans, M., Verhoef, L., Duizer, E., Aidara-Kane, A., Sprong, H., Opsteegh, M., Langelaar, M., Threfall, J., Scheutz, F., Van Der Giessen, J., & Kruse, H. Food-borne diseases—the challenges of 20 years ago still persist while new ones continue to emerge. *Int. J. Food Microbiol.*, **2010**, *139*(1), S3–S15.
4. Potasman, I., Paz, A., & Odeh, M. Infectious outbreaks associated with bivalve shellfish consumption: A worldwide perspective. *Clin. Infect. Dis.*, **2002**, *35*, 921–928.
5. Ochlenschlager, J., & Rehbein, H., Basic facts and figures. In: Rehbein, H., Ochlenschlager, J., (eds.), *Fishery Products: Quality, Safety, and Authenticity* (pp. 1–18). Wiley-Blackwell Inc., USA, **2009**.
6. Majowicz, S. E., Musto, J., Scallan, E., Angulo, F. J., Kirk, M., O'Brien, S. J., Jones, T. F., Fazil, A., & Hoekstra, R. M. International collaboration on enteric disease 'burden of illness' studies. The global burden of nontyphoidal *Salmonella* gastroenteritis. *Clin. Infect.*, **2010**, *50*, 882–889.
7. Smith, B., & Oliver, J. D. *In situ* and *in vitro* gene expression by *Vibrio vulnificus* during entry into, persistence within, and resuscitation from the viable but nonculturable state. *Appl. Environ. Microbiol.*, **2006**, *72*, 1445–1451.
8. Gulig, P. A., Bourdage, K. L., & Starks, A. M. Molecular pathogenesis of *Vibrio vulnificus*. *J Microbiol.*, **2005**, *43*, 118–131.
9. FDA. *Quantitative Risk Assessment on the Public Health Impact of Pathogenic Vibrio parahaemolyticus in Raw Oysters*. US Food and Drug Administration, Washington, DC, **2005**.
10. WHO/FAO. *Risk Assessment of Vibrio vulnificus in Raw Oysters*. Interpretive summary and technical report, World Health Organization/Food and Agriculture Organization of the United Nations, Geneva, Switzerland, **2005**.

11. Hofreuter, D., Tsai, J., Watson, R. O., Novik, V., Altman, B., Benitez, M., et al. Unique features of a highly pathogenic *Campylobacter jejuni* strain. *Infect Immun.*, **2006**, *74*, 4694–4707.

12. Chatzidaki-Livanis, M., Hubbard, M. A., Gordon, K., Harwood, V. J., & Wright, A. C. Genetic distinctions among clinical and environmental strains of *Vibrio vulnificus*. *Appl. Environ. Microbiol.*, **2006**, *72*, 6136–6141.

13. Guo, X., Liu, B., Chen, M., Wang, Y., Wang, Y., Chen, H., Wang, Y., Tu, L., Zhang, X., & Feng, L. Genetic and serological identification of 3 *Vibrio parahaemolyticus* strains as candidates for novel provisional O serotypes. *Int. J. Food Microbiol.*, **2017**, *245*, 53–58.

14. Colwell, R. R. Infectious disease and environment: Cholera as a paradigm for waterborne disease, *Int. Microbiol.*, **2004**, *7*, 285–289.

15. Butt, A. A., Aldridge, K. E., & Sanders, C. V. Infections related to the ingestion of seafood Part I: Viral and bacterial infections. *Lancet Infect. Dis.*, **2004**, *4*, 201–212.

16. Centers for Disease Control and Prevention (CDC). Surveillance for food-borne disease outbreaks. United States, 1998–2002. *Morb Mortal Wkly Rep Surveill Summ*, **2006**, *55*(10), 1–42.

17. Hurley, C. C., Quirke, A., Reen, E. J., & Boyd, F. E. Four genomic islands that mark post-1995 pandemic *Vibrio parahaemolyticus* isolates, *BMC Genomics*, **2006**, *7*(1), 104.

18. Nordstrom, J. L., Vickery, M. C. L., Blackstone, G. M., Murray, S. L., & Depaola, A. Development of a multiplex real-time PCR assay with an internal amplification control for the detection of total and pathogenic *Vibrio parahaemolyticus* bacteria in oysters. *Appl. Environ. Microbiol.*, **2007**, *73*, 5840–5847.

19. Zimmerman, A. M., Depaola, A., Bowers, J. C., Krantz, J. A., Nordstrom, J. L., Johnson, C. N., & Grimes D Variability of total and pathogenic *Vibrio parahaemolyticus* densities in northern Gulf of Mexico water and oysters. *Appl. Environ. Microbiol.*, **2007**, *73*, 7589–7596.

20. Meibom, K. L., Blokesch, M., Dolganov, N. A., Wu, C. Y., & Schoolnik, G. K. Chitin induces natural competence in *Vibriocholerae*. *Science*, **2005**, *310*, 1824–1827.

21. Gonzalez-Escalona, N., Blackstone, G. M., & Depaola, A. Characterization of *a Vibrio alginolyticus* strain, isolated from Alaskan oysters, carrying a hemolysin gene similar to the thermostable direct hemolysin-related hemolysin gene (trh) of *Vibrio parahaemolyticus*. *Appl. Environ. Microbiol.*, **2006**, *72*, 7925–7929.

22. Formiga-Cruz, M., Tofino-Quesada, G., Bofill-Mas, S., Lees, D. N., Henshilwood, K., Allard, A. K., et al. Distribution of human virus contamination in shellfish from different growing areas in Greece, Spain, Sweden and the United Kingdom. *Appl. Environ. Microbiol.*, **2002**, *68*, 5990–5998.

23. Dore, W. J., Mackie, M., & Lees, D. N. Levels of male-specific RNA bacteriophage and *Escherichia coli* in molluscan bivalve shellfish from commercial harvesting areas. *Lett. Appl. Microbiol.*, **2003**, *36*, 92–96.

24. Pruzzo, C., Huq, A., Colwell, R. R., & Donelli, G. Pathogenic vibrio species in the marine and estuarine environment. In: Belkin, S., & Colwell, R. R., (eds.), *Ocean and Health Pathogens in the Marine Environment* (pp. 217–252). Springer, Verlag, New York, **2005**.

25. Mahmud, Z. H., Neogi, S. B., Kassu, A., Mai Huong, B. T., Jahid, I. K., Islam, M. S., & Ota, F. Occurrence, seasonality and genetic diversity of *Vibrio vulnificus* in coastal seaweeds and water along the Kii Chanel, Japan. *FEMS Microbiol. Ecol.*, **2008**, *64*, 209–218.

26. Martinez-Urtaza, J., Lozano-Leon, A., Varela-Pet, J., Trinanes, J., Pazos, Y., & Garcia-Martin, O. Environmental determinants of the occurrence and distribution of *Vibrio*

parahaemolyticus in the rias of Galicia, Spain. *Appl. Environ. Microbiol.,* **2008**, *74,* 265–274.

27. Sousa, O. V., Macrae, A., Menezes, F. G., Gomes, N. G., Vieira, R. H., & Mendonca-Hagler, L. C. The impact of shrimp farming effluent on bacterial communities in mangrove waters, Ceara, Brazil. *Mar. Pollut. Bull.,* **2006**, *52,* 1725–1734.

28. Eiler, A., Johansson, M., & Bertilsson, S. Environmental influences on *Vibrio* populations in northern temperate and boreal coastal waters (Baltic and Skagerrak Seas). *Appl. Environ. Microbiol.,* **2006**, *72,* 6004–6011.

29. Gugliandolo, C., Carbone, M., Fera, M. T., Irrera, G. P., & Maugeri, T. L. Occurrence of potentially pathogenic vibrios in the marine environment of the straits of Messina (Italy)', *Mar. Pollut. Bull.,* **2005**, *50,* 692–697.

30. Le Hello, S., Hendriksen, R. S., Doublet, B., Fisher, I., Nielsen, E. M., Whichard, J. M., et al. International spread of an epidemic population of *Salmonella* enterica serotype Kentucky ST198 resistant to ciprofloxacin. *JID,* **2011**, *204,* 675–684.

31. Foley, S. L., & Lynne, A. M. Food animal-associated *Salmonella* challenges: Pathogenicity and antimicrobial resistance. *J. Anim. Sci.,* **2008**, *86*(E. Suppl.), E173–E187.

32. Bhaftopadhyay, P. Fish-catching and handling. In: Robinson, R. K., (ed.), *Encycl. Food Microbiol.,* (Vol. 2, p. 1547). Academic Press, London, **2000**.

33. El-Barbary, M. I. First recording of *Shewanella putrefaciens* in cultured *Oreochromis niloticus* and its identification by 16Sr RNA in Egypt. *Egypt. J. Aqua. Res.,* **2017**, *43,* 101–107.

34. Al-Harbi, A. H., & Uddin, N. Bacterial diversity of tilapia (*Oreochromis niloticus*) cultured in brackish water in Saudi Arabia. *Aquaculture,* **2005**, *250,* 566–572.

35. Korun, J., Akgun-Dar, K., & Yazıcı, M. Isolation of *Shewanella putrefaciens* from cultured European sea bass, (*Dicentrarchus labrax*) in Turkey. *Rev. Med. Vet.,* **2009**, *11,* 532–536.

36. Altun, S., Büyükekiz, A. G., Duman, M., Özyiğit, M. Ö., Karataş, S., & Turgay, E. Isolation of *Shewanella putrefaciens* from Goldfish (*Carassius auratus auratus*). *Isr. J. Aquacult. Bamid, IJA,* **2014**, *66*(956), 7.

37. Al-Harbi, A. H., & Uddin, M. N. Seasonal variation in the intestinal bacterial flora of hybrid tilapia (*Oreochromis niloticus, Oreochromis aureus*) cultured in earthen ponds in Saudi Arabia. *Aquaculture,* **2004**, *229,* 37–44.

38. Qin, L., Zhu, M., & Xu, L. First report of *Shewanella* sp. and *Listonella* sp. infection in freshwater cultured loach, *Misgurnus anguillicaudatus. Aquacult. Res.,* **2012**, *45*(4), 602–608.

39. Pezkala, A., Kozinska, A., Pazdzior, E., & Głowacka, H. Phenotypical and genotypical characterization of *Shewanella putrefaciens* strains isolated from diseased freshwater fish. *J. Fish Dis.,* **2015**, *38,* 283–293.

40. Lu, S., & Levin, R. E. *Shewanella* in a tilapia fish farm. *J. Fish Sci. Commun.,* **2010**, *4*(2), 159–170.

41. Altun, S., Duman, M., Buyukekiz, A., Ozyigit, M., & Karatas, S. First isolation of *Citrobacter braakii* from rainbow trout (*Oncorhynchus mykiss*). *Isr. J. Aquacult. Bamid.,* **2013**, *65*(915), 7.

42. El-Barbary, M. I. Some clinical microbiological and molecular characteristics of *Aeromonas hydrophila* isolated from various naturally infected fishes. *Aquacult. Int.,* **2010**, *18,* 943–954.

43. EL-Barbary, M. I. Pathogenic characteristics and molecular identification of *Aeromonas hydrophila* isolated from some naturally infected cultured fishes. *Egypt. J. Aquat. Res.,* **2010**, *36*(2), 345–356.

44. Al-Afify, F. D., Osman, M. E., & Elnady, M. *Ecological Studies on El-Serw Fish Farm, Egypt: Drainage Water Quality, Sediment Analysis and Heavy Metals Pollution in Nile Delta* (p. 200). Farms Paperback-LAP LAMBERT Academic Publishing, **2014**.

45. El-Barbary, M. I., & Hal, A. M. Molecular identification and pathogenicity of *Citrobacter* and *Serratia* species isolated from cultured *Oreochromis niloticus*. *Egyptian J Aquatic Res.*, **2017**, *43*, 255–263.

46. Chuang, Y. M., Tseng, S. P., Teng, L. J., Ho, Y. C., & Hsueh, P. R. Emergence of cefotaxime resistance in *Citrobacter freundii* causing necrotizing *fasciitis* and osteomyelitis. *J. Infect.*, **2006**, *53*, 161–163.

47. Hassan, A. H. M., Noor-El Deen, A. E., Galal, H. M., Sohad, M. D., Bakry, M. A., & Hakim, A. S. Further characterization of *Enterobacteriaceae* isolated from cultured freshwater fish in Kafr El Shiek Governorate: Clinical, biochemical and histopathological study with emphasis on treatment trials. *Glob Vet.*, **2012**, *9*(5), 617–629.

48. Chan, X. Y., Chang, C. Y., Hong, K. W., Tee, K. K., Yin, W. F., & Chan, K. G. Insights of biosurfactant producing *Serratiam arcescens* strain W2.3 isolated from diseased tilapia fish: A draft genome analysis. *Gut. Pathog.*, **2013**, *5*, 29.

49. Thanigaivel, S., Vijayakumar, S., Gopinath, S., Mukherjee, A., Chandrasekaran, N., & Thomas, J. *In vivo* and *in vitro* antimicrobial activity of *Azadirachta indica* (Lin) against *Citrobacter freundii*, isolated from naturally infected Tilapia (*Oreochromis mossambicus*). *Aquaculture*, **2015**, *437*, 252–255.

50. Jeremic, S., Jaki-Dimi, D., & Veljovic, L. J. *Citrobacter freundii* as a cause of disease in fish. *Acta Vet. Beograd*, **2003**, *53*(5/6), 399–410.

51. Vazquez, L., Alpuche, J., Maldonado, G., Agundis, C., Pereyra-Morales, A., & Zanteno, E. *Innate Immunity*, **2009**, *15*(3), 179–188.

52. Wanka, K. M., Damerau, T., Costas, B., Krueger, A., Schulz, C., & Wuertz, S. Isolation and characterization of native probiotics for fish farming. *BMC Microbiol.*, **2018**, *18*, 119–131.

53. Fredrick, W. S., & Ravichandran, S. Hemolymph proteins in marine crustaceans. *Asian Pacific J. Tropical Biomed.*, **2012**, *2*(6), 496–502.

54. Smith, V. J., & Chisholm, J. R. Antimicrobial proteins in crustaceans. *Adv. Exp. Med. Biol.*, **2001**, *484*, 95–112.

55. Gross, P. S., Bartlett, T. C., Browdy, C. L., Chapman, R. W., & Warr, G. W. Immune gene discovery by expressed sequence tag analysis of hemocytes and hepatopancreas in the Pacific white shrimp, *Litopenaeus vannamei*, and the Atlantic white shrimp, *L. setiferus*. *Dev. Comp. Immunol.*, **2001**, *25*, 565–577.

56. Muñoz, M., Vandenbulcke, F., Garnier, J., et al. Involvement of penaeidins in defense reactions of the shrimp *Litopenaeusstylirostris* to a pathogenic *Vibrio*. *Cell Mol. Life Sci.*, **2004**, *61*, 961–972.

57. Kang, C. J., Wang, J. X., Zhao, X. F., Yang, X. M., Shao, H. L., & Xiang, J. H. Molecular cloning and expression analysis of Ch-penaeidin, and antimicrobial peptide from Chinese shrimp, *Fenneropenaeus chinensis*. *Fish Shellfish Immunol.*, **2004**, *16*, 513–525.

58. Chiou, T. T., Wu, J. L., Chen, T. T., & Lu, J. K. Molecular cloning and characterization of cDNA of penaeidin-like antimicrobial peptide from tiger shrimp (*Penaeus monodon*). *Mar. Biotechnol.*, **2005**, *7*, 119–127.

59. Bartlett, T. C., & Guillanume, J. Nutrition of *Litopenaeus vannamei* and *Litopenaeus setiferus*. *Marine Biotechnol.*, **2004**, *4*, 278–293.

60. Rosa, R. D., & Barracco, M. A. Antimicrobial peptides in crustacean. *Invertebrate Survival Journal*, **2010**, *195*, 262–284.

61. Omojowo, F. S., & Omojasola, P. F. Microbiological quality of fresh catfish raised in ponds fertilized with raw and sterilized poultry manures. *Am. J. Res. Commun.*, **2013**, *1*, 42–45.

62. El-Naggar, G. O., NasrAlla, A. M., & Kareem, R. O. Economic analysis of fish farming in Behera governorate of Egypt. In: Elghobashy, H., Fitzsimmons, K., & Diab, A. S., (eds.), *From the Pharaohs to the Future, Proceedings of 8th International Symposium on Tilapia in Aquaculture* (pp. 693–708), **2008**.

63. Enyidi, U., Pirhonen, J., Kettunen, J., & Vielma, J. Effect of feed protein: Lipid ratio on growth parameters of african catfish *Clarias gariepinus* after fish meal substitution in the diet with bambaranut (*Voandzeia subterranea*) meal and soybean (*Glycine max*) meal. *Fishes*, **2017**, *2*, 1.

64. Embaby, E. M., Ayat, N. M., Abd El-Galil, M. M., Allah, A. H. N., & Gouda, M. M. Mycoflora and mycotoxin contaminated chicken and fish feeds. *Middle East J. Appl. Sci.*, **2015**, *5*(1), 1044–1054.

65. Mahfouz, M. E., & Sherif, A. H. A multiparameter investigation into adverse effects of aflatoxin on *Oreochromis niloticus* health status. *J. Basic Appl. Zool.*, **2015**, *71*, 48–59.

66. Fallah, A. A., Pirali-Kheirabadi, E., Rahnama, M., Saei-Dehkordi, S. S., & Pirali-Kheirabadi, K. Mycoflora, aflatoxigenic strains of *Asperigillus* section Flavi and aflatoxins in fish feed. *Qual. Assur. Saf. Crop. Foods,* **2014**, *6*, 419–424.

67. Sofie, M., Van Poucke, C., Detavernier, C., Dumoultn, F., Van Velde, M. D. E., Schoeters, E., et al. Occurrence of mycotoxins in feed as analyzed by a multi-mycotoxin LC-MS/MS method. *J. Agric. Food Chem.*, **2010**, *58*, 66–71.

68. Bryden, W. L. Mycotoxin contamination of the feed supply chain: Implications for animal productivity and feed security. *Anim. Feed Sci. Technol.*, **2012**, *173*, 134–158.

69. Charo-Karisa, H., Opiyo, M., Munguti, J., Marijani, E., & Nzayisenga, L. Cost-benefit Analysis and growth effects of pelleted and unpelleted on-farm feed on African catfish (*Claries Gariepinus* Burchell 1822) in earthen ponds. *Afr. J. Food Agric. Nutr. Dev.*, **2013**, *13*, 8019–8033.

70. Njobeh, P. B., Dutton, M. F., Åberg, A. T., & Haggblom, P. Estimation of multimycotoxin contamination in South African compound feeds. *Toxins*, **2012**, *4*, 836–848.

71. Elsaidy, N., Abouelenien, F., & Kirrella, G. A. K. Impact of using raw or fermented manure as fish feed on microbial quality of water and fish. *Egyptian J. of Aquatic Res.*, **2015**, *41*, 93–100.

72. Sahlstrom, L. A review of survival of pathogenic bacteria in organic waste used in biogas plants. *Bioresour. Technol.*, **2003**, *87*(2), 161–166.

73. World Bank. *The World Bank Annual Report – 2013*. Washington, DC. © World Bank, **2013**. https://openknowledge.worldbank.org/handle/10986/16091 (accessed on 6 January 2020).

74. Meric, I., Wuertz, S., Kloas, W., Wibbelt, G., & Schulz, C. Cottonseed oilcake as a protein source in feeds for juvenile tilapia (*Oreochromis niloticus*): Antinutritional effects and potential detoxification by iron supplementation. *Isr. J. Aquacult-Bamid.*, **2011**, *63*, 568–576.

75. Slawski, H., Adem, H., Tressel, R. P., Wysujack, K., Koops, U., Wuertz, S., & Schulz, C. Replacement of fish meal with rapeseed protein concentrate in diets fed to Wels catfish (*Silurusglanis* L.). *Aquac. Nutr.*, **2011**, *17*, 605–612.

76. IARC. Formaldehyde: Wood dust and formaldehyde. *IARC Monographs on the Evaluation of the Carcinogenic Risks to Humans* (Vol. 62, pp. 1–405). World Health Organization, **1995**.

77. IARC. Chemical agents and related occupations. *IARC Monogr. Eval. Carcinog. Risks Hum.*, **2012**, *100F*, 401–430.
78. Bhowmik, S., Begum, M., Hossain, M. A., Rahman, M., & Alam, N. A. K. M. Determination of formaldehyde in wet marketed fish by HPLC analysis: A negligible concern for fish and food safety in Bangladesh. *Egyptian J Aquatic Res.*, **2017**, *43*, 245–248.
79. El-Hariri, O., Bouchriti, N., & Bengueddour, R. Risk assessment of histamine in chilled, frozen, canned, and semi-preserved fish in morocco, implementation of risk ranger and recommendations to risk managers. *Foods*, **2018**, *7*, 157.
80. Food and Agriculture Organization and World Health Organization FAO (The Food and Agriculture Organization). *Assessment and Management of Seafood Safety and Quality, the Food and Agriculture Organization* (p. 574). Roma, Italy, **2014**.
81. Espelund, M., & Klaveness, D. Botulism outbreaks in natural environments: An update. *Frontiers in Microbiology*, **2014**, *5*, 1–7.
82. Solomon, H. M., & Lilly, Jr. T. BAM: *Clostridium botulinum. Bacteriological Analytical Manual*, **2001**.
83. World Health Organization. Cholera – 2017. *Weekly Epidemiological Record*, **2018**, *93*(38), 489–500.
84. European Food Safety Authority and European Centre for Disease Prevention and Control (EFSA and ECDC). The European Union summary report on trends and sources of zoonoses, zoonotic agents and food-borne outbreaks in 2017. *EFSa Journal*, **2018**, *16*(12), e05500.
85. European Food Safety Authority. *Foodborne Outbreaks*. http://www.efsa.europa.eu/en/microstrategy/food-borne-outbreaks (accessed on 20 January 2020).
86. Center for Disease and Control Prevention (**2019**). Foodborne Germs and Illness. https://www.cdc.gov/foodsafety/diseases/index.html (accessed on 6 January 2020).
87. Baker-Austin, C., Oliver, J. D., Alam, M., Ali, A., Waldor, M. K., Qadri, F., & Martinez-Urzata, J. *Vibrio* spp. infections. *Nat. Rev. Dis. Primer*, **2018**, *4, 8.*
88. Gooch, J. A., DePaola, A., Bowers, J., & Marshall, D. L. Growth and survival of *Vibrio parahaemolyticus* in postharvest American oysters. *J. Food Prot.*, **2002**, *65*(6), 970–974.
89. Kaysner, C. A. *Vibrio* species. In: Lund, B. M., Baird-Parker, T. C., & Gould, G. W., (eds.), *The Microbiological Safety and Quality of Foods* (pp. 1336–1362). Aspen Publishers Inc., Gaithersberg, Maryland, USA, **2000**.
90. Jones, M. K., & Oliver, J. D. *Vibrio vulnificus*: Disease and pathogenesis. *Infect Immun.*, **2009**, *77*(5), 1723–1733.
91. Chiavelli, D. A., Marsh, J. W., & Taylor, R. K. The mannose-sensitive Hemagglutinin of *Vibrio cholerae* promotes adherence to zooplankton. *Appl. Environ. Microbiol.*, **2001**, *67*, 3220–3225.
92. Oliver, J. D., & Kaper, J. B. *Vibrio* species. In: Doyle, M. P., Beuchat, L. R., & Montville, T. J., (eds.), *Food Microbiology. Fundamentals and Frontiers* (pp. 228–264). ASM Press, Washington DC, USA, **1997**.
93. Huq, A., Sack, R. B., Nizam, A., Longini, I. M., Nair, G. B., Ali, A., et al. Critical factors influencing the occurrence of *Vibrio cholerae* in the environment of Bangladesh. *Appl. Environ. Microbiol.*, **2005**, *71*, 4645–4654.
94. Farber, J. M., & Peterkin, P. I. *Listeria monocytogenes*. In: Lund, B. M., Baird-Parker, T. C., & Gould, G. W., (eds.), *The Microbiological Safety and Quality of Foods* (pp. 1178–1232). Aspen Publishers, Inc. Gaithersberg, Maryland, USA, **2000**.

95. Denny, J., & McLauchlin, J. Human Listeria monocytogenes infections in Europe—an opportunity for improved European surveillance. *Euro Surveillance*, **2008**, *13*, 8082.
96. Gram, L. Potential hazard in cold-smoked fish: *Listeria monocytogenes*: Special supplement. *J. Food. Sci.,* **2001**, *66*, S1072–S1081.
97. FAO. *Expert Workshop on the Application of Biosecurity Measures to Control Salmonella Contamination in Sustainable Aquaculture FAO Fisheries and Aquaculture Report No. 937 ISSN 2070-6987* (pp. 19–21). Mangalore, India, **2010**.
98. Amagliani, G., Brandi, G., & Schiavano, G. F. Incidence and role of *Salmonella* in seafood safety. *Food Res. Internat.*, **2012**, *45*, 780–788.
99. Wachsmuth, K., & Morris, G. K. *Shigella*. In: Doyle, M. P., (ed.), *Foodborne Bacterial Pathogens* (pp. 447–462). Marcel Dekker Inc., **1989**.
100. Wan, B., Zhang, Q., Ni, J., Li, S., Wen, D., Li, J., Xiao, H., He, P., Ou, H. Y., Tao, Q., Lu, J., Wu, W., & Yao, Y. F. Type VI secretion system contributes to Enterohemorrhagic *Escherichiacoli* virulence by secreting catalase against host reactive oxygen species (ROS). *PLOS Pathog.*, **2017**, *13*(3), e1006246.
101. EU. Commission Regulation (EC) No. 2074/2005 of 5 December 2005 laying down implementing measures for certain products under regulation (EC) No. 853/2004 of the European Parliament and of the Council and for the organization of official controls under Regulation (EC) No. 854/2004 of the European Parliament and of the Council and Regulation (EC) No. 882/2004 of the European Parliament and of the Council, derogating from Regulation (EC) No. 852/2004 of the European Parliament and of the Council and amending. *Official Journal of the European Commission*, **2005**, *338*, 27–58.
102. Jablonski, L. M., & Bohach, G. A. *Staphylococcus aureus*. In: Doyle, M. P., Beuchat, L. R., & Montville, T. J., (eds.), *Food Microbiology: Fundamentals and Frontiers* (pp. 353–375). ASM Press, Washington DC, USA, **1997**.
103. Maftuch, K. I., Adam, A., & Zamzami, I. Antibacterial effect of *Gracilaria verrucosa* bioactive on fish pathogenic bacteria. *Egyptian J Aquatic Res.*, **2016**, *42*, 405–410.
104. Wan-Yusof, W. R., Badruddin-Ahmad, F., & Swamy, M. A brief review on the antioxidants and antimicrobial peptides revealed in mud crabs from the genus Scylla. *J. Marine Biol.,* **2017**, ID 1850928.
105. Centers for Disease Control and Prevention (CDC). Preliminary food net data on the incidence of infection with pathogens transmitted commonly through food—10 states, United States – 2006. *Morb. Mortal. Wkly. Rep.*, **2007**, *56*, 336–339.
106. Iwamoto, M., Huang, J. Y., Cronquist, A. B., Medus, C., Hurd, S., Zansky, S., et al. Bacterial enteric infections detected by culture-independent diagnostic tests—Food Net, United States – 2012–2014. *MMWR. Morbidity and Mortality Weekly Report*, **2015**, *64*(9), 252–257.
107. European Food Safety Authority and European Centre for Disease Prevention and Control. The European Union summary report on antimicrobial resistance in zoonotic and indicator bacteria from humans, animals, and food in 2017. *EFSA Journal*, **2018**, *16*(2), e05000.
108. Hao, X., Zhang, X., Duan, B., Huo, S., Lin, W., Xia, X., & Liu, K. Screening and genome sequencing of deltamethrin-degrading bacterium ZJ6. *Curr. Microbiol.*, **2018**, *75*(11), 1468–1476.
109. Lovecka, P., Pacovska, I., Stursa, P., Vrchotova, B., Kochankova, L., & Demnerova, K. Organochlorinated pesticide degrading microorganisms isolated from contaminated soil. *New Biotechnol.*, **2015**, *32*(1), 26–31.

110. Robodonirina, S., Rasolomampianina, R., Krier, F., Drider, D., Merhaby, D., Net, S., & Oudddane, B. Degradation of fluorine and phenanthrene in PAHs-contaminated soil using *Pseudomonas* and *Bacillus* strains isolated from oil spill sites. *J. Environ Management*, **2019**, *232*, 1–7.

111. Kim, D. W., Lee, K., Lee, D. H., & Cha, C. J. Comparative genomic analysis of pyrene-degrading *Mycobacterium* species: Genomic island and ring-hydroxylating dioxygenases involved in pyrene degradation. *J. Microbiol.*, **2018**, *56*(11), 798–804.

112. Mishra, G. K., Barfidokht, A., Tehrani, F., & Mishra, R. K. Food safety analysis using electrochemical biosensors. *Foods*, **2018**, *7*, 141.

113. Mishra, G. K., Sharma, V., & Mishra, R. K. Electrochemical aptasensors for food and environmental safeguarding: A review. *Biosensors,* **2018**, *8,* 28.

114. Gangola, S., Bhatt, P., Chaudhary, P., Khati, P., Kumar, N., & Sharma, A. Bioremediation of industrial waste using microbial metabolic diversity. In: Bharagava, R. N., & Saxena, G., (eds.), *Bioremediation of Industrial Pollutants*. IGI Global, Pensilvania, USA, **2016**.

115. Altamirano, G. R. I., & Marquez-Rocha, F. J. PAHs-bacterial biodegradation capability of hydrocarbon-contaminated sediments. *Curr. Res. Microbiol. Biotechnol.*, **2016**, *4*(3), 867–873.

116. Yang, R., Liu, G., Chen, T., Zhang, W., Zhang, G., & Chang, S. The complete genomic sequence of a novel cold-adapted bacterium, *Planococcus maritimus* Y42, isolated from crude oil-contaminated soil. *Std. Genomic. Sci.,* **2018**, *13*, 23.

CHAPTER 4

Microbiological Quality and Food Safety Challenges in the Meat Industry

DANIELA SÁNCHEZ-ALDANA, TOMAS GALICIA-GARCÍA, and
MARTHA YARELY LEAL-RAMOS

*School of Chemical Sciences, Autonomous University of
Chihuahua, 31125 Chihuahua, Chih, Mexico, Tel.: +52 614 236 6000,
E-mail: dsancheza@uach.mx (D. Sánchez-Aldana)*

ABSTRACT

Meat products are rich in nutrients, so spoilage microorganisms easily contaminate these food products. Also, pathogenic microorganisms such as *Listeria monocytogenes* and *Escherichia coli* can be found in not well-cooked products. As such, there are several challenges for food safety that the meat industry needs to take into account, in order to provide safe products to the consumer. The present review will address four important aspects of the microbiological quality and food safety of meat products. First, the aspects related to the production of raw materials, including meat and additives. Second, the monitoring of the parameters for microbial quality, by identification of critical control points (CCP), through meat processing; then, the storage conditions and the cold chain assurance during the transport and distribution of meat products will be described, and finally, the importance of the quality control systems will be addressed.

4.1 INTRODUCTION

The quality of a food product is related to its sensory, nutritional, and functional characteristics, as well as to the presence of pathogenic microorganisms or any other contaminant (physical or chemical) that could represent a potential health risk. Nevertheless, microbiological quality and food safety are a crucial part of guaranteeing the final quality of a product.

Meat processing is one of the most challenging industries for food safety, due to the wide variety of raw materials involved (both meat and additives), the high variability in the product quality, and the large amount of process variables that need to be controlled. Ready to eat meat (RTE) products from pork, turkey, chicken, beef or poultry meat are industrialized to meet the consumer demands so that there is no further preparation required. RTE meats preparation involves a large sequence of processes that goes from raw material reception, through grinding, emulsification, cooking, and packaging until storage and distribution; all those processes need to be controlled to guarantee the safety of the final product. Pathogen such *Listeria monocytogenes* must be absent in the food product, other pathogens must also be absent or a controlled maximum permitted, including *Salmonella*, *Escherichia coli*, and *Clostridium botulinum;* the presence or a specified quantity of those pathogens are set in rigorous international standards, and vary according to the country where the product will be marketed. Some examples of RTE meats are cured meats; sliced, cooked, and boneless hams; sliced cooked loaves; snack sticks; jerky; barbecued meats poultry rolls and breasts; roast beef; and other delicatessen items [1].

It is important in the meat industry to have traceability of the production processes, from the origin of their raw materials before processing, as well as during storage and distribution. Within the production processes, it is impending to know the physical and taxonomic characteristics of the animal's breed, feeding mode, age at slaughter, a method of slaughter, etc. Also, it is important to maintain reliable information on quality throughout the production process to ensure consumers of the high quality of the meat products. Another parameter to consider is its stability after production, during packaging, storage, transport, and distribution operations, before the acquisition by the consumer and its consumption.

4.2   PRODUCTION PROCESSES OF RAW MATERIALS

Despite the consumer tendencies on plant-based diets, meat, and meat products production and consumption are expected to increase, especially in developing countries, where consumers will expand their consumption of beef and other meat products [2]. On the other hand, meat, and meat products deterioration generate significant economic losses (up to 40% of production) in the meat industry [3], so that producers are always looking for strategies to guarantee the safety of meat products, from the acquisition of raw materials, during processing and after storage, until reaching the final consumer.

The meat industry uses criteria based on raw material and finished products, to assess their food products quality; the specifications include physical hazards such as bone or metal fragments, chemical hazards such as hormones or antibiotics, and microbial hazards, including *Clostridium botulinum, Listeria monocytogenes,* and *Salmonella.* [4].

The quality of a food product starts with the consistent quality of raw materials, and need to be controlled through the food processing steps. Subsequent manufacturing operations are more likely to be successful if they are standardized and supported by quality control methods, during, and at the end of processing. Thus, typical, and profitable food processes include freezing, cooling, dehydration, and several thermal approaches. Other common approaches include the use of hurdle technology, involving modulation of pH and water activity (A_w), antimicrobials, spices, and active packaging [5]. The use of these processes depends on the type of meat product to be developed, the nature of the ingredients and the profile of the consumer to whom the product is focused. These operations include the selection of raw meat and additive materials, their supply, thawing of meat raw materials, trimming, brining, injection, tenderization, mixing of ingredients, emulsification, tumbling, curing, molding, resting, cooking, and/or smoking, cooling, unmolding, slicing, packaging, storage, distribution, etc. In Figure 4.1, the Commonly Meat Industry Unit Operations and the identification of CCP are presented.

4.2.1 MEAT RAW MATERIALS

Pathogenic microorganisms such as *Campylobacter* spp., *Salmonella* spp., *Clostridium botulinum, Clostridium perfringens,* and *Bacillus cereus,* are part of the normal microbiota usually present in raw meat. Some could also be introduced during food preparation, such as *Staphylococcus aureus.* However, these microorganisms can persist in the finished product, and continue its development during the storage and distribution, with a detrimental effect on the food product quality, and therefore, in its safety and shelf life. Comminuting processes of raw meat include mechanical separation; the carcass must be boned and broken down into specific cuts depending on the type of manufacturing operation. Hazards associated with the handling of raw materials include temperature abuse, that can promote microbial

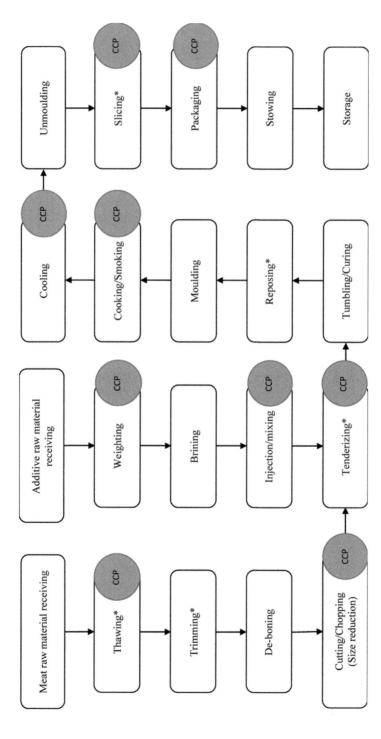

FIGURE 4.1   Commonly meat industry unit operations and critical control points. Operations marked with * are optional unit operations in ready to eat meat products. Unit operations marked with CCP are critical control points in manufactured process.

growth, as well as metal particles from equipment, or large quantities of bone fragments [6]; thus, this process depend on the training and skill of the operators and supervision is the best control strategy. Thus, it is important to avoid the contamination of meat raw materials, including even contamination in the live animals before sacrifice and during slaughtering, with special attention to equipment, to good sanitation procedures by the employees, and the environment [6].

The raw meat ingredients must be weighed before processing according to the list of ingredients and the formulation process. After weighing, these materials are placed in correctly labeled containers with the name of the raw material, name of the product to be processed, quantity, assigned batch, manufacturing order number, etc. Meat raw materials that require pre-treatment before their incorporation into a formulation, such as deboning, removes leather or some grinding, must be processed using good manufacturing practices (GMP) in order to reduce the incidence in microbial contamination and to avoid cross-contamination.

4.2.2 ADDITIVE RAW MATERIALS

Additives are any permitted substance that, without having nutritional properties, is integrated into the formulation of the products and acts as a stabilizer, preservative, or modifier of its sensorial characteristics. It is added to favor stability, preservation, appearance, or acceptability of the final product [7]. It is important to mention that the corresponding authorities of each country based on toxicological, microbiological, and nutritional studies regulate the addition of certain additives. Therefore, the use and incorporation of additives must be carried out under strict controlled standards. It is necessary to establish control parameters for the reception of raw materials incorporated into meat products as additives, such as compliance to physicochemical specifications by the supplier, identification, storage, and their subsequent disposal, as described below.

In particular, additives classified as allergens and/or sensitives must have special treatment. Allergen additives (such wheat, gluten, milk, and its derivate, soy, peanuts among others) are substances that cause a hypersensitivity reaction in some patients; this reaction involves the recognition of the allergen as a foreign substance. On the other hand, sensitive additives that contain ingredients such as sulfites can potentially cause severe reactions.

The reception of additives is the first step to ensure the safety of meat products. Once the material is incorporated into the additives warehouse, the quality inspector must inspect them. The inspection needs to agree with operating instructions for sampling, in order to prevent cross-contamination. The inspection needs to be documented and the information captured in administration systems so that the quality inspector can analyze that later. In case of a rejection of a material for not meeting quality standards or product specifications, it must be sent to a specific area so that an operator does not use it. The storage of allergen and/or sensitive additives must be carried out with strict control of identification and location in the specially assigned spaces within the additive store. Allergen and sensitive ingredients should be placed at lower levels of racks separated by a group of allergens to avoid cross-contamination. The ingredients must be controlled by registration records, where the description of the ingredient, quantity, code, reception date, date of departure, minimum stock, maximum stock, and current stock. Each space with allergen and sensitive material must be identified with color labels defined according to the operating procedures. The weighing of allergenic and sensitive products is carried out by the weighing operator, who must carry out the operation in a designated area, with specially identified protective and uniform equipment. Once the weighing operation has been completed, the operator must leave the equipment and unite in an assigned place in order to avoid cross-contamination; the utensils used should be exclusive for the sensitive material. Operational cleaning procedure should be performed with disposable wet materials at each change of allergen or sensitive additives, which should involve the entire area, including direct contact areas such as ladles, working tables, and floor, as well as the changing of gloves and protective equipment.

The registration of formulations must be submitted by the supervisor of the area and should differentiate or separate allergenic and sensitive materials as a group, leaving space for the necessary cleaning, in order to avoid cross-contamination. The operator in charge of weighing must notify the quality supervisor, who must validate compliance with the cleaning, at each change of product, signing in the space programmed for cleaning. The quality supervisor must ensure compliance with the procedure for the control of allergenic and sensitive ingredients through a general audit in the dry storage area and in the formulation process. The weighted ingredients must be placed in the assigned storage area, or they are directed towards their final destination.

4.3   MEAT PRODUCTS UNIT OPERATIONS

The meat industry produces a large variety of meat products ranging from whole muscle to ground and comminuted products, and RTE meat products;

each requires different equipment and process steps. These processes of meat products are usually long and complex operations, where various transformations as physicochemical and sensory changes occur, which require constant monitoring as well as the establishment and monitoring of CCP. RTE meat products are considered those products that are pre-cleaned, precooked, mostly packaged, and ready for consumption without prior preparation or cooking [8]. RTE meat and meat products are manufactured from meat obtained from the carcass and edible offal (noncarcass parts such as liver, kidney, and heart) [1], and that are subjected to a cooking process that can eliminate microorganisms of public health concern. Most of the processed fresh meat is prepared by high-speed and high technology equipment, but some particular products are still prepared by the traditional methods used for many years, as some artisanal meat products including dry-cured ham.

After verification of raw materials compliance, it is indispensable to guarantee the quality of the intermediate products is maintained during the processing operations, in order to obtain a safe and secure final product. Before starting with the elaboration of a product, it is necessary to count on the material lists and process specifications. The process variables implicit in each of the operations mentioned above must be controlled to the same extent. These include temperature, humidity, time, speed, pressure, etc. The lack of control in these variables can significantly affect the quality not only microbiological but also of physicochemical, sensory, and nutritional attributes.

The unit operations are described below, with particular attention to points where more control must be exercised in order to guarantee the maintenance of food safety at the industrial level.

4.3.1 RAW MEAT THAWING

When raw meat is received frozen, it must be thawing before using it in the manufacturing process. Thawing is a critical phase in the process, since the melted water from ice crystals, can promote microbial reactivation. On the other hand, if heat is applied to a frozen product, the heat can be transferred to the inside of the product, so that an appropriate environment for microbial growth will also be present. Thawing temperatures below 5°C reduce the risk of microbial growth and produces a slow thawing rate which guarantees efficient reabsorption of melted water [9]. Defrosting can be carried out in air or water; it is important to control an above 4°C temperature in the entire product. Industrially tempered or thawed tumblers are also used to accelerate the thawing process of meat raw materials. This equipment must be clean, sanitized, and operated at refrigerated temperatures under the strict control of GMP to guaranty food safety.

4.3.2 TRIMMING (SIZE REDUCTION)

Bacteria present in grown meat and poultry, are derived from the carcasses, as well as the contamination that occurs during trimming. Griding will also distribute the microbial cells present in the surface, to the entire product. Another source of contamination is the equipment used for trimming and griding, if not properly sanitized. Cutting, slicing, and dicing of fresh meat is also done in processing industries, either manually by operators or by automated equipment. In either case, cross-contamination needs to be avoided [10].

4.3.3 TUMBLING, MASSAGING, AND INJECTION

Mixing ground meat or poultry pieces with muscle is used in the preparation of some meat products, to incorporate additives and enhance their textural properties. The processing of tumbling, massaging, and injection can be used to incorporate brine into the meat piece; also, the brine can be incorporated by a low-pressure injection system [1]. In some cases, the tumbler equipment is supplemented with paddles or blades, which help on the tumbling action. All these processes are helpful in meat curing [10].

Another important ingredient in the brine mixture is sodium nitrite. It is responsible for the cured meat color, adds flavor, and can help prevent the development of spores from microorganisms such as *C. botulinum*. The injection process is susceptible to microbiological contamination and growth in the brine/marinade and equipment components. Therefore, thorough microbiological testing and sanitation protocol is important to maintain product quality and integrity.

4.3.4 CURING

A large proportion of all processed meat products are cured. Most sausages and hams RTE are cured. Thus, curing is one of the most important operations in processed meat production [11]. Nitrate was originally used in curing as a form to stabilize color in the food product, but it was also found to have a significant antimicrobial activity [1], since demonstrated to be effective in the control of *Clostridium botulinum* growth; its use has also been related to flavor enhancement and growth inhibition of spoilage-related microbial groups [11].

The prepared meat should be transferred immediately to the curing room and maintained at 1–4°C to prevent a rise in meat temperature. Failure to maintain temperature may lead to defects and loss of quality [6]. Most cured meat products are also heated or cooked in a smokehouse to enhance the meat flavor and surface color. To maintain a good color of the cured meat products, packaging the meat with oxygen-impermeable packaging materials is essential [1].

4.3.5 COOKING

Cooking involves thermal processing, where the meat product undergoes a series of physicochemical modifications, which are related to the final sensorial properties of the product. The objective of the cooking procedure is to develop the sensory characteristics (color, flavor, structure, texture, etc.) and microbiological stabilization of the product [12]. Cooking time and temperature should assure that the procedure or its combination with other preservation methods is sufficient to eliminate pathogenic microorganisms and toxins.

Before cooking, meat can become microbiologically contaminated, and thermal processing is intended to reduce this contamination, to diminish microbiological risks and increase shelf life. Therefore, time and temperature conditions should consider the suspected initial contamination or microbial load; but the effect on the sensorial parameters are also important [6]. For cured-cooked meat products, the center point in the product needs to be at a constant temperature of 68–70°C for 30–60 min; this temperature must be measured using a suitable hand-held probe thermometer of hygienic design, previously disinfected. Another factor to consider is the response of bacteria to stress conditions that can give them the capacity of thermal resistance; therefore, the products cannot be exposed for a long time to temperatures that favor thermal resistance (40–50°C) [12].

Temperature control is frequently overlooked, but many bacteriological problems could be avoided if the meat temperature was controlled throughout this critical process. After cooking, it is necessary a cooling process, to cool down products up to packing temperature. The cooling process can be carried out by water as a shower or using airflow. The cooling process in just-cooked meat products can affect properties such as cohesion but is also fundamental to identify this process as an integral part of product pasteurization [12]. The lower internal temperature of the meat products (50–60°C), can prevent excessive heat taken into chilling rooms, with the consequent lowering of energy costs.

4.3.6   SLICING AND PACKING

It is important to consider that cooked meat products can be recontaminated with pathogenic bacteria that can be found in the environment (*Listeria monocytogenes, Bacillus*), or that can be present in surfaces that are in contact with the food product before the final packing procedure. Contamination of RTE meat products can also be related to slicing machines that can be responsible for cross-contamination with *L. monocytogenes* among other pathogenic bacteria [13].

Meat packaging must be an effective barrier in preventing undue external-internal interactions that could detract from a meat product's viability. This is a core requirement but is often complicated by the inherently physical nature of meat processing, distribution, and trade. In response, much effort has been dedicated to improving the durability of meat packaging and identifying when its barrier function has been compromised [14].

The correct choice of packaging material to be used for meat and meat products will depend on the chemical, biochemical, and microbiological knowledge of the product, and take in consideration the changes that can take place after packaging. In the selection of meat and meat products, the consumer will be influenced largely by their color. The color of fresh meat is predominantly due to oxymyoglobin, and it is only maintained by the high partial pressure of oxygen in the surrounding atmosphere. Conversely, the color of cured meat is best maintained in the absence of oxygen. Thus, in considering packaging requirements meat and meat products may be divided into two groups, unprocessed, and cured products [1, 6].

4.3.7   POST-PACKING THERMAL TREATMENT

It is well recognized that post-thermal exposure of meat products to surfaces, can be a cause of microbial recontamination with *L. monocytogenes* and other microbial contaminants; including contact with equipment, utensils, and personnel, which should be avoided [8]. Thermal pasteurization is usually used in combination with antimicrobial agents [8, 10] to achieve maximum reduction of bacterial counts and retention of product quality. One form to avoid post-contamination is to apply heat treatment after packing. In-package pasteurization serves as the final step in product processing to eliminate vegetative pathogens such as *L. monocytogenes*, pathogenic *E. coli,* and *Salmonella* spp. in RTE foods. [8].

4.4 STORAGE, DISTRIBUTION, AND PRESERVATION OF THE COLD CHAIN

The preservation of the cold chain is one of the main principles and basic requirements to guarantee food safety in meat products. The cold chain should not be interrupted at all times along the meat storage and distribution chain since inappropriate preservation conditions can lead to shorter shelf life and can increase the probability of health-associated risks [15].

Cold chain preservation is necessary for the safety of meat products, several points need to be carefully controlled, including chilling before transportation and the control of temperature along the distribution channels [16]. Meat products are vulnerable to failure in the cold chain, especially in the present day, where global trade is complex and sometimes involves the mobilization in more than one country or continent [15]. The record of the meat product temperature, which needs to be monitored along the distribution pathway, can control the preservation of the cold chain and assure the quality of the meat product.

4.4.1 IDENTIFICATION AND TRACEABILITY

As has been reviewed, the quality of meat products is determined by sensorial, technological, and hygienic quality indexes. Therefore, it is necessary to establish the methodology for the identification and lot number of the different raw materials, finished product and products for direct sale through the supply chain (reception, processing, and finished product), in order to maintain traceability.

For product traceability, it is important to maintain the records of all materials used in food preparation, from the purchase of raw material to finished product that will be identified by a unique number. In the food industry, traceability is intended to create a line that unites all the steps in the food chain, the so-called "from farm to fork." In the meat processing industry, traceability needs to include animal production, processing for preparation of raw material and additives, processing for preparation of the meat products, distribution to wholesalers and food stores, to the point where the product is on the consumer's table [17].

Identification is defined as the differentiation of an article, input, single finished product from another article, input or finished product within the supply chain. Lot number refers to establishing an identification code for raw materials, materials in process, finished product or for direct sale, which can be numeric or alphanumeric in order to identify it within the supply chain. The assigned lot of a finished product must be automatically generated according to the lot number program. This lot consists of a series of digits and must be consecutive. To ensure traceability, each finished product must be divided, assigning a label that contains the name of the product, the lot number, production line where the product was manufactured, work shift, expiration date, date of manufacture, time, weight, etc. This lot must be used until the product reaches the customer. An example of lot number is presented in Figure 4.2.

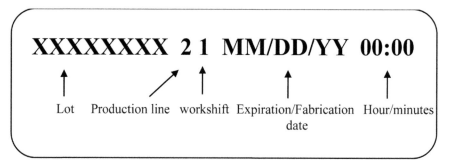

FIGURE 4.2   Example of identification used in meat products.

A traceability system must be sufficiently clear in order to identify the product of interest and follow up along the supply chain. For this, the product identifier should ideally: (a) uniquely identify the manufacturing lot; (b) be sure; (c) be permanent; (d) preserve the identity of the entire life cycle of the product; and (e) be easy to read and capture identification data [17].

4.5   QUALITY CONTROL PROGRAMS FOR FOOD SAFETY

The main purpose of a quality control program in the food industry is to assure the safety of the final product, by the identification and reduction of associated hazards, including chemical, physical, and biological risks. To guarantee that the products are safe regardless of the production and processing location, national, and international criteria are established for food quality, particularly in relation to the presence of foodborne pathogens. In meat products, there have been identified risks associated with several foodborne pathogens that can be present in raw meat and processed products. Special care needs to be taken in regards to storage temperature and humidity of the storage facilities since those factors can contribute to microbial growth [6]. Thus, the control of temperature, humidity, and air distribution in workrooms and cold stores, although frequently overlooked, play an important part in the quality control of meat products [18].

In general, quality control is necessary in order to: (a) safeguard that the product composition is uniform and does not fall below the limits established in standards; (b) comply with legislation; and (c) maintain quality at levels and tolerances, which are acceptable while minimizing the cost of production [19]. Wherever possible, methods of quality control should be objective, rapid, simple, and low in cost about labor, materials, and equipment. Laboratory analyses are frequently time-consuming and expensive; and although they are a necessary part of quality control, a careful study of production methods and the use of strategically placed "in-plant" control can often lead to a considerable reduction in the volume of analyses without any loss of control [20].

The objective of quality control programs is to maintain a sanitary/clean environment necessary for the production of safe food. The frequency of these programs can be established monthly, quarterly, semi-annually or annually or when a significant change occurs in a production line or work area and must be supervised by a multidisciplinary team. This type of program is also used when there is a deviation in the process or a complaint of affection to a consumer that directly involves the health program. In the case of training, all new personnel must be trained, and this training must be constantly reinforced.

4.5.1 GOOD MANUFACTURING PRACTICES (GMP) AND SANITATION STANDARD OPERATING PROCEDURES (SSOP)

The GMP program is aimed at assuring the sanitary quality of the food products and includes requirements for the processing plant, strict rules of personal hygiene and cleanliness of the workplace, and the description in writing form of all procedures involved in the product production [21].

GMPs also provide the operating and environmental conditions that are needed to protect food products during processing and storage. The control program includes raw material, and the control of the quality of the material used in food processing, and includes the development of quality control programs with the suppliers. Therefore, GMP will cover raw material to storage, transportation, and distribution down to consumers [20]. GMP in the food processing plants include the cleaning and sanitization of equipment, but also the employee's hygienic practices; training is required and is recommended to have continuous training programs, so that food handlers have a strong sense of their responsibility in the production of safe food products [21].

Sanitation standard operating procedures (SSOP) are written procedures that an establishment develops and implements to prevent contamination of the food product. The SSOP need to be monitored and recorded to assure its implementation; the procedures should include corrective actions, in case there is a failure in the procedure that can contaminate the food product. The corrective actions should include the proper disposition of the product, the restoration of hygienic conditions, and preventing measures to avoid further accidents [22].

During production, the hygienic conditions in the plant need to be maintained, and the employees are fundamental for this objective since they constitute a potential source of contamination for either biological, chemical or physical hazards. Due to the constant movement of the personnel through the plant, they might contribute to the distribution of hazardous material; therefore, it is fundamental to promote hygienic practices in all the production areas [23]. GMP and SOP programs are the basis for other quality control programs, such as the hazard analysis and critical control point (HACCP) system.

4.5.2   *HAZARD ANALYSIS AND CRITICAL CONTROL POINT (HACCP)*

The HACCP system, which has a scientific and systematic basis, can identify specific hazards and measures for their control in order to ensure the safety of food; the system focuses on preventive more than in corrective actions. The entire HACCP system is susceptible to changes that may derive from advances in the design of the equipment, the manufacturing procedures, or technological sector [18]. Successful HACCP programs depend upon prerequisite programs, such as GMPs to ensure the production of safe food products.

The HACCP system can guarantee the quality and safety of the products endorsed by control from suppliers, reception, and storage of raw materials, packaging, as well as in the process of preparation and distribution; in order to offer the consumer products and services of the highest quality [24, 25]. The effectiveness of the HACCP system is based on the results of audits, monitoring of control points (CP), corrective, and preventive actions on the entire food production and processing steps, in order to reduce potential risks to consumers [26].

A HACCP program maintains the safety of meat products since the process is analyzed to detect and control possible hazardous points. A hazard is defined as any biological, physical, or chemical property that could cause a product to be unsafe for consumption [24].

The HACCP system consists of five preliminary stages and seven fundamental principles; before beginning to develop a HACCP, it is necessary to perform five preliminary tasks, as enlisted below [25]:

1. Assemble the HACCP Team. The first task is to develop a HACCP plan is to assemble the equipment; the team participants have specific knowledge and experience with the product and the process since it is your responsibility to develop the HACCP plan. The team must be multidisciplinary, must know enough about the process that is evaluated and must be trained.
2. Describe the food and its distribution. The HACCP team should formulate a complete product description that includes relevant information on its safety, for example, composition, important physicochemical characteristics (pH, Aw), packaging, durability, storage conditions, and distribution system.
3. Describe the intended use and consumers of the food. The use to which it has to be allocated should be based on the intended uses of the product by the user or final consumer (in some instances, such as in the food in institutions, it will be necessary to take into account if it is vulnerable groups of the population).
4. Develop a flow diagram which describes the process. The flow chart should be prepared by the HACCP team and cover all phases or stages of the operation. The diagram should also consider the inputs and outputs of rework, reprocessing, waste, services (water, steam, and liquids). Temperatures and storage points must also be indicated in the CP and CCP.

5. Verify the flow diagram. Check the flow chart. The HACCP team must compare the flow diagram with the processing operation in all its stages and moments, and amend it when appropriate.

The seven principles are enumerated as follows:

1. Analyze the hazards in the plant. Make a list of all processing steps where a hazard might occur;
2. Identify all CCP's in the process. CCP's are critical to the safety of the product;
3. Establish a critical limit for each of the identified CCP's;
4. Establish CCP monitoring requirements;
5. Establish corrective action to be taken if the CCP deviates from the critical limit;
6. Establish effective record-keeping procedures to document the HACCP program; and
7. Establish a procedure to verify that the HACCP program is working.

Corrective actions are taken to eliminate the causes of a detected risk or non-desired situation. On the other hand, preventive action is taken to eliminate the cause of a potential non-desired situation. [24]. the identification of the CCP along the process of elaboration of meat products is imminent; an example of these critical points is schematized in Figure 4.1.

Based on what has been previously described, the HACCP plans of each process that is elaborated in the plants must be elaborated, trained, communicated, and implemented, these plans must have the five preliminary steps and seven principles within the same document (the hazard analysis of the process and raw materials must be documented separately).

KEYWORDS

- critical control points
- good manufacturing practices
- hazard analysis and critical control point
- meat products
- quality control programs
- sanitation standard operating procedures

REFERENCES

1. Shmidt, Ronald, H., & Rodrick, G. E. *Food Safety Handbook*. John Wiley & Sons, New Jersey, USA, **2003**.
2. Lonergan, S. M., Topel, D. G., & Marple, D. N. Fresh and cured meat processing, and preservation. *Sci. Animal Growth Meat Technol., * **2019**, 205–228.
3. Lorenzo, J. M., Batlle, R., & Gómez, M. Extension of the shelf-life of foal meat with two antioxidant active packaging systems. *LWT-Food Sci. Technol.*, **2014**, *59*(1), 181–188.
4. OECD/FAO. *OECD-FAO Agricultural Outlook*. OECD agriculture statistics (database), **2018**. http://dx.doi.org/10.1787/agr-outl-data-en (accessed on 6 January 2020).
5. Barbosa-Cánovas, G. V., Medina-Meza, I., Candoğan, K., & Bermúdez-Aguirre, D. Advanced retorting, microwave assisted thermal sterilization (MATS), and pressure assisted thermal sterilization (PATS) to process meat products. *Meat Sci.*, **2014**, *98*(3), 420–434.
6. Herschdoerfer, S. *Quality Control In The Food Industry* (p. 2). Elsevier. Amsterdam, Netherland, **2012**.
7. Mexican Official Standard NOM-213-SSA1–2002. *Products and Services, Processed Meat Products, Sanitary Standards, Test Methods*. (Norma Oficial Mexicana 213 (NOM-213-SSA1-2002). Productos y servicios, Productos cárnicos procesados, Especificaciones sanitarias, Métodos de prueba), 2002.
8. Huang, L., & Hwang, C. A. Package pasteurization of ready-to-eat meat and poultry products. *Advances in Meat, Poultry, and Seafood Packaging*, **2012**, pp. 437–450.
9. Cano-Muñoz, G. Manual on meat cold store operation and management. *Food and Agri. Org.*, **1991**, *92*.
10. Kerry, J. P., John, F. K., & Ledward, D. *Meat Processing: Improving Quality*. Elsevier, Amsterdam, Netherland, **2002**.
11. Pearson, M. A., & Dutson, T. R. *HACCP in Meat, Poultry, and Fish Processing* (p. 10). Springer Science & Business Media, Berlin, Germany, **2012**.
12. Lagares, J. Manufacturing process for whole muscle cooked meat products V: Cooking. *Metalquimia Tech Articles*, **2010**, *1*, 161–169.
13. Kurpas, M., Kinga, W., & Jacek, O. Ready-to-eat meat products as a source of *Listeria monocytogenes*. *J. Veterinary Res.*, **2018**, *62*(1), 49–55.
14. Holman, W. B., Kerry, J. P., & Hopkins, D. L. Meat packaging solutions to current industry challenges: A review. *Meat Sci.*, **2018**.
15. Nastasijević, I., Lakićević, B., & Petrović, Z., Cold chain management in meat storage, distribution and retail: A review. *IOP Conference Series: Earth and Environmental Science*. IOP Publishing, **2017**, *85*(1).
16. Raab, V., Petersen, B., & Kreyenschmidt, J. Temperature monitoring in meat supply chains. *British Food J.*, **2011**, *113*(*10*), 1267–1289.
17. Yordanov, D., & Angelova, G. Identification and traceability of meat and meat products. *Biotechnol. Biotechnol. Equip.*, **2006**, *20*, 3–8.
18. Forsythe, S. J. The microbial flora of food. In: *The Microbiology of Safe Food*. John Wiley & Sons. New Jersey, USA, **2011**.
19. Toldrá, F. *Safety of Meat and Processed Meat*. Springer. New York, USA, **2009**.
20. Joint FAO/WHO Codex Alimentarius Commission, and World Health Organization. *Codex Alimentarius, Hygienic Conditions of Food: Basic Texts*, **2018**. http://www.fao.org/docrep/012/a1552s/a1552s00.htm (accessed on 6 January 2020).

21. Paiva, C. L. Quality management: Important aspects for the food industry. *Food Industry*, **2013**.

22. FAO. Hazard Analysis and Critical Control Point System (HACCP) and rules for its application, **2018**. http://www.fao.org/docrep/005/y1579s/y1579s03.htm#bm3.2 (accessed on 6 January 2020).

23. Park, W. H., Chung, D. H., & Yi, S. H. SSOP program development for HACCP application in fresh raw fish manufacturing. *J. Food Hygiene Safety,* **2004**.

24. Northcutt, J. K., & Scott, R. M. *General Guidelines for Implementation of HACCP in a Poultry Processing Plant*, **2010**.

25. Unnevehr, L. J., & Jensen, H. H. HACCP as a regulatory innovation to improve food safety in the meat industry. *Amer. J. Agr. Eco.,* **1996**, *78*(3)*,* 764–769.

26. Kvenberg, J., Stolfa, P., Stringfellow, D., & Garrett, E. S. HACCP development and regulatory assessment in the United States of America. *Food Control*, **2000**, *11*(5), 387–401.

CHAPTER 5

Fermentation as a Preservation Strategy in Foods

MARÍA GEORGINA VENEGAS-ORTEGA,[1]
VÍCTOR EMMANUEL LUJÁN-TORRES,[2]
ADRIANA CAROLINA FLORES-GALLEGOS,[1]
JOSÉ LUIS MARTINEZ-HERNÁNDEZ,[1] and
GUADALUPE VIRGINIA NEVÁREZ-MOORILLÓN[2]

[1]*Research Group of Bioprocesses and Bioproducts,
Department of Food Research, School of Chemistry,
Autonomous University of Coahuila, 25280 Saltillo, Coahuila, México*

[2]*School of Chemical Sciences, Autonomous University of Chihuahua,
31125 Chihuahua, Chihuahua, Mexico, Tel.: +52 614 236 6000,
Ext: 4248, E-mail: vnevare@uach.mx*

ABSTRACT

Food-fermentation is an economic, highly used preservation method that prevents food contamination but can also increase the functionality of the final product due to the microbial metabolites released during the fermentative process. Food-fermentation is industrially implemented due to its wide application, which included animal and plant-derived food-products, where sensorial food characteristics are improved. Among the microorganisms used for this process, lactic acid bacteria (LAB) are one of the most used group; their feasibility includes their safety for consume status and the production of high spectrum antimicrobials. Some of the most important benefits of preservative-fermentation, including economical aspects, are revised in this chapter.

5.1 INTRODUCTION

Food contamination is the process whereby microorganisms or chemical are in contact and absorbed on the surface of the food matrix. It is a condition of high concern for food industry and consumers; therefore, there is a constant research on the determination of contamination causes, as well as on implementation of prevention methods. Food microbial contamination includes the presence and growth of bacteria, molds, yeast, parasites, and virus. Foodborne microbial outbreaks are a worldwide problem, with differences in the causal agent due to prevailing climate and environmental conditions. However, not all microbial groups found in food are related to infectious processes or food deterioration. Beneficial microorganisms can be established in food superficies and provide a protective effect against deleterious microorganisms. The metabolic activities of many beneficial microorganisms are associated in foods, with natural fermentation processes, whereby there are chemical and physical changes in the food matrix due to the growth and metabolic activity of fermenting microorganisms. Fermentation can be carried out by one or more microbial strain; usually a relationship is formed among the beneficial microorganisms responsible for fermentation; in many cases, a synergistic effect can be observed. Depending on the microorganisms involved in fermentation, and the metabolic processes involved, chemical, and physical changes in foodstuff, can be associated with beneficial effects in human health. These beneficial aspects include the antimicrobial capacity of the fermented product; therefore, fermentative processes have been known as a food preservation method. Fermentation is a traditional, inexpensive method that generated textural and sensorial changes in the food product that in many cases is better accepted by the consumer than the previous raw food. The consumption of fermented food and beverages has increased recently, due to the high demand for naturally processed products. Hence, in this chapter, the advantages of fermentation as preservation strategy compared to other preservation methods will be described; an introduction in how an ancient process attract the attention of the industry and became a billion-dollar business is also included.

5.2 FOOD MICROBIAL-CONTAMINATION

Food is usually exposed to chemical or microbial contamination, causing an important number of illnesses, with the consequent economic impacts. Global estimates report 31 foodborne agents responsible of food-related diseases, including bacteria, protozoa, viruses, and helminths; in a lesser extent, chemical contaminants are also included [1]. Diarrheal infections are among the top ten global causes of death related to food contamination [2]. Foodborne pathogens include *Bacillus cereus, Campylobacter jejuni, Clostridium botulinum, Clostridium perfringens, Cryptosporidium, Cyclospora cayetanensis, Escherichia coli* producing toxin, *E. coli* O157:H7, Hepatitis A, *Listeria monocytogenes*, Noroviruses, *Salmonella* sp., *Shigella* sp., *Staphylococcus aureus, Vibrio parahaemolyticus,* and *V. vulnificus* [3] and their presence is related to different food sources. The most recent information on each causal agent is followed in the outbreaks reports; in 2018, the Center for Disease Control and Prevention (CDC) report a high incidence of *Salmonella* outbreaks in ready for consumption products [4].

The microbial contamination of food products can occur at different stages of production, processing, and commercialization. Environmental contamination is related to food spoilage in the first steps of the process, raw food is most affected by this problem; chemical contaminants such as agrochemicals, industrial or urban emissions are potential contaminants of fruits and vegetables [5]. On the other hand, microbial contamination of raw food is most commonly caused by enteropathogenic bacteria, and is associated to the use of residual water for irrigation, fertilizer residues, and excreta from wild animals, and poor hygiene practices of field workers [6]. Food transport, the next step in the food supply chain is most commonly related to cross-contamination, due to the lack of hygiene in the retail cabinets and the combination of different products in a single trip. Also, the long-term transportation increases the probability of microbial contamination by the absence of refrigeration and aeration of the raw food products [7]. Food storage as the final step in the food-retailing process is a crucial point for food quality and safety. The microbial load in that the food product is related to the previous steps and conditions [8]. The challenge once the product is commercialized is the maintenance of the quality, flavor, and safety during the product shelf life. Consumer management of food products to maintain the quality, flavor, and safety is an extensive topic that is related to their hygienic practices; this final step will not be covered in this chapter.

FDA establishes a series of mandatory preventive controls that include the following specifications:

1. Evaluation of the threats that can be affecting the food safety;
2. Establishment of a preventive plan that will be put in place to minimize or prevent the hazards;

3. Establishment of the monitoring controls to confirm the prevention plan;
4. Keeping routine registration of the monitoring plan;
5. Establishment of the actions that correct future problems [3].

5.3 FOOD-PRESERVATION STRATEGIES

During the processing steps for food production, a series of strategies can be developed to safeguard food preservation. Those strategies are discussed below, but also some of the disadvantages that can have a harmful effect on human health will be commented. In general, examples of physical and chemical methods that participate in the reduction of the microbial load in a food matrix, by using a combination of more than one method are excluded.

5.3.1 PHYSICAL METHODS

Some of the physical methods have been used since ancient times, but there are also some examples of the newest technologies like non-thermal methods, which has minimal effects on sensorial food properties. Despite the advantages of using physical methods, the antimicrobial effectiveness of the technology can be at risk if the working conditions, the time of exposure, and the material of choice for processing are not adequately handled. High levels of resistance have been reported in a large number of foodborne pathogens due to the stress responses to the application of physical methods. Also, some methods are expensive or require large quantities of energy costs to be implemented in the food industry.

1. **Temperature:** One of the oldest and most used methods used for food preservation. Pasteurization is a continuous thermal processing that uses different temperature with variations of time in order to achieve microbial inactivation or sterilization, depending on the stress adaptation of the microorganisms. The expression of heat shock proteins (HSPs) induced as a consequence of damage caused by environmental physical or chemical stressful conditions, like high temperatures, can facilitate cellular recovery [9]. *Bacillus cereus, Bacillus subtilis, Escherichia coli,* and *Listeria monocytogenes* are among the most common pathogenic bacteria present in food that are examples of heat stress-adaptive microorganisms by

the production of HSPs [10, 11]. The HSPs system is only involved in heat treatments although cold temperatures are also employed as a food preservation strategy; in this later case, other types of adaptation are involved. For example, it has been demonstrate that proteins implicated in metabolic pathways of *L. monocytogenes* also participate in their survival at refrigeration temperatures, and under those conditions, the microorganism has an increased energy demand for growth and survival [12]. Therefore, high or low temperatures have disadvantages as compared with other methods that are not affected by microbial resistance mechanisms.

2. **High Pressure:** The application of high pressure can be as effective as temperature treatments in microbial load reduction, but is a softer process that does not cause changes in appearance or sensorial properties in foods. This technology uses high hydrostatic pressures (HHP) (>350 MPa) inside of a vessel that contains a pressure transmission fluid in low temperatures [13, 14], and can inactive vegetative cell and spores. As in the case of high and low temperatures, it has been observed resistance mechanisms for high pressure in foodborne pathogens, including *E. coli* [15] or even in *Clostridium botulinum* spores [16]. Genes involved in stress response to high pressure are involved in the regulation of the barotolerance mechanisms [17]. If the conditions of the process are altered, the resistance to high pressure could be induced [18]; for that reason, the process has to be highly understood in order to avoid bigger problems caused by the following contamination with resistant bacteria.

3. **Ultrasound:** This method is based on the use of ultrasonic waves of high intensity, which can penetrate cells and denature enzymes. This property has been exploited not only for food processing but also for food preservation [19]. Ultrasound promotes a better mass and energy transfer process; therefore, this technology is used in combination with heat or with organic acid treatments, with the consequent synergistic effect that destroys cells, spores, and enzymes [20–22]. The utilization of ultrasound requires the use of specific frequencies during an optimal period, due to possible negative side effects on the treated food matrix [23] and even more when simultaneous treatment is being used.

4. **Pulse Light:** Also called as intense light pulses [24] and high intensity broad spectrum pulsed light [25]. It rapidly inactivates bacteria, fungi, and viruses from surfaces in addition to solid and liquid foods. What constitute this technology are continuous and instant

light pulses with a broad wavelength spectrum that goes from UV to infrared. Pulse-light does not act at the microbial membrane level, instead causes molecular injury in the pyrimidine nucleotides of DNA, thus, cell replication is affected [26]. The disadvantages from pulse light technology where discussed by Heinrich et al. [27] including high cost, negative ecological impact, lack of standardized and regulated methods, among others.

5. **Pulsed Electric Fields (PEF):** PEF processing system requires a treatment chamber with a high-voltage pulse generator that delivers short explosions to the food matrix, which in most cases are liquids. It is known that electric pulses cause a break through the biological membranes for the liberation of intracellular material [28]. It is not fully understood how the microbial inactivation is achieved. In normal condition, bacteria exhibited sensitivity to PEF, but after exposure to external stress, the microorganism can be resistant by expression of stress-regulation genes [29]. PEF is not effective with inactivation of viral capsid or spores [30, 31].

6. **Packaging:** The use of a physical barrier against deterioration in foods is used even at home, but is important to take in consideration, possible reactions between the food components and the material used for packaging. The material should prevent not only sensorial changes, but also must prevent light transmittance, and it should be resistant to storage conditions. There are many polymeric materials combined with other compounds that can be used for packaging; the novelty in the case of protects barriers as preservation strategy is active packaging. The materials considered as active packing, are functionalized, or combined with antimicrobials to support the antimicrobial compound and at the same time interacting with the food-matrix [32].

5.3.2 CHEMICAL METHODS

Chemical methods include effective substances against foodborne pathogens, but also have an impact on health by using them in food products. In the last few years, there is a growing necessity to use fewer chemical products and in lower concentrations.

1. **Essential Oils (EOs):** EO is plant secondary metabolites that are obtained by extraction processes in form of aromatic and volatile

liquids, many of them with antimicrobial properties [33]. The functions of EOs as additives in foods are still controversial, because usually their active components cause sensorial changes in food, that are not always desirable [34]. The antimicrobial spectrum of EOs includes pathogenic bacteria, fungi, and virus but some have also shown cytotoxic effects.

2. **Chlorine Dioxide:** Chemical compounds that acts as an oxidizing agent for the disinfection of water, fruits, and vegetables [35]. Aqueous chlorine dioxide (ClO_2) has the same effectiveness than other sanitizers like sodium hypochlorite and did not induce the toxicity in foods by using it [36]. Chlorine dioxide gas can be more effective than aqueous and have been proved against biofilm formation of *L. monocytogenes* [37]. Both presentations of ClO_2 are unstable substances that are inactivated by sunlight.

5.4 FERMENTATION PROCESSES AND INDUSTRIALIZATION

The fermentation process is known in general, as chemical and physic transformation of raw materials to obtain a product or sub-products with a specific role or purpose. This transformation can be achieved thanks to the addition of a microorganism in the materials in controlled-culture conditions. The role of microorganisms consists of the decomposition of the substrate contained in the raw materials for their metabolic necessities, with the generation of microbial biomass, enzymes, and metabolites. There are several types of fermentation processes; the most common are batch and continuous processes, which can be carried out in liquid or solid-state; selection of the fermentation process selected depends on the microorganism used (e.g., bacteria, fungus or yeast) and the final product of interest.

Historically fermentation processes have been applied in food production; the production of cheese was made through an accidental fermentation in which bladders made from ruminants were used to transport milk, after a long journey of transportation, the milk separate in two phases by the action of the native microorganisms in the transporter vessel. Recently, industrialization has complicated the development of some fermentation process, so that the study of critical variables (e.g., temperature, pressure, mass, and energy transfer, pH, time of growth, etc.) became important for optimization. Some of the industrialized fermentation-processes of foods include the production of alcoholic beverages (e.g., beer, wine, etc.), bread, dairy products (e.g., cheese, yogurt, etc.), sauces, and tofu, among others.

However, the real importance of a successful process in the food industry is not just for the capacity to create a simple food product, but trough fermentation the sensorial characteristics such as flavor, as well as textural properties can be enhanced for consumer satisfaction. Also, the presence of live microorganisms or de addition of supplements will provide a healthier food product, with a longer shelf life.

The augmented shelf life in food and preservation by fermentation is the result of the metabolic activity of microbial communities (bacteria, fungi, and yeast) either one type or the combination of them; a practical example of interactions between more than one microbial community and how they growth together with specifics roles is cheese production [38]. The principal role of a bacterial community is the utilization of oxidized carbohydrates to produce alcohol, carbon dioxide, and organic acids; these metabolites prevent the growth of spoilage microorganisms in food [39].

Lactic acid bacteria (LAB) are a heterogeneous group of fermentative microorganisms, which are naturally found or intentionally added to nutrient-rich environments. LAB is industrially important due to the biosynthetic capacities and the production of high-interest metabolites. These products have several applications in the food industry; for example, lactic acid is an inhibitor of pathogenic bacteria, but also it is often used in the malting process to enhance beer quality, contributing in the brewing process and subsequent operations such as filtering or separation [40].

The interest in fermentation has lead food industry to improve scale-up processes on products that are traditionally manufactured in an artisanal way. Scaling up requires the combination of factors to develop a successful process with high production yields. These factors depend on the result seek the type of fermentation, the equipment, and the microorganism used since control parameters are variable. The most common parameters to consider include physical characteristics; mixing, shear, pH, heat, and mass transfer by aeration, working volume, as well as their interactions. Escalation requires the design of mathematical models that can explain the whole process and control its variation. Kombucha is a traditional fermented beverage; Cvetković et al. [41] introduce a method for scaling-up this product taking into consideration the interphase area, which explains the relationship between the ratio of the cross-section reaction and the volume of the reacting liquid phase. Another criterion analyzed in alcoholic fermentation is the relation between the amount carbohydrates (or sugars) and the gas flow rate (or oxygen levels) in an early stage of the process, because a high concentration of sugars and a high concentration of oxygen, promote biomass production instead of alcoholic concentration.

Alcoholic beverages are products that have been consumed for a long time in human history, and have been studied and improved over the years. Nowadays, is an industry with a global value of billions of dollars (USD). Generally, the production of ethanol by fermentation is considered as a simple procedure, but to obtain a product with the exact amount of ethanol (alcohol), a balanced flavor with the quality and safety parameters necessary for its consumption, is not an easy task. Therefore, the industry is currently looking for the improvement of their production systems; most of these processes are based in yeast fermentation; thus, selection of yeast strains as inoculum can help on the prediction of the product quality, repeatability of the process, and the predominance over indigenous yeast or molds present in the raw materials. Industrially, yeast strains have been categorized in two classes: top-fermenting yeasts, that are capable of create a foam layer on the top of the liquid and grow rapidly at 20°C; *Saccharomyces cerevisiae* is the most common in alcoholic fermentation and is part of this category. The second class is the bottom-fermenting yeasts, suitable for low temperature fermentation (10°C to 15°C) and slow growth; an example is *Saccharomyces pastorianus*, formerly known as *S. carlsbergensis*, mainly used to produce some types of beer [40].

5.5 NATIVE MICROORGANISMS IN FOOD AND THEIR PROTECTIVE ROLES

Native microorganisms in food systems can enrich a natural environment where they can be metabolic active and reproduce. Due to their biosynthetic capability, microorganisms are excellent candidates for the solution of a variety of environmentally-related problems [42]. Natural niches are selective for particular types of microorganisms, where they play a particular role; in foods, microorganisms can intervene in natural fermentation. There is a particular interest in the microbial identification of traditional foods, which not only represent the culture of a region, but can also harbor a huge genetic potential of probable undiscovered microbial strains [43]. Lately, metagenomics studies have helped not only on the identification of microbial communities in different environments, but also on the genetic characterization to recognized potential microorganisms [44]. In many cases, native microorganisms develop a protective effect in food products; however, the same strain could participate as food-spoilage microorganism. One strategy to identify the beneficial or deleterious effect of a microorganism in a food system is the determination of the production of biogenic amines, which are toxicological nitrogenous compounds that some microorganisms can generate by the decarboxylation of amino acids [45].

5.5.1 ANIMAL-DERIVED PRODUCTS

Animal-derived products are considered food material obtained from the body of an animal, including meat, milk, and eggs. Contrary to what consumers think, livestock, and poultry are not in a sterile environment; animal production is carried out in an exposed environment, subjected to the presence of a variety of microorganisms from the air, feces, insects, and humans with a high probability of cross contamination [46]. Due to the high nutritional content of animal products, they can be contaminated by pathogens and deteriorating contamination; once the product is obtained, the hygienic practices are much needed during the distribution and retail for adequate shelf life. Dairy products are the most common naturally fermented products since their microbial population includes LAB, and the fermentation products are important for food preservation. Table 5.1 summarizes microbial strains found in animal-derived products, where fermentation has proved its protective effect.

5.5.2 VEGETABLE-DERIVED PRODUCTS

Vegetable-derived products are the ones that growth directly from the ground in a form of a plant including fruits, vegetables, legumes, and cereals. Plant-associated microorganisms are more variable than animal products since they include rhizosphere and soil microbes [51]. Plants exhibit natural defenses for microbial contamination and the microbiota present act more as endophytes (plant-microbe association that involve mutual interactions) [52]. Table 5.2 includes some isolated microbial strains found in vegetable-derived products, where fermentation has proved its protective effect.

5.6 LACTIC ACID BACTERIA (LAB) AS FOOD DEVELOPERS AND FUNCTIONAL ADDITIVES

LAB can improve the health benefits associated with fermented foods, as detoxifiers in food, by enhanced digestion of polymers, due to removal of intestinal-swelling compounds, and enrichment of food-substrates with essential nutrients like folate [56], an important vitamin from the B-group,

TABLE 5.1 Native Microorganisms Isolated from Animal-Derived Foods and Their Protective Effect Against Food Spoilage Microorganisms

Isolated Microbial Strains	Food Product	Proved Protective Effect	References
Lactococcus lactis subsp. *lactis* BGMN1–5	Traditionally home-made cheeses	Three narrow spectrum class II heat-stable bacteriocins	[47]
L. lactis subsp. *lactis*BGSM1–19		Low molecular mass (7 kDa) bacteriocin SM19 that showed antimicrobial activity against *Staphylococcus aureus, Micrococcus flavus* and partially against *Salmonella paratyphi*	
Lactobacillus paracasei subsp. *paracasei* BGSJ2–8		Heat-stable bacteriocin SJ (approx. 5kDa) polypeptide	
L. sake, L. curvatus, L. divergens, L. carnis, L. sanfrancisco, L. plantarum, L. casei, L. brevis, E. faecium, L. mesenteroides, P. pentosaceous, B. subtilis, B. mycoides, B. thuringiensis, B. lentus, B. licheniformis, Micrococcus, Staphylococcus, D. hansenii, D. polymorphus, D. pseudopolymorphus, P. burtonii, P. anomala, Candida famata, and *Rhizopus*	Himalayas meat products	LAB strains inhibited *K. pneumoniae* subsp. *pneumonia,* and *E. agglomerans*	[48]
B. licheniformis Me1 and *B. flexus* Hk1	Milk and cheese	Antioxidant activity and inhibition of *L. monocytogenes*	[49]
Lactobacillus pobuzihii	Fish food product	Bacteriocinogenic potential against *S. typhi, B. cereus, K. pneumoniae, E. coli* and *B. licheniformis*	[50]

TABLE 5.2 Native Microorganisms Isolated from Vegetables/Vegetable-Derived Foods and Their Protective Effect Against Food Spoilage Microorganisms

Isolated Microbial Strains	Food Product	Metabolite Production for a Protective Effect	References
E. faecium, L. lactis, E. hiraeand, and *E. canis*	Swiss chard and spinach	Temperature and storage resistant bacteriocin-like compounds and antagonistic effect against *E. coli* and *A. xylosoxidans*	[53]
A. hydrophila, P. fluorescens, P. aeruginosa, A. salmonicida spp. *salmonicida*	Carrots, green peppers, iceberg lettuce, green cabbage, celery, purple cabbage, and yellow onion	Inhibition of *E. coli* O157:H7, *L. monocytogenes, Salmonella Montevideo* and *S. aureus*	[54]
B. siamensis	Ripened *Doejang* (Fermented Soybean)	Not proved	[55]
B. licheniformis			
E. faecalis, E. faecium, and *T. halophilus*			
B. subtilis Bn1	Fermented beans	Antioxidant activity and inhibition of *L. monocytogenes*	[49]

which participates in metabolic pathways and cannot be produced in mammalian cells [57]. Different genera of microorganisms comprise the group of LAB: *Lactobacillus, Lactococcus, Leuconostoc, Pediococcus, Streptococcus, Aerococcus, Carnobacterium, Enterococcus, Lactobacillus, Lactococcus, Leuconostoc, Pediococcus, Streptococcus, Tetragenococcus, Vagococcus,* and *Weissella.* These taxonomic groups differ between micro and macro-morphology, mode of glucose fermentation (homo and hetero-fermentation), meso or thermo-philic growth, the configuration of the lactic acid production, halophilic ability, and acid or alkaline tolerance [58].

Preservation through fermentation with LAB includes the formation of inhibitory metabolites such as organic acids, bacteriocins, and biosurfactants. Not all the LAB can produce these metabolites; however, fermentation also promotes a competition for space and food with another microorganism, which results in their growth inhibition.

5.6.1 ORGANIC ACIDS

Lactic, acetic, and propionic acids are the main products obtained by carbohydrate utilization of LAB. Organic acids are potential bio-preservers as they can cause destabilization in the cellular processes by trespassing the cellular membrane and decreasing intracellular pH. Organic acids in food preservation are inexpensive and effective ways of microbial growth inhibition; however, their utilization has caused an emergence in acid-tolerance pathogens [59]. The antimicrobial effectiveness includes Gram-negative bacteria but Gram-positive but also against yeast and molds [60].

5.6.2 BIOSURFACTANTS

Biosurfactants are amphipathic molecules produced by several microorganisms, including some LAB. The combination of hydrophobic and hydrophilic moieties of these molecules is responsible for their antimicrobial and anti-biofilm formation capacities. The mechanism of action is not clearly understood; however, it has been hypothesized that biosurfactants can infiltrate into the plasma membrane to form pores, adhesion of microbial surfaces that block the nutrient acquisition, or insertion into the cell membrane that causes structural changes [61]. Their antimicrobial spectrum includes bacteria, fungi, and viruses [62].

5.6.3 BACTERIOCINS

Among all the antimicrobial metabolite produced by LABs, bacteriocins comprise the most important and studied compounds. Bacteriocins are ribosomally synthesized peptides, which exhibit antimicrobial activities against foodborne pathogens. The first discovered bacteriocin was nisin that has been used for more than 50 years and is approved in 40 countries and in the United States of America [63]. The search for bacteriocins with new properties is still an ongoing research area. The mechanism of LAB bacteriocins is based on the capacity of peptides to bind into the cellular membrane by interactions with specific receptors or direct contact to disrupt the membrane potential and cause the microbial cell death.

5.7 ECONOMICAL ADVANTAGE OF PRESERVATIVE FERMENTATION

Preservative fermentation improves food quality, desirable sensorial changes with better acceptance by the consumer and the simultaneous production of metabolites with health benefits, as compared with other food preservation methods. Also, there is becoming a common practice, to manufacture your food products, for recreation or to start a small business; one of the most popular products is artisanal beer. Raw materials and equipment are accessible in the market, and the process could be a little bit less long than wine to obtain a good quality beer.

Large investments in the food industry are dedicated to the development of methods for increasing the product shelf life, but with the preservation of aspects such as appearance, nutritional value, and flavor [64]. The reduction in quality is directly related to a lower commercialization value. Food preservation by traditional fermentation processes can be inexpensive because of the unnecessary high technological equipment, infrastructure, or highly trained personal. Fermentation products represent one-third of food consumed worldwide [65]. An example of how these products play an important economic role is in the fermentation of cassava in African, where linamarin is eliminated. Linamarin is a toxic substance present in some varieties of cassava; the inoculation of *Lactobacillus plantarum* A6 (amylolytic) and *Lactobacillus plantarum* (nonamylolytic) in cassava pulp, achieved a total reduction of the linamarin in less than five hours [66]. Fermentations are more effective than the physical treatment for the removal of linamarin; this is an example of the implementation of the fermentative process in developing countries, with social and economic impact [67]. However, the fermentative process is also implemented in countries like the United States and the majority of European countries either artisanal (also called handmade or craft products) or by industrialized processes.

Fermentation can provide a better method of food preservation, compared with heat temperature treatments, that have an effect on flavor and texture of the food product; on the other hand, refrigeration or freezing, in some cases cause physical damage and bad taste due to the increase in solute concentration by ice formation [68]. More sophisticated process of conservation such as HHP, ultrasound, high voltage pulse and irradiation, are expensive technologies. Irradiation as preservation method does not cause major changes in the food matrix, but the use of gamma rays required a label with the legend "treated with radiation" or "treated by irradiation" in addition to the radula symbol; even when scientific evidence demonstrates that irradiated foods does not become radioactive and there is no radiation residue remains in the food, this process is not attractive to the consumer [69]. On the contrary, fermentative process for food preservation is considered as a natural way of conservation, with additional benefits of flavor enhancing in the product that make them more attractive to the consumer.

5.8   FUTURE PERSPECTIVES IN FOOD PRESERVATION TECHNOLOGY

In recent years, there are some fermentative processes of food products that have been industrialized; however, a large number of fermented products are still obtained by artisanal production, by native microorganisms. The main problem with artisanal products is lack of standardized processes, and the microorganisms present are not characterized, which can lead to the presence of pathogens or the production of fermented foods that are not suitable for consumption. Metagenomic studies have helped on the characterization of the microbial population of selected environments, including fermented foods. Also, by the isolation and extensive characterization of microbial populations, a better understanding of the fermentative processes has been obtained. The knowledge acquired and the microbial strains isolated, can promote the use of fermentation as a way to preserve food, with the required microbial and chemical safety.

KEYWORDS

- **chlorine dioxide**
- **essential oils**
- **food-contamination**
- **functional benefits**
- **industrial fermentation**
- **lactic acid bacteria**

REFERENCES

1. World Health Organization (WHO). *The Top 10 Causes of Death*, **2014**.
2. World Health Organization. *WHO Estimates of the Global Burden of Foodborne Diseases: Foodborne Disease Burden Epidemiology Reference Group 2007–2015*, **2015**.
3. US Food and Drug Administration. *Background on the FDA Food Safety Modernization Act (FSMA)*, **2018**. https://www.fda.gov/newsevents/publichealthfocus/ucm239907.htm (accessed on 20 January 2020).
4. Centers for Disease Control and Prevention. List of selected multistate foodborne outbreak investigations. *Centers for Disease Control and Prevention*, **2018**. https://www.cdc.gov/foodsafety/outbreaks/multistate-outbreaks/outbreaks-list.html (accessed on 6 January 2020).
5. Kaushik, G., Satya, S., & Naik, S. N. Food processing a tool to pesticide residue dissipation: A review. *Food Res. Int.*, **2009**, *42*(1), 26–40.
6. Heaton, J. C., & Jones, K. Microbial contamination of fruit and vegetables and the behavior of enteropathogens in the phyllosphere: A review. *J. Appl. Microbiol.*, **2008**, *104*(3), 613–626.
7. Nychas, G. J. E., Skandamis, P. N., Tassou, C. C., & Koutsoumanis, K. P. Meat spoilage during distribution. *Meat Sci.*, **2008**, *78*(1/2), 77–89.
8. Rodas-González, A., Narváez-Bravo, C., Brashears, M. M., Rogers, H. B., Tedford, J. L., Clark, G. O., Brooks, J. C., Johnson, R. J., & Miller, M. F. Evaluation of the storage life of vacuum packaged Australian beef. *Meat Sci.*, **2011**, *88*(1), 128–138.
9. Beere, H. M. The stress of dying: The role of heat shock proteins in the regulation of apoptosis. *J. Cell Sci.*, **2004**, *117*(13), 2641–2651.
10. Periago, P. M., Van Schaik, W., Abee, T., & Wouters, J. A. Identification of proteins involved in the heat stress response of *Bacillus cereus* ATCC 14579. *Appl. Environ. Microbiol.*, **2002**, *68*(7), 3486–3495.
11. Chastanet, A., Derre, I., Nair, S., & Msadek, T. clpB, a novel member of the *Listeria monocytogenes* CtsR regulon, is involved in virulence but not in general stress tolerance. *J. Bacteriol.*, **2004**, *186*(4), 1165–1174.

12. Cacace, G., Mazzeo, M. F., Sorrentino, A., Spada, V., Malorni, A., & Siciliano, R. A. Proteomics for the elucidation of cold adaptation mechanisms in *Listeria monocytogenes*. *J. Proteomics,* **2010**, *73*(10), 2021–2030.

13. Considine, K. M., Kelly, A. L., Fitzgerald, G. F., Hill, C., & Sleator, R. D. High-pressure processing-effects on microbial food safety and food quality. *FEMS Microbiol. Lett.,* **2008**, *281*(1), 1–9.

14. Heinz, V., & Buckow, R. Food preservation by high pressure. *J. Verbraucherschutz und Lebensmittelsicherheit.,* **2010**, *5*(1), 73–81.

15. Vanlint, D., Mitchell, R., Bailey, E., Meersman, F., McMillan, P. F., Michiels, C. W., & Aertsen, A. Rapid acquisition of Giga Pascal-high-pressure resistance by *Escherichia coli*. *Mbio.,* **2011**, *2*(1), e00130-10.

16. Margosch, D., Ehrmann, M. A., Gaenzle, M. G., & Vogel, R. F. Comparison of pressure and heat resistance of *Clostridium botulinum* and other endospores in mashed carrots. *J. Food Prot.,* **2004**, *67*(11), 2530–2537.

17. Malone, A. S., Chung, Y. K., & Yousef, A. E. Genes of *Escherichia coli* O157: H7 that are involved in high-pressure resistance. *Appl. Environ. Microbiol.,* **2006**, *72*(4), 2661–2671.

18. Vanlint, D., Mitchell, R., Bailey, E., Meersman, F., McMillan, P. F., Michiels, C. W., & Aertsen, A. Rapid acquisition of Giga Pascal-high-pressure resistance by *Escherichia coli*. *Mbio.,* **2011**, *2*(1), e00130–10.

19. Chemat, F., & Khan, M. K. Applications of ultrasound in food technology: Processing, preservation and extraction. *Ultrasonics Sonochemistry,* **2011**, *18*(4), 813–835.

20. Villamiel, M., & De Jong, P. Influence of high-intensity ultrasound and heat treatment in continuous flow on fat, proteins, and native enzymes of milk. *J. Agric Food Chem.,* **2000**, *48*(2), 472–478.

21. Piyasena, P., Mohareb, E., & McKellar, R. C. Inactivation of microbes using ultrasound: A review. *Int. J. Food Microbiol.,* **2003**, *87*(3), 207–216.

22. Sagong, H. G., Lee, S. Y., Chang, P. S., Heu, S., Ryu, S., Choi, Y. J., & Kang, D. H. Combined effect of ultrasound and organic acids to reduce *Escherichia coli* O157: H7, *Salmonella typhimurium*, and *Listeria monocytogenes* on organic fresh lettuce. *Int. J. Food Microbiol.,* **2011**, *145*(1), 287–292.

23. Režek, J. A., Lelas, V., Herceg, Z., Badanjak, M., Batur, V., & Muža, M. Advantages and disadvantages of high power ultrasound application in the dairy industry. Mljekarstvo: Journal for the Advancement of Milk Production and Processing, **2009**, *59*(4), 267–281.

24. Gómez-López, V. M., Devlieghere, F., Bonduelle, V., & Debevere, J. Factors affecting the inactivation of micro-organisms by intense light pulses. *J. Appl. Microbiol.,* **2005**, *99*(3), 460–470.

25. Roberts, P., & Hope, A. Virus inactivation by high intensity broad spectrum pulsed light. *J. Virological Methods,* **2003**, *110*(1), 61–65.

26. Oms-Oliu, G., Martín-Belloso, O., & Soliva-Fortuny, R. Pulsed light treatments for food preservation: A review. *Food Bioprocess Technol.,* **2010**, *3*(1), 13–23.

27. Heinrich, V., Zunabovic, M., Varzakas, T., Bergmair, J., & Kneifel, W. Pulsed light treatment of different food types with a special focus on meat: A critical review. *Crit. Rev. Food Sci. Nutr.,* **2016**, *56*(4), 591–613.

28. Barba, F. J., Parniakov, O., Pereira, S. A., Wiktor, A., Grimi, N., Boussetta, N., Saraiva, J. A., Raso, J., Martin-Belloso, O., Witrowa-Rajchert, D., & Lebovka, N. Current applications and new opportunities for the use of pulsed electric fields in food science and industry. *Food Res. Int.,* **2015**, *77*, 773–798.

29. Yun, O., Liu, Z. W., Zeng, X. A., & Han, Z. *Salmonella typhimurium* resistance on pulsed electric fields associated with membrane fluidity and gene regulation. *Innov. Food Sci. Emerg. Technol.*, **2016**, *36*, 252–259.

30. Cserhalmi, Z., Vidács, I., Beczner, J., & Czukor, B. Inactivation of *Saccharomyces cerevisiae* and *Bacillus cereus* by pulsed electric fields technology. *Innov. Food Sci. Emerg. Technol.*, **2002**, *3*(1), 41–45.

31. Khadre, M. A., & Yousef, A. E. Susceptibility of human rotavirus to ozone, high pressure, and pulsed electric field. *J. Food Prot.*, **2002**, *65*(9), 1441–1446.

32. Appendini, P., & Hotchkiss, J. H. Review of antimicrobial food packaging. *Innov. Food Sci. Emerg. Technol.*, **2002**, *3*(2), 113–126.

33. Hyldgaard, M., Mygind, T., & Meyer, R. L. Essential oils in food preservation: Mode of action, synergies, and interactions with food matrix components. *Frontiers Microbiol.*, **2012**, *3*, 12.

34. Solórzano-Santos, F., & Miranda-Novales, M. G. Essential oils from aromatic herbs as antimicrobial agents. *Curr. Op. Biotechnol.*, **2012**, *23*(2), 136–141.

35. Gómez-López, V. M., Rajkovic, A., Ragaert, P., Smigic, N., & Devlieghere, F. Chlorine dioxide for minimally processed produce preservation: A review. *Trends Food Sci. Technol.*, **2009**, *20*(1), 17–26.

36. López-Gálvez, F., Allende, A., Truchado, P., Martínez-Sánchez, A., Tudela, J. A., Selma, M. V., & Gil, M. I. Suitability of aqueous chlorine dioxide versus sodium hypochlorite as an effective sanitizer for preserving quality of fresh-cut lettuce while avoiding by-product formation. *Postharvest Biol. Technol.*, **2010**, *55*(1), 53–60.

37. Trinetta, V., Vaid, R., Xu, Q., Linton, R., & Morgan, M. Inactivation of *Listeria monocytogenes* on ready-to-eat food processing equipment by chlorine dioxide gas. *Food Control.*, **2012**, *26*(2), 357–362.

38. Imran, M., Bré, J. M., Guéguen, M., Vernoux, J. P., & Desmasures, N. Reduced growth of *Listeria monocytogenes* in two model cheese microcosms is not associated with individual microbial strains. *Food Microbiol.*, **2013**, *33*(1), 30–39.

39. Caplice, E., & Fitzgerald, G. F. Food fermentations: Role of microorganisms in food production and preservation. *Int. J. Food Microbiol.*, **1999**, *50*(1/2), 131–149.

40. Sarma, S. J., Verma, M., & Brar, S. K. Industrial fermentation for production of alcoholic beverages. In: *Fermentation Processes Engineering in the Food Industry* (pp. 322–345). CRC Press, **2013**.

41. Cvetković, D., Markov, S., Djurić, M., Savić, D., & Velićanski, A. Specific interfacial area as a key variable in scaling-up Kombucha fermentation. *J. Food Engin.*, **2008**, *85*(3), 387–392.

42. Kumar, B. L., & Gopal, D. S. Effective role of indigenous microorganisms for sustainable environment. *3 Biotech.*, **2015**, *5*(6), 867–876.

43. Tamang, J. P., Watanabe, K., & Holzapfel, W. H. Diversity of microorganisms in global fermented foods and beverages. *Frontiers Microbiol.*, **2016**, *7*, 377.

44. Jung, J. Y., Lee, S. H., Kim, J. M., Park, M. S., Bae, J. W., Hahn, Y., Madsen, E. L., & Jeon, C. O. Metagenomic analysis of kimchi, the Korean traditional fermented food. *Appl. Environ. Microbiol.*, **2011**, *77*(7), 2264–2274.

45. Spano, G., Russo, P., Lonvaud-Funel, A., Lucas, P., Alexandre, H., Grandvalet, C., et al. Biogenic amines in fermented foods. *Eur. J. Clin. Nutr.*, **2010**, *64*(S3), S95.

46. Cunningham, F. *The Microbiology of Poultry Meat Products*. Elsevier, **2012**.

47. Topisirovic, L., Kojic, M., Fira, D., Golic, N., Strahinic, I., & Lozo, J. Potential of lactic acid bacteria isolated from specific natural niches in food production and preservation. *Int. J. Food Microbiol.*, **2006**, *112*(3), 230–235.

48. Rai, A. K., Tamang, J. P., & Palni, U. Microbiological studies of ethnic meat products of the Eastern Himalayas. *Meat Sci.*, **2010**, *85*(3), 560–567.

49. Nithya, V., & Halami, P. M. Evaluation of the probiotic characteristics of Bacillus species isolated from different food sources. *Ann. Microbiol.*, **2013**, *63*(1), 129–137.

50. Rapsang, G. F., Kumar, R., & Joshi, S. R. Identification of *Lactobacillus pobuzihii* from tungtap: A traditionally fermented fish food, and analysis of its bacteriocinogenic potential. *African J. Biotechnol.*, **2011**, *10*(57), 12237–12243.

51. Gunatilaka, A. L. Natural products from plant-associated microorganisms: Distribution, structural diversity, bioactivity, and implications of their occurrence. *J. Nat. Prod.*, **2006**, *69*(3), 509–526.

52. Reinhold-Hurek, B., & Hurek, T. Living inside plants: Bacterial endophytes. *Current Opinion in Plant Biology*, **2011**, *14*(4)., 435–443.

53. Ponce, A. G., Moreira, M. R., Del Valle, C. E., & Roura, S. I. Preliminary characterization of bacteriocin-like substances from lactic acid bacteria isolated from organic leafy vegetables. *LWT-Food Sci. Technol.*, **2008**, *41*(3), 432–441.

54. Schuenzel, K. M., & Harrison, M. A. Microbial antagonists of foodborne pathogens on fresh, minimally processed vegetables. *J Food Prot.*, **2002**, *65*(12), 1909–1915.

55. Jeong, D. W., Kim, H. R., Jung, G., Han, S., Kim, C. T., & Lee, J. H. Bacterial community migration in the ripening of doenjang, a traditional Korean fermented soybean food. *J. Microbiol. Biotechnol.*, **2014**, *24*(5), 648–660.

56. Mfc, M. F. C. Safety demonstration of microbial food cultures (MFC) in fermented food products. *Bulletin of the International Dairy Federation*, **2012**, *455*.

57. LeBlanc, J. G., De Giori, G. S., Smid, E. J., Hugenholtz, J., & Sesma, F. Folate production by lactic acid bacteria and other food-grade microorganisms. *Communicating Current Research and Educational Topics and Trends in Applied Microbiology*, **2007**, *1*, 329–339. FORMATEX.

58. Rattanachaikunsopon, P., & Phumkhachorn, P. Lactic acid bacteria: Their antimicrobial compounds and their uses in food production. *Annals Biol. Res.*, **2010**, *1*(4), 218–228.

59. Ricke, S. C. Perspectives on the use of organic acids and short chain fatty acids as antimicrobials. *Poultry Sci.*, **2003**, *82*(4), 632–639.

60. Dalié, D. K. D., Deschamps, A. M., & Richard-Forget, F. Lactic acid bacteria: Potential for control of mould growth and mycotoxins: A review. *Food Control*, **2010**, *21*(4), 370–380.

61. Sharma, D., Saharan, B. S., & Kapil, S. *Biosurfactants of Lactic Acid Bacteria*. Switzerland: Springer, **2016**.

62. Inès, M., & Dhouha, G. Lipopeptide surfactants: Production, recovery and pore forming capacity. *Peptides*, **2015**, *71*, 100–112.

63. Cleveland, J., Chikindas, M., & Montville, T. J. Multi-method assessment of commercial nisin preparations. *J. Ind. Microbiol. Biotechnol.*, **2002**, *29*(5), 228–232.

64. Johnson, M. E., & Steele, J. L. Fermented dairy products. In: *Food Microbiology* (pp. 825–839). American Society of Microbiology, **2013**.

65. Soccol, C. R., Pandey, A., & Larroche, C. *Fermentation Processes Engineering in the Food Industry*. CRC Press, **2013**.

66. Savadogo, A. The role of fermentation in the elimination of harmful components present in food raw materials. In: Mehta, B. M., Kamal-Eldin, A., Iwanski, R. Z., (eds.), *Fermentation: Effects on Food Properties* (pp. 169–179). CRC Press, Taylor & Francis Group. NY, **2012**.

67. Oyewole, O. A, & Isah, P. Foods fermented locally in Nigeria and its importance for the national economy: A review. *J. Recent Adv Agric.*, **2012**, *1*(4), 92–102.

68. Vaclavik, V. A., & Christian, E. W. Food preservation. In: *Essentials of Food Science* (pp. 323–342). Springer, New York, NY, **2014**.

69. Gould, G. W. *New Methods of Food Preservation.* Springer Science and Business Media, **2012**.

CHAPTER 6

Food Preservation Using Plant-Derived Compounds

SOFIA DEL ROSARIO ROMERO-RAMOS, DIANA B. MUÑIZ-MÁRQUEZ,
PEDRO AGUILAR-ZÁRATE, and JORGE E. WONG-PAZ

*Master Program in Engineering, Technological Institute of Ciudad Valles,
National Technological Institute of Mexico 79010, Ciudad Valles,
S.L.P., México, Tel.: +52 481 381 2044, E-mail: jorge.wong@tecvalles.mx
(J. E. Wong-Paz)*

ABSTRACT

The growing need for more natural products, as well as the demand of consumers to eat fresh, minimally prepared, or ready to eat (RTE) food products, presents an enormous challenge for the food industry, and to meet this demand, new methodologies are developed continuously for the preservation of food. In this context, the use of compounds derived from plants that have preservative properties results in an alternative. Nevertheless, the use of these compounds has the challenge of maintaining the quality and safety of food by also providing functional properties. At the same time, it is necessary to develop new technologies for more processes for the extraction, application, and conservation of bioactive compounds (BCs). This chapter is focused on the present use of plant-derived compounds in the preservation of food, and pays particular attention to the primary agents that cause spoilage and represents a risk for food products.

6.1 INTRODUCTION

The ability to conserve food is one of the most important advances in the history of humanity because they ensure quality and nutritional content and also represents an evolution in the development of techniques and resources for food security [1, 2]. Because of this development, it was possible the establishing human communities without the need to move around to look for fresh food. The first techniques were employed with empiric knowledge using several methods to preserve the food (Figure 6.1) [2].

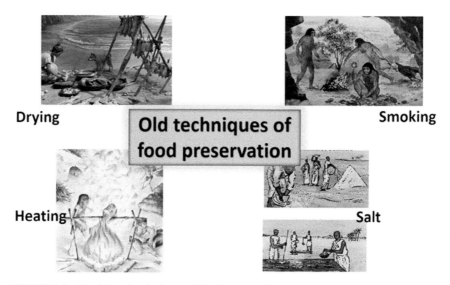

FIGURE 6.1   Traditional techniques of food preservation.

Food preservation techniques were evolving, and after pasteurization, it was possible to reduce the number of microorganisms in food products, to improve their quality and extend its shelf life. The addition of several substances, such as salt and spices, has allowed the preservation methods to improve, in order to offer a better quality product, and avoid the undesirable changes in the food [2]. The evolution of the techniques of food preservation was due to the understanding of the causes of deterioration and damage to food products [3]. These causes are mentioned in Figure 6.2.

Physical agents are not, on many occasions, the direct responsible for food spoilage, but they represent the entrance to other alterations [3]. Besides being unpleasant, this type of affectations, produce a rupture of the cells and damage of the tissue which causes the loss of water and a rapid increase in the respiration of the damaged tissue. The mechanical damages are caused by the incorrect manipulation of products, causing cuts and bruises which increase the physiological modifications and risk the integrity of the product by allowing the entry of pathogens [4].

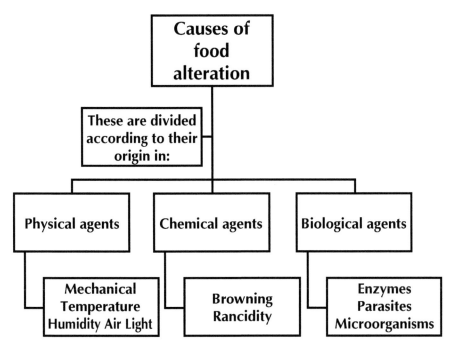

FIGURE 6.2 Causes of food alteration in aliments.

Temperature is one of the most critical factors of food deterioration because the chemical and enzymatic activities double their speed for each 10°C increase; therefore, decomposition processes are accelerated. Also, several of the nutrients present in food are sensitive to the increase in temperature coupled with the heat causes changes in the state of emulsions or mixtures containing water due to drying [3]; furthermore, low temperatures affect the fresh products, causing freeze damage [4]. On the other hand, high humidity levels favor microbial development, and atmospheric conditions such as the concentration of oxygen in the air, are determinant for the oxidation processes of fresh products, since this oxygen is used for oxidation reactions of starches, and in turn, these reactions produce undesirable changes in food [3, 4]. Light conditions should also be controlled because many foods contain several photosensitive compounds. Among the microbiological agents of food deterioration, are bacteria, molds, and yeasts. These are considered the most important factors that affect the food quality, because their presence represents a risk to the integrity and safety of food [1].

Regarding extrinsic factors, those that are not related to the food itself, but the environmental conditions, two main factors are considered:

temperature and atmospheric composition. Depending on the interactions of these factors and the requirements of the microbiota present in the food environment, the microbial growth will be fast or slow. Therefore, the knowledge about the microbiological requirements is fundamental to secure safe foods [5, 6].

Food preservation is a continuous effort to assure that microbial responsible for unsafe food, will be minimized, or destroyed to guarantee the safety and quality of food products. The use of preservation techniques has been changing through time in the food industry. Nowadays, the research has focused on the replacement of traditional food preservation techniques by new preservation techniques, due to the increased consumer demand for tasty, nutritious, natural, and easy-to-handle food products. The primary research objective is centered on non-thermal technologies, due to the preservation of food attributes [7].

This chapter is focused on the present use of plant-derived compounds in the preservation of food, especially on the main agents that cause spoilage and represent a risk for food products.

6.2 CONVENTIONAL AND EMERGING TECHNOLOGIES IN FOOD PRESERVATION

Food preservation processes are designed to avoid alterations produced by physical, chemical, and biological agents [2, 3]. Currently, there are different treatments to preserve food. In general, terms, the main conventional methods that use the temperature process are describing in Figure 6.3.

Water activity (A_w) of the food is one of the intrinsic factors that affect the quality of the food product and makes its preservation difficult because high values of A_w it favors the proliferation of microorganisms. Different processes are involved in the modification of the amount of water in a product, among which are dehydration, lyophilization, and concentration. On the other hand, the chemical methods are techniques that modify the sensorial properties, for example, flavor odor and color. Figure 6.4 shows what the principal chemical methods used in food preservation are.

Non- thermal food treatments for food preservation are methods that are of interest for scientists, manufacturers, and consumers because they have a minimal impact on the nutritional and sensorial food properties and extend their shelf life by the control of microbial growth. Moreover, the non-thermal process represents an efficient method to preserve food quality attributes while being environmentally efficient, since they use less energy than traditional methods. These methods include ultrasound, ultra-high pressure, ionizing radiation, pulsed X-ray, magnetic fields, pulsed light and pulsed electric fields (PEF), high-voltage arc discharge, dense phase carbon dioxide and hurdle technologies [8]. Regardless of their benefits, thermal processes are still the main used methods for food industrial food preservation; although they can, because undesirable changes in the food characteristics, as shown in Figure 6.5.

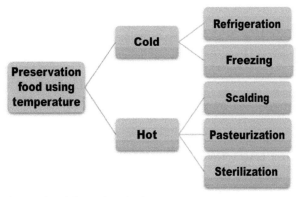

FIGURE 6.3   Conventional thermal methods.

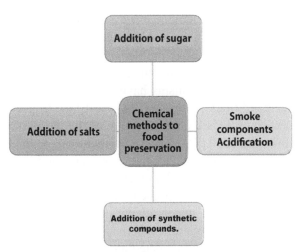

FIGURE 6.4   Chemical methods of food preservation.

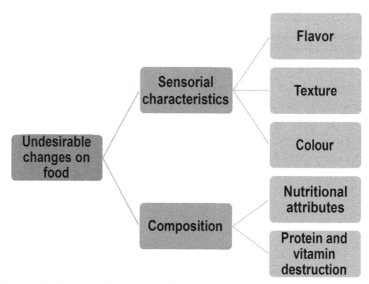

FIGURE 6.5   Undesirable changes in food.

6.3   USE OF PLANT-DERIVED COMPOUNDS IN FOOD PRESERVATION

Microorganisms have an essential role in all the processes in the food industry. Spoilage and foodborne pathogens are the main risk factor for food quality and safety; when the security of the aliments is threatened, it is possible that the deterioration caused by microorganisms will have an effect in human health. This food safety problem is the main reason for food products to be subjected to several treatments to reduce their microbial load. Nonetheless, some microorganisms have developed resistance to adverse environmental conditions, including the ability to reduce the effects of external attacks, by reversible modification of the morphology and physiology of the microbial cell [9]. Also, the current trend on the bacterial resistance to multiple treatments has become a problem for food treatment [10]. On the other hand, chemical food preservatives that have been conventionally used in the food industry to prevent the spoilage caused by pathogenic microorganisms have been under review due to their adverse effects on consumers, and to environmental concerns as well. Moreover, the growing demand for products minimally treated and the need to reduce the risk on the human health have opened the door to the new preservation methods, enabling the preservation of food with minimal impact on their quality of them [11]. Biopreservation is the use of naturally derived antimicrobial and antioxidant agents or controlled microbiota, capable of high antimicrobial potential to preserve food, thus extending their shelf lives and increasing the food safety. Several studies have examined and reported on promising natural antimicrobial agents, such as natamycin, nisin, pediocin, reuterin, bacteriocins, lactoferrins, lysozymes, alkaloids, dienes, flavonols, flavones, glycosides, lactones, organic acids, phenolic compounds, and essential oils (EOs) [12–14].

Natural products to control of pests and to prevent damages in food products have been used for centuries, but recently, the search for natural products to control parasitic and bacterial agents has been intensified [15]. Preservation processes are the main techniques to guarantee the quality in the food technology. This process requires an integrated and interdisciplinary approach, taking into account that each part of the food production process is important [1].

Natural substances from plants are commonly used as antioxidant and ant microbiological biopreservatives, due to their natural composition, these products are biocompatible, biodegradable, safety, and have low toxicity and cost efficiency [1]. The majority of these compounds are naturally present in plants, where they have a role as agents of protection against infections or injuries. Some examples of these compounds are shown in Table 6.1 [11, 12, 16].

6.3.1 MECHANISMS OF THE BIOLOGICAL ACTIVITY OF PLANT-DERIVED COMPOUNDS

Plants are pluricellular organisms with the ability to generate action mechanisms that allow them to protect against external aggressions; furthermore, their proprieties are very useful and are employed in several forms, as shown in Figure 6.6.

These mechanisms have been exploited for centuries, especially in traditional medicine. In recent years, the compounds obtained through these mechanisms are used with a different focus and has been demonstrated the potential of plants extracts, complete or parts of them, including stems, roots, leaves, flowers, and fruits scientifically, to affects the microbial growth and guaranty food safety and human health. Antimicrobial compounds that are derived and isolated naturally from several sources, including microorganisms, animal and plant extracts, and EOs [17, 18].

TABLE 6.1 Natural Compounds to Food Preservation Processes

Phenolic Derivatives	Essential Oils (Rich in Monoterpenoids)	Aromatic Derivatives	Organosulfur Compounds	Bacteriocins	Others
Simple phenols, polyphenolcarboxylic acids, floroglucinols, flavonoids, tannins	Geranial, 1,8-cineole, α-terpineol, linalool, borneol, camphor, carvacrol, thymol, limonene, *p*-cymene, α-pinene	Eugenol, isoeugenol, eugenone, cinnamaldehyde	Thiosulfinates allyl isothiocyanate	Nisin, pediocins, reuterin, sakacins, lacticin S, lacticin 3147, enterocins	Lysozyme, lactoperoxidase system, lactoferrin, chitosan, vitamins, sugar, polysaccharides, organic acids

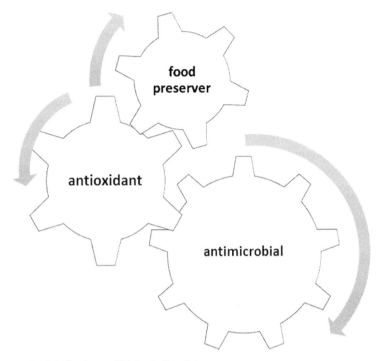

FIGURE 6.6   Mechanisms of biological activity.

Microbial growth, lipid oxidation, and color are important factors to shelf-life and consequently to consumer acceptance of the food [19].

Naturally occurring antimicrobial and antioxidant compounds derived from plant sources have been preferably employed for addition to the food products, because of their potential health benefits and safety compared to synthetic preservatives [19–22]. The potential damage on human health associated with the use of synthetic compounds for food preservation, have impulse new research focused on the development of natural antimicrobial agents and antioxidants, as a secure alternative to synthetic compounds, and could help avoid the excessive physical and chemical processing of food to ensure microbial safety and eco-preservation. Further, the use of natural preservatives to improve shelf life is a promising technology; plant-derived substances have antioxidant and antimicrobial properties as food preservatives [13, 23].

Antioxidants as food preservatives have increasing importance in food preservation, since they prevent food deterioration by oxidation, reduce the loss of nutritional value and energy contents, allows freshness by ensuring flavors, odors, color pigments, taste, and texture preservation. Consequently, numerous health benefits in cardiovascular and neurological diseases as well as anti-inflammatory, antibacterial, anti-allergic, anti-hypertensive, and antiviral and skin wound healing effects, and even reduction in cancer incidence have been attributed to the role of dietary antioxidants. Potent natural antioxidants are obtained of different natural resources; some examples are included in Figure 6.7. There are several applications for each compound; for example, it has been reported the use of plant phenolic extracts as perseverative of seafood, meat, and oils. The ascorbic acid founded in citric fruits is employed to preserve juices, jams, cured meats, and some canned foods. Tocopherols are used in the preservation of cereals, meat, and poultry products, butter, oils, and dairy products.

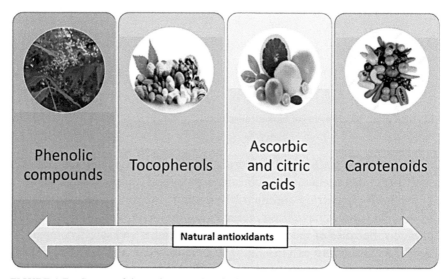

Phenolic compounds | Tocopherols | Ascorbic and citric acids | Carotenoids

Natural antioxidants

FIGURE 6.7   Source of the main natural antioxidants.

6.3.2 *IMPLICATIONS FOR FOOD SAFETY, NUTRITIONAL, AND QUALITY PROPERTIES*

The growing demand for ready-to-eat products has also increased the need to ensure the safety and quality of these foods. Additionally, there is a desire of consumers for natural, healthier, and better foodstuffs, with sensory characteristics similar to fresh food, and the consensus that functional food has an important role in the prevention of diseases. This requirement has resulted in the reduction of thermal processes, or the use of lower treatment temperatures, the decrease in the use of conservatives, and changes in packaging materials. Nevertheless, these changes have implicated a reduction in the microbiological margins of safety; food stability is reduced and improved control in manufacturing and distribution is required to ensure food safety. Perishable foods are mostly exposed to microbial contamination and the need to employ antioxidant compounds, and the incorporation of bioactive antioxidant compounds derived of plans to the formulations to preserve the aliments or coated on it sources is a secure alternative [13, 16, 21, 24].

In recent decades, many natural bioactive compounds (BCs) whit demonstrated proprieties such as antioxidant, immunoregulatory, and antimicrobial had been identified from plants. Nevertheless, unfavorable characteristics of these compounds such as poor solubility, undesirable flavors, low bioavailability and instability during food processing and storage of food, represent a challenge to be solved. It is necessary to develop technologies that allow the extraction and improvement of those compounds, as well as the processes for its application, in such a way that such undesirable characteristics disappear without affecting their beneficial properties [25]. In the last years, the need to develop functional products using biotechnological processes, specifically, applying BCs derived from plants has become a promising field of research both in the food industry and in the academia.

It should be noted that BCs can be degraded by factors such as light, oxygen, temperature, humidity, and the presence of unsaturated bonds in their molecular structures. Taking into account the susceptibility of biocomposites, their microencapsulation, appropriate materials, is an alternative to improve their stability and storage and provides the opportunity to mask unpleasant characteristics such as bitter taste and astringency of polyphenols. Consequently, an area of study is the research on the effects of microencapsulation on food antioxidants. There are several techniques of micro and nanoencapsulation, as well as different materials used; each one has advantages and disadvantages. The main challenge is to determine what is the better process of encapsulation (micro or nanoencapsulation) of each bioactive compound to be used in food preservation, keeping the quality of the products where they are applied, while maintaining the effectiveness of these compounds [14, 26].

Besides, nanotechnology has transformed many rules in food science, particularly, those that involve the processes, packaging, storage, shipment, functionality, and other safety issues related to food. The mixing of nanostructured materials with other existing polymers or biomolecules in aggregate form can result in a particle size of >100 nm, which leads to the formation of nanocomposites. These nanomaterials show superior physicochemical characteristics, such as solubility, magnetism, color, and thermodynamics. Its unbeatable ability to increase solubility and bioavailability and to protect bioactive components while they are processed and stored is one of the areas that need to improve in biocomposites. Also, due to the physicochemical nature and the antimicrobial potential of nanomaterials, they are widely used for food preservation. However, the nanoparticles have advantages and disadvantages, and the first ones include economic benefits due to the speed and efficiency of the processes as well as improvement of the sensory characteristics of the products, along with the retention of most of the properties of the bioactive compound. On the other hand, some disadvantages include the release of liquid compounds with high humidity, a limited number of cover materials, and high-temperature conditions required during the immobilization process; some of those conditions are not compatible with the stability of the BCs [26, 27].

6.4   CONCLUDING REMARKS AND FUTURE PERSPECTIVES

Currently, there is evidence of the security and viability of the natural preservatives. The recent discoveries and development of new natural antimicrobials and antioxidants have shown an efficient alternative to reduce the use of synthetic products to food preservation. Phenolic compounds, organic acids, phytoalexins, and other products derived from plants, represents a relevant field of investigation for better recovery techniques, as well as on the quality of the compounds recovered, increasing the stability of the food products. New techniques are emerging in the extraction and preservation of the properties of BCs. However, industrial equipment is yet needed. Also, the development of standardized techniques to ensure that these substances do not lose the potential effect on food safety. In this sense, advances in the micro and nanoencapsulation of BCs derived from plants are a feasible alternative.

ACKNOWLEDGMENTS

Romero-Ramos is grateful with her advisors by the support this project.

KEYWORDS

- **bioactive compounds**
- **functional additives**
- **nanoencapsulation**
- **phytoalexins**
- **plant-derived preservatives**
- **water activity**

REFERENCES

1. Mogoşanu, G. D., Grumezescu, A. M., Bejenaru, C., & Bejenaru, L. E. 11 – Natural products used for food preservation BT – Food preservation. In: *Nanotechnology in the Agri-Food Industry* (pp. 365–411). Academic Press, **2017**.

2. Introduction. In: Zeuthen, P., & Bøgh-Sørensen, L. B. T. F. P. T., (eds.), *Woodhead Publishing Series in Food Science, Technology and Nutrition* (pp. 1–2). Woodhead Publishing, **2003**.

3. Juliarena, P., & Gratton, R. *Technology, Environment and Society Unicen,* **2013**, 1–12 (Tecnología, Ambiente y Sociedad).

4. Castellanos, M., Vejarano, I., & Flores, E. Harvest and marketing manual. In: Pitty, A., & Valladares, P., (eds.), *Integrated Pest Management Program in Central America.* Secretaría de Educación, Eds., República de Honduras, **2012**.

5. Augusto, P. E. D., Soares, B. M. C., & Castanha, N. Chapter 1. In: Francisco, J., Sant'Ana, A. S., Orlien, V., & Koubaa, M. B. T. I., (eds.), *Conventional Technologies of Food Preservation A2-Barba* (pp. 3–23). Academic Press, **2018**.

6. Berk, Z. Chapter 16 – Spoilage and preservation of foods BT – food process engineering and technology (3rd edn.). In: *Food Science and Technology* (pp. 395–398). Academic Press, **2018**.

7. Devlieghere, F., Vermeiren, L., & Debevere, J. New preservation technologies: Possibilities and limitations. *Int. Dairy J.,* **2004**, *14*(4), 273–285.

8. Morris, C., Brody, A. L., & Wicker, L. Non-thermal food processing/preservation technologies: A review with packaging implications. *Packag. Technol. Sci.,* **2007**, *20*(4), 275–286.

9. Ferro, S., Amorico, T., & Deo, P. Role of food sanitizing treatments in inducing the 'viable but nonculturable' state of microorganisms. *Food Control.,* **2018**, *91*, 321–329.

10. Solórzano-Santos, F., & Miranda-Novales, M. G. Essential oils from aromatic herbs as antimicrobial agents. *Curr. Opin. Biotechnol.,* **2012**, *23*(2), 136–141.

11. Botella, J. R. 13-Biotechnology and reduced spoilage. In: Zeuthen, P., & Bøgh-Sørensen, L. B. T. F. P. T., (eds.), *Woodhead Publishing Series in Food Science, Technology and Nutrition* (pp. 243–262). Woodhead Publishing, **2003**.

12. Panja, P. *Minimal Processing of Fruits and Vegetables*, **2017**.

13. Yusuf, M. *Natural Antimicrobial Agents for Food Biopreservation* (pp. 409–438). Food Packaging and Preservation Academic Press, **2018**.

14. Ozkan, G., Franco, P., Marco, I. De, Xiao, J., & Capanoglu, E. A Review of microencapsulation methods for food antioxidants: Principles, advantages, drawbacks and applications. *Food Chem.,* **2019**, *272*, 494–506.

15. Antruejo, A., & Viola, M. N. *Evaluation of the Use of Plant Polyphenols as a Measure of Biosecurity for the Control of Flies in Sheds of Laying Hens*, **2015**.

16. Rodríguez, E. Use of natural antimicrobial agents in the conservation of fruits and vegetables. *Ra Ximhai,* **2011**, *7*(1), 153–170.

17. Siwe-Noundou, X., Ndinteh, D. T., Olivier, D. K., Mnkandhla, D., Isaacs, M., Muganza, F. M., et al. Biological activity of plant extracts and isolated compounds from Alchornea Laxiflora: Anti-HIV, Antibacterial and cytotoxicity evaluation. *South African J. Bot.,* **2018**, 6–11.

18. Difonzo, G., Russo, A., Trani, A., Paradiso, V. M., Ranieri, M., Pasqualone, A., Summo, C., Tamma, G., Silletti, R., & Caponio, F. Green extracts from coratina olive cultivar leaves: Antioxidant characterization and biological activity. *J. Funct. Foods,* **2017**, *31*, 63–70.

19. Hayes, J. E., Stepanyan, V., Allen, P., O'Grady, M. N., & Kerry, J. P. Effect of Lutein, sesamol, ellagic acid and olive leaf extract on the quality and shelf-life stability of packaged raw minced beef patties. *Meat Sci.,* **2010**, *84*(4), 613–620.

20. Ciriano, M. G. I., De García-Herreros, C., Larequi, E., Valencia, I., Ansorena, D., & Astiasarán, I. Use of natural antioxidants from lyophilized water extracts of borago officinalis in dry fermented sausages enriched in ω-3 PUFA. *Meat Sci.,* **2009**, *83*(2), 271–277.

21. Nikmaram, N., Budaraju, S., Barba, F. J., Lorenzo, J. M., Cox, R. B., Mallikarjunan, K., & Roohinejad, S. Application of plant extracts to improve the shelf-life, nutritional and health-related properties of ready-to-eat meat products. *Meat Sci.,* **2018**, *145*, 245–255.

22. Al-Hijazeen, M., & Al-Rawashdeh, M. Preservative effects of rosemary extract (*Rosmarinus officinalis* L.) on quality and storage stability of chicken meat patties. *Food Sci. Technol.,* **2017**, 1–8.

23. Lopez-Malo, A., Guerrero, S., & Alzamora, S. M. Probabilistic modeling of *Saccharomyces cerevisiae* inhibition under the effects of water activity, PH, and potassium sorbate concentration. *J. Food Prot.,* **2000**, *63*(1), 91–95.

24. Speranza, P., Lopes, D. B., & Martins, I. M. *Chapter 16 – Development of Functional Food from Enzyme Technology: A Review* (pp. 263–286). Enzymes in Food Biotechnology, Academic Press, **2019**.

25. Zhang, C., Feng, F., & Zhang, H. Emulsion electrospinning: Fundamentals, food applications and prospects. *Trends Food Sci. Technol.,* **2018**, *80*, 175–186.

26. Bajpai, V. K., Kamle, M., Shukla, S., Mahato, D. K., Chandra, P., Hwang, S. K., Kumar, P., Huh, Y. S., & Han, Y. K. Prospects of using nanotechnology for food preservation, safety, and security. *J. Food Drug. Anal.,* **2018**, *26*(4), 1201–1214.

27. Ray, S., Raychaudhuri, U., & Chakraborty, R. An overview of encapsulation of active compounds used in food products by drying technology. *Food Biosci.,* **2016**, *13*, 76–83.

CHAPTER 7

Lactic Acid Bacteria in Preservation and Functional Foods

IVAN SALMERÓN, SAMUEL B. PÉREZ-VEGA,
NÉSTOR GUTIÉRREZ-MÉNDEZ, and ILDEBRANDO PÉREZ-REYES

School of Chemical Sciences, Autonomous University of Chihuahua, 31125 Chihuahua, Chih, Mexico, Tel.: +52 (614) 236 6000, E-mail isalmeron@uach.mx (I. Salmerón)

ABSTRACT

Fermentation as a food conservation method has an ancestral history, and some of those products are highly consumed nowadays. With the rise in the awareness of health improvement through the consumption of functional foods, the segment of fermented dairy-based products such as probiotics for the attention of gastrointestinal diseases made a significant market breakthrough. Furthermore, the use of lactic acid bacteria (LAB) with probiotic characteristics was incorporated as the most important microorganisms in these products. Nevertheless, consumer demands for non-dairy probiotics have motivated the need to explore other potential substrates like fruits, vegetables, legumes, and cereals for the formulation of probiotic products. Cereals could be used to design new cereal-based fermented beverages with probiotic characteristics if these formulations fulfill probiotic requirements and have acceptable physicochemical characteristics and organoleptic properties. Thus, food technology encounters several challenges during the development of novel products. Therefore, the incorporation of multidisciplinary fields of research will aid the creation of novel fermented products within the segment of functional foods.

7.1   INTRODUCTION

Fermented food has its ancestral roots in the uncontrolled transformation of food by wild species of bacteria and yeast that provoked changes in the texture and flavor that were well accepted and thus fermentation was converted into a conventional practice of food preservation by many civilizations. In the past decades, consumers became more interested in the link between nutrition and good health; these gave way to the creation of functional foods that are foods that claim to promote human health over and above the provision of basic nutrition. The largest segment of this market comprises foods designed to improve gut health such as probiotics, prebiotics, and synbiotics [1, 2]. Probiotic products have gained widespread popularity and acceptance throughout the world with established markets in Japan, the USA, Europe, and Canada [3]. The global success in the sales of these products has its roots in the market campaign and launch of dairy probiotic products by Yakult, Danone, and Valio over the past twenty years [4]. Food technologists and scientists have the challenge to search for other raw materials for the development of new probiotic products. In this respect, interest in the use of cereals has increased for designing these new health improvement foods, as these are abundant and form an essential part of the human diet. During the process of developing non-conventional probiotic foods, any decrease in the organoleptic properties concerning the traditional products will not be acceptable by the consumer as health is important but not at the expense of flavor. Therefore, in order to see rise new probiotics of great market success products with novel features and outstanding flavor characteristics should be developed.

In this chapter, a review of the literature associated with the aspects of probiotics and the development of new food products has been provided. The most important technological features and current status of probiotic foods are given. An insight into the physiology of lactic acid bacteria (LAB) and the genus *Lactobacillus* has been performed as these are the main microorganisms used in the formulation of probiotic foods. The importance of cereals and their fermented foods has been carried out, as these could be the platform for the development of non-conventional probiotics. The impact of flavor in functional food fermented by LAB has also been reviewed, as this is a crucial feature behind the creation of successful food products.

7.2 PRESERVATION OF FOODS BY FERMENTATION

Fermentation is an ancient technique used for the elaboration and preservation of foods. Civilizations have described different fermentation methods that they have applied on a variety of raw materials, and there is a register of these since 6000 BC in the Middle East region. Fermentation is considered as a desirable effect of the microbial biochemical activity in foods. Microbial enzymes break down carbohydrates, lipids, proteins, and other food components that can improve food digestion in the human gastrointestinal tract (GIT) improving the consumption of nutrients [5]. As well, it reduces the volume of material to be stored, increases the nutritive value and appearance of the food and the most important inhibits the growth of several microorganisms that cause its alteration and spoilage [6].

7.2.1 THE CULTURE OF FERMENTED PRODUCTS

Bread, beer, wine, and cheese were created long before Christ. Despite that food, technology advances have developed a high standard of quality and hygiene of fermented foods, the principles of old processes have hardly changed. In industrialized societies, a variety of fermented foods are very popular with consumers because of their attractive flavor and their nutritional value. In developing countries, this is one of the main options for processing foods, as it serves as an affordable and manageable technique for food preservation. Fermented foods have a high demand because of their added value regarding consistency, color, and especially flavor and taste. The nutritional value of fermented food products is also greatly enhanced, compared with raw food material, as fermentation contributes to the nutritional value of the fermented food by increasing the bioavailability of nutrients or by the production of vitamins and other nutritional components by the fermenting microorganisms [7–9].

7.2.2 THE ROLE OF LACTIC ACID BACTERIA (LAB) IN FERMENTATION

There are several types of fermented food produced throughout the world using diverse manufacturing techniques, raw materials and microorganisms (i.e., mold, yeast, bacteria). Thus, only four types of fermentation processes exist; alcoholic, lactic acid, acetic acid, and alkali fermentation [6]. The second process is the most common; this relays on LAB to mediate the fermentation process. Also, these bacteria cause rapid acidification of the raw material through the production of organic acids, mainly lactic acid. Throughout this mechanism, they promote the shelf life and microbial safety, improve texture, and contribute to the pleasant sensory profile of the end product [5, 10, 11]. Finally, the LAB has a long and safe history of application in the production of fermented foods and beverages. As a consequence, they have been given the status of "generally recognized as safe" (GRAS) within the food industry [12].

Fermentation carried out by LAB is the process that at the moment, of most significant interest to food safety and public health. Scientific studies have demonstrated that lactic acid fermentation inhibits the growth, survival, and toxin production of several pathogenic bacteria. In this context, study results have demonstrated that lactic acid fermented products have veridical and antitumor effects [7, 13]. Functional starter cultures can be employed in the fermentation industry to improve the use of LAB in food processes. These starter cultures can contribute to food safety and offer one or more organoleptic, technological, nutritional, or health advantages. The implementation of carefully selected LAB strains as starter cultures or co-cultures in fermentation processes can help to maintain a perfectly natural and healthy product. Finally, these could LAB with health-promoting properties, referred to as probiotic strains. Therefore, this represents an alternative for the substitution of chemical additives by natural compounds, at the same time providing the consumer with new, attractive food products [14–16].

7.3 APPLICATION, PHYSIOLOGY, AND METABOLISM OF LACTIC ACID BACTERIA (LAB)

The concept of LAB as a group of organisms developed at the start of the 19th century, continued by revolutionary scientific and technical improvements during the latter part of the same century. The relations of LAB in foods caused the attention of scientists and resulted in the considerable contribution by Pasteur on lactic acid fermentation in 1857, followed by the first isolation of a pure bacterial culture, *Bacterium lactis*, by Lister in 1857. The use of starter cultures for cheese and sour milk production was introduced almost simultaneously in 1890 by Weigrnann and Storch respectively; this opened the way for the industrialization of fermented food [17].

7.3.1   THE USE OF LACTIC BACTERIA BY HUMANS

Humans have acknowledged the role of LAB in the preservation of organic products throughout history [18]. The mention of fermented dairy products (cheese, yogurt, butter) is stated in ancient texts from Uruk/Warka (Iraq) which dates about 3200 B.C. LAB have also been employed to produce other fermented milk products, such as leiben, dahi, kefir, and koumiss. Furthermore, procedures for the fermentation of meat, throughout the use of LAB, were developed as early as the 15th century B.C. in Babylon and China, as well as methods for the fermentation of vegetables, was applied in China in the 3rd century B.C. [17, 19]. In many African countries, lactic fermentation has been used to produce traditional sorghum, maize, and millets beers [20]. In these processes, a stable product is produced due to the rapid lactic fermentation, which reduces the pH to levels inhibitory to the growth of spoilage microorganisms. Also, this fermentation enhances some organoleptic characteristics of food products, such as flavor and texture [18]. The conventional food associated fermentations are known for a long time, but now only the metabolic pathways are understood so that it is possible to apply selected strains and controlled conditions for carrying out a particular fermentation [21].

7.3.2   CLASSIFICATION OF LACTIC ACID BACTERIA (LAB)

During the 18th century, the term LAB were used synonymously with "milk-souring organisms." Significant progress in the classification of these bacteria was made when the similarity between milk-souring bacteria and other lactic acid-producing bacteria of other habitats was recognized [22]. Early definitions of LAB as a group, based on the ability to ferment and coagulate milk, included coliform bacteria. The description of *Lactobacillus* organisms by Beijerink in 1901 as Gram-positive bacteria separated the coliforms from the LAB [17]. Though, confusion was still prevalent when the monograph of Orla-Jensen appeared in 1919. This work had a significant impact on the systematic of LAB and, although revised to a considerable extent, the classification basis is remarkably unchanged. Thus, the LAB is a group of Gram-positive bacteria united by a constellation of morphological, metabolic, and physiological characteristics. The general description of the bacteria included in the group is Gram-positive, non-sporing, catalase-negative, devoid of cytochromes, of nonaerobic habitat but aerotolerant cocci or roods, fastidious, acid-tolerant, and strictly fermentative which produce lactic acid as the major end product during the fermentation of carbohydrates [22]. The genera that fit the general description of the typical LAB in most respects are (as they appear in the Bergey's Manual from 1986): *Aerococcus* (A.), *Lactobacillus* (Lb.), *Leuconostoc* (Ln.), *Pediococcus* (P.), and *Streptococcus* (S.). The genus *Bifidobacterium* is historically also considered to belong to the LAB group. Although *Bifidobacterium* species do fit the general description above, they are phylogenetically more related to the *Actinomycetaceae* group of Gram-positive group. Besides, they have a special pathway for sugar fermentation unique to the genus [22].

7.3.2.1 THE GENUS LACTOBACILLUS

The genus *Lactobacillus* is by far the largest of the genera included in LAB. It is also very heterogeneous, encompassing species with a large variety of phenotypic, biochemical, and physiological properties [22]. The classical division of the lactobacilli is based on their fermentative characteristics:

1. obligate homofermentative;
2. facultative heterofermentative; and
3. obligate heterofermentative [17].

The physiological basis employed for the division is (generally) the presence or absence of the key enzymes of homo- and heterofermentative sugar metabolism, fructose-1,6-diphosphate aldolase, and phosphoketolase, respectively [22]. The lactobacilli are strictly fermentative and have complex nutritional requirements. They are widespread in nature, and many species have found applications in the food industry. They are generally the most acid-tolerant of the LAB and will, therefore, terminate many spontaneous lactic fermentations such as silage and vegetable fermentations. They can produce a pH 4.0 in foods containing a fermentable carbohydrate. As a result, they often suppress the growth or kill other bacteria. It is generally accepted that lactobacilli grow up to a maximum pH of 7.2, although exceptions concerning substrate and strain exist. Lactobacilli are also associated with the oral cavity, GIT, and vagina of humans and animals. They can also be found in plants and plant materials, soil, water, sewage, and manure. Lactobacilli are used as starter cultures for several varieties of cheese, fermented plant foods, fermented meats, in wine and beer production, sourdough bread and silage [17, 22].

7.3.3 FERMENTATION METABOLISM OF LAB

The foremost characteristic of LAB metabolism is their efficient carbohydrate fermentation coupled to substrate-level phosphorylation. The generated ATP is subsequently used for biosynthetic purposes. LAB as a group displays a vast capacity to degrade different carbohydrates and related compounds. Generally, the predominant end-product is, of course, lactic acid. It is clear, however, that LAB adapt to various conditions and change their metabolism accordingly [22]. The taxonomy of LAB for many decades heavily relied on the type of sugar fermentation. In order to deal with a large number of species being described, Orla Jensen proposed a classification of LAB which was based on morphology, temperature range of growth, nutritional characteristics, carbon sources utilization, and agglutination effects [23]. The subdivision of lactobacilli into three major fermentation groups for taxonomic reasons was maintained until the late 1970s. The development and application of advanced molecular techniques brought new insights in the taxonomy of the genus with currently 113 different species described. Besides, the accumulated knowledge on their sugar fermentation patterns created a solid basis on which further research was carried out, including other metabolic properties of the lactobacilli, such as proteolytic and lipolytic activities, which are equally important in foods applications [23].

7.3.3.1 SUGARS CATABOLISM

The conversion of carbohydrates to lactate by LAB may well be considered as the most important fermentation process employed in food technology [24]. However, it is clear that LAB adapts to various conditions and changes their metabolism accordingly leading to different end-product patterns. As mentioned above, there are two major pathways for hexose fermentation occurring within LAB. The Embden-Meyerhof pathway (glycolysis) generates lactic acid practically as the only end-product. The second major pathway is known as the pentose phosphate pathway characterized by the creation of 6-phosphogluconate which is further divided into glyceraldehyde-3-phosphate (GAP) and acetyl phosphate. GAP is metabolized to lactic acid while acetyl phosphate is reduced to ethanol and the important flavor compound acetaldehyde [22]. Furthermore, through this heterofermentative process equimolar amounts of CO_2, lactate, and acetate or ethanol are formed from hexose fermentation. The ratio acetate/ethanol depends on the oxidation-reduction potential of the system [24]. During the fermentation of complex substrates such as fruit juices and vegetables that contain compounds other than hexoses, e.g., pentose or organic acids the production of lactate, acetate, and CO_2 will be of different ratios. In this sense, pyruvate may, moreover, not only be reduced to lactate but also converted to several other products by alternative mechanisms depending on the growth conditions and properties of the particular microorganisms [24].

7.3.3.2 *PYRUVATE AND LACTATE CATABOLISM*

Pyruvate is mainly reduced to lactate, catalyzed by lactate dehydrogenase (LDH) in LAB. Therefore, pyruvate can alternatively be converted by LAB to other flavor compounds such as diacetyl, acetoin, acetaldehyde, acetic acid, formate, ethanol, and 2,3-butanediol [25, 26]. Although lactate is the end-product of lactic acid fermentation, it can be catabolized under aerobic conditions by lactate oxidase or NAD^+ independent LDH in some LAB to produce pyruvate, which is further catabolized [24]:

$$\text{Lactate} \xrightarrow[\text{lactate oxidase}]{O_2} \text{Pyruvate} \xrightarrow[\text{pyruvate oxidase}]{} \text{Acetate} + CO_2$$

$$\text{Lactate} \xrightarrow[\text{LDH}]{O_2} \text{Pyruvate} \xrightarrow[\text{pyruvate oxidase}]{} \text{Acetate} + CO_2$$

Under anaerobic conditions, lactate can also be catabolized via NAD^+ independent LDH by some lactobacilli. Such lactate degrading lactobacilli use other electron acceptors (e.g., oxaloacetate derived from citrate and 3-hydroxypropionaldehye derived from glycerol) for anaerobic catabolism of lactate as below [26]:

$$\text{Lactate} \longrightarrow \text{pyruvate} \longrightarrow \text{acetate} + CO_2$$

or

$$\text{Lactate} \longrightarrow \text{pyruvate} \longrightarrow \text{acetate} + \text{formate}$$

The products formed during pyruvate degradation are dependent upon the presence of a particular enzyme(s) in a particular LAB, pyruvate oxidase or pyruvate-formatelyase or both [26].

7.4 MICROBIAL AND FUNCTIONAL ASPECTS OF PROBIOTICS

Consumers around the globe are more aware of the relationship between nutrition and good health; this has lead scientific studies to identify food and food components that have special health benefits. With these efforts, probiotic products have come into the market and are an important category within 'functional foods [2]. The last ones are defined as the foods that in addition to nutrients, supply the organism with components that contribute to the cure of diseases or to reduce the risk of developing them [27]. Thus, a food marketed as functional contains added, technologically developed ingredients with a specific health benefit [28]. During the last decade in Japan and Europe, the market of functional food was dominated by gut health products, in particular, probiotics which accounted for 379 product launches in 2005 [28]. This demand is due mainly to the growing appreciation for the important role of commensal microbes in human and animal health by the mediation of the intestinal development and innate immunity, or digestion of foods for protection of the host against disease. Also, the amount of scientific publications on probiotics has significantly risen due to scientific and clinical developments using well-documented probiotic organisms [27, 29].

The concept of probiotics was in use in the early 1990s. Even so, the term was created since 1965 by Lilly and Stillwell and subsequently developed [30]. An Expert committee technically defined the term probiotic as "live microorganisms which upon ingestion in certain numbers exert health benefits beyond inherent general nutrition" [31]. This leads to the conception that all probiotic products must have high numbers of live microorganisms at the time of the intake in order to impart beneficial effects, it is also a requirement that will allow these microorganisms to survive the passage throughout the harsh conditions of the gut [32]. There is limited information about the effective dose for particular probiotic strains, so large numbers of viable bacteria are recommended in probiotic foods. Scientific evidence recommends that probiotic bacteria consumed at a range of 10^9–10^{11} CFU/day can lower the severity of some intestinal disorders [33–35]. The food industry has generally marked a minimum dose of 10^6 CFU/mL at the time of the intake. Also, it has been recommended that probiotic food should be consumed on a daily basis by portions of 100/g day in order to deliver high cell viability into the intestine [36].

Currently, most of the individual probiotic foods belong to the dairy products like yogurt, fermented milk, and cheese. Nevertheless, some new products based on cereals, fruits, vegetables, and meat are being developed mainly for the treatment of lactose intolerance and control of cholesterol levels that is a drawback in milk-based products. The application of new technologies such as encapsulation could facilitate the use of non-conventional dairy products as vehicles to deliver the probiotic organism in an active form to the GIT. This technology promises to be of great help in increasing the survival rate of probiotics during process and storage [1, 2]. The alternative application of probiotics in animal health has also been exploited as they could replace feed antibiotics, which have been commonly used as supplements in the diets of farm animals. This comes into place as possible residues in animal products and cross-resistance with human pathogens might result in health risks encouraging the development and use of non-antibiotic products [37].

7.4.1 PROBIOTICS MECHANISMS OF ACTION

With the understanding of the probiotic mechanistic, it could be demonstrated that many gastrointestinal disorders are based on malfunctions of intestinal microflora [38]. The human GIT is the habitat of a complex microbiota consisting of either facultative or obligate anaerobic microorganisms including streptococci, lactobacilli, and yeasts. These are settled in different quantities throughout the GIT and posse a broad spectrum of differences (Figure 7.1). In this respect, the human being could be considered as a traveling bioreactor, as the number of bacteria that colonize the human body is so large that researchers have estimated that the human body contains 10^{14} cells, only 10% of which are not bacteria and belong to the human body [39, 40]. Two groups of bacteria that form part of the microbial environment in the gut have been of interest in recent years for their beneficial effects on the host, these species are Bifidobacterium and *Lactobacillus*, which promote the correct "balance" of bacteria, to achieve an optimally intestine function [19, 41–43]. Some of the probiotic mechanisms of action for the prevention of gastrointestinal disorders are the suppression of harmful bacteria and viruses, stimulation of local and systemic immunity and the modification of gut microbial metabolic activity [38].

The production of bacteriocins is one of the mechanisms that probiotic bacteria use against pathogenic bacteria. Fecal bacteria enzymes, such as β-glucuronidase, nitroreductase, and azoreductase produced by the autochthonous microflora, are thought to convert procarcinogens in the colon to carcinogens. It has been proposed that probiotic bacteria decrease such enzyme activities since they produce acids, peroxides, and bacteriocins which suppress the growth of these fecal bacteria [44].

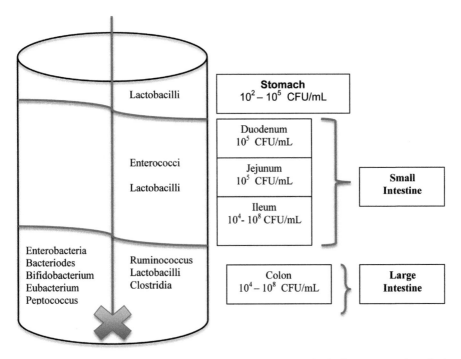

FIGURE 7.1 The human bioreactor: Major microbiota present in the human gastrointestinal tract (GIT).

Another factor such as the decrease of pH by organic has bactericidal and bacteriostatic effects that act directly on the growth inhibition of pathogenic organisms. Probiotic bacteria competing against harmful pathogens for a position in the intestinal mucosa during the colonization of the GIT can endorse the state of well-being in the host [33, 45]. Further metabolic activities of the GIT microbiota that beneficially affect the host include continued degradation of food components (mainly by fermenting dietary carbohydrates that have escaped digestion in the upper GIT such as resistant starch, cellulose, hemicelluloses, pectins, and gums), vitamin production, and production of short-chain fatty acids (FA) [46].

7.4.1.1 DEVELOPMENT OF THE GUT BACTERIA

The normal colonization of the sterile newborn intestine is a complex process, which begins immediately after birth when the newborn is in contact with the surroundings such as the maternal vaginal and fecal bacteria flora [47]. These first bacteria that colonize the GIT of the newly arisen include species such as *Enterobacter, Streptococcus,* and *Staphylococcus,* which have a high reductive potential as well they metabolize oxygen encouraging the growth of anaerobic bacteria such as *Lactobacilli,* and *Bifidobacteria* is one of the predominant anaerobic bacteria in the intestinal microbiota from early infancy until the old age [40].

The initial microbe colonization of the GIT plays an important role in the development of the newborns' microbiota. In contrast, this process of microflora maturation is poorly manifested in cesarean born infants who are colonized by microbes from the hospital environment. These children who are given birth throughout abdominal delivery are not exposed to the maternal flora; therefore, the colonization process by bacteria such as lactobacilli and bifidobacteria is held back by periods of ten days or even one month. In the same context, these colonization delays are alike when there is an early use of antibiotics or when the newly arisen is situated under sterile environments such as incubators. Studies have shown that early colonization of the intestinal tract is a process required to activate the efferent limb of the mucosal immune response [40, 41].

7.4.2 MICROORGANISMS USED AS PROBIOTICS

The microbes used as probiotics have broad genera origins; however, the principal strains used belong to the heterogeneous group of LAB; lactobacilli, enterococci, and bifidobacteria from which lactobacilli are most commonly used during the formulation of probiotic products [48]. These types of LAB have a long and safe history of application in the formulation of fermented foods and beverages [10]. Despite this, several microbial organisms have been claimed to poses probiotic properties only a reduced number of these have been industrialized and are used in the formulation of food and beverages (Table 7.1). Amongst these strains the most frequently used are *Lactobacillus acidophilus, Lactobacillus casei, Lactobacillus plantarum, Lactobacillus reuteri, Lactobacillus plantarum, Bifidobacterium bifidum,* and *Bifidobacteria lactis* which are all of human origin and are recognized by distinct brand names [49–51]. However, other species belonging to the genera

Sacaromyces and *Propionibacterium* yeasts, *Enterococcus, Lactococcus,* and filamentous fungi (e.g., *Aspergillus oryzae*) are also used as probiotics due to their health improvement properties [36].

TABLE 7.1 Commercial Strains Used in Probiotics Market

Company (Owned by)	Probiotic Strain
Chr Hansen A/S	*L. acidophilus* La5
	B. bifidum BB12
	L. reuteri RC-14
Danisco A/S	*L. acidophilus* NCFM (HOWARU™ Dophilus)
	L. rhamnosus HN001 (HOWARU™ Rhamnosus)
	B. lactis HN019 (HOWARU™ Bifido)
Danone	*Bifidobacterium* Actifessensis
	B. regularis (Activia-proprietary)
DSM	*L. acidophilus* LAFTI L10
	B. lactis B94
Yakult Honsha Co. Ltd	*L. casei* Shirota
SofarSpA	*L. casei* DG
Puleva Biotech SA	*L. coryne* Hereditum
Centro Sperimentale delLatte S.p.A.	*L. delbrueckii* subsp. bulgaricus
Nestle SA	*L. fortis*
Medipharm AB	*L. paracasei*
Siffra FarmaceuticiSrl	*L. paracasei*
Probi AB (& Institut Rosell)	*L. plantarum* 299v
BioGaia AB	*L. reuteri* ATCC 55730
Valio	*L. rhamnosus* ATCC 53103 (Gefilus)
Dicofarm	*L. rhamnosus* GG
Institut Rosell	*Saccharomyces boulardii*
Biocodex	*Saccharomyces boulardii* (ULTRA-LEVURE/ Florastor)
BLIS Technologies	*Streptococcus salivarius* K12
Sanofi Winthrop	*Bacillus clausii*

Source: Adapted from Refs. [49–51].

7.4.3 DESIRABLE PROBIOTIC CHARACTERISTICS

A microorganism should meet several predefined criteria, and it must enclose the status of GRAS before being considered for the development of a probiotic product [19, 52]. The theoretical basis for the selection of probiotic microorganisms including safety, functionality, and technological aspects are shown in Figure 7.2. A probiotic microorganism besides promoting human health should fulfill several technological characteristics such as be metabolically stable, demonstrate high viability during processing conditions and storage, pose adequate sensorial properties, and it should survive the passage through the harsh conditions of the gut arriving in sufficient numbers to its site of action [36, 53, 54]. Also, reliable techniques are required to establish the effect of the probiotic strain in other members of the intestinal microbiota and most important the host [53].

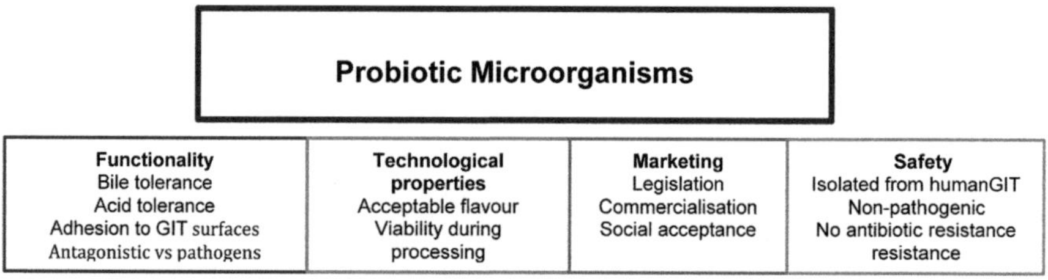

Probiotic Microorganisms			
Functionality Bile tolerance Acid tolerance Adhesion to GIT surfaces Antagonistic vs pathogens	**Technological properties** Acceptable flavour Viability during processing	**Marketing** Legislation Commercialisation Social acceptance	**Safety** Isolated from humanGIT Non-pathogenic No antibiotic resistance resistance

FIGURE 7.2   Theoretical basis for the selection of probiotic microorganisms.
Source: Adapted from Refs. [5] and [19].

7.4.4   *IMPROVE HEALTH EFFECTS INDUCED BY PROBIOTICS*

There is increasing scientific research to sustain the concept that the maintenance of healthy gut microflora may ensure protection against gastrointestinal disorders including gastrointestinal infections, inflammatory bowel diseases (IBDs), and cancer [41]. Furthermore, it has been established that the beneficial effects of probiotics can extend outside the GIT as they could improve health conditions such as the metabolic syndrome, allergies, diabetes, obesity, and autism [55]. The use of probiotic bacteria cultures induce the health of the gut by several mechanisms like the reinforcement of intestinal barrier function, regulation of the innate and acquired immune response of the host, improvement of digestion and uptake of dietary nutrients, production of antimicrobial substances (bacteriocins, antibiotics, and microcins) and competition with pathogens for binding sites [56].

There are important health benefits attributed to the ingestion of these bacteria, as enhancement of the immune response, prevention, and treatment of gastrointestinal disorders (antibiotic associated-diarrhea, IBDs, irritable bowel syndrome and so forth), allergies, decrease of necrotizing enterocolitis (NEC) in infants, intolerances, reduction in acute pediatric diarrhea, reduction of cholesterol levels, lowering blood levels, as well they exhibit anti-carcinogenic effects. These health properties are strain-specific and are influenced by several of the mechanisms stated previously [56–58].

7.5 PROBIOTIC PRODUCTS

In Europe, the probiotic food and beverage market has developed with outstanding success within the functional foods industry. Since the formation of this market about mid-1980, it has been growing and reaching major achievements. It has been reported that the application of bacterial strains for consumer health improvement was carried out around the 1920s; nevertheless, the first serious attempts at commercializing probiotics were performed in Japan in the 1950s with the foundation of Yakult Honsha Co. Ltd. At present, due improved technologies in micro-encapsulation and coatings have encouraged numerous food producers to integrate probiotic strains in their food products [49, 59]. Some of these commercial probiotics are shown in Table 7.2.

The leading market division for probiotics is the dairy division which accounted for sales of around 1.35 billion US$ in 1999 and about 56% of functional foods total 31.1 billion US$ global sales in 2004. Germany, France, the United Kingdom, and the Netherlands accounted for about two-thirds of functional dairy products in Europe [28]. Commercially, yogurt drinks are the most popular within this market and due to their general positive image among consumers, they account for the success exhibited by dairy-based probiotic products that have the greatest market shares among the bacteria friendly products [49].

7.5.1 NON-DAIRY AND CEREAL-BASED PROBIOTIC PRODUCTS

Fermented milk and yogurts are the most successful probiotic products. Though, lactose intolerance and the cholesterol content are two major downsides encountered by these products. Also, with an increase in the consumer vegetarianism, there is also a demand for the vegetarian probiotic products [60, 61]. The fortification of fruit juices with probiotic bacteria has been studied as these are already positioned as healthy food products, and are consumed regularity by a large percentage of the consumer population [62]. Nevertheless, sensorial studies performed in these products have shown that

TABLE 7.2   Commercial Examples of Probiotic Products

Producer/Brand	Description
Danone, France/Actimel	Probiotic drinking yogurt with *L. casei* Imunitass® cultures
Danone, France/Activia	Creamy yogurt containing Bifidus ActiRegularis®
Valio, Finnland/Gefilus	A wide range of LGG products
Bioferme, Finland/SOYosa	Range of products based on soy and oats and includes a refreshing drink and a probiotic yogurt-like soy-oat product
Bioferme, Finland/Yosa	Yogurt-like oat product flavored with natural fruits and berries containing probiotic bacteria (*Lactobacillus acidophilus, Bifidobacterium lactis*)
Ingman Foods, Finland/Rela	Yogurts, cultured milks, and juices with *L. reuteri*
Attune Foods, USA/Attune	Probiotic chocolates containing *Lactobacillus acidophilus, Lactobacillus casei*
Healthy Delights, USA/DelightsTM	Healthy DelightsTM ProBiotic Bites milk chocolates
Wysongs Corporation, USA/ OrizinsTM	Orizins TM bars, Chocolate Therapy™ containing *Lactobacillus acidophilus, Lactobacillus bifidus, Lactobacillus plantarum, Enterococcus faecium*
Lifeway, USA/Soytreat	Kefir type product with six probiotics
Wysongs Corporation, USA/ OrizinsTM	OrizinsTM bars, Chocolate Therapy™ containing *Lactobacillus acidophilus, Lactobacillus bifidus, Lactobacillus plantarum, Enterococcus faecium*
Aviva Natural Health Solutions, Canada	Healthy Digestives-ProBiotic Dark Chocolate Bites containing Bacillus coagulans
Tallinna Piimatoööstuse AS, Estonia	Hellus
H&J Bruggen, Germany/Jovita Probiotisch	Blend of cereals, fruit, and probiotic yogurt
Ohso, Belgium/Ohso	Ohso Probiotic chocolate containing *Lactobacillus helveticus, Bifidobacterium longum*
Valašské Meziříčí Dairy, Czech Republic/Pohadka	Yogurt milk with probiotic cultures
Skånemejerier, Sweden/ProViva	Refreshing natural fruit drink and yogurt in many different flavors containing *Lactobacillus plantarum*
Olma, Czech Republic/Revital Active	Yogurt and drink yogurt with probiotics
Celigüeta, Spain/Snack Fibra	Snacks and bars with natural fibers and extra minerals and vitamins
Yakult, Japan/Yakult	Milk drink containing *Lactobacillus casei* Shirota
Müller, Germany/Vitality	Yogurt with pre- and probiotics and omega-3
Campina, the Netherlands/Vifit	Drink yogurts with LGG, vitamins, and minerals

Source: Adapted from Refs. [28, 59].

these products developed off-flavors such as dairy, medicinal, and savory which could negatively affect their performance within the market of functional foods [63–65].

Cereals present a further alternative for the creation of non-dairy functional foods with probiotic features. The various health-promoting effects of these raw materials can be exploited in different ways leading to the development of well-being products design for specific groups of population [66]. Cereals could be used as fermentable substrates for the growth of probiotic microorganisms. Additionally, cereals can be applied as sources of non-digestible carbohydrates that besides promoting several beneficial physiological effects can also selectively stimulate the growth of lactobacilli and bifidobacteria in the colon and act as prebiotics [66, 67]. Furthermore, there are a wide variety of traditional non-dairy fermented beverages produced around the world. Much of them are non-alcoholic products manufactured with cereals as principal raw material. Cereal-based fermented gruels and porridges are extensively accepted in many cultures [27]. This implies that cereals can be used for the creation of novel cereal beverages embracing probiotic characteristics and launched within the market of non-dairy probiotic foods and beverages.

7.5.2 CEREAL-BASED FUNCTIONAL PRODUCTS

Cereals are the most important components of the human diet as they are an important source of carbohydrates and proteins [68]. They supply a variety of micronutrients; in particular, certain vitamins (B and E), minerals (iron, in the case of wheat), and significant amounts of dietary fiber. The principal cereal crops grown in the world are maize, wheat, barley, rice, oats, rye, and sorghum. In 2007 the Food and Agriculture Organization of the United Nations (FAO) estimated that the global cereal output was about 2.129 million tons of this it was estimated that the main uses were human consumption and animal feed [69]. Since the past decade, there have been researched carried out regarding cereals as potential sources for the development of functional products [70]. The possible applications of cereals or cereal constituents in functional food formulations could be stated:

- As fermentable substrates for growth of probiotic microorganisms.
- As dietary fiber promoting several beneficial physiological effects.
- As prebiotics due to their content of specific non-digestible carbohydrates.
- As encapsulation materials for probiotic in order to enhance their stability.

7.5.3   CEREAL-BASED FUNCTIONAL PRODUCTS IN THE MARKET

A current challenge for the food industry is the development of non-dairy functional products, in its attempt to applying natural resources such as cereals. In this respect, some cereal-based functional foods have been developed and are currently in the market such as "Adavena," a food base derived from oats, another product "Mill Milk," which is a non-dairy milk substitute, has been reported to have high acceptance among consumers. In the same context, "Yosa" is a yogurt-like oat product, is at the market in Finland. Other non-dairy, yogurt-like products have been developed from rice and mixtures of cereals and legumes [71]. Also, the probiotic strain product *Lactobacillus casei* F-19®, manufactured by Medipharm AB, has been introduced into a porridge cereal product for children in Sweden and Norway by the Swedish baby food producer Semper AB. This strain, *L. casei* F-19®, was selected during development from a tested range of 400 strains, and Medipharm AB also state that it has been approved as the only strain for use in NASA's space program [49]. In Great Britain, it was estimated that breakfast cereals fortified with extra fiber and minerals, to aid digestion and address bone health had a market value of £325 m in the year ending 8 September 2007 [72].

7.5.4   THE USE OF CEREALS FOR THE DEVELOPMENT OF PROBIOTICS

Probiotic bacteria such as lactobacilli and bifidobacteria require several nutrients for their metabolism such as carbohydrates, amino acids, vitamins, and metal ions. An alternative could be the use of cereals as a source of nutrients for these bacteria; these raw materials have within their structure simple sugars, complex oligosaccharides such as fructans, and polysaccharides [68]. For such reasons, interest in the use of cereals for the development of non-diary probiotic products has increased enormously in recent years.

In previous works, malt, wheat, barley, and oat extracts were studied in order to evaluate their performance as substrates for the growth of human-derived LAB, the results of such investigations indicated that cereals were able to support bacterial growth and increase their acid and bile tolerance exhibiting a significant protective effect of the lactobacilli strains under the harsh conditions tested [66, 73, 74]. In addition, further potential benefits of cereal extracts and cereal fiber for the development of new functional products have been studied showing that these substrates can improve the tolerance of a *Lactobacillus plantarum* strain to simulated gastric and bile juice [75]. The potential use of mixed cultures for the fermentation of cereal substrates and the evaluation of the fermentability of cereal fractions by lactobacilli strains has been carried out. The results showed that cereal-based products with a greater concentration of non-volatile components and higher numbers of LAB can be achieved and that by the use of different fractions from cereal grains functional formulations with higher probiotic levels can be reached than those levels obtained with the use of the whole grain [67, 76]. Therefore, cereals could be used to design cereal-based fermented beverages with probiotic characteristics if these formulations fulfill probiotic requirements and have acceptable physicochemical characteristics and organoleptic properties.

7.6 SIGNIFICANCE OF THE FLAVOR IN NOVEL PROBIOTIC FOODS

Flavor is among the key factors that determine the success of a product as consumers consider it is one of the critical factors in the selection and acceptance of a particular food [77]. Fruit juices, vegetables, and cereals constitute an emerging segment of nonconventional dairy substrates for the design of probiotic foods that will require sensory and physicochemical characterization for quality control and further product design. Most of the current work in the area of product development in non-dairy probiotic food such as soy dessert, cantaloupe melon juice, an oat flakes beverage, soy-based petit-Suisse cheese, cashew apple juice, and peanut soy milk have focused in sensory, color, and rheological evaluations [78–82]. Nevertheless, the analysis of volatile flavor compounds in novel non-dairy probiotic products has not been studied in depth. The flavor of fermented food is due to a delicate equilibrium among volatile and non-volatile compounds present in the raw materials and metabolites synthesized through biological pathways. Thus, the flavor and aroma of fermented foods are composed of many volatile compounds. However, only a few are designated as aroma-impact compounds. LAB is widely used during the formulation of probiotic products. These microorganisms have complex enzyme systems through which they produce a range of metabolites that provide important flavor attributes. Acetaldehyde and diacetyl metabolites synthesized by LAB have been described as flavor active compounds in yogurt [83, 84]. Alternatively, in Bushera and Togwa, which are traditional African beverages produced by the fermentation of cereal substrates with mix-cultures containing LAB, volatile flavor compounds such as acetaldehyde, diacetyl, ethanol, and acetone have been detected [85, 86]. In sourdough, an important modern fermentation method of cereal flours and water inoculated with yeasts and LAB strains, the non-volatiles acetic and lactic acid are important flavor compounds produced during this fermentation procedure [87]. The evaluation of the flavor volatiles acetaldehyde, acetone, ethyl acetate, diacetyl, and the primary metabolites lactic and acetic acid which have been reported to have important flavor attributes in fermented products could be of great interest during the development of novel probiotic beverages formulated with lactobacilli and non-dairy substrates. Consequently, during the development of novel probiotic food, these compounds should be carefully examined in order to understand their possible sensorial attributes and effect on the flavor quality in order to design unique probiotic beverages and foods with an acceptable taste and aroma.

KEYWORDS

- **cereals**
- **fermentation**
- **gastrointestinal tract**
- **lactic acid bacteria**
- *Lactobacillus*
- **probiotics**

REFERENCES

1. Stanton, C., Ross, R. P., Fitzgerald, G. F., & Van Sinderen, D., Fermented functional foods based on probiotics and their biogenic metabolites. *Curr. Opin. Biotechnol.,* **2005,** *16,* 198–203.
2. Agrawal, R. Probiotics: An emerging food supplement with health benefits. *Food Biotechnol.,* **2005,** *19*(3), 227–246.
3. Kumar, H., Salminen, S., Verhagen, H., Rowland, I., Heimbach, J., Bañares, S., Young, T., Nomoto, K., & Lalonde, M., Novel probiotics and prebiotics: Road to the market. *Current Opinion in Biotechnology,* **2015,** *32,* 99–103.
4. Reid, G. The growth potential for dairy probiotics. *International Dairy Journal,* **2015,** *49,* 16–22.
5. Nout, M. J. *Foods and Their Production* (pp. 1–18). Aspen Publishers: Gaithersburg, Maryland, **2001.**
6. Blandino, A., Al-Aseeri, M. E., Pandiella, S. S., Cantero, D., & Webb, C. Cereal-based fermented foods and beverages. *Food Research International,* **2003,** *36,* 527–543.

7. Motarjemi, Y., Asante, A., Adams, R. M., & Nout, M. J. *Practical Applications: Prospects and Pitfalls* (pp. 253–271). Aspen Publishers: Gaithersburg, Maryland, **2001**.

8. Sybesma, W. F., & Hugenholtz, J. *Food Fermentation by Lactic Acid Bacteria for the Prevention of Cardiovascular Disease* (pp. 448–473). Woodhead Publishing Limited: Cambridge, **2004**.

9. Heydanek, M. G., & McGorrin, R. J., Gas chromatography-mass spectroscopy investigations on the flavor chemistry of oat groats. *J. Agric. Food Chem.,* **1981**, *29*, 950–954.

10. Leroy, F., & De Vuyst, L. Lactic acid bacteria as functional starter cultures for the food fermentation industry. *Trends in Food Science and Technology,* **2004**, *15*, 67–78.

11. Salminen, S. *Lactic Acid Bacteria* (pp. 1–138). Mercer Deckker: New York, **1998**.

12. Adams, M. R., & Marteau, P. On the safety of lactic acid bacteria from food. *Int. J. Food Microbiol.,* **1995**, *27*, 263–264.

13. Imhof, R., Glattli, H., & Bosset, J. O. Volatile organic compounds produced by thermophlic and mesophlic single strain dairy starter cultures. *LWT-Food Sci. Technol.,* **1995**, *28*, 78–86.

14. Chandan, R. C. Enhancing market value of milk by adding cultures. *J. Dairy Sci.,* **1999**, *82*, 2245–2256.

15. De Vuyst, L. Technology aspects related to the application of functional starter cultures. *Food Technology and Biotechnology,* **2000**, *38*, 105–112.

16. Ross, R. P., Stanton, C., Hill, C., Fitzgerald, G. F., & Coffey, A. Novel cultures for cheese improvement. *Trends in Food Science and Technology,* **2000**, *11*, 96–104.

17. Stiles, M. E., & Holzapfel, W. H. Lactic acid bacteria of foods and their current taxonomy. *Int. J. Food Microbiol.,* **1997**, *36*(1), 1–29.

18. Martin, A. M. Role of lactic acid fermentation in bioconversion of wastes. In: Bozoğlu, T. F., & Ray, B., (eds.), *Lactic Acid Bacteria: Current Advances in Metabolism, Genetics and Applications* (pp. 219–252). Springer: Berlin, **1996**.

19. Farnworth, E. R. *Handbook of Fermented Functional Foods*. CRC Press: Boca Raton, Florida, **2003**.

20. Haggblade, S., & Holzapfel, W. H. Industrialization of Africa's indigenous beer brewing. In: Steinkraus, K. H., (ed.), *Industrialization of Indigenous Fermented Foods* (pp. 191–283). Marcel Dekker: New York, **1989**.

21. Singh, S. K., Ahmed, S. U., & Pandey, A. Metabolic engineering approaches for lactic acid production. *Process Biochem.,* **2006**, *41*, 991–1000.

22. Axelsson, L. Lactic acid bacteria: Classification and physiology. In: Salmínen, S., & Von Wright, A., (eds.), *Lactic Acid Bacteria: Microbiology and Functional Aspects* (2ⁿᵈ edn., pp. 1–72). Marcel Dekker, Inc: New York, **1998**.

23. Pot, B., & Tsakalidou, E. Taxonomy and metabolism of *Lactobacillus*. In: Ljungh, Å., & Wadström, T., (eds.), *Lactobacillus Molecular Biology From Genomics to Probiotics* (pp. 3–58). Caister Academic Press: Norfolk, UK, **2009**.

24. Kandler, O. Carbohydrate metabolism in lactic acid bacteria. *Antonie Van Leeuwenhoek,* **1983**, *49*, 209–224.

25. Smit, G., Smit, B. A., & Engels, W. J. M. Flavor formation by lactic acid bacteria and biochemical flavor profiling of cheese products. *Federation of European Microbiological Societies (FEMS) Microbiology Reviews,* **2005**, *29*, 591–610.

26. Liu, S. Q. Practical implications of lactate and pyruvate metabolism by lactic acid bacteria in food and beverage fermentations. *Int. J. Food Microbiol.,* **2003**, *83*, 115–131.

27. Prado, F. C., Parada, J. L., Pandey, A., & Soccol, C. R. Trends in non-dairy probiotic beverages. *Food Research International,* **2008**, *41*(2), 111–123.

28. Siró, I., Kápolna, E., Kápolna, B., & Lugasi, A. Functional food. Product development, marketing and consumer acceptance: A review. *Appetite,* **2008**, *51*(3), 456–467.

29. Li, Q., & Zhou, J. M. The microbiota-gut-brain axis and its potential therapeutic role in autism spectrum disorder. *Neuroscience,* **2016**, *324,* 131–139.

30. Sindhu, S. C., & Khetarpaul, N. Development, acceptability, and nutritional evaluation of an indigenous food blend fermented with probiotic organisms *Nutrition and Food Science,* **2004**, *35*(1), 20–27.

31. FAO/WHO. *Health and Nutritional Properties of Probiotics in Food Including POWDER milk with Live Lactic Acid Bacteria, Food and Agriculture Organization of the United Nations and World Health Organization Expert Consultation Report.* Cordoba, Argentina, **2001**.

32. Kandylis, P., Pissaridi, K., Bekatorou, A., Kanellaki, M., & Koutinas, A. A. Dairy and non-dairy probiotic beverages. *Current Opinion in Food Science,* **2016**, *7,* 58–63.

33. Fooks, L., Fuller, R., & Gibson, G. R. Prebiotics, probiotics, and human gut microbiology. *International Dairy Journal,* **1999**, *9,* 53–61.

34. Tannock, G. W. Studies of the intestinal microflora: A prerequisite for the development of probiotics. *Int. Dairy J.,* **1998**, *8,* 527–533.

35. Zubillaga, M., Weill, R., Postaire, E., Goldman, C., Caro, R., & Boccio, J. Effect of probiotics and functional foods and their use in different diseases. *Nutr. Res.,* **2001**, *21,* 569–579.

36. Tripathi, M. K., & Giri, S. K. Probiotic functional foods: Survival of probiotics during processing and storage. *Journal of Functional Foods,* **2014**, *9*(0), 225–241.

37. Nousiainen, J., & Setälä, J. Lactic acid bacteria as animal probiotics. In: Salminen, S., & Von Wright, A., (eds.), *Lactic Acid Bacteria: Microbiology and Functional Aspects,* (2nd edn., Vol. 2). Marcel Dekker Inc.: New York, USA, **1998**.

38. Gismondo, M. R., Drago, L., & Lombardi, A. Review of probiotics available to modify gastrointestinal flora. *Int. J. Antimicrob. Agents,* **1999**, *12,* 287–292.

39. Puupponen-Pimiä, R., Aura, A. M., Oksman-Caldentey, K. M., Myllärinen, P., Saarela, M., Mattila-Sandholm, T., & Poutanen, K. Development of functional ingredients for gut health. *Trends in Food Science and Technology,* **2002**, (13), 3–11.

40. Teitelbaum, J. E., & Walker, A. W., Nutritional impact of pre- and probiotics as protective gastrointestinal organisms. *Annu. Rev. Nutr.,* **2002**, *22,* 107–138.

41. Saarela, M., Lahteenmaki, L., Crittenden, R., Salminen, S., & Mattila-Sandholm, T. Gut bacteria and health foods: The European perspective. *Int. J. Food Microbiol.,* **2002**, *78,* 99–117.

42. Holzapfel, W. H., Haberer, P., Snel, J., Schillinger, U., & Huisint, V. J. H. J. Overview of gut flora and probiotics. *Int. J. Food Microbiol.,* **1998**, *41,* 85–101.

43. Shanahan, F. The host-microbe interface with the gut. *Best Practice and Research Clinical Gastroenterology,* **2002**, *16*(6), 915–931.

44. Scheinbach, S. Probiotics: Functionality and commercial status. *Biotechnology Advances,* **1998**, *16,* 581–608.

45. Tuohy, K. M., Probert, H. M., Smejkal, C. W., & Gibson, G. R. Using probiotics and prebiotics to improve gut health. *Therapeutic Focus,* **2003**, *15,* 692–700.

46. Salminen, S., Bouley, C., Boutron-Ruault, M. C., Cummings, J. H., Franck, A., Gibson, G. R., Isolauri, E., Moreau, M. C., Roberfroid, M., & Rowland, I. Functional food science and gastrointestinal physiology and function. *Br. J. Nutr.,* **1998**, *I*(80), S147–Sl71.

47. Limdi, J. K., O'Neill, C., & McLaughlin, J. Do probiotics have a therapeutic role in gastroenterology? *World J. Gastroenterol.,* **2006**, *12*(34), 5447–5457.
48. Ouwehand, A., Salminen, S., & Isolauri, E. Probiotics: An overview of beneficial effects. *Antonie van Leeuwenhoek, 82,* **2002**, *82,* 279–289.
49. Frost, S. *Strategic Analysis of the European Food and Beverage Probiotics Markets (#B956–88).* Frost & Sullivan Ltd.: London, **2007**.
50. Lippolis, R., Siciliano, R. A., Mazzeo, M. F., Abbrescia, A., Gnoni, A., Sardanelli, A. M., & Papa, S. Comparative secretome analysis of four isogenic *Bacillus clausii* probiotic strains. *Proteome Science,* **2013**, *11*(28).
51. Succi, M., Tremonte, P., Pannella, G., Tipaldi, L., Cozzolino, A., Coppola, R., & Sorrentino, E. Survival of commercial probiotic strains in dark chocolate with high cocoa and phenols content during the storage and in a static in vitro digestion model. *Journal of Functional Foods,* **2017**, *35,* 60–67.
52. Nout, M. J. R. Accelerated natural lactic fermentation of cereal-based formulas at reduced water activity *Int. J. Food Microbiol.,* **1992**, *16,* 313–322.
53. Saarela, M., Mogensen, G., Fodén, R., Mättö, J., & Mattila-Sandholm, T. Probiotic bacteria: Safety, functional and technological properties. *J. Biotechnol.,* **2000**, *84,* 197–215.
54. Martín, M. J., Lara-Villoslada, F., Ruiz, M. A., & Morales, M. E. Microencapsulation of bacteria: A review of different technologies and their impact on the probiotic effects. *Innovative Food Science and Emerging Technologies,* **2015**, *27,* 15–25.
55. Sanders, M. E., Guarner, F., Guerrant, R., Holt, P. R., Quigley, E. M. M., Sartor, R. B., Sherman, P. M., & Mayer, E. A. An update on the use and investigation of probiotics in health and disease. *Gut.,* **2013**, *62*(5), 787–796.
56. Mahajan, B., & Singh, V. Recent trends in probiotics and health management: A review. *International Journal of Pharmaceutical Sciences and Research,* **2014**, *5,* 1643.
57. Gao, J., Wu, H., & Liu, J., Importance of gut microbiota in health and diseases of new born infants (review). *Exp. Ther. Med.,* **2016**, *12,* 28.
58. Sanders, M. E. Probiotics and microbiota composition. *BMC Med.,* **2016**, *14*(1), 82.
59. Patel, A. Current trend and future prospective of functional probiotic milk chocolates and related products: A review. *Czech Journal of Food Sciences,* **2015**, *33,* 295–301.
60. Heenan, C. N., Adams, M. C., Hosken, R. W., & Fleet, G. H. Survival and sensory acceptability of probiotic microorganisms in a nonfermented frozen vegetarian dessert. *LWT-Food Sci. Technol.,* **2004**, *37,* 461–466.
61. Yoon, K. Y., Woodams, E. E., Hang, Y. D., & Ziemer, C. J. Production of probiotic cabbage juice by lactic acid bacteria. *Bioresource Technology,* **2006**, *97,* 1427–1430.
62. Tuorila, H., & Cardello, A. V., Consumer response to an off-flavor in juice in the presence of specific health claims. *Food Qual. Prefer.,* **2002**, *13,* 561–569.
63. Luckow, T., & Delahunty, C. Consumer acceptance of orange juice containing functional ingredients. *Food Research International,* **2004**, *37*(8), 805–814.
64. Luckow, T., Sheehan, V., Delahunty, C., & Fitzgerald, G. Determining the odor and flavor characteristics of probiotic, health-promoting ingredients and the effects of repeated exposure on consumer acceptance. *J. Food Sci.,* **2005**, *70,* S53–S59.
65. Luckow, T., Sheehan, V., Fitzgerald, G., & Delahunty, C. Exposure, health information and flavor-masking strategies for improving the sensory quality of probiotic juice. *Appetite,* **2006**, *47*(3), 315–323.
66. Charalampopoulos, D., Pandiella, S. S., & Webb, C. Growth studies of potentially probiotic lactic acid bacteria in cereal-based substrates. *J. Appl. Microbiol.,* **2002**, *92*(5), 851–859.

67. Salmerón, I. Fermented cereal beverages: From probiotic, prebiotic and synbiotic towards nano science designed healthy drinks. *Letters in Applied Microbiology,* **2017,** *65*(2), 114–124.

68. Henry, R. J. Biotechnology, cereal, and cereal products quality. In: Owens, G., (ed.), *Cereals Processing Technology* (pp. 53–72). Woodhead Publishing Limited: Cambridge, **2001.**

69. FAO. *Crop Prospects and Food Situation.* Food and Agriculture Organization of the United Nations (www.fao.org/worldfoodsituation) Rome, **2008.**

70. Charalampopoulos, D., Pandiella, S. S., Wang, R., & Webb, C. Application of cereals and cereal components in functional foods: A review. *Int. J. Food Microbiol.,* **2002,** *79,* 131–141.

71. Angelov, A., Gotcheva, V., Hristozova, T., & Gargova, S. Application of pure and mixed probiotic lactic acid bacteria and yeast cultures for oat fermentation. *Journal of the Science of Food and Agriculture,* **2005,** *85,* 2134–2141.

72. KeyNote. *Functional Foods* (pp. 26–30). Key Note Ltd.: Hampton, Middlesex, **2008.**

73. Patel, H. M., Pandiella, S. S., Wang, R. H., & Webb, C. Influence of malt, wheat, and barley extracts on the bile tolerance of selected strains of lactobacilli. *Food Microbiol.,* **2004,** *21*(1), 83–89.

74. Rozada-Sánchez, R., Sattur, A. P., Thomas, K., & Pandiella, S. S. Evaluation of *Bifidobacterium* spp. for the production of a potentially probiotic malt-based beverage. *Process Biochem.,* **2008,** *43* (8), 848–854.

75. Michida, H., Tamalampudi, S., Pandiella, S. S., Webb, C., Fukuda, H., & Kondo, A. Effect of cereal extracts and cereal fiber on viability of Lactobacillus plantarum under gastrointestinal tract conditions. *Biochemical Engineering Journal,* **2006,** *28*(1), 73–78.

76. Rozada, R., Vázquez, J. A., Charalampopoulos, D., Thomas, K., & Pandiella, S. S. Effect of storage temperature and media composition on the survivability of Bifidobacterium breve NCIMB 702257 in a malt hydrolisate. *International Journal of Food Microbiology,* **2009,** *133*(1/2), 14–21.

77. Fisher, C., & Scott, T. R. *Food Flavors Biology and Chemistry.* The Royal Society of Chemistry (RSC) Cambridge, **1997.**

78. Chattopadhyay, S., Raychaudhuri, U., & Chakraborty, R. Optimization of soy dessert on sensory, color, and rheological parameters using response surface methodology. *Food Sci. Biotechnol.,* **2013,** *22*(1), 47–54.

79. Fonteles, T., Costa, M., De Jesus, A., Fontes, C., Fernandes, F., & Rodrigues, S. Stability and quality parameters of probiotic cantaloupe melon juice produced with Sonicat juice. *Food Bioprocess Technol.,* **2013,** *6*(10), 2860–2869.

80. Luana, N., Rossana, C., Curiel, J. A., Kaisa, P., Marco, G., & Rizzello, C. G. Manufacture and characterization of a yogurt-like beverage made with oat flakes fermented by selected lactic acid bacteria. *Int. J. Food Microbiol.,* **2014,** *185*(0), 17–26.

81. Matias, N. S., Bedani, R., Castro, I. A., & Saad, S. M. I. A probiotic soy-based innovative product as an alternative to petit-Suisse cheese. *LWT-Food Sci. Technol.,* **2014,** *59*(1), 411–417.

82. Pereira, A., Almeida, F., de Jesus, A., Da Costa, J., & Rodrigues, S. Storage stability and acceptance of probiotic beverage from cashew apple juice. *Food Bioprocess Technol.,* **2013,** *6*(11), 3155–3165.

83. Imhof, R., Glättli, H., & Bosset, J. O. Volatile organic aroma compounds produced by thermophilic and mesophilic mixed strain dairy starter cultures. *LWT-Food Sci. Technol.,* **1994,** *27*(5), 442–449.

84. Ott, A., Fay, L. B., & Chaintreau, A. Determination and origin of the aroma impact compounds of yogurt flavor. *J. Agric. Food Chem.,* **1997**, *45*(3), 850–858.
85. Mugula, J. K., Narvhus, J. A., & Sørhaug, T. Use of starter cultures of lactic acid bacteria and yeasts in the preparation of togwa, a Tanzanian fermented food. *International Journal of Food Microbiology,* **2003**, *83*(3), 307–318.
86. Muyanja, C. M. B. K., Narvhus, J. A., Treimo, J., & Langsrud, T. Isolation, characterization and identification of lactic acid bacteria from bushera: A Ugandan traditional fermented beverage. *Int. J. Food Microbiol.,* **2003**, *80*(3), 201–210.
87. Pozo-Bayón, M. A., G-Alegría, E., Polo, M. C., Tenorio, C., Martín-Álvarez, P. J., Calvo De La Banda, M. T., Ruiz-Larrea, F., & Moreno-Arribas, M. V. Wine volatile and amino acid composition after malolactic fermentation: Effect of *Oenococcus oeni* and *Lactobacillus plantarum* starter cultures. *J. Agric. Food Chem.,* **2005**, *53*(22), 8729–8735.

CHAPTER 8

Fermented Milks: Quality Foods with Potential for Human Health

BLANCA ESTELA GARCÍA-CABALLERO,[1,2]
OLGA MIRIAM RUTIAGA-QUIÑONES,[2]
SILVIA MARINA GONZÁLEZ-HERRERA,[2] CRISTÓBAL NOÉ AGUILAR,[1]
ADRIANA CAROLINA FLORES-GALLEGOS,[1] and
RAÚL RODRÍGUEZ HERRERA[1]

*[1]Department of Food Research, Faculty of Chemical Sciences,
Autonomous University of Coahuila, Saltillo Coahuila, Mexico,
E-mail: raul.rodriguez@uadec.edu.mx (R. R. Herrera)*

*[2]National Technological Institute of Mexico, Technological Institute of
Durango Department of Chemical and Biochemical Engineering,
Felipe Pescador. Blvd. 1830 Ote., C.P. 34080 Durango, Durango, Mexico*

ABSTRACT

Lactic acid fermentation helps to preserve the milk, improves its nutritional value, and intensifies its sensory properties, as well as to preserve the consumers health. For a long time, consumption of this type of milk has been considered beneficial for health, because of the microorganisms that it contains, different compounds released during the fermentation process, among them, metabolites, and biologically active molecules, which confer it antimicrobial, prebiotic, antioxidant, antihypertensive, and hypocholesterolemic properties, among others. Some of the best-known fermented kinds of milk are Yogurt, Kefir, Buttermilk, and Acidophilus Milk or Koumis. Currently, the consumption of these foods has skyrocketed in both, developing, and developed countries. This document aims to discuss the most up-to-date information on fermented milk.

8.1   INTRODUCTION

Fermented milks are products where lactic acid fermentation technology is used, as a means for its elaboration, which helps to preserve these foods, improve their nutritional value, and intensify their sensory properties, as well as to preserve the consumer health, since they are a source of probiotics [1]. For a long time, consumption of this type of milk has been considered beneficial for human health; this has been attributed to the microorganisms involved during its fermentation. Some studies showed that these benefits are also due to the different compounds released during the fermentation process, such as specific metabolites and biologically active molecules [2, 3]. Both milk and fermented milk products are foods that humans have consumed for millennia around the world; these foods have components with antimicrobial, probiotic, prebiotic, antioxidant, antihypertensive, hypocholesterolemic characteristics [4]. Among the best-known fermented milk types are Yogurt, Kefir, Buttermilk, and Acidophilus milk or Kumis. These products in recent years have become very popular all over the world. One of the main reasons is because consumers have become more aware of the relationship between diet and health [5].

Different investigations support the hypothesis that while the diet satisfies the nutritional needs, it is also capable of modulating several physiological functions and can play a role in the benefit or detriment of some diseases [6]. Additionally, fermented milk has become key for human health, so their consumption has skyrocketed; one of the reasons is because they act as vehicles of healthy elements. Nowadays, it is not strange that each country or region has traditional fermented milk of higher consumption, autochthonous, and elaborated mostly by hand. In Mexico, fermented foods have been known since pre-Hispanic times, the fermentation process being one of the most used ancestral techniques to conserve, produce or transform food and beverages. This practice involves modification of raw materials structure [7, 8].

8.2   ORIGIN OF FERMENTED MILKS

Transformation of the nomadic life to a sedentary lifestyle of our ancestors occurred as a process that included the development of agriculture and the domestication of animals, later, the need for food conservation arose. Perhaps, the simplest ways for food conservation were sun-drying and fermentation; the latter as a complicated process, since the transformation of the food is produced by action of different microorganisms from the environment or artificially added, which transform flavor, color, smell, texture, and even the nutritional value of the product in which they are growing and fermenting [7]. Microbiological studies of fermented beverages indicate that they contain a high concentration of beneficial microorganisms such as lactic bacteria, which are the first to develop and are present throughout the process. These microorganisms are responsible for biomass acidification (pH value could be close to 4) since these microorganisms produce lactic acid, which imparts a fresh and pleasant taste to the product) [7].

Before, use of food was emphasized on survival, hunger satisfaction, and prevention of adverse effects on health. Currently, they are considered as promoters of a state of wellness, health enhancers, and coadjutants for risk reduction of diseases. These concepts are particularly important, because of cost increases of health maintaining, an increase in life expectancy, and the desire of older people to improve their quality of life [9]. Historically, nutritional status of human populations has been affected by a high intake of sugars, salt, saturated trans fats, and low levels of fiber, vitamins, and essential minerals. These habits are the leading cause of chronic-degenerative non-transmissible diseases; to reduce this risk, it has been proposed the development of new products containing biologically active substances [10].

The term "Functional foods" was initially defined in Japan in the 1980s as a specific food for health "FOSHU" (Food for Specified Health Uses) and these were described as foods or nutrients that when ingested, produce important physiological changes in the human organism, which are distinct from others associated with their role as nutrients [11]. All foods are functional at some physiological level because they contain nutrients or other substances that provide energy, sustain growth, or maintain/repair vital processes. However, functional foods also provide additional benefits that can reduce the risk of disease, and promote optimal health. Such foods include conventional, modified (fortified, enriched, or improved) foods, medicinal foods, and foods for a special-use diet [12].

Fermented milks are central functional foods. The best known worldwide is the fermented thermophilic milk, known as yogurt. It is reported that a large proportion of fermented milk products seem to have their origin in nomadic cattle ranches from Asia, which transported fresh milk, in bags usually made of goatskin. Heat and contact of milk with the goatskin promoted multiplication of lactic acid bacteria (LAB), which fermented the milk. Thus, milk was converted into a semi-solid mass by coagulation. Galen, in the second century, highlighted its beneficial effect on stomach problems [4]. It is at the beginning of the 20th century when yogurt begins to be part of food habits of the general population. Elie Metchnikoff, a member of the Pasteur Institute and Nobel Prize winner in 1908, demonstrated the benefits of yogurt bacteria on diarrhea in infants. In 1917, Isaac Carasso produced yogurt following industrial processes, as a product sold exclusively in pharmacies. In the 50s, yogurt began to be distributed in dairies and later, in food stores [4].

Currently, in almost all countries, there are one or several native fermented products such as cheeses or Bulgarian milk. In Mexico, there is a wide variety of traditional fermented dairy products, produced in an artisanal way [13]; cream cheese [14], or fresh goat cheese [15] for talking about some of the most consumed fermented foods. All of them are obtained through lactic acid fermentation, thanks to its content of native or inoculated lactic bacteria. Through fermentation, different metabolites with a beneficial effect on health are produced. It is important to note that this type of traditional foods reflect the history, culture, idiosyncrasy, and lifestyle of their place of origin [14], which is why they are called "identitary" products, which confer them additional value.

8.3   LACTIC ACID FERMENTATION

Fermented milks are the product of a lactic acid fermentation process, which is one of the cheapest and oldest processes to elaborate and preserve food, using different types of microorganisms. Among the changes that occur in the milk during fermentation is the decrease in pH (up to 4.0–4.6), a factor that contributes to the maintenance of a low pH in the stomach after consuming milk. Also, inhibition of microbial growth by non-dissociated acids (lactic acid), and by other metabolites such as H_2O_2 and other substances with antibiotic activity; a low oxide-reduction potential and consumption by lactic bacteria of components that is vital for other microorganisms [16, 17]. The use of microorganisms to prepare food has been known around the world for thousands of years; these can come from the natural microbiota or starter cultures [18].

In Mexico, some fermented foods are produced industrially, but most are still made by hand [8]. During the elaboration process, the metabolic activity of the microorganisms improves nutritional quality and sensory properties of different raw materials, including dairy products [19, 20]. These foods are appreciated for their attributes of flavor, aroma, and texture, improving their sensory properties during cooking, and their processing [21]. In our country other traditional fermented beverages are also consumed, some are made with corn (atole sour, pozol, and tesgûino), fruit (tepache and colonche), and those obtained by fermenting plants sap (pulque, tuba, and tavern) [8].

Around the world, there is a wide variety of fermented milk typical of the regions where they are produced, most of them are elaborated and consumed at the artisanal level. Table 8.1 shows some types of fermented milk originating from different countries and their microorganisms (native or inoculated), imparting functional properties, depending on the types they contain. In South Africa, as in Mexico and other countries, spontaneous fermentation is used as one of the oldest forms of food processing technology, the knowledge that has been passed on from generation to generation [22]. Most of this knowledge, it has not been documented, and there is a danger that the technologies involved will be lost and that families will not be able to preserve traditional food processing methods [23].

Recent studies have documented the quality and diversity of traditional African fermented milk [24–29]. Milk produced in rural communities in Africa is consumed fresh or sour [30], with only a small fraction produced entirely by the commercial sector [22, 30].

8.4 LACTIC ACID BACTERIA (LAB)

LAB comprise a heterogeneous group of Gram-positive microorganisms of bacillary or coccoid form, whose main characteristic is the production of lactic acid and other compounds such as acetate, ethanol, CO_2, formate, and succinate from fermentable carbohydrates. This group of bacteria includes genera such as *Bifidobacterium*, *Lactococcus*, *Lactobacillus*, *Streptococcus*, and *Pediococcus*, which can be isolated from fermented foods, acid masses, beverages, plants, and the respiratory, intestinal, and vaginal tracts of warm-blooded animals, among others. These bacteria are considered probiotic since they contribute beneficially to the balance of the intestinal microbiota [65, 66] and are useful for the prevention and control of gastrointestinal disorders [67].

LAB has been inadvertently used for thousands of years for the production of fermented foods such as cheese and yogurt. However, it was not until the mid-nineteenth century that Louis Pasteur demonstrated that production of lactic acid during fermentation was due to the action of microorganisms (lactic ferments). LAB strains are used in the food industry, not only for its ability to acidify and therefore preserve food from spores, but also for its involvement in texture, flavor, smell, and aroma development of fermented foods [68].

The LAB are widely distributed in different ecosystems, besides being generated on a large scale in processes for the commercial production of

TABLE 8.1 Fermented Milk Types Produced in Different Countries, Some Microorganisms That Intervene in Their Fermentation and Nutritional or Functional Properties

Dairy Product	Origin Country	Microorganisms	Functional Potential According to Its Microorganisms	References
Amasi	Zimbawe, South Africa	*Lc. lactis* subsp. *lactis* (dominating), *Lc. lactis* subsp. *cremoris*, *Lactobacillus*, *Enterococcus*, and *Leuconostoc* spp.	Probiotic, antagonistic activity to pathogens, antioxidant activity	[28, 31, 32]
Airag	Mongolia	*Lb. helveticus*, *Lb. kefiranofaciens*, *Bifidobacterium mongoliense*, *Kluyveromyces marxianus*	Probiotic, protection of infants against nosocomial diarrhea	[33–36]
Chhurpi	India, Nepal, China	*Lb. farciminis*, *Lb. brevis*, *Lb. alimentarius*, *Lb. salivarius*, *Lact. lactis*, *Candida* sp. *Saccharomycopsis* sp.	Antimicrobial activity, resistance to antibiotics, cholesterol assimilation	[37, 38]
Dadih	Indonesia	*Leuc. mesenteroides*, *Ent. faecalis*, *Strep. Lactis* supsp. *lactis*, *Strep. cremoris*, *Lb. casei* subsp. *casei*, and *Lb. casei* subsp. *rhamnosus*	Resistance to antibiotics, antagonist activity against pathogens	[32, 39]
Dahi	India	*Strep. Salivarius* subsp. *thermophilus*, *Lb. acidophilus*, *Lb. delbruecki* subsp. *bulgaricus*, *Lc. lactis* subsp. *lactis*, *Sacch. cerevisiae*	De-inflammatory, antioxidant activity, resistance to antibiotics, inhibition of *Helicobacter pylori*	[32, 40–42]
Ergo	Ethypia	*Lb. mesenteroides*, *Lc. lactis* subsp. *cremoris*, *Lc. lactis* subsp. *lactis*, *Lc. cremoris*, *S. thermophilus*, *Lb. delbrueckii*, *Lb. homi*, *Micrococcus* spp.	Protection against infections, protection against nosocomial diarrhea	[36, 43–45]
Kefir	Russia	*Lb. brevis*, *Lb. caucasicus*, *Strep. thermophilus*, *Lb. bulgaricus*, *Lb. plantarum*, *Lb. casei*, *Lb. brevis*, *Tor. holmii*, *Tor. delbruechii*	Antimicrobial activity against *E. coli*, *Staphylococcus aureus*, and *Salmonella*	[46, 47]
Kumiss	Russia, Mongolia	*Lb. brevis*, *Lb. caucasicus*, *Strep. thermophilus*, *Lb. bulgaricus*, *Lb. plantarum*, *Lb. casei*, *Lb. brevis*, *Tor. holmii*, *Tor. delbruechii*	Protection against diarrhea, de-inflammatory antimicrobial activity	[36, 48, 49]

TABLE 8.1 *(Continued)*

Dairy Product	Origin Country	Microorganisms	Functional Potential According to Its Microorganisms	References
Laban raveb	Egypt	*Lb. casei, Lb. plantarum, Lb. brevis, Lact. lactis, Leuconostoc* sp., *Sacch. Kéfir*	Antimicrobial activity against *Listeria monocitogenes* and de-inflammatory	[46, 50]
Leben	North, East Central Africa	*Candida* sp., *Saccharomyces* sp., *Lactobacillus* sp., *Leuconostoc* sp.	Inhibition of *Salmonella tiphimorium,* prevention of diarrhea rotavirus	[29, 51]
Acidophilus milk	USA	*Lactobacillus acidophilus*	Inflammatory activity and antimicrobial	[52]
Nono	Nigeria, Ghana	*Lb. fermentum, Lb. plantarum, Lb. helveticus, Leuc. mesenteroides, Ent. faecium, Ent. italicus, Weissella confusa, Candida parapsilosis, C. rugosa, C. tropicalis, Galactomyces geotrichum, Pichia kudriavzevii, Sacch. Cerevisiae*	Antimicrobial activity, anti-inflammatory activity	[50, 53, 54]
Goat cheese	Mexico	*Leu. Mesenteroides, Leu desxtranicum, Lactobacillus plantarum, lactococcus lactis subsp. lactis*	Probiotic, anti-inflammatory, antimicrobial activity	[15]
Ricotta Cheese	Tunicia	*Lb plantarum*	Antimicrobial activity against *Listeria monocitogenes*	[50]
Cream Cheese	México	*Lc. lactis* subsp. *lactis, Lc. lactis* subsp. *cremoris,*	Protection against infections	[14, 43]
Sua-Chua	Vietnam	*Lb. bulgaricus, Strep. Thermophilus*	Probiotic	[43, 55]
Tarag	Mongolia	*Lb. delbrueckii* subsp. *bulgaricus, Lb. helveticus, Strep. thermophilus, Sacch. cerevisiae, Issatchenkia orientalis, Kazachstania unispora*	De-inflammatory, reduced diarrhea induced by antibiotics	[33, 56–59]

TABLE 8.1 (Continued)

Dairy Product	Origin Country	Microorganisms	Functional Potential According to Its Microorganisms	References
Viilli	Finland	*Lc. lactis* subsp. *lactis*, *Lc. lactis* subsp. *cremoris*, *Lc. lactis* subsp. *lactis* biovar. *diacetylactis*, *Leuc. mesenteroides* subps. *cremoris*, *G. candidum*, *K. marxianus*, *P. fermentans*	Probiotic, inhibition of *Salmonella tiphimorium*, inhibition of *Helicobacter pylori* (gastritis)	[42, 51, 60–62]
Yogurt	Europe, America and Australia	*Strep. thermophilus*, *Lb. delbrueckii* subsp. *bulgaricus*, *Lb. acidophilus*, *Lb. casei*, *Lb. rhamnosus*, *Lb. gasseri*, *Lb. johnsonii*, *Bifidobacterium* spp.	Probiotic, prevention of diarrhea rotavirus, inhibition of *Helicobacter pylori* (gastritis)	[42, 43, 63, 64]

fermented foods, alcoholic beverages, silages, yeasts for production of beer, wines, and in vegetables or meat fermentations [69], cheese, butter, yogurt, sausages, olives, grapes, and cereals like bread, preserving, and providing sensory and nutritional properties to these food products [70, 71]. The extensive use of LAB in food preservation is because they produce organic acids and a wide variety of antimicrobial substances such as hydrogen peroxide (H_2O_2), carbon dioxide (CO_2), and diacetyl and bacteriocins [72, 73].

Different factors affect LAB growth during fermentation. In addition to the nutritional requirements, the temperature is one of the most important factors that influence LAB growth. There is an optimal temperature at which the growth rate is higher and depends on the microorganism characteristics, as well as, environmental conditions [74]; most species need B vitamins (lactoflavin, thiamine, biotin, nicotinic acid, pantothenic acid, folic acid) and several amino acids [75]. They are chemo-organotrophs and only grow on complex media. Fermentable carbohydrates and alcohols can be used as energy sources to form mainly lactic acid, through degradation of hexoses to lactate (homofermentative) and additional products such as acetate, ethanol, CO_2, formate, or succinate (heterofermentative) [71]. Some of the LAB functions are: acid taste formation, inhibition of pathogenic organisms, milk gelling, reduction of lactose content, aroma formation, gas production required for formation of "eyes" in cheeses, and proteolysis required for cheeses maturation [76], LAB have also been widely used as probiotics [77].

LAB Taxonomy was initiated in 1919 by Orla Jensen. LAB comprises a diverse group of organisms [71]. They are cocci and bacilli of variable length and a thickness between 0.5–0.8 μm. It is a group of physiologically uniform, Gram-positive wall bacteria [75], facultative anaerobes, and catalase negative and non-spore-forming bacteria [78]. They lack respiratory activity because of the absence of cytochrome catalase, LAB contain a hemin group, which allows them to start the respiratory chain with oxygen as an electron acceptor. Despite their anaerobic metabolism, they are tolerant anaerobes and in solid culture, media form colonies in the presence of air [75].

LAB belong to the *Firmicutes* phylum which comprises around 20 genera and *Lactococcus, Lactobacillus, Streptococcus, Leuconostoc, Pediococcus, Aerococcus, Carnobacterium, Enterococcus, Oenococcus, Tetragenoccus, Vagococcus,* and *Weisella* are the main members of LAB, being *Lactobacillus* the largest of these genera [79–81]. Type and characteristics of the initiating organisms that are used for the production of fermented milks are the two most important factors that determine final product quality. The essential criteria for selection of initiators include acidification, fragrance, flavor, stability, and texture [82]. LABs can be classified in several ways, depending on their shape, growth temperature, and functions [83].

According to the fermentation of lactose, LABs are classified as homofermentative (produce only lactic acid) and heterofermentative (produce lactic acid and other metabolites) and according to growth temperature in mesophiles and thermophiles [84]. The homofermentative group composed of *Lactococcus, Pediococcus, Enterococcus,* and *Streptococcus,* use the Embden-Meyerhoff-Parnas pathway to convert 1 mole of glucose to two moles of lactic acid [69], and more than 85% of lactic acid from glucose. In contrast, heterofermentative bacteria produce equimolar concentrations of lactate, CO_2, and ethanol from glucose using the hexose monophosphate or the pentose pathway, and thus only generating half the energy of the homofermentative group [69]. Lactic acid is the main product of LAB fermentation. The bacteria belonging to this group possess the enzymes aldolase and hexose isomerase, but lack of the phosphoketolase. Within this group is *Lactobacillus* of long rods, isolated or in short chains, thermophiles, very energetic acidifiers and with important caseinolytic activity. *Streptococcus,* with spherical shapes in chains, rapid acidification and with little caseinolytic activity [76].

The obligate heterolactic species use only the phosphoketolase dependent pathway to metabolize sugars, and in addition to lactic acid, they produce significant amounts of acetic acid and ethanol with the generation of carbon dioxide; D-galactose can be metabolized through the tagatose 6-phosphate or Leloir pathway [69]. These fermentative species metabolize hexoses through the Emden-Meyerhoff glycolytic pathway, but pentoses and some other substances are metabolized via phosphoketolase to produce lactic acid and other products (typically acetic acid and ethanol).

LAB members can be subdivided into two different groups based on their carbohydrate metabolism. In the heterolactic fermentation, *Lactobacillus* species are: *plantarum, ramnosus, coryneformis, curvatus, casei, paracasei, brevis, buchneri, fermentun, kefir, reuteri,* and *leuconostoc.* While, in the homolactic group the *Lactobacillus* species are: *acidophilus, helveticus, delbrueckii subsp delbrueckii, debrueckii subsp lactis, delbrueckii subsp bulgaricus, lactis,* and *thermophilus* [85].

The homofermentative LAB, as they do not possess pyruvate decarboxylase, transfer the hydrogen formed by the phosphatase-dehydrogenase action to pyruvic acid with the help of the nicotinamide-adenine-dinucleotide (NAD), and transform it into lactic acid. Initiative LABs produce intracellular enzymes (peptidases, lipases, and amino acid catabolism enzymes), which play an important role in the development of cheese flavor during ripening.

After that, the initial breakdown of caseins by renin, endogenous milk proteases and bacterial proteases of the cell wall, peptidases are able to degrade the resulting peptides into free amino acids [86]. These can be subsequently catabolized to volatile aroma components by various enzymatic routes. Also, esterases and lipases catalyze the hydrolysis of triglycerides from milk fat into free fatty acids (FFAs) that are later converted to aromatic components. LAB can also synthesize esters from alcohols and glycerides and are attributed the imparting of nutritional and therapeutic attributes to fermented dairy products, such as improved digestion and viability of milk constituents, inhibition of harmful bacteria from the gastrointestinal tract (GIT), suppression of cancer, alleviation of lactose intolerance and effect on hypocholesterolemia [87, 88].

8.5 FUNCTIONAL PROPERTIES OF FERMENTED MILK PRODUCTS

Currently, with an increase of human life expectancy and the exponential growth of medical costs, human society needs a new point of view to bring about changes through the development of new scientific knowledge, and technologies that result on important modifications to improve people's lives style [89]. This trend and advances in food science and technology will improve the physical structure and chemical composition of foods, creating functional foods that have superior attributes beyond nutritional properties [90].

Some functional foods have been associated with risk reduction and protection against hypertension, diabetes, cancer, osteoporosis, and heart disease [91]. These foods have potential for promotion of good health by maximizing the physiological functions of a person and do not to cure the disease [9, 92], since they contain one or more beneficial components such as prebiotics, probiotics, antioxidant polyphenols and sterols, carotenoids, and others [93, 94].

The role of live microorganisms in fermented milk products has gained considerable interest by producers and consumers. LAB are important in the conservation, nutritional, and therapeutic properties of food, which helps to make human life longer [95]. LAB is natural inhabitants of the GIT. During fermentation of lactose to lactic acid, the milk acidity increases and the conditions become unfavorable for the growth of other microorganisms. In addition to producing most lactic acid metabolites, fermentation also produces bacteriostatic compounds. It has been suggested that ingestion of LAB can counteract the effect of *Escherichia coli*, through possible mechanisms such as anti-*E. coli* metabolites, detoxification of enterotoxins, prevention of synthesis of toxic amines and adhesion to the intestine, thus preventing colonization by bacterial pathogens. The homofermentative LAB convert lactose almost entirely to lactic acid, this can help the removal of toxic or antinutritive factors, such as lactose and galactose, from fermented milk to prevent lactose intolerance and galactose accumulation. Galactose metabolism through Leloir pathway can be employed as a catabolic pathway for utilization of galactose as carbon and energy source. Also, contributes with structural units needed for the production of lipopolysaccharides, components of cell wall and exopolysaccharides. In the absence of an external source of galactose, the Leloir pathway is the unique mean to obtain this sugar through the interconversion of glucose to galactose [96].

The antimicrobial effects of LAB have been attributed to the production of antibacterial substances such as lactic acid, hydrogen peroxide, and bacteriocins [97, 98]. Some physiological benefits of functional foods are cholesterol reduction, vitamin E production, detoxification of carcinogens, increased nutrient viability, immunostimulation, treatment against diarrhea, reduction of incidence and number of colon tumors, reduction of toxic compounds, regularization of intestinal flow, improvement of absorption of Fe, Ca, and Mg, and reduction of intestinal pathogens and increase of tolerance to lactose [94].

Fermented foods are also a source of probiotics, which have been defined by experts as "living microorganisms that, when administered in adequate concentrations, confer benefits to host health." On the other hand, yeasts can also be used as probiotics; however, even with its wide distribution and importance for the production of food and beverages around the world, its probiotic potential has been ignored [99]. Those that are produced industrially (wine, cheese, beer, bread, and yogurt) are accessible to many people. However, there are also, the so-called "traditional fermented foods that are produced in smaller quantities, by certain social or ethnic groups" [100], which are also part of the gastronomic culture of these groups. Several authors point out these foods as an example of "biological ennoblement," due to their bioenrichment with essential nutrients during fermentation. They are a source of beneficial microorganisms, which have an important role in food preservation, palatability, and nutrient viability [101].

Probiotics are considered as functional foods [102]; the dairy sector is strongly linked to them; it is the most abundant functional food market with about 33% of the current market [103]. In recent years, *per capita* consumption of yogurt has increased drastically, because many consumers associate it with good health [104].

The concept of probiotic was introduced at the beginning of the 20th century. Fooks et al. [105] indicate that the word probiotic derives from two Greek words that mean "for life." Also, this term was used to describe a substance that stimulated microorganisms growth [65] or tissue extracts that promote microbial growth [106] but did not receive general acceptance. Parker [107] was the first author to use the probiotic word in the context of animal supplementation, and this was defined as organisms and substances that contribute to the balance of the intestinal microbiota. Fuller [108] defined probiotics as supplements containing live microorganisms that affect the host in a healthy balance of the intestinal microbiota.

Supplementation of infant formulas with *Bifidobacterium lactis* and *S. thermophilus* has been reported to protect infants against nosocomial diarrhea [36]. Pediatric beverages containing *B. animalis, L. acidophilus,* and *L. reuteri* have been reported to be used for the prevention of rotavirus diarrhea. Treatment with *Lactobacillus* and local systemic GG promoters of the rotavirus immune response may be necessary for protection against infection [43]. Probiotics have been reported to reduce diarrhea induced by antibiotics, including *B. longum, B. lactis, Lactobacillus* GG, *L. acidophilus* LA5, *Streptococcus faecium,* and the *Saccharomyces boulardii* yeast [56–59]. Several strains of *Lactobacillus* have been reported to inhibit *Salmonella tiphimorium* [51]. *Helicobacter pylori* is recognized as the major cause of gastritis. Inhibition of *H. pylori* NCTC 11637 by *L. casei* subsp. *Rhamnosus* and several strains of *L. acidophilus* and production of antihelicobacter factors by this LAB have been reported [42, 61].

For more than ten years, products known as Symbiotic containing *Bifidobacteria* and *Lactobacillus,* fortified with FOS (fructo-oligosaccharides), inulin (prebiotics) have emerged. Also, food supplements containing probiotic bacteria. The purpose of these additives is to confer to food products, beneficial effects, ensuring the correct balance in order to perform an excellent job of the intestinal microbiota, regulation of the intestinal immune system and strengthening of the intestinal barrier. In this context, the link among food, health, probiotics, and prebiotics has been the subject of numerous studies, demonstrating its therapeutic effectiveness in the intestinal and systemic tract. Parallel to this, the industry is taking advantage of research results, promoting its products in the market; suggesting that probiotics and prebiotics intake will benefit human health.

8.6 HEALTHY BENEFITS OF FERMENTED DAIRY PRODUCTS

Different scientific studies are reporting that probiotic microorganisms can improve health, however, it is through media, and marketing that human population is informed of these discoveries. An informed population of these benefits will increase the demand for probiotic foods. To meet the growing demand of fermented dairy products, the food industry has developed new probiotic products. Some specific LAB strains, such as *L. acidophilus, L. casei, B. longum, L. fermentum, L. rhamnosus, L. reuteri, L. crispatus, L. plantarum, B. animalis,* and *B. lactis* are in the market and produce good profits for various companies. In this same context, other probiotic beverages and yogurts have been developed, but in the academic and industrial sectors, new sources of improved products for the consumer in Brazil are offered. However, this market still has tremendous potential growth, since most of the products available are yogurts or milk fermented with *B. animalis* and *L. acidophilus.*

Different health benefits are attributed to ingestion of probiotic microorganisms contained in some foods, some of them have been scientifically tested, and others still require further studies in humans. Some of these benefits are: antimicrobial and antimutagenic activities [102], anticarcinogenic properties [66], antihypertensive properties [109], beneficial effects on mineral metabolism, especially considering bones stability [110], attenuation of symptoms of bowel diseases and Crohn's syndrome [66], reduction of allergic symptoms [111] and reduction of LDL-cholesterol levels [112]. Strains of *Lactobacillus* have also shown suppression of pathogenic microorganisms such as *Salmonella enteritidis, Escherichia coli, Shigella sonnei,* and *Serratia marcescens* [113]. In addition to the probiotic effects, some bacteria have been shown to be promoters of endogenous host defense mechanisms.

Numerous probiotic microorganisms (*Lactobacillus rhamnosus* GG, *L. reuteri, L. acidophilus,* bifidobacteria, and certain strains of the *L. casei* group) are used in foods, particularly for fermented milk products [114]. Different health benefits have been attributed to probiotic bacteria, among them: transient modulation of host gut microbiota and ability to interact with the immune system, through direct basic mechanisms or with autochthonous microbiota. These results have been supported by *in vivo* and *in vitro* experiments using conventional biological and molecular methods. Additionally, a limited number of well-controlled, selected trials have been reported [115]. Probiotic effects that have been well documented include: (1) Prevention and/or reduction of diseases duration induced by rotavirus or antibiotics associated with diarrhea, as well as alleviation of diseases due to lactose intolerance. (2) Reduction of concentration of cancer-promoting enzymes and/or putrefactive bacterial metabolites in the intestine. (3) Prevention and relief of nonspecific and irregular diseases of the GIT of healthy people. (4) Beneficial effects on anomalies of microbial origin, inflammation, and other disorders related to inflammatory diseases of the GIT, infection by

Helicobacter pylori or excessive bacterial growth [62]. (5) Normalization of bowel movements and stool consistency in subjects suffering from constipation or irritable bowel syndrome. (6) Prevention or alleviation of allergies and atopic diseases in infants. (7) Prevention of infections of the respiratory tract (cooling, influenza) and other infectious diseases, as well as the treatment of urogenital infections. However, there is still insufficient or very preliminary evidence regarding probiotics effect on prevention of cancer, the hypocholesterolemic effect, improvement of oral microbiota and prevention of caries or therapy of ischemic diseases of the heart or reduction of autoimmune diseases (arthritis) [116].

8.7 ANTI-INFLAMMATORY EFFECT OF PROBIOTIC AND SYMBIOTIC MICROORGANISMS CONTAINED IN MILK, ON CHRONIC INTESTINAL DISEASES

Probiotics and symbiotics are consumed in numerous and various forms, such as yogurt and fermented milk, cheese, and other fermented foods. The use of probiotics and symbiotic in preventive medicine to maintain healthy intestinal function is well documented. Additionally, both probiotics and symbiotic have been proposed as therapeutic agents for gastrointestinal disorders and other pathologies [117, 118] and are being used for the treatment of chronic diseases, due to their role on modulation of the immune system and anti-inflammatory responses. Also, effects *in vitro* of probiotics and symbiotic have been reported on chronic intestinal diseases, in studies with humans and animals, particularly in selected clinical trials [119]. The selected probiotics exhibited anti-inflammatory properties *in vitro*. The probiotic strains and supernatant free cells reduced the expression of pro-inflammatory cytokines, an action that is the principal intermediary between similar receptors [120]. Administration of probiotics improved clinical symptoms, histological alterations, and mucus production, in most of the animals evaluated, but some results suggest caution, because, during the administration of these agents, there may be a relapse of chronic inflammatory bowel diseases (IBDs), necrotizing enterocolitis (NEC), and malabsorption syndrome [121]. IBDs are a term used to describe four pathologies: ulcerative colitis (UC), Crohn's disease (CD), pouchitis, and microscopic colitis. These conditions are systemic disorders that affect the GIT and have frequent extraintestinal manifestations, in which the function of the epithelial barrier is a critical factor for the attack. Likewise, the native immunity of commensal enteric bacteria also plays an important role.

8.8 PREBIOTICS

This type of food has been defined as food ingredients, which beneficially affect the host's immunity by stimulating the growth, and/or activity of one or more bacterial species resident in the colon [122]. Currently, they have been described as short-chain carbohydrates, with a degree of polymerization between two and seven, and are usually non-digestible by enzymes from humans or animals. However, they can be digested by the natural metabolism of *Bifidobacteria* and *Lactobacillus* [123].

8.9 BIOACTIVE PEPTIDES

It has been reported that food proteins can also exert other functions *in vivo*, through their peptides with biological activity [124]. Milk is one of the primary sources of these peptides, which can be released from the protein sequence by the action of digestive enzymes during intestinal transit, or during food processing, such as milk fermentation or maturation of cheese [125]. Once released, these peptides can exert a physiological effect on the organism [126]. LAB has the ability to hydrolyze milk proteins through their proteolytic system.

Currently, food proteins are investigated not only from a nutritional or functional point of view but as a raw material for obtaining peptides [127], since any source of food proteins is capable of providing functional peptides. However, it must be critical regarding its use, although in some cases its functionality is not something new, since the existence of proteins in breast milk which contributes to maturation and function of the immune system in the newborn, is a fact known for more than 50 years [128, 129].

Bioactive peptides are defined as inactive amino acid sequences within the precursor protein, which exert certain biological activities once, released using chemical or enzymatic hydrolysis. They are small-sized peptides, from 3 to 20 amino acids, which are released during the industrial processing of food, or during gastrointestinal digestion [130, 131]. After being ingested, bioactive peptides can exert their effect on the cardiovascular, digestive, immune, and nervous systems [132]. It has been proven that these peptides can cross the intestinal epithelium and reach peripheral tissues via systemic circulation, being able to exert specific functions locally, in the GIT, and at a systemic level. Therefore, bioactive peptides could influence cellular metabolism and act as vasoregulator, growth factors, hormonal inducers, and neurotransmitters [133].

8.9.1   *BIOACTIVE PEPTIDES IN FOOD*

The bioactive peptides are a small sequence of amino acids encrypted in proteins, so due to the wide diversity of protein foods existing in nature and the market, their intake is ensured with a balanced diet. However, their bioavailability is not so clear, since they have to be released from the proteins in which they are, after suffering the action of gastric and intestinal proteases and must be able to cross the intestinal epithelium and reach the peripheral tissues through of blood circulation to be able to exert its action [134, 135]. Due to the relevance that bioactive peptides have obtained in the market, different techniques have been developed for obtaining new bioactive peptides from food proteins by enzymatic digestion *in vitro*, using proteolytic enzymes of microbial origin [136]. Also, modified peptides, designed from natural peptides, have been obtained in order to increase the activity of the latter. In this way, proteins of different origin (animal and vegetal) have been used for isolation of peptides from enzymatic hydrolysates [134, 137]. As for proteins of animal origin, milk, and other dairy products are precursors of the most studied bioactive peptides, although these peptides have also been identified in egg ovalbumin, meat, fish muscle (sardine, tuna, bonito) [138], and royal jelly [139].

8.9.2   *BENEFICIAL PROPERTIES OF BIOACTIVE PEPTIDES*

Among the biological activities of the peptides are included: the inhibitory activity of the angiotensin-converting enzyme (ACE), the antithrombotic, the antimicrobial, the antioxidant, the immunomodulatory, the antihypercholesterolemic, the opioid, and activity on the cardiovascular system, among others [124]. Bioactive peptides can produce local effects on GIT or they can be absorbed, entering the bloodstream and reaching intact the organ where they exert their effect. The most studied are those that exert an antihypertensive effect through inhibition of the ACE [140].

8.10   BACTERIOCINS

Bacteriocins are antimicrobial peptides produced by a large number of bacteria, including those of the LAB group. They usually act against unwanted microorganisms, closely related or responsible for the deterioration of food and causing diseases. For this reason, they are used in several applications, such as biopreservation, the extension of life span, clinical anti-microbial action and for control of fermentation [141]. The term "bacterio-cins" was first proposed by Jacob et al. [142] to refer to protein substances with antimicrobial activity of bacterial origin; then, Tagg et al. [143] defined them as "a group of antimicrobial substances of bacterial origin, character-ized by having a biologically active protein component and by exerting a bactericidal mode of action" [144].

The bacteriocins comprise a large and diverse group of proteins or anti-microbial peptides which are synthesized in the ribosomes. They are classi-fied according to their chemical characteristics and activity spectra into four classes. The class I peptides have a size between 2 and 6 kDa and are modi-fied post-translationally, by dehydrating serine and threonine and the subse-quent addition of cysteine, to form the amino acids lanthionine and methyl-lanthionine. Class II corresponds to unmodified post-translational peptides; they are thermostable and have a relatively limited range of activity. Class III contains proteins with a weight greater than 30 kDa and sensitive to heat. Class IV is composed of a heterogeneous group of proteins that have addi-tions of carbohydrates and lipids [71, 145].

Peptides with post-translational modifications have a bactericidal or bacteriostatic effect on other bacteria, either of the same species (narrow spectrum) or other genera (broad-spectrum) [141]. The cell synthesizes a molecule that immunizes it against its Bacteriocin. The production occurs naturally during the logarithmic phase of bacterial development or at the end of it, keeping a direct relationship with the biomass produced [146]. Several bacterial species produce these peptides, those that come from the LAB, are of particular interest for the food industry since they have the status of QPS (qualified presumption of safety). This quality is attributed to peptides produces by LAB, because these microorganisms are considered as safe for health since its metabolites have been consumed in foods fermented by countless generations without any adverse effects on the population [147].

The most relevant Bacteriocins are:

1. **Nisin:** It is produced by strains of *Lactococcus lactis* subsp. Lactis. It was the first bacteriocin isolated from this strain and was discovered by Rogers in 1928, who observed that, during maturation of some cheeses, certain strains of *Lactococcus lactis* inhibited the growth of other pathogenic lactic bacteria, and that this too was not harmful to health [148]. The action of nisin in sensitive bacteria is carried out in the cytoplasmic membrane. Nisin forms pores that affect the motive power of protons and pH balance causing loss of ions and

hydrolysis of ATP, resulting in cell death. This natural preservative is used mainly to: prolong the shelf life of various pasteurized dairy products, prevent alteration of canned by thermophilic microorganisms, and reduce the intensity of heat treatment of canned foods. In addition, it could contribute to the substitution of nitrites as an anti-botulinic agent and inhibit *Listeria* sp., since it can control alteration of alcoholic beverages, participate on the preservation of low pH foods, in bakery products with high humidity, as well as inhibit the growth of *Clostridium* sp. in cheeses [149].

2. **Pediocin:** It is produced by strains *Pediococcus acidilactici, P. parvulus, P. pentosaceus, P. damnosus*, and a strain of *Lactobacillus plantarum* isolated from cheese. It is active against a narrow range of Gram-positive bacteria such as some species of the *Lactococcus, Lactobacillus*, and *Enterococcus* genera. It also has activity against other Gram-positive bacteria farther phylogenetically, among which are responsible for some food toxin-infections such as *Bacillus, Brochotrix, Listeria* and *Staphylococcus*. [150]. Concerning Gram-negative bacteria, their cell covers prevent access of bacteriocins to the plasma membrane, however, as with nisin, it is sufficient to alter the permeability of their outer membranes to allow the antimicrobial action of these bacteriocins [151]. It is used as a preservative in vegetal and meat products, and because of its high activity against *Listeria* species, it has a high potential to be used as a preservative in dairy foods [152]. For example, the company Danisco formulated a lyophilized culture of *P. acidilactici*, called CHOOZIT®, which is suggested as an adjunct for cheddar cheese and semi-soft cheeses [153].

3. **Plantaricin:** They are bacteriocins produced by different *Lactobacillus plantarum* strains [154, 155], belonging to classes I and II. Its activity depends on the synergistic action of two different plantaricin in peptides that act in equal amounts, such as plantaricin E with plantaricin F, and plantaricin J with Plantaricin K, which leads to an increase on its action [156, 157]. These bacteriocins show relatively narrow inhibition spectra, being mostly active against *Lactobacillus* species (for example, *L. plantarum, L. casei, L. sakei, L. curvatus*), *Listeria monocytogenes* [154, 158, 159] and other Gram-positive bacteria closely related to *L. plantarum* (*P. pentosaceus* and *P. acidilactici*) [160]. Plantaricins occur in fermentations in which *Lactobacillus plantarum* is used, such as fermentation of olives for extraction of olive oil (OO) [155], wine production [161], fermentation of cereals and legumes [162], among others.

8.11 CONCLUSIONS

There is a worldwide spectrum of fermented milk products, including different types of fermented milk, cheeses, etc. Different types of fermented milk have been developed in different countries, which are mostly elaborated as traditional products at a handicraft level, especially in developing countries, since they can be more affordable and are considered as "identitary" products. On the other hand, in developed countries, the products are manufactured at an industrial level, through standardized processes and with a high-quality final product from the microbiological and sensory point of view, but at a higher cost. It is observed in most of the countries that the interest in the intake of fermented products is increasing. Several reasons can be considered, including: a higher level of consumer awareness, high costs in traditional medicine and the capacity and characteristics of dairy products to serve as vehicles that transport functional components, thus giving it a health potential to the product.

ACKNOWLEDGMENTS

BEGC expresses recognition to the National Council of Science and Technology of Mexico (CONACYT) for the financial support during her postgraduate studies under the scholarship agreement number: 6075.17. Financial support was received by The National Technologic of Mexico through the project: "Jocoque un producto lácteo fermentadotradicional y sus características microbiológicas, fisicoquímicas y sensoriales" 621938.

KEYWORDS

- carbon dioxide
- fermented dairy products
- functional foods
- Jocoque
- lactic acid bacteria
- probiotics

REFERENCES

1. Gadaga, T. H., Mutukumira, A. N., Narvhus, J. A., & Feresu, S. B. A review of traditional fermented foods and beverages of Zimbabwe. *Int. J. Food Microbiol*, **1999**, *53*(1), 1–11.
2. Leroy, F., & De Vuyst, L. Lactic acid bacteria as functional starter cultures for the food fermentation industry. *Trends Food Sci. Technol.*, **2004**, *15*(2), 67–78.
3. Figueroa-González, I., Hernández-Sánchez, H., Rodríguez-Serrano, G., Gómez-Ruiz, L., García-Garibay, M., & Cruz-Guerrero, A. Antimicrobial effect of *Lactobacillus casei* Shirota variety co-cultivated with *Escherichia coli* UAM0403. *Rev Mex Ing. Quim.*, **2010**, *9*(1), 11–16.
4. Juárez, I. M. *The Fermented Milks* (pp. 31–35). The white book of dairy products, **2008**.
5. Mark-Herbert, C. Innovation of a new product category – functional foods. *Technovation*, **2004**, *24*(9), 713–719.
6. Koletzko, B., Aggett, P. J., Bindels, J. G., Bung, P., Ferre, P., Gil, A., Lentze, M. J., Roberfroid, M., & Strobel, S. Growth, development and differentiation: A functional food science approach. *Brit. J. Nut.*, **1998**, *80*(S1), S5–S45.
7. Wacher-Rodarte, C. The antique food biotechnology: The fermented foods. *Rev. Dig. Univ.*, **2014**, *15*(8), 64.
8. Romero-Luna, H. E., Hernández-Sánchez, H., & Dávila-Ortiz, G. Traditional fermented beverages from Mexico as a potential probiotic source. *Annals Microbiol.*, **2017**, *67*(9), pp. 577–586.
9. Roberfroid, M. B. Concepts and strategy of functional food science: The European perspective. *Am. J. Clin. Nutr.*, **2000**, *71*(6), 1660S–1664S.
10. Roberfroid, M. Functional food concept and its application to prebiotics. *Digestive Liver Dis.*, **2002**, *34*, S105–S110.
11. Food and Drug Administration (FDA). *Probiotics*, **2004**. Available from: http://www.webdietitians.org/Public/GovernmentAffairs/92_adap1099.cfm (accessed on 6 January 2020).
12. American Dietetic Association (ADA). Position of the American dietetic association: Functional foods. *J Am Diet Assoc.*, **2009**, *109*, 735–46.
13. García-Garibay, M. Fermented milks as vehicle for probiotics. *Arch Invest. Pediatr. Mex.*, **2000**, *2*, 327–341.
14. Lozano, M. O., & Villegas, D. G. A. Symbolic valorization of cream cheese from Chiapas, a Traditional Mexican cheese with quality. *PASOS. Rev Turismo Patrimonio Cultural*, **2016**, *14*(2), 459–473.
15. Ramírez-López, C., & Vélez-Ruiz, J. F. Isolation, characterization and selection of autochthonous lactic bacteria of milk and fresh artisanal cheese from goat. *Informacion Tecnologica*, **2016**, *27*(6), 115–128.
16. Walstra, P., Geurts, T. J., Noomen, A. C., Jellema, A. C., & Van Boekel, M. A. J. *Milk Science and Technology of Dairy Products (No. 637.1 W169c Ej. 1 019044)*. Acribia, **2001**.
17. Varnam, A. H., & Sutherland, J. P. *Beverages, Technology, Chemistry and Microbiology Acribia*. Zaragoza, **1997**.
18. Sangwang, S., Kumar, S., & Goyal, S. Maize utilization in food bioprocessing: An overview. In: Chaudhary, D. P., Kumar, S., & Langyan, S., (eds.), *Maize: Nutrition Dynamics and Novel Uses* (pp. 119–134). Springer India, New Delhi, **2014**.

19. Sanni, A., Franz, C., Schillinger, U., Huch, M., Guigas, C., & Holzapfel, W. Characterization and technological properties of lactic acid bacteria in the production of "sorghurt," a cereal-based product. *Food Biotechnol.,* **2013**, *27*(2), 178–198.

20. Chilton, S., Burton, J. P., & Reid, G. Inclusion of fermented foods in food guides around the world. *Nutrients,* **2015**, *7*, 390–404.

21. Holzapfel, W. H. Appropriate starter culture technologies for small-scale fermentation in developing countries. *Int. J. Food Microbiol.,* **2002**, *75*, 197–212.

22. Walshe, M. J., Grindle, J., Nell, A., & Bachmann, M. Dairy development in sub-Saharan Africa: A study of issues and options. *World Bank Technical Paper No. 135*, Africa Technical Department Series, The World Bank, Washington, DC ISBN 0-8213-1781-4, **1991**.

23. Kebede, A., Viljoen, B. C., Gadaga, T. H., Narvhus, J. A., & Lourens-Hattingh, A. The effect of container type on the growth of yeast and lactic acid bacteria during production of Sethemi, South African spontaneously fermented milk. *Food Res. Int.,* **2007**, *40*(1), 33–38.

24. Abdelgadir, W. S., Ahmed, T. K., & Dirar, H. A. The traditional fermented milk products of the Sudan. *Int. J. Food Microbiol.,* **1988**, *44*(1/2), 1–13.

25. El-Gendy, S. M. Fermented foods of Egypt and the Middle East. *J Food Protect,* **1983**, *46*, 358–367.

26. FAO. *The Technology of Traditional Milk Products in Developing Countries*. FAO Animal production and health paper (85) 92-5-102899-0. Rome: Food and Agriculture Organization of the United Nations, **1990**.

27. Gran, H. M., Wetlesen, A., Mutukumira, A. N., & Narvhus, J. A. Smallholder dairy processing in Zimbabwe: The production of fermented milk products with particular emphasis on sanitation and microbiological quality. *Food Control,* **2002**, *13*, 161–168.

28. Mutukumira, A. N. Properties of amasi, a natural fermented milk produced by smallholder milk producers in Zimbabwe. *Milchwissenschaft,* **1995**, *50*, 201–205.

29. Odunfa, S. A., & Oyewole, O. B. African fermented foods. In: Wood, B. J. B., (ed.), *Microbiology of Fermented Foods* (2nd edn., pp. 713–752) Blackie Academic and Professional, **1998**.

30. Ganguly, B. K., Bandopadhyay, P., & Kumar, S. Processed milk products. In: Falvey, & Chantalakhana, (eds.), *Smallholder Dairying in the Tropics ILRI* (pp. 462–481). International. Livestock Research Institute, Nairobi, Kenya, **1999**.

31. Osvik, R. D., Sperstad, S., Breines, E. M., Godfroid, J., Zhou, Z., Ren, P., Geoghegran, C., Holzapfel, W., & Ringø, E. Bacterial diversity of amasi, a South African fermented milk product, determined by clone library and denaturing gradient gel electrophoresis analysis. *African J. Microbiol. Res.,* **2013**, *7*, 4146–4158.

32. Leite, A. M., Miguel, M. A. L., Peixoto, R. S., Ruas-Madiedo, P., Paschoalin, V. M. F., Mayo, B., & Delgado, S. Probiotic potential of selected lactic acid bacteria strains isolated from Brazilian kefir grains. *J. Dairy Sci.,* **2015**, *98*(6), 3622–3632.

33. Watanabe, K., Fujimoto, J., Sasamoto, M., Dugersuren, J., Tumursuh, T., & Demberel, S. Diversity of lactic acid bacteria and yeasts in Airag and Tarag, traditional fermented milk products of Mongolia. *World J. Microbiol. Biotechnol.,* **2008**, *24*, 1313–1325.

34. Watanabe, K., Fujimoto, J., Tomii, Y., Sasamoto, M., Makino, H., Kudo, Y., & Okada, S. *Lactobacillus kisonensis* sp. nov., *Lactobacillus otakiensis* sp. nov., *Lactobacillus rapi* sp. nov. and *Lactobacillus sunkii* sp. nov., heterofermentative species isolated from sunki, a traditional Japanese pickle *Int. J. Syst. Evol. Microbiol.,* **2009**, *59*, 754–760.

35. Yu, J., Wang, W. H., Menghe, B. L. G., Jiri, M. T., Wang, H. M., Liu, W. J., Bao, Q. H., Lu, Q., Zhang, J. C., Wang, F., & Xu, H. Y. Diversity of lactic acid bacteria associated with traditional fermented dairy products in Mongolia *J. Dairy Sci.,* **2011,** *94,* 3229–3241.

36. Saavedra, J. M., Bauman, N. A., Oung, I., Perman, J. A., & Yolken, R. H. Feeding of *Bifidobacterium bifidum* and *Streptococcus thermophilus* to infants in hospital for prevention of diarrhea and shedding of rotavirus. *Lancet,* **1994,** *344,* 1046–1049.

37. Dewan, S., & Tamang, J. P. Dominant lactic bacteria and their technological properties isolated from the Himalayan ethnic fermented milk products. *A Van Leeuw,* **2007,** *92,* 343–352.

38. Syal, P., & Vohra, A. Probiotic potential of yeasts isolated from traditional Indian fermented foods. *Int. J. Microbiol. Res.,* **2013,** *5*(2), 390–398.

39. Hosono, A., Wardoyo, R., & Otani, H. Microbial flora in "dadih," a traditional fermented milk in Indonesia. *Lebensm Wiss Technol.,* **1989,** *22,* 20–24.

40. Ghosh, J., & Rajorhia, G. S. Selection of starter culture for production of indigenous fermented milk product (*Misti dahi*). *Lait.,* **1990,** *70,* 147–154.

41. Gupta, R. C., Bimlesh, M., Joshi, V. K., & Prasad, D. N. Microbiological, chemical and ultrastructural characteristics of Mishti Doi (Sweetened Dahi). *J. Food Sci. Technol.,* **2000,** *37,* 54–57.

42. Kawamura, T., Ohnuki, K., & Ichida, H. A clinical study on a *Lactobacillus casei* preparation (LBG-01) in patients with chronic irregular bowel movement and abdominal discomfort. *Jpn. Pharmacol. Ther.,* **1981,** *9,* 4361–4370.

43. Kalia, M., Isolauri, E., Soppi, E., Virtanen, E., Laine, S., & Arvilimmi, H. Enhancement of circulating antibody secreting cell response in human diarrhea by a human *lactobacillus* strain. *Ped. Res.,* **1992,** *32,* 141–144.

44. Gonfa, A., Fite, A., Urga, K., & Gashe, B. A. Microbiological aspects of Ergo (Ititu) fermentation. *SINET: Ethiop. J. Sci.,* **1999,** *22,* 283–290.

45. Almaze, G., Foster, H. A., & Holzapfel, W. H. Field survey and literature review on traditional fermented milk products of Ethiopia. *Int. J. Food Microbiol.,* **2001,** *68*(3), 173–186.

46. Bernardeau, M., Guguen, M., & Vernoux, J. Beneficial lactobacilli in food and feed: Long-term use, biodiversity, and proposals for specific and realistic safety assessments. *FEMS Microbiol. Rev.,* **2006,** *30,* 487–513.

47. Li, Q., Liu, X., Dong, M., Zhou, J., & Wang, Y. Adhesion and probiotic properties of *Lactobacillus plantarum* isolated from Chinese traditional fermented soybean paste. *Glob Adv Res J. Food Sci. Technol.,* **2015,** *4*(1), 001–009.

48. Wu, R., Wang, L., Wang, J., Li, H., Menghe, B., Wu, J., Guo, M., & Zhang, H. Isolation and preliminary probiotic selection of lactobacilli from koumiss in Inner Mongolia *J. Basic Microbiol.,* **2009,** *49,* 318–326.

49. Hao, Y., Zhao, L., Zhang, H., & Zhai, Z. Identification of the bacterial biodiversity in koumiss by denaturing gradient gel electrophoresis and species-specific polymerase chain reaction. *J. Dairy Sci.,* **2010,** *93,* 1926–1933.

50. Ben Slama, R., Kouidhi, B., Zmantar, T., Chaieb, K., & Bakhrouf, A. Anti-listerial and anti-biofilm activities of potential probiotic lactobacillus strains isolated from Tunisian traditional fermented food. *J. Food Safety,* **2013,** *33*(1), 8–16.

51. Hitchins, A. D., Wells, P., McDonough, F. E., & Wong, N. P. Amelioration of the adverse effect of a gastrointestinal challenge with *Salmonella enteritidis* on weanling rats by a yogurt diet. *Am. J. Clin. Nutr.,* **1985,** 41(1), 92–100.

52. Rettger, L. F., & Cheplin, H. A. A. Treatise on the transformation of the intestinal flora: With special reference to the implantation of *Bacillus Acidophilus* (Vol. 13). Yale Univ. Press, **1921**.

53. Obadina, A. O., Akinola, O. J., Shittu, T. A., & Bakare, H. A. Effect of natural fermentation on the chemical and nutritional composition of fermented soymilk nono. *Nigerian Food J.,* **2013**, *31*(2), 91–97.

54. Akabanda, F., Owusu-Kwarteng, J., Tano-Debrah, K., Glover R. L., Nielsen K., & Jespersen L. Taxonomic and molecular characterization of lactic acid bacteria and yeasts in *nunu*, a Ghanaian fermented milk product. *Food Microbiol.,* **2013**, *34*, 277–283.

55. Alexandraki, V., Tsakalidou, E., Papadimitriou, K., & Holzapfel, W. H. *Status and Trends of the Conservation and Sustainable Use of Microorganisms in Food Processes.* Commission on genetic resources for food and agriculture. *FAO* Background Study Paper No. 65, **2013**.

56. Borgia, M., Sepe, N., Brancato, V., & Borgia, R. A. A controlled clinical study on *Streptococcus faecum* preparation for the prevention of side reactions during long term antibiotic therapy. *Curr. Ther.,* **1982**, *31*, 265–271.

57. Colombel, J. F., Cortot, A., Neut, C., & Romond, C. Yoghurt with *Bifidobacterium longum* reduces erythromycin induced gastrointestinal effects. *Lancet,* **1987**, *2*, 43.

58. Surawicz, C. M., Elmer, G. W., Speelman, P., Mcfaraland, L., Chinn, J., & Belle, G. Prevention of antibiotic associated diarrhea by *Saccharomyces boulardii*: A perspective study. *Gastroenterol.,* **1989**, *84*, 2072–2078.

59. Nord, C. E., Lidbeck, A, Orrhage, K., & Sjstedt, S. Oral supplementation with lactic acid bacteria during intake of clindamycin. *Clin. Microbiol. Infect.,* **1997**, *3*, 124–132.

60. Kahala, M., Mäki, M., Lehtovaara, A., Tapanainen, J. M., Katiska, R., Juuruskorpi, M., Juhola, J., & Joutsjoki, V. Characterization of starter lactic acid bacteria from the Finnish fermented milk product viili. *J. Appl Microbiol.,* **2008**, *105*(6), 1929–1938.

61. Lambert, J., & Hull, R. Upper gastrointestinal tract disease and probiotics. *Asia Pacific J. Clin. Nutr.,* **1996**, *5*, 31–35.

62. Mukai, T., Asasaka, T., Sato, E., Mori, K., Matsumoto, M., & Ohori, H. Inhibition of binding of *Helicobacter pylori* to the glycolipid receptors by probiotic *Lactobacillus reuteri. FEMS Immunol Med Microbiol.,* **2002**, *32*, 105–110.

63. Tamime A. Y., & Robinson R. K. *Yoghurt Science and Technology.* Cambridge: Woodhead Publishing Ltd., **2007**.

64. Angelakis, E., Million, M., Henry, M., & Raoult D. Rapid and accurate bacterial identification in probiotics and yoghurts by MALDI-TOF mass spectrometry. *J. Food Sci.,* **2011**, *76*, M568–M572.

65. Lilley, D. M., & Stillwell, R. H. Probiotics: Growth promoting factors produced by microorganisms. *Science,* **1965**, *147*, 747–748.

66. Marteau P. R., Vrese, M., Cellier C. J., & Schrezenmeir, J. Protection from gastrointestinal diseases with the use of probiotics. *Am. J. Clin. Nutr.,* **2001**, *73*, 430–436.

67. Guandalini, S., Pensabene, L., & Zikri, M. A. *Lactobacillus* GG administered in oral rehydration solution to children with acute diarrhea: A multicenter European trial. *J. Pediatr. Gastroenterol Nutr.,* **2000**, *30*, 54–60.

68. Axelsson, L. Lactic acid bacteria: Classification and physiology. In: Salminen, S., Von Wright, A., & Marcel, D., (eds.), *Lactic Acid Bacteria*, (pp. 1–63). New York: USA, **2002**.

69. Almanza, F., & Barrera, E. *Technology of Milk and Its Derivatives.* Bogotá Unisur, **1991**.

70. Hugenholtz, J. Review The lactic acid bacterium as a cell factory for food ingredient production. *Int. Dairy J.*, **2008**, *18*, 466–475.
71. Savadogo, A., Quatara, A. T., Bassole, H. N., & Traore S. A. Bacteriocins and lactic acid bacteria a mini review. *African J. Biotechnol.*, **2006**, *5*(9), 678–683.
72. Podolak P. K., Zayas, J. F., Kastner, C. L., & Fung, D. Y. C. Inhibition of *Listeria monocytogenes* and *Escherichia coli* O157:H7 on beef by application of organic acids. *J. Food Prot.*, **1996**, *59*(4), 370–373.
73. Smulders, F. J. M., Barendsen, P., Van Logstestijn, J. G., Mossel, D. A. A., & Van Der Marel, G. M. Lactic acid: Considerations in favor of its acceptance as a meat decontaminant. *J. Food Technol.*, **1986**, *21*, 419–436.
74. Guerra, N., Rúa, M., & Pastrana, L. Nutritional factors affecting the production of two bacteriocins from lactic acid bacteria on whey, *Int. J. Food Microbiol.*, **2001**, *70*, 267–281.
75. Ekinci, F., & Gurel M. Effect of using propionic acid bacteria as an adjunct culture in yogurt production. *J. Dairy Sci.*, **2007**, *91*, 892–899.
76. Hernández. B. Preparation of a whey-bases probiotic product with *Lactobacillus reuteri* and *Bifidobacterium bifidum*, *J. Food Technol. Biotechnol.*, **2007**, *45*, 27–31.
77. Hill, C., Okeeffe, T., & Ross P. Antimicrobial factors produced by lactic acid bacteria. *Encyclopedia Food Sci. Nutr.*, **2002**, *14*, 273–285.
78. Duboc, P., & Mollet, B. Applications of exopolysaccharides in the dairy industry. *Int. Dairy J.*, **2001**, *11*, 759–768.
79. Bouzar, F., Cerning, J., & Desmazeaud, M. Exopolysaccharide production and texture-promoting abilities of mixed-strain starter cultures in yogurt production. *J. Dairy Sci.*, **1997**, *80*, 2310–2317.
80. Galvez, H. Bacteriocin based strategies for food biopreservation. *Int. J. Food Microbiol.*, **2007**, *120*, 51–70.
81. Jagnow, G., & Wolfang, D. In: Zaragoza, (ed.), *Introduction with Model Experiments*. Acribia, **1991**.
82. Laws, Y., Gu, Y., & Marshall, V. Biosynthesis, characterization, and design of bacterial exopolysaccharides from lactic acid bacteria. *Biotechnol. Adv.*, **2001**, *19*, 597–625.
83. Neira, E., & López, J. Technical guide for dairy products elaboration Bogotá-Colombia: *Litografía Enzas Ltda*, **2001**.
84. Bertrand, C. Evolution of Lactoglobulin and D-lactoalbumin content during yogurt fermentation, *Int. Dairy J.*, **2003**, *13*, 39–45.
85. Blanco, S., Delahaye, P., & Fragenas, N. Physical and nutritional evaluation of a yogurt with tropical fruits and low in calories. *Revi. Fac. Agron. (Maracay) Venezuela*, **2006**, *32*, 131–144.
86. Veisseyre, R. *Technical Dairy Products* (2nd edn.). Editorial Acribia. Zaragoza, España, **1990**.
87. Mathur, S., & Singh, R. Antibiotic resistance in food lactic acid bacteria-a review. *Int. J. Food Microbiol.*, **2005**, *105*, 281–295.
88. Pancsar, P. S. Fermented dairy products: Starter cultures and potential nutritional benefits. *Food Nutr. Sci.*, **2011**, *2*, 47–51.
89. Kwak, N., & Jukes, D. J. Functional foods. Part 1: The development of a regulatory concept. *Food Control*, **2001**, *12*, 99–107.
90. Behrens, J. H., Roig, S. M., & Da Silva, M. A. A. P. Functional aspects, regulations of oil extracted from hydrolyzed fermented soybean and probiotic dairy products. *Cien. Tec. Alim.*, **2001**, *34*, 99–106.

91. Arihara, K., Nakashima, Y., Ishikawa, S., & Itoh, M. Antihypertensive activities generated from porcine skeletal muscle proteins by lactic acid bacteria [abstract]. In: *Abstracts of 50ᵗʰ International Congress of Meat Science and Technology* (p. 236). Helsinki, Finland, Elsevier Ltd., **2004**.

92. Sanders, M. E. Overview of functional foods: Emphasis on probiotic bacteria. *Int. Dairy J.,* **1998**, *8,* 341–347.

93. Andlauer, W., & Fűrst, P. Nutraceuticals: A piece of history, present status, and outlook. *Food Res. Int.,* **2002**, *35,* 171–176.

94. Granato, D., Branco G. F., Cruz A. G., Faria, J., & Shah N. P. Probiotic dairy products as functional foods. *Compr. Rev. Food Sci. F.,* **2010**, *9*(5), 455–470.

95. Ram, C., & Bhavadasan, M. K. Probiotic dairy foods-present status and future perspectives. *Indian Dairy Man.,* **2002**, *54,* 53–57.

96. Frey, P. A. The Leloir pathway: A mechanistic imperative for three enzymes to change the stereo-chemical configuration of a single carbon in galactose. *FASEB J.,* **1996**, *10,* 461–470.

97. Kodama, R. Studies on lactic acid bacteria: Lactolin a new antibiotic substance produced by lactic acid bacteria. *J. Antibiot,* **1952**, *5,* 72–74.

98. Kansal, V. K. Probiotic application of culture and culture containing milk products. *Indian Dairy Man.,* **2001**, *53,* 49–55.

99. Zhu, K., Hölzel, C. S., Cui, Y., Mayer, R., Wang, Y., Dietrich, R., Didier, A., Bassitta, R., Märtlbauer, E., & Ding, S. Probiotic *Bacillus cereus* strains, a potential risk for public health in China. *Frontiers Microbiol.,* **2016**, *7,* 718.

100. Olivares-Illana, V., Wacher-Rodarte, C., Le Borgne, S., & López-Mungía, A. Characterization of a novel cell-associated levansucrase from a *Leuconostoc citreum* strain isolated from pozol, a fermented corn beverage of Mayan origin, *J. Ind. Microbiol. Biotechnoly,* **2002**, *28,* 112–117.

101. Champagne, C. P., Gardner, N. J., & Roy, D. Challenges in the addition of probiotics cultures to foods. *Crit. Rev. Food Sci.,* **2005**, *45*(1), 61–84.

102. Lourens-Hattingh, A., & Viljoen, B. C. Yogurt as probiotic carrier food. *Int. Dairy J.,* **2001**, *11,* 1–17.

103. Leatherhead Food International. The international market for functional foods. In: *Functional Food Market Report.* London, U.K.: Leatherhead Food International Publication, **2006**.

104. Hekmat, S., Soltani, H., & Reid, G. Growth and survival of *Lactobacillus reuteri* RC-14 and *Lactobacillus rhamnosus* GR-1 in yogurt for use as a functional food. *Innov. Food Sci. Emerg. Technol.,* **2009**, *10,* 293–296.

105. Fooks, L. J., Fuller, R., & Gibson, G. R. Prebiotics, probiotics and human gut microbiology. *Int. Dairy J.,* **1999**, *9,* 53–61.

106. Sperti, G. S. *Probiotics.* West Point, CT: Avi Publishing Co., **1971**.

107. Parker, R. B. Probiotics, the other half of the antibiotic story. *Anim. Nutr. Health,* **1974**, *29,* 4–8.

108. Fuller, R. Probiotics in man and animals. *J. Appl. Bacteriol.,* **1989**, *66,* 365–378.

109. Liong, M. T., Fung, W. Y., Ewe, J. A., Kuan, C. Y., & Lye, H. S. The improvement of hypertension by probiotics: Effects on cholesterol, diabetes, renin, and phytoestrogens. *Int. J. Mol. Sci.,* **2009**, *10,* 3755–3775.

110. Arunachalam, K. D. Role of bifidobacteria in nutrition, medicine, and technology. *Nutr. Res.,* **1999**, *19,* 1559–97.

111. Salminen, S., Ouwehand, A. C., & Isolauri, E. Clinical applications of probiotic bacteria. *Int. Dairy J.,* **1998**, *8,* 563–572.
112. Sindhu, S. C., & Khetarpaul, N. Effect of feeding probiotic fermented indigenous food mixture on serum cholesterol levels in mice. *Nutr. Res.,* **2003**, *23,* 1071–1080.
113. Drago, L., Gismondo, M. R., Lombardi, A., Haen, C., & Gozzoni, L. Inhibition of entero-pathogens by new *Lactobacillus* isolates of human intestinal origin. *FEMS Microbiol. Letters,* **1997**, *153,* 455–463.
114. Vinderola, G. Probiotic bacteria in fermented dairy products. *Anales Academia Nacional de Ciencias. Ex., Fís. y Nat.,* **2014**, *66,* 5–21.
115. Martín, R., Langa, S., Reviriego, C., Jiménez, E., Marín, M. L., Olivares, M., et al. The comensal microflora of human milk: New perspectives for food bacteriotherapy and probiotics. *Trends Food Sci. Technol.,* **2004**, *15,* 121–127.
116. De Vrese, M., & Schrezenmeir, J. Probiotics, prebiotics andsymbiotic. *Adv. Biochem. Eng. Biotechnol.,* **2008**, *111,* 1–66.
117. Tojo, R., Suárez, A., Clemente M. G., De los Reyes-Gavilán C. G., Margolles, A., Gueimonde, M., & Ruas-Madiedo P. Intestinal microbiota in health and disease: Role of bifidobacteria in gut homeostasis. *World J. Gastroenterol.,* **2014**, *20,* 15163–15176.
118. Upadhyay N., & Moudgal V. Probiotics: A review. *J. Clin. Outcomes Manag.,* **2012**, *19*(2), 76–84.
119. Arribas, A. M. B. Probiotics: A new strategy in modulation of immune system. *Doctoral Thesis.* Universidad de Granada. Facultad de Farmacia. Departamento de Farmacología, **2009**.
120. Coppack, S. W. Pro-inflammatory cytokines and adipose tissue. *Proc. Nutr. Soc.,* **2001**, *60,* 349–356.
121. Plaza-Díaz, J., Ruiz-Ojeda, F., Vilchez-Padial, L., & Gil, A., Evidence of the anti-inflammatory effects of probiotics and synbiotics in intestinal chronic diseases. *Nutrients,* **2017**, *9*(6), 555.
122. Gibson, G. R., & Roberfroid, M. B. Dietary modulation of the human colonic microbiota: Introducing the concept of prebiotics. *J. Nutr.,* **1995**, *125,* 1401–1412.
123. Cummings, J. H., & Macfarlane, G. R. Gastrointestinal effects of prebiotics. *Brit, J, Nutr.,* **2002**, *87,* S145–151.
124. Korhonen, H., & Ve Philanto, A. Bioactive peptides: Production and functionality. *Int. Dairy J.,* **2009**, *16,* 945–960.
125. Hata, Y., Yamamoto, M., Ohni, M., Nakajima, K., Nakamura, Y., & Takano, T. A placebo-controlled study of the effect of sour milk on blood pressure in hypertensive subjects *Am. J. Clin. Nutr.,* **1996**, *64,* 767–771.
126. Foltz, M., Meynen, E. E., Bianco, V., van Platerink, C., Koning, T. M. M. G., & Kloek, J. Angiotensin converting enzyme inhibitory peptides from a lactotripeptide-enriched milk beverage are absorbed intact into the circulation. *J. Nutr.,* **2007**, *137,* 953–958.
127. Korhonen, H. Technology options for new nutritional concepts. *Int. J. Dairy Tech.,* **2002**, *55,* 79–88.
128. Devlieghere, F., Vermeirenl, J., & Debevere, J. New preservation technologies: Possibilities and limitations. *Int. Dairy J.,* **2004**, *14,* 273–285.
129. Lonnerdal, B. Nutritional and physiologic significance of human milk proteins. *Am. J. Clin. Nutr.,* **2003**, *77,* 1537S–1543S.
130. Pihlanto-Leppälä, A. Bioactive peptides derived from bovine whey proteins: Opioid and ace-inhibitory peptides. *Trends Food Sci. Technol.,* **2000**, *11,* 347–356.

131. Shahidi, A., & Zhong, B. Bioactive peptides. *Jour. Aoac. Int.*, **2008**, *91,* 914–931.
132. Fitzgerald, R. J., & Meisel, H. Milk protein derived inhibitors of angiotensin-I-converting enzyme. *Brit. J. Nutr.*, **2000**, *84,* S33–37.
133. Roberts, P. R., & Zaloga, G. P. Dietary bioactive peptides. *New Horizons,* **1994**, *2*(2), 237–243.
134. Korhonen, H., & Pihlanto, A. Food-derived bioactive peptides opportunities for designing future foods. *Curr. Pharm. Design,* **2003**, *9*(16), 1297–1308.
135. Vermeirssen, V., Van Camp, J., & Verstraete, W. Bioavailability of angiotensin I converting enzyme inhibitory peptides. *Br. J. Nutr.*, **2004**, *92,* 357–366.
136. Meisel, H. Multifunctional peptides encrypted in milk proteins. *Biofactors,* **2004**, *21,* 55–61.
137. Dziuba, J., Niklewicz, M., Iwaniak, A., Darewicz, M., & Minkiewicz, P. Structural properties of proteolytic-accessible bioactive fragments of selected animal proteins. *Polimery,* **2005**, *50*(6), 424–428.
138. Yamamoto, N., Ejiri, M., & Mizuno, S. Biogenic peptides and their potential use. *Curr. Pharm. Des.*, **2003**, *9,* 1345–1355.
139. Matsui, T., Yukiyoshi, A., Doi, S., Sugimoto, H., Yamada, H., & Matsumoto, K. Gastrointestinal enzyme production of bioactive peptides from royal jelly protein and their antihypertensive ability in SHR. *Nutr. Biochem.*, **2002**, *13,* 80–86.
140. Domínguez-González, K., Cruz-Guerrero, A., González-Márquez, H., Gómez-Ruiz, L. C., García-Garibay, M., & Rodríguez-Serrano. G. M. The antihypertensive effect of fermented milks. *Rev. Argent Microbiol.,* **2014**, *46*(1), 58–65.
141. Marcos, E., Castillo, F. A., Dimitrov S. T., Gombossy de Melo B. D., & De Souza R. P. Novel biotechnological applications of bacteriocins: A review. *Food Control,* **2013**, *32,* 134–142.
142. Jacob, F., Lwoff, A., Siminovitch, A., & Wollman, E. L. Definition of some terms related to lysogeny. *Ann. Inst. Pasteur*, **1953**, *84,* 222–224.
143. Tagg, J. R., Dajani, A. S., & Wannamaker, L. W. Bacteriocins of Gram positive bacteria. *Bacteriol. Rev.*, **1976**, *40,* 722–756.
144. Cristóbal, R. L. Lactobacilli producers of bacteriocins isolated from artisanal cheeses from Lima and provinces. *MSc Thesis.* Lima, Perú: Universidad Nacional Mayor de San Marcos, **2008**.
145. Jack, R. W., Tagg, J. R., & Ray, B. Bacteriocins of Gram-positive bacteria. *Microbiol. Rev.,* **1995**, *59,* 171–200.
146. Vázquez, S. M., Suárez, H., & Zapata, B. S. Utilization of antimicrobial substances produced by lactic acid bacteria during meat conservation. *Revista Chilena de Nutrición,* **2009**, *36*(1), 64–71.
147. Joerger, R. Alternatives to antibiotics: Bacteriocins, antimicrobial peptides and bacteriophages. *Poultry Sci.*, **2003**, *82*(4), 640–647.
148. Suarez, A. M. Antibody production front nisin: Strategies of immunization and development of immunoassays. *PhD Thesis.* Madrid, España: Universidad Complutense de Madrid, **1997**.
149. Martínez, B. Bacteriocins of *Lactococcus lactis* isolated from Asturian artisanal cheese: Nisin Zy Lactococin *972.* PhD Thesis. Oviedo, España: Universidad de Oviedo, **1996**.
150. Heredia-Castro, P. Y., Hernández-Mendoza, A., González Córdova, A. F., & Vallejo-Córdoba, B. Bacteriocins from lactic acid bacteria in action mechanisms and antimicrobial activity against pathogen in cheese. *Interciencia,* **2017**, *42*(7), 340–346.

151. Díez, L. Effects of enological agents and pediocin PA-1 against wine lactic bacteria. *PhD Thesis*. Logroño, España: Universidad de la Rioja, **2011**.

152. González, B. E., Gómez, M., & Jiménez, Z. Bacteriocins of probiotics. *RESPYN,* **2003**, *4*(2).

153. Papagianni, M., & Anastasiadou, S. Pediocins: The bacteriocins of pediococci. Sources, production, properties and applications. *Microb. Cell Fact,* **2009**, *8*(3), 1–16.

154. Chen, Y. S., Wang, Y. C., Chow, Y. S., Yanagida, F., Liao, C. C., & Chiu, C. M. Purification and characterization of plantaricin Y, a novel bacteriocin produced by *Lactobacillus plantarum* 510. *Arch Microbiol,* **2014**, *196*(3), 193–199.

155. Doulgeraki, A., Paraskevopoulos, N., Nychas, G., & Panagou, E. An *in vitro* study of *Lactobacillus plantarum* strains for the presence of plantaricin genes and their potential control of the table olive microbiota. *A Van Leeuw,* **2013**, *103*(4), 821–832.

156. Atrih, A., Rekhif, N., Moir, A. J. G., Lebrihi, A., & Lefebvre, G. Mode of action, purification and amino acid sequence of plantaricin C19, an anti-Listeria bacteriocin produced by *Lactobacillus plantarum* C19. *Int. J. Food Microbiol.,* **2001**, *68*(1/2), 93–104.

157. Soliman, W., Wang, L., Bhattacharjee, S., & Kaur, K. Structure-Activity relationships of an antimicrobial peptide plantaricin S from two-peptide class IIb bacteriocins. *J. Med. Chem.,* **2011**, *54*(7), 2399–2408.

158. Zhang, H., Liu, L., Hao, Y., Zhong, S., Liu, H., Han, T., & Xie, Y. Isolation and partial characterization of a bacteriocin produced by *Lactobacillus plantarum* BM-1 isolated from a traditionally fermented Chinese meat product. *Microbiol. Immunol.,* **2013**, *57*(11), 746–755.

159. Tiwari, S. K., & Srivastava, S. Purification and characterization of plantaricin LR14: A novel bacteriocin produced by *Lactobacillus plantarum* LR/14 *Appl. Microbiol. Biotechnol.,* **2008**, *79*(5), 759–767.

160. Diep, D., Straume, D., Kjos, M., Torres, C., & Nes, I. An overview of the mosaic bacteriocin pln loci from *Lactobacillus Plantarum*. *Peptides,* **2009**, *30*(8), 1562–1574.

161. Yanagida, F., Srionnual, S., & Chen, Y. Isolation and characteristics of lactic acid bacteria from koshu vineyards in Japan. *Lett. Appl. Microbiol.,* **2008**, *47*(2), 134–139.

162. Jama, Y., & Varadaraj, M. Antibacterial effect of plantaricin LP84 on foodborne pathogenic bacteria occurring as contaminants during idli batter fermentation. *World J. Microbiol. Biotechnol.,* **1999**, *15*, 27–32.

CHAPTER 9

Training in Food Safety Practices and Food Manufacturing for Safe Food Production

JOSÉ RAFAEL LINARES-MORALES,[1] ARELY PRADO-BARRAGÁN,[2] and GUADALUPE VIRGINIA NEVÁREZ-MOORILLÓN[1]

[1]*School of Chemical Sciences, Autonomous University of Chihuahua, 31125 Chihuahua, Chih, Mexico, Tel.: +52 614 236 6000, ext 4248, E-mail: vnevare@uach.mx (G. V. Nevárez-Moorillón)*

[2]*Autonomous Metropolitan University, Iztapalapa Campus, 09340 CDMX, Mexico*

ABSTRACT

The incidence of foodborne infections remains as one of the most critical problems for the food industry. The main focus on food preservation techniques is the increase in shelf life of foodstuff, as well as to produce safe and nutritious products. Food preservation technologies are developed for the reduction of the microbial load in food products, but as necessary as the processing methods, the control of hygienic conditions in the food industry will contribute to the production of safe food. The importance of training for all employees involved in food production in good manufacturing practices (GMP) will also contribute to safer food. This chapter will cover some important considerations for training in the food industry.

9.1 INTRODUCTION

The incidence of foodborne illness is a problem that has persisted for many years threatening public health, affecting millions of people all over the world and causing an enormous monetary loss. [1–3]. Among the causes of foodborne outbreaks is inadequate food handling practices, including cross-contamination, abuse of temperatures during cooking and storage, as well as unsuitable hygiene [1, 3, 4]. Food Safety training, imparts knowledge and abilities to those employees who direct or indirectly handle food, to enrich their skills [2]. Over the years, training has been applied to prevent inappropriate food handling practices. Nevertheless, it is not always possible to improve food handler's behavior through the communication of knowledge [1, 4].

9.2 TRAINING PROCESS

9.2.1 TRAINING DEFINITION

According to Uddin et al. [5] training is a plan consciously organized by an establishment to make the learning process more comfortable for employees, improving key job capabilities like knowledge and behavior, both crucial to achieve a production goal, on either quantity or quality.

Astegiano et al. [6] defined training as "*a planned process to modify attitude or skill behavior through learning experience to achieve effective performance in an activity or range of activities.*" Development of Human resources using a training program is a long-term approach, and its application needs an acute analysis and a suitable training method [5]. Training is a complex task that has been turning more difficult; it needs continuous replication and creativity to deliver the message. The company can organize it through expertise trainers or by hiring external companies in the area. Smaller establishments would get better results in hiring an external company because their experience with other customers can make it easier for them to find out vulnerabilities. The evaluation should comprise an exhaustive round through all the processes, a complete review of previous performances and knowledge assessments. Larger businesses need more personalized and frequent training; also, they could finance their own internal trainers, which would be more in contact with processes and food handlers and consequently, more capable of finding and correcting unconformities [7].

Well trained employees are a key factor when industries try to get or to maintain certification; thus, training becomes an obligation. Continuous training, besides being a mandatory requisite refreshes personnel on the effect of recall, defects, nonconformities, and more. There are some approaches to arrange training throughout the year. Training may be carried out due to a reduction on the scores during an evaluation, as part of induction training for a new employee or an employee that was changed to a new area or simply, can be done corresponding to a schedule. In all cases, the employee has a deficiency in ability, knowledge, or incentive that should be improved with the training course activities [8]. Soon et al. [9] pointed out that focalized training and communication of risks have to be applied in order to achieve an adjustment in the attitudes of food handlers towards a harmless food handling. Also, an upgrade in the knowledge of the food safety practices can be achieved, since those practices are the most frequent causes of foodborne illnesses, including cross-contamination, temperature control, and personal hygiene.

Adesokan et al. [1] mentioned that food safety training courses have to emphasize the needs of each food handler, instead of giving a complete course that encompasses all the training topics that probably are not needed; furthermore, it is probable that food handlers have different needs of knowledge and practices.

9.2.2   TRAINEE CHARACTERISTICS

Cognitive ability is a desirable quality for trainees. Even though it is an attribute that cannot be easily controlled, companies need to take this into account before defining which workers are to assist in a food safety course. However, reducing the cognitive load necessary for the course would help those employees to be included [10]. Education level is another desirable quality, Khanal and Poudel [11] pointed that education level is critical for a food handler, because to be able to read and comprehend a subject, makes it easier to uptake all that helpful information imparted in a food safety training programs.

9.2.3   TRAINING DESIGN

Design of a training course and the method of transmission affect information uptake. A food safety program should be adapted to the needs of the personnel of each facility. A general program will not succeed in all food processing plants. The success of the program will depend on how much knowledge is disseminated. Training sessions must be kept in short practical presentations, tailored for the actual job function of the employees, scheduled yearly, quarterly, monthly, and even weekly, and measuring objectives and results [10]. Adesokan et al. [1] recommend that an updating and brief time training (maximum of two weeks in a row) are very important characteristics of a suitable training course for improved food safety practices.

Moreover, refresher training should be carried out periodically as a preventive action against food safety failures and in order to enable food handlers to enhance their practices. Araújo et al. [12] developed and applied a checklist for good hygiene practices during the production of fresh fruit and vegetables. They demonstrated the lack of food safety knowledge and practices among food manipulators during the production of fresh fruit and vegetables. Food handlers do not consider that humans can be a threat to food safety since there is mandatory to disinfect fresh produce before eating them. They concluded that because of the lack of knowledge in food safety, it is fundamental periodic training.

Pre and post-evaluations are used to determine the efficiency of the course by tracing enhancement through the results [7]. Tyler et al. [2] evaluated variations in knowledge of participants in the Level One Food Safety Certification Program at Dayton and Montgomery Counties. Question on a survey was analyzed concerning the correctness of the answers using pre- and post-tests. Also, the relationship between the records and job responsibilities were determined. Results from 692 participants were collected and assessed with a paired t-test. A considerable improvement between both tests was observed. Questions related to temperature had higher wrong answers; nevertheless, these questions had the highest upgrading. Therefore, Level One Food Safety Certification course demonstrated improvement in the knowledge of applicants.

Khanal and Poudel [11] assessed the knowledge of meat handling and hygiene practices among slaughterers in Nepal. They interviewed the butchers using a semi-structured close-ended survey. A checklist was designed based on the standard guidelines for slaughterhouse and meat inspection regulations of the Codex Alimentarius Commission. Knowledge and practice levels were evaluated. Most of the participants were unconscious of good hygiene practices. A total of 63.2% of the respondents had a deficient level of knowledge on meat hygiene, and 52.6% demonstrated an unacceptable level of practice on meat hygiene. Authors suggest that among the reasons for these low qualifications are participants with a second job, a reduced education level, burden of work and facilities.

Pilling et al. [4] pointed out that variation in food safety knowledge and behavior among food handlers could be the result of a deficient course; a short session will not be enough to explain a complete food-handling course widely, and nonetheless, participants will receive their certification. On the other hand, food handlers and managers would consider the course as satisfactorily completed, and they would be complying with a requisite. Local health departments impart 2-hour session courses while private companies offer up to 8-hour programs.

9.2.4 CHARACTERISTICS OF TRAINING

Machado and Cutter [13] explain that a variation in knowledge is the initial stage in the model knowledge-attitude-behavior (KAB) since that variation may cause modifications in attitude and consequently modifications in behavior. However, Schrader, and Lawless [14] indicate that attitudes can influence a person's perception, with the consequent modification of his knowledge. Likewise, knowledge cannot automatically predict a behavior by itself. For those reasons, the three concepts have a dynamic and at times reciprocal correlation. Therefore, training programs focused only on knowledge issues usually are not effective. An appropriate way to manage food safety training programs is to integrate issues of knowledge, attitudes, and behavior *in situ.*

Behavioral modeling, error management, and realistic training environments are factors that should be considered for training design. Behavioral modeling let the trainees observe and practice determined behaviors enhancing their capability to absorb and maintain information. Allowing trainees to incur in mistakes during an in-place course session provide opportunities to deliver instructions based on those mistakes, improving the appropriate application of focused knowledge and abilities in the workplace. Managing errors during training have been more useful for post-training new assignments [10]. Food safety programs have been approached by improving knowledge. Nevertheless, it should be imparted towards the reasons to adopt improved behavior during food handling. Thus, food safety training should be restructured and based considering that a determined behavior has to be modified after an acute observation of the working process and in this way, employees will be more concerned with the consequences of improper actions during food handling [3]. Trainers have the mission of identifying which aspects are considered by food handlers as obstacles to developing their tasks, to comply with good hygiene practices. Thus, trainers can design the program fostering on how to eliminate those barriers, mostly directed to managers rather than food handlers, because of their responsibility [4]. Roberts et al. [3] determined the influence of training on food safety knowledge and behavior. They applied a knowledge test before training. This test comprised the conducts that mostly contribute to foodborne diseases like cross-contamination, temperature abuse, and personal hygiene.

Once the test was applied, the researchers monitored the workers during a working period with a checklist including those three main issues. Later, the workers assisted to a ServSafe training session and finally answered again the same knowledge test. This research group reported that an enhancement of knowledge is not a guaranty of an improvement of behavior since the participants obtained high grades on knowledge but low percentages on behaviors regarding food safety. Machado and Cutter [13] designed an adapted counter-top food safety and sanitation-training program for farmstead cheesemakers. Firstly, a test about food safety knowledge was applied, later, participants received a one-hour-long counter-top training focused on their shortcomings, and three weeks later, another test was applied. Both tests comprised food safety knowledge, attitudes, and behavior, and a handwashing skills test. The operators achieved a substantial improvement in knowledge and handwashing skills; nevertheless, the researchers could not identify any modification in attitude and behavior.

Soon et al. [9] executed a meta-analysis to evaluate whether food-safety training and other approaches foster a strengthening of knowledge and behaviors about hygienic practices in handwashing and control. This analysis proved that food safety training helps to reinforce knowledge in food handlers related to proper hand hygiene. Nevertheless, the behavior of food handlers was less positively affected since almost all the participants included in the meta-analysis provided these results themselves, without the supervision of any trainer. Results self-reported are subjected to bias since people, in general, give answers that they consider correct and acceptable hindering the actual situation. As a result, modification of knowledge and behavior of food operators should be examined by an observer that can suggest modification, and provide an updated course, related to the particular process step. Variations in results reported, may be the consequence of a long space of time between the training course and the exam since it could make difficult to the participants the answering of the most challenging questions. This inference is supported by the transfer of training theory, which states that cycles of training have to be accompanied by additional learning options like an update training [1].

Rowell et al. [15] reported the necessity to develop robust food safety resources, which include a method of training *in situ* as well as the knowledge of the food safety practices recognized as the sources of foodborne infections like cross-contamination, temperature control, and personal hygiene. During food safety courses, it is necessary to present with palpable experiences, the risks of deficient food handling, as an excellent way to sensitize manipulators. A good example is a demonstration in which participants rub on an invisible powder, wash their hands, and put them through a black light. The powder shines when it is exposed to black light, revealing those spots where microorganisms would remain if the hand wash were inefficient [4].

Another important aspect to take into account is to gain the attention of the participants and ask them to use their food safety knowledge. Gallant explains that an effective way to apply training programs is to place the topics of the course in the form of games. Jeopardy, board games, guessing games, e-learning activities, crossword puzzles, are commonly used, but there are still many more. Even games created by trainers can be effectively applied [16]. Carrying out the training sessions and practices in a place similar to the workroom increases the probability of effective delivery of information. [10]. Araújo et al. [12] recommend that fresh fruit and vegetable producers must hold a specific facility to impart food safety courses. Furthermore, this training must be periodically imparted and must be focused on those habits that the operators ought to apply throughout their working day.

9.2.4.1 WORK ENVIRONMENT

The work environment should foster the application of targeted behaviors. A suitable work environment has to include characteristics like transfer climate, support, and opportunities:

1. **Transfer Climate:** It refers to those circumstances recognized by trainees in the facility that obstruct or enable the application of learned abilities. A favorable transfer climate should include keys that persuade trainees to practice new abilities [10]. Pilling et al. [4] explain that no suitable assets or materials, the absence of training and behavior reminders and time limitations are some of the common aspects reported by food handlers as hurdles. Creating encouragement programs by which exceptional employees will be rewarded can be a great help to establish a food safety culture in the food processing facility [7]. Rowell et al. [15] propose to form a food safety culture system that could be able to encourage the skills of the managers and employees enhancing their understanding on the food safety risks in their facilities and stimulating a successful food safety system.
2. **Support:** In order to be successful, a food safety programs must include all levels of the organization. The general manager has to be aware of the food safety program and emphasize how important is food safety for the organization, and that this program has to update it regularly [7]. Superiors and coworkers support significantly reassure trainees to employ learned abilities. Once the training cycle finishes, trainees should take advantage of the opportunity to apply

and debate in their workplace everything they have learned [10]. Knowledgeable restaurant managers could be an ideal resource of information if they are willing to transmit food safety training to their staffs. Even if it were executed in an informal environment or a theory-practice like a workshop, food safety behaviors would not be related to mandatory training classes. Hence, this informal approach would make effortless for the personnel to understand, accept, and modify (K-A-B model) their perception of food safety [4]. Pilling et al. [4] evaluated two clusters of food handlers: personnel of restaurants where food safety training was obligatory for all and personnel from restaurants where food safety training was mandatory only for shift managers. Results demonstrated that a well-trained manager was related to food handlers with superior knowledge and better behaviors since restaurants where training course was mandatory only for managers, had food handlers with more robust knowledge and more proper behaviors than those food handlers with obligatory training. Additionally, training at all levels of the corporation should be measurable [7].

3.   **Opportunities:** Presence of personnel that is not bothered by inappropriate food safety practices may interfere in the work environment. Production personnel are always well trained, while maintenance personnel are not taken into account for food safety training. There are many repairs and maintenances actions in the facilities that can become sources of contamination if they are not carried out properly. Khanal and Poudel [11] evaluated the food safety knowledge and practices among butchers in Nepal found that only 17% had an official certificate that authorizes them for slaughtering practice. This fact creates a negative work environment.

9.3   INFLUENCE OF GOOD MANUFACTURING PRACTICES (GMP) IN QUALITY OF PRODUCTS

For Tavolaro and Oliveira [17], microbiological analysis serves as a support for GMP assessment but cannot be taken as a standalone measure. Astegiano et al. [6] explained that inappropriate practices throughout food operation are determinant in the incidence of foodborne illness. Microorganisms are disseminated by cross-contamination, insufficient cooking procedures, and application of improper storage temperatures. In consequence, the spreading of microorganisms directly affects the quality of the final products. The microbiological analysis of food products during processing and as a finished product, when it is accompanied by the evaluation of food safety knowledge and practice, are a useful resource for the development of training programs. Tavolaro and Oliveira [17] evaluated the application of a good manufacturing practices (GMP) training course directed to milkers in dairy goat farms in Brazil. To fulfill this evaluation, milkers received a one-hour course that employed an educational technique known as "content-based communicative approach." Information of the course was transmitted through illustrated cards with short sentences about GMP and guidelines directed to the milking process. This information was focused on the deficiencies detected during a previous GMP evaluation and the observation of the milking process. At the end of the course, easy-to-implement proposals, aimed to improve the milking process were presented by the trainers and debated and approved by the farm participants. Correction of GMP deficiencies and application of the proposals to improve the milking process were evaluated by a new assessment one or two months later.

Furthermore, raw milk samples were taken in the first and the second appointment to evaluate aerobic mesophilic, psychrophilic, and total coliform microbial counts, as well as *Staphylococcus aureus* counts, and presence of *Salmonella* spp. Although mesophilic counts were reduced in samples taken a post-GMP course, the training was not considered as helpful, since the proposed items were not incorporated in the process. The reduction of mesophilic microbial counts can be due to extrinsic factors and not to the knowledge acquired in the course and applied in their practice. Astegiano et al. [6], who evaluated the hygiene during the production process of Piedmonte cheese obtained similar results. Personnel of the of the Istituto Zooprofilattico Sperimentale of Piemonte, Liguria, and Valle d'Aosta (IZSPLV) technically checked and advised cheese producers through the entire process: from milking until the end of the ripening period, with the aim of designing a training program based on detected deficiencies and the execution of corrective actions. The trainers explained the correct processes to avoid cross-contamination during milking, cheese manufacturing, and instrument cleaning. They stated that correcting improper handling practices was more advantageous to accomplish and remember proper behavior.

Moreover, in order to improve the process, trainers were able to determine problems in the system and apply corrective actions according to the results. Finally, samples were taken at different stages of the process. Microbiological results were in general acceptable. Nonetheless, the isolation of hygiene quality indicator microorganisms such as *E. coli* and *Enterobacteriaceae* during the first steps of process and information obtained during the observation of the process suggested that there was a need of strengthening complying with GMP.

One of the reasons why trainees are not successful in transferring knowledge is the lack of association between GMP training and the activities they carry out. Trainees should find a way to make the participants visualize the consequences of poor food safety practices. Moreover, it is widespread to employ the same information resources and the same training techniques, so that participants get bored and do not give importance to the information. Furthermore, if there is not management committed to food safety programs, training programs are applied occasionally, and personnel is not asked to apply what they have learned; therefore, GMP will be perceived as a waste of time that will not deliver good outcomes and profits [16].

Training in good hygiene practices and GMP is an indispensable tool for every food handler and even more when they live in poverty and have a total or partial knowledge of food safety. This situation should be considered to design a suitable food safety training course. Also, industries need skilled personnel that facilitate the production of safe food [17]. Ramón et al. [18] evaluated the impact of a GMP course for small producers on the quality of goat milk cheese. The authors assessed the current conditions of the food processing facilities, with a GMP checklist, to identify those areas that needed improvement. Training course included modular courses using a workshop methodology, which included participating activities that helped on the determination of the level of capabilities learned by participants. At the end of the course, another evaluation was applied. Population evaluated belonged to a rural-marginal zone where milk producers had very limited food safety knowledge. In fact, the evaluation found deficiencies like inadequate working materials and very low educational levels, which creates a favorable environment favorable for the growth of foodborne pathogens. This evaluation had a very positive impact on the producers since their product had better sensorial and microbiological characteristics after their training.

Microbiological analysis is also important to determine which aspects of GMPs need adjustments to reduce sources of contamination. Carpentier and Cerf [19] explain that microorganisms like *L. monocytogenes* have no exceptional properties that let them persist in facilities and equipment. Therefore, these bacteria can remain mostly due to an inability to eradicate them from the food processing facilities, besides the capacity of *L. monocytogenes* to grow at low temperatures. Consequently, they recommend an adjustment in the design of food processing facilities and equipment, as well as the use of disinfectants to guaranty an effective application.

Daelman et al. [20] determined that hands and gloves of employees and equipment could retain high concentrations of microbial load. A total of 39.8% of production surfaces analyzed in a facility of cooked, chilled foods had total psychrophilic aerobic counts higher than 3.0 logs CFU/g. Hence, they propose that those surfaces must be monitored to avoid the generation of a contamination point that can raise the risk of contamination of the final products, and highlighted the importance of keeping clean and disinfected the production zones.

It is necessary to find ways to encourage the application of GMP. Benchmarked schemes of the Global Food Safety Initiatives (GFSI) are significant for companies that need to fulfill with the U.S. 2011 Food Safety Modernization Act (FSMA). Crandall et al. [21] surveyed to evaluate the view of food suppliers on the decision of Walmart Stores (Bentonville, AR) to call for their suppliers to become GFSI compliant. Almost all the suppliers agree to modify their food safety management systems to accomplish the GFSI requirements. These modifications were taken as upgrades to the safety of their food products. In consequence, the capability to generate and provide safe food was increased. Moreover, training for workers in food safety was adopted, suggesting that there was a tremendous educational opportunity to familiarize retailers and suppliers worldwide with the GFSI.

The stimulation for learning and application of good hygienic practices (GHP) and GMP should be carried out in all types of food processing facilities, and need to be applied to all the personnel. This action will provide an improvement in the facility and the quality of the products as well as on the wellbeing of the staff, along with the reduction of risk of foodborne diseases associated with food products.

KEYWORDS

- **Food Safety Modernization Act**
- **Global Food Safety Initiatives**
- **good hygienic practices**
- **Good Manufacturing Practices in Quality of Products**
- **Knowledge-attitude-behavior**
- **psychrophilic aerobic counts**

REFERENCES

1. Adesokan, H. K., Akinseye, V. O., & Adesokan, G. A. Food safety training is associated with improved knowledge and behaviors among foodservice establishments' workers. *Int. J. Food Sci.,* **2015**, 1–8, ID 328761.
2. Tyler, M. M., Khalil, N., & Paton, S. Increasing knowledge with food safety training at public health-Dayton & Montgomery county. *Food Prot. Trends,* **2015**, *35*(4), 262–269.
3. Roberts, K., Barrett, B., Howells, A., Shanklin, C., Pilling, V., & Brannon, L. Food safety training and foodservice employees' knowledge and behavior. *Food Prot. Trends,* **2008**, *28*(4), 252–260.
4. Pilling, V., Brannon, L., Shanklin, C., Roberts, K., Barrett, B., & Howells, A. Food safety training requirements and food handlers' knowledge and behaviors. *Food Prot. Trends,* **2008**, *28*(3), 192–200.
5. Uddin, S., Khan, M. A., & Solaiman, M. human resource development through training on business teachers in Bangladesh. *Int. J. Business Technopreneurship.,* **2014**, *4*(1), 83–96.
6. Astegiano, S., Bellio, A., Adriano, D., Bianchi, D. M., Gallina, S., Gorlier, A., et al. Evaluation of hygiene and safety criteria in the production of a traditional piedmont cheese. *Ital. J. Food Saf.,* **2014**, *3*(3), 1705.
7. Hisey, P. *Establishing an Effective Food Safety Training Program.* Food safety magazine, **2008**. [online] https://www.foodsafetymagazine.com/magazine-archive1/junejuly-2008/establishing-an-effective-food-safety-training-program/ (accessed on 6 January 2020).
8. Welty, G. Developing a continuing CGMP training program. *J. GXP Compliance,* **2009**, *13*(4), 86–96.
9. Soon, J. M., Baines, R., & Seaman, P. Meta-analysis of food safety training on hand hygiene knowledge and attitudes among food handlers. *J. Food Prot.,* **2012**, *75*(4), 793–804.
10. Grossman, R., & Salas, E. The transfer of training: What really matters. *Int. J. Training and Development,* **2011**, *15*(2), 103–120.
11. Khanal, G., & Poudel, S. Factors associated with meat safety knowledge and practices among butchers of ratnanagar municipality, Chitwan, Nepal: A Cross-sectional study. *Asia Pac. J. Public Health,* **2017**, *29*(8), 683–691.
12. Araújo, J., Esmerino, E. A., Alvarenga, V. O., Cappato, L. P., Hora, I. C., Silva, M. C., Freitas, M. Q., Pimentel, T. C., Walter, E. H., Sant'Ana, A. S., & Cruz, A. G. Development of a checklist for assessing good hygiene practices of fresh-cut fruits and vegetables using focus group interviews. *Foodborne Pathog. Dis.,* **2018**, *15*(3), 132–140.
13. Machado, R., & Cutter, C. Training hard-to-reach Pennsylvanian cheese makers about food safety, using a low-tech training tool. *Food Prot. Trends,* **2018**, *38*(4), 266–283.
14. Schrader, P. G., & Lawless, K. The Knowledge, attitudes and behaviors approach how to evaluate performance and learning in complex environments. *Performance Improvement,* **2004**, *43*, 9, 8–15.
15. Rowell, A. E., Binkley, M., Thompson, L., Burris, S., & Alvarado, C. The impact of food safety training on employee knowledge of food safety practices for hot/cold self-serve bars. *Food Prot. Trends,* **2013**, *33*(6), 376–386.
16. Gallant, J. Life science training institute. *Power up Your GMP Training: How To make it More Relevant, Meaningful, and Engaging,* **2016**. https://www.lifesciencetraininginstitute.com/power-up-your-gmp-training-how-to-make-it-more-relevant-meaningful-engaging/ (accessed on 20 January 2020).

17. Tavolaro, P, Oliveira, C. Evaluation of a GMP training of milkers in dairy goat farms in Sao Paulo, Brazil. *Int. J. Environ. Health Research,* **2006**, *16*(1), 81–88.
18. Ramón, A. N., De La Vega, S. M., Ferrer, E. C., Cravero Bruneri, M. P., Millán, M. P., Gonçalvez De Oliveira, E., Borelli, M. F., Villalva, F. J., & Paz, N. F. Training small producers in good manufacturing practices for the development of goat milk cheese. *Food Sci. Technol. Campinas,* **2018**, *38*(1), 134–141.
19. Carpentier, B., & Cerf, O. Persistence of *Listeria monocytogenes* in food industry equipment and premises. *Int. J. Food Microbiol.,* **2011**, *145*, 1–8.
20. Daelman, J., Jacxsens, L., Lahou, E., Devlieghere, F., & Uyttendaele, M. Assessment of the microbial safety and quality of cooked chilled foods and their production process. *Int. J. Food Microbiol.,* **2013**, *160*, 193–200.
21. Crandall, P., Van Loo, E. J., O'Bryan, C. A., Mauromoustakos, A., Yiannas, F., Dyenson, N., & Berdnik, I. Companies' opinions and acceptance of global food safety initiative benchmarks after implementation. *J. Food Protect,* **2012**, *75*(9), 1660–1672.

PART II
Food Biotechnology

CHAPTER 10

Production, Recovery, and Application of Invertases and Lipases

DEICY YANETH LÓPEZ ACUÑA,[1] JOSÉ D. GARCÍA-GARCÍA,[1]
ANNA ILINÁ,[1] MÓNICA L. CHÁVEZ GONZÁLEZ,[1]
AYERIM HERNÁNDEZ-ALMANZA,[2] REBECA GALINDO BETANCOURT,[3]
CRISTÓBAL NOÉ AGUILAR,[2] and JOSÉ LUIS MARTÍNEZ-HERNÁNDEZ[1]

[1]Nanobioscience Group, Food Research Department,
School of Chemistry, Autonomous University of Coahuila,
Saltillo Campus, 25280 Coahuila, México, Tel.: (844) 4161238,
E-mail: jose-martinez@uadec.edu.mx (J. L. Martínez-Hernández)

[2]Bioprocesses and Bioproducts Group, Food Research Department,
School of Chemistry, Autonomous University of Coahuila,
Saltillo Campus, 25280 Coahuila, Mexico

[3]Reseach Center for Applied Chemistry, 25294 Saltillo, Coahuila, Mexico

ABSTRACT

Fermentative systems have allowed the production and the recovery of a great variety of enzymes. The design of bioprocesses to produce enzymes of importance in diverse industrial areas has been possible by means the use of the existing microbial richness and the abilities of these to grow in different substrates such is the case of the enzyme invertase and lipase. This chapter describes fermentation systems to produce these enzymes, as well as a detailed analysis of their reaction mechanisms, characterization, and their current and potential uses.

10.1 SOLID STATE AND SUBMERGED FERMENTATION (SF) SYSTEMS

The production and the recovery of various enzymes produced by the microbial route have been carried out through fermentative processes like solid-state fermentation (SSF) or submerged fermentation (SF). The main difference between these two bioprocesses is the amount of free water in the system. Solid-state fermentation (SSF) or fermentation in a solid substrate, could be defined as a fermentative process which occurs in the absence or near the absence of free water, using natural or inert media that must serve as a source of nutrients for the microorganism growth on them [1–3]. The SSF is characterized by carrying out the bioprocess on a solid matrix, which has a low moisture content (limit ≈ 12%) and occurs in an aseptic and natural state. A more suitable definition of SSF is proposed, due to the extensive use of inert supports in which there can be an excess of free water: the SSF process can be any fermentative process that is carried out in a solid or semi-solid substrate, or already in support of natural origin, which can be used as a source of carbon and energy for the microorganism [4]. Most industrial enzymes are produced by SSF since this process requires low energy yields and produces a lower amount of wastewater compared to SF processes [1, 5]

The SF, also named liquid fermentation, is a system in which the substrates are dissolved or suspended in an aqueous medium and are stirred to preserve the homogeneity of the system [6]. Compared to SSF, SF has the following advantages: the product obtained is more homogeneous, it is easier to control the fermentation factors such as temperature, aeration, agitation, and pH, direct measurement of the biomass can be carried out [7].

The filamentous fungi growth in liquid medium is highly recommended. SF not only allows an easy quantification of the material obtained throughout the culture but also the study of the main physiological characteristics of the fungus. Extracellular enzymes produced by fungus are present in sufficient quantities in the liquid culture (LC) and can be detected and easily characterized by electrophoresis.

10.2   SURFACE ADHESION FERMENTATION (SAF)

The surface adhesion fermentation (SAF) is an alternative category of fermentation described by Gutiérrez and Villena [8]. It is a process where the spores are adsorbed on an inert solid support and cells grow uniformly to form a biofilm in the presence of a considerable amount of liquid phase [9].

The biofilm concept presumes either a population or a community of microorganisms that live together on a surface [10]. This form of growth supposes a differential expression of genes that is reflected in a greater production of enzymes [8, 11]. The formation of biofilms by fungi is a complex process that begins with the adhesion of spores to a surface, which is influenced by physical, chemical, and environmental factors [12]. According to Gutiérrez and Villena [8], the formation of fungal biofilms generally occurs in three phases:

1. adhesion;
2. initial growth and development; and
3. maturation.

Figure 10.1 shows a diagram in which spores reach the substrate surface and adhere to initiate germination, which is highly influenced by the presence of hydrophobic proteins called hydrophobins. These proteins are produced exclusively by mycelial fungi and present cysteine residues in their sequences that form intramolecular disulfide bonds [13, 14]. The characteristic property of hydrophobins is the formation of an amphipathic membrane in contact with a hydrophilic-hydrophobic interface which allows changing the nature of the surface.

It can be stated that the initial adhesion of the spores to the support surface determines the structure and stability of the biofilm. From this point of view, cell adhesion can be considered as the initial step necessary to trigger the signaling and mechanisms of differential gene expression that is absent in conventional SF, which may explain the higher enzyme productivity found in fungal biofilms [15].

After the adhesion to the surface, the initial growth and development phase occurs, which is very similar to the formation of biofilms by bacteria where there is an initial formation of micro-colonies formed as a consequence of the generation of new cells. Subsequently, the germination of the spores and the production of the germinative tube begin, thus promoting the colonization of the surface and continuing to perceive the almost total colonization of the same. Finally, the maturation phase, in which the biomass density is increased by the formation of the biofilm, allowing the differential expression of genes which leads to greater production of enzymes and metabolites [8].

10.3 SUBSTRATES USED IN SSF

In addition to the structural aspect, biofilms are important because they promote a physiological change in microorganisms. These can be developed on inert and synthetic supports, generally showing a higher production of metabolites and enzymes than SF cultures [9, 16, 17].

It has been used inert supports in SSF like polyester [8] and polyurethane, also cellulosic materials, for example, residuals of papaya trunk, coyonoxtle trunks [18, 19] and more recently, fibrous networks of *Opuntia ficus indica* [20]. The advantages of such biostructures are that they are generally reusable; there is an absence of toxicity problems and the open spaces within the matrix for the growth of cells thus avoiding diffusion problems [21, 22].

Nowadays, numerous efforts have focused on the development of novel routes that allow the development of immobilized systems, through the use of porous materials different from the polymers of natural origin that have been used until now. As an example, nanomaterials (Figure 10.1) have recently become one of the most prolific research areas because they are presented as an interesting alternative to control the stabilization and homogeneous dispersion of such nano-supports in the system. Also, they have the additional advantage of their ease of recycling and reuse an essential and desired property in many of the applications of these nanomaterials.

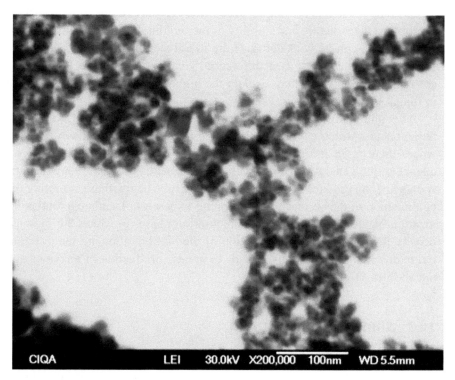

FIGURE 10.1 Microphotography of magnetic nanoparticles in scanning electron microscope (Own source).

10.4   POLYSTYRENE (PS)

Polystyrene (PS) is a thermoplastic polymer obtained from the polymerization of styrene. This type of material is an alternative to be used as support in SSF due to it exhibits characteristics that offer excellent conditions for surface adhesion such as its hydrophobic character, its surface roughness, high porosity, and low density (Figure 10.2). The PS-divinyl benzene resins are produced by the copolymerization of styrene and divinyl benzene as a crosslinking agent. These matrices have a low water absorption capacity in their pores as well [23].

FIGURE 10.2   Microphotograph of a polystyrene sample in a scanning electron microscope.

It should be noted that there are very few reports about the use of this material as support in a fermentative process. As it is an inert material, it offers a support that is used only as a binding structure for the microorganism, and it does not contribute nutritionally in the growth of the microorganism. From the engineering point of view, inert supports are better because their geometric and physical characteristics do not change due to microbial growth, which allows better control of heat and mass transfer as well as obtaining products free of contamination [24, 25].

The PS-divinyl benzene resins are a type of material designed in separation operations of aromatic compounds, peptides, steroids, etc. They are also used for the packaging of columns used in high performance liquid chromatography.

10.5   MAGNETIC NANOPARTICLES (MNP)

Magnetic nanoparticles (MNP) have been the subject of numerous studies in recent decades owing to their high activity and specificity of interaction, as well as interesting properties like high surface/volume ratio combined with small sizes [26]. One of the main advantages of metallic nanoparticles is the existence of a series of energy states compared to the conventional state of energy in metals, which also increase as the nanoparticle size decreases. This phenomenon involves a decrease in the density of electronic states that can facilitate the mobility of nanoparticles between states and therefore a high specificity. However, this small size and high surface gives them great instability due to their high surface energies, so they tend to aggregate to stabilize.

Various methods have been described in order to stabilize NPM, among which are the addition of organic ligands [27], coating agents [27, 28], the use of ionic liquids, as well as colloids and soluble polymers [27, 29]. Within these nanostructured materials, magnetite (Fe_3O_4) is a magnetic material commonly used for its strong magnetic properties, superparamagnetism, and low toxicity, which favors its application in medicine and biotechnology field [30]. Besides being magnetic, magnetite can be manipulated by an external magnetic field gradient, and their surface can be modified with biological agents as well [31].

Despite all these advantages, magnetite nanoparticles tend to aggregate due to strong magnetic attractions between dipole-dipole particles, so for its stabilization have been reported various studies of nanoparticle synthesis coated with different materials as chitosan [32], PS [33], carbon [34], among others.

Recently, MNP coated with chitosan have attracted considerable attention in various fields due to their physicochemical and biological properties. In contrast to other biopolymers, chitosan is a positively charged hydrophilic polyamino saccharide that comes from weak basic groups that give it special characteristics from the technological point of view. Due to the presence of free amino groups in chitosan, it has a high degree of solubility and reactivity, also, chitosan has excellent biological importance and excellent chemical properties including biocompatibility, bioactivity, it is biodegradable, non-toxic, and polycationic [35, 36]. Therefore, the use of MNP coated with chitosan represents an alternative material to be used as a support in fermentation systems.

10.6 FILAMENTOUS FUNGI USED IN SSF

Various fungi have been used in SSF, mainly those that adhere through the production of hydrophobins where adhesion can be considered as a metabolically active process involving signaling mechanisms [9]. Among them, *Mucor griseocyanus* is an organism that has allowed the production of secondary metabolites of industrial interest in good production levels. Mitra et al. [37] used this technology in the production of cellulase and xylanase enzymes from two filamentous fungi: *Chaetomium crispatum* and *Gliocladium viride,* respectively, reporting high levels of enzymatic activity.

The genus *Aspergillus* is a genus of fungal microorganism widely used in the production of various enzymes such as lipases [38–42], invertases [43], proteases, and glucose oxidase, among others; which are of great importance and demand in industries such as the food, pharmaceutical, chemical and industry. Among the species of interest, *Aspergillus niger* shows a high capacity of adherence to different solid supports [40, 41], as well as shows some advantages to produce enzymes due to it can be grown in defined and low-cost media, which allows the production of its metabolites on a large scale. Its products are generally considered as GRAS (generally recognized as safe) [44], this allows its application in the food industry for human being, animals, and pharmaceutical industry [45].

In this same context, the importance of filamentous fungi in biotechnological processes lies in the metabolites produced in the different stages of their growth and maturation. Metabolites like enzymes play an important role because they allow the reduction of production costs in industrial processes, improve the quality of products, and participate in the development of more efficient processes and generation of profits and higher yields [44].

10.7 INVERTASE

The β-D-fructofuranosidase fructohydrolase, commonly called invertase and known as β-D-fructofuranosidase, β-fructosidase, β-fructosylinvertase and sucrose, was first discovered by Berthelot in 1860 [45, 46]. Sucrose is the most adequate substrate for invertase; however it can catalyze the hydrolysis of raffinose and stachyose [47] producing a mixture of fructose and glucose. This mixture is known as invert sugar and its sweetening power is higher than sucrose, and it is used in the production of chocolate, marshmallows, syrup, synthetic honey, jam, and jellies. Invert sugar is not only the main product that produces invertase. The production and application of fructo-oligosaccharides (FOS) formed by invertase obtained from strains of *A. niger* have gained great commercial importance due to its functional properties. These include improvement of the intestinal microflora, ability to alleviate constipation, reduce cholesterol and lipids, promote animal growth, and are low-calorie sweeteners [48, 49].

10.7.1 MECHANISM OF ACTION OF INVERTASE

Invertase also called β-D-fructofuranosidase (EC.3.2.1.26) catalyzes the hydrolysis of non-reducing terminal residues of β-D-fructofuranoside in β-D-fructofuranosides. The result of the hydrolysis reaction (Figure 10.3) is a mixture of glucose and fructose called "invert sugar," due to the reversal of its optical properties from a positive rotation of sucrose (+66°) to a negative rotation of the glucose (+52) and fructose (−92°) [44, 50].

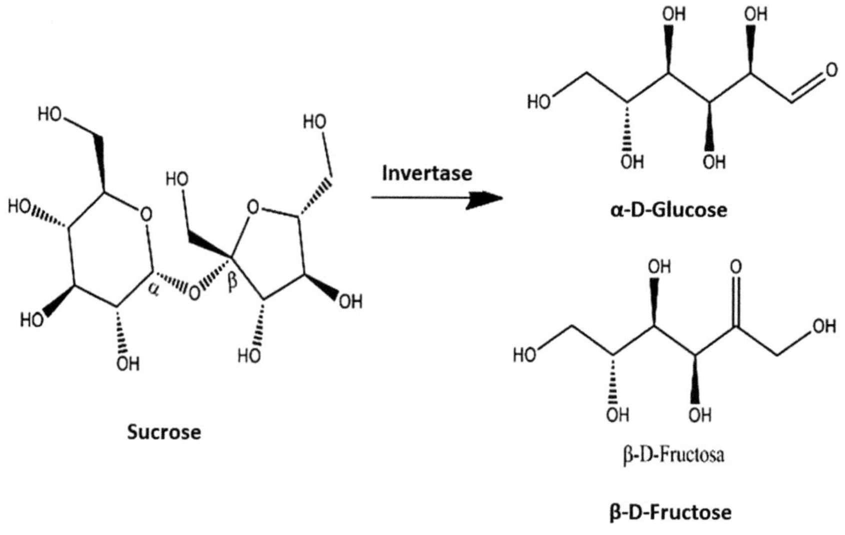

FIGURE 10.3 Hydrolysis reaction catalyzed by the invertase enzyme.

10.7.2 MICROORGANISMS PRODUCERS OF INVERTASE

It has been reported the presence of invertase by bacteria, fungus, higher plants and animal cells [48, 51]. Most of the reports about invertase production have been carried out using *Saccharomyces cerevisiae*, which is an interesting microorganism since other yeasts consume saccharose to synthesize just one form of invertase intra or extracellular. *S. cerevisiae* can synthesize both forms of invertase [52]; however, even within the same yeast culture there are different isoforms of invertase, for example, the intracellular invertase has a weight of 135 KDa, while the extracellular invertase weighs 270 KDa [53, 54].

The invertase is commercially produced mainly by *Saccharomyces cerevisiae* and some strains of fungi of the genus *Aspergillus* such as *A. niger, A. oryzae,* and *A. ficcum.* In these fungi, the invertase can be intra and extracellular. In *A. niger,* the presence of multiple forms of this enzyme has been observed, and approximately 40–60% of it is intracellular [55]. Contrary to most enzymes, the invertase is very tolerant to changes in pH, presenting a relatively high activity between pH 3.5 to 5.5 with an optimum pH approximately 4.5 K_m values vary significantly from one species to another, but most are between 2 mM to 5 mM, with a typical value of 3 mM for the free enzyme [56, 57].

10.7.3 INVERTASE PRODUCTION

Researches works about the production of invertase have been developed UN SF using sucrose and molasses as main source of carbon. Belcarz et al. [52] observed that 1% sucrose concentration the synthesis of invertase is induced by *Candida utilis.* Several attempts have been performed with filamentous fungus to increase the production of invertase. It has been reported the selection of *A. fumigatus* strain by means mutagenesis techniques, finding the higher activity of invertase intracellular and extracellular after 96 hours of incubation. Mukherjee et al. [58] report the production of invertase by *Termitomyces clypeatus* when fungi were growth by 5 days in synthetic media with 1% of sucrose.

The production of invertase by *Aspergillus niger* has been explored as well with residues of pineapple peals in SF, reaching maximum yields of 24.20 U/mL of invertase activity at 120 hours in pH condition of 5.0 and 35°C [49]. Another viable option to produce invertase is the fermentation processes in solid-state. Lowest catabolic repression, the high productivity of enzymes, and the highest concentration of enzymes stand out are some advantages of the SSF in comparison of the FS [59, 60]. More recently, Romero-Gómez et al. conducted a study comparing four xerophilic fungal strains (*Penicillium pinophilum EH2, P. purpurogenum GH2, P. citrinum ESS, Aspergillus niger GH1*) to produce invertase in SF, finding that *A. niger GH1* strain showed higher growth rate than *Penicillium* strains. This strain produced extracellular invertase with the highest enzymatic activity (8.625 U/L) at 48 hours [60].

Dinarvand et al. [61] optimized the production of inulinase and invertase by *A. niger* ATCC 20611; these authors demonstrated that the optimal production conditions were at pH 6.5, temperature 30°C and agitation at 150 rpm, and under these conditions the production increases 10–32 times. Chand Bhalla et al. [48] evaluated the production and characterization of invertase produced by *S. cerevisiae* SAA-621, besides they studied the application of invertase in the synthesis of FOS. The authors report that the optimal production conditions were at a pH of 6.0 and a temperature of 40°C and enzyme showed stability at a temperature range of 30 to 40°C. Furthermore, authors indicate that the maximum production of FOS was observed with 2.5 U of invertase in 1 mL of reaction, pH 5.5 and 40°C.

Romero-Gómez [60] compared the invertase production by *A. niger* in SF and in SSF with polyurethane as support and *A. niger* C28B25, N-402 and Aa20 strains. They observed that the growth is faster and with a higher level of biomass in SSF than SF. A microscopic analysis of the polyurethane cubes showed that the mycelium is adhered to the polyurethane generating spaces filled with air; this suggests that the culture in SSF is not limited by the transfer of oxygen as in the case of SF [62].

10.8 LIPASES

Lipases (EC 3.1.1.3) belong to the family of hydrolases that act on carboxylic bonds. Their physiological role is to hydrolyze triglycerides to diglycerides, monoglycerides, fatty acids (FA), and glycerol; additionally lipases can also carry out esterification, interesterification, and transesterification in non-aqueous media, among other reactions [63]. These enzymes are widely distributed in nature and are found in microorganisms, plants, and animals [64]. There is a report of the presence of lipases since 1901 by *Bacillus prodigiosus, B. pyocyaneus,* and *B. fluorescens* [65].

It has been found that most lipases share a common structure, a polypeptide folding composed of 8 β-sheets, connected by 6 α-helices [66]. An important quality of lipases is that the catalytic triad Ser-His-Asp/Glu is entirely covered by a lid that must be completely open to access the substrate. This catalytic triad is embedded in a Glycon consensus region X-Ser-X-Gly [66–68].

Therefore, lipases can exist in two forms: in one of them, the active center of the lipase is hidden by a polypeptide chain that forms the lid, under this form the enzyme is inactive (form closed). In the open or active form, the polypeptide chain is displaced, and the active center is exposed to the reaction medium. In an aqueous solution, lipases can exist in equilibrium between the two forms. This exchange between the closed and open form of the enzyme provided by the water-oil interface is accompanied by conformational changes [64, 67]. Recently, it has been shown that many of the lipases tend to form molecular aggregates, by adsorption of the open form of the lipase-mediated by the hydrophobicity of the active center. These aggregates present different catalytic properties when compared with an individual molecule [69, 70].

10.8.1 MECHANISM OF ACTION OF LIPASES

Regardless of the environment in which the enzyme is located, the catalytic mechanism currently accepted to describe the action of lipases is centered on the catalytic residue serine, located in its active center. The oxygen atom present in the serine acts as a nucleophile, forming a tetrahedral hemiacetal with the triglyceride. Then, the ester bond of hemiacetal is hydrolyzed and diacylglyceride is released forming an acyl enzyme complex. Subsequently, an attack on acyl-enzyme complex occurs by a nucleophilic reagent, which may be a diglyceride, an alcohol or water, obtaining finally a triglyceride, an ester, or a carboxylic acid respectively, and releasing the enzyme [71, 72]. All these stages are outlined in Figure 10.4.

10.8.2 INTERFACIAL ACTIVATION

Lipases develop a unique mechanism of action, called interfacial activation, when the lipase is in a polar medium, the lid is closed; this causes the enzyme to be protected and can only act on the water-oil interface generated by an emulsion [73, 74]. In general, lipases are enzymes that require inter-facial activation to maximize their catalytic activity [75, 76]. The interface is understood as the imaginary surface that separates two homogeneous and physically different portions of space. At the molecular level, an interface consists of a set of two adjacent layers of ordered molecules with different hydrophobic-hydrophilic character [75]. When lipase comes into contact with an interface, the dielectric environment on the protein surface is modi-fied, which enhance electrostatic interactions. It is then possible for the cover of the active center to move [77, 78], and a restructuration occurs in the conformation of the molecule [74]. Thus, the catalytic amino acids are exposed to the solvent in a suitable orientation, and around them, the oxyanion cavity is formed, by exposure of certain hydrophobic residues and internalization of other hydrophilic ones [76]. All this increases the affinity of the enzyme for its lipid substrates and contributes to the stabilization of the transition state during the catalytic cycle [75, 76, 79].

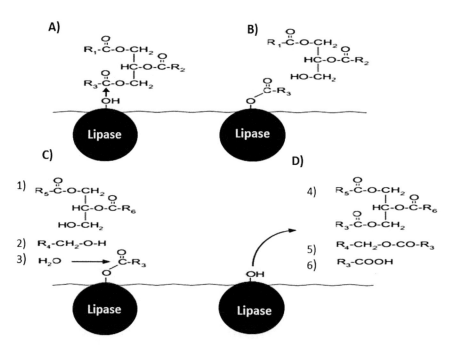

FIGURE 10.4 Reaction stages catalyzed by lipases: (A) Nucleophilic attack on carbon atom of the carbonyl group of triglyceride, (B) formation of acyl-enzyme complex and alcohol or diglyceride, (C) attack of reagents nucleophiles, such as diglyceride (1), alcohol (2) or water (3), and (D) release of the respective products of nucleophilic attacks: triglyceride (4), ester (5) or carboxylic acid (6).

10.8.3 MICROORGANISMS THAT PRODUCE LIPASE

Lipases differ in their properties depending on their origin. They catalyze the hydrolysis or synthesis of a large variety of carboxylic esters and release organic acids and glycerol. All lipases show a high specificity on substrates. Microorganisms with a high potential to produce lipases can be found in different habitats, mainly in waste or residues of vegetable oils used in the preparation of fried foods, dairy products industries, soils contaminated with oils and deteriorated foods.

These aspects indicate that the same natural environment offers a wide potential to isolate new sources of lipases with novel properties. Bacteria, filamentous fungi, yeasts, and actinomycetes have been isolated, among which stand out the genera *Pseudomonas, Chromobacterium Bacillus, Rhodococcus, Staphylococcus, Rhizopus, Mucor, Candida, Aspergillus,* and *Geotrichumsp* for their ability to produce extracellular lipases, facilitating in this way, the recovery of these enzymes from the culture medium [67, 80]. Due to lipases are among the most widely used enzymes in biotechnological processes and in chemical processes, research has focused on the search for new microorganisms with different lipolytic properties.

Likewise, different species of fungi such as *Aspergillus* and *Penicillium* have been evaluated for the production of extracellular lipases in solid and LC media using olive oil (OO) as the inducing agent. From these studies it has been found that the species *A. alliaceus, A. candidus, A. carneus, A. fischeri, A. niger, A. ochraceus, A. parasiticus, A. sundarbanii, A. terreus, A. versicolor, P. Aurantiogriseum, P. brevicompactum, P. camemberti, P. chrysogenum I, P. coryrnbiferum I, P. crustosum, P. egyptiacum, P. expansum,* and *P. spiculisporum* present the highest levels of lipase production, representing an important source and great potential for its biotechnological production [81].

10.8.4 INDUSTRIAL APPLICATIONS OF LIPASES

Under certain conditions, lipases can catalyze chemical reactions different hydrolysis, such as esterification, interesterification, alcoholysis, acidolysis, and aminolysis [84], which allow numerous applications in various industries such as: food industry, textiles, detersive, paper, medicine, pharmaceutical, cosmetics, skins treatment, as well as in environmental treatments and in the design of biosensors [82]. The expectation is to position lipases in the market as widely used and easily accessible enzymes, such as proteases and carbohydrases.

10.8.4.1   LIPASES IN THE DETERGENT INDUSTRY

Due to its ability to hydrolyze fats, lipases find great use as additives in the laundry industry and household detergents. Sahay and Chouhan [83] demonstrated the potential use of isolated lipases from *Penicillium canesense* and *Pseudogymnoascus roseus* to be incorporated into detergent formulations. Outcomes showed that lipases are resistant to a wide range of temperature; therefore they can be used in cold washing and in this way, generate less environmental impact.

10.8.4.2   LIPASES IN THE FOOD INDUSTRY

Fats and oils are important components of food. Nutritional, sensory value and physical properties of a triglyceride are greatly influenced by factors such as the position of the fatty acid in the glycerol backbone, the chain length of the fatty acid, and its degree of establishment. Lipases make it possible to modify the properties of lipids, altering the location of fatty acid chains in the glyceride and replacing one or more of FA with new ones [43, 66, 84, 85]. Also, lipases have been used to develop cheese maturation flavors, bakery products, and beverages. Lipases are used to eliminate fats from meat and fish products [74, 86]. Lipase has been used to improve the characteristics of products of the baking industry, improvements in the taste of various products of the dairy industry such as cheeses and butter and to produce foods that are directed to specific sectors such as babies and older adults. Besides, in recent years, lipases have been tested for the generation of biodiesel [82, 87, 88].

10.8.4.3   LIPASES IN THE PAPER INDUSTRY

The hydrophobic components of wood (mainly triglycerides and waxes) make pulp manufacture and paper more complicated. Lipases are used to eliminate the pulp tone produced for papermaking [82]. Nippon Paper Industries, Japan, have developed a tone control method that uses lipase from the *Candida rugosa* fungus to hydrolyze up to 90% of wood triglycerides.

10.8.4.4   LIPASES IN ORGANIC SYNTHESIS

The use of lipases in organic chemical synthesis is becoming an increasingly important challenge. Lipases are used to catalyze a wide variety of chemo-, regio-, and stereoselective transformations [89, 90].

10.8.4.5   LIPASES IN THE RESOLUTION OF RACEMIC ACIDS AND ALCOHOLS

The stereoselectivity of lipases has been used to resolve various racemic mixtures of organic acids in immiscible biphasic systems [91]. Racemic alcohols can also be developed in enantiomerically pure forms by trans-esterification catalyzed trough lipase [92].

10.8.4.6   LIPASES AS FOOD SUPPLEMENTS IN ANIMAL FEED

Owing to its hydrolysis, capacity the application of lipases in animal feed particularly in monogastric animals has increased. Fats are an important source of energy for their high caloric intake; however, this is not well used due to the lack of enzymes capable of hydrolyzing triglycerides releasing FA which are absent in young monogastric animals [93]. In the pancreatic lipase insufficiency, enzyme replacement therapy with microbial lipases has been the subject of attention in recent years because they show a wide range of characteristics including greater stability in acidic conditions.

10.9   MICROENCAPSULATED LIPASES APPLICATION AS SUPPLEMENTS IN POULTRY FEED

The use of enzymes as additives for the feeding of monogastric animals has achieved significant advances in recent years, which explain a large amount of scientific research. The reason why enzymes are used in monogastric nutrition is because of the possibility of degradation of compounds in raw materials that animal with its digestive system is not able to degrade it efficiently, as well as the ability to increase the availability of nutrients that exist inside cells by breaking cell wall. Likewise, the use of enzymes reduces adverse effect of anti-nutritional factors found in many raw materials (such as soy) that affect digestion and absorption processes. Enzymes can be considered as a complement of the digestive system of young animals when enzymes are limited [94, 95]. Enzymes such as proteases, amylases, lipases, cellulases, xylanase, phosphatases, and pectinases have been used in different formulations of monogastric diets [93] and in pigs. Monogastric animals produce all the enzymes necessary for the degradation and absorption of nutrients, but in certain stages as in young animals, especially under stress conditions, the appropriate enzymes are not always present in sufficient quantity such is the case of absorption and fat degradation where lipases play an important role, however, very little research has been developed.

Fats are important ingredients in poultry diets. The incorporation of fats in diets has received increasing attention in recent years. Fats are the most potent source of energy and provide many other positive effects in poultry diets. Based on energy content and price compared to other energy sources, it is often economical to add fats to poultry diets. However, many physiological functions related to digestion and absorption of fats is not ripe at hatching and need to continue to develop over several weeks.

Some aspects of the process of digestion and absorption of fats have been diagnosed as deficient in young birds but the exact reason for the malabsorption of fats is unknown. However, the lack of maturity in pancreatic secretions of the enzyme lipase can be an important factor that causes low utilization of FA in small chickens. In the pancreatic lipase insufficiency, enzymatic substitution therapy with microbial lipases is practiced.

Several fungi produce a lipase resistant to acids. The clinical efficacy of a fungal lipase depends, at least to some degree, on its characteristic survival and activity in the stomach [96]. The enzymes produced by the fungi *Rhizopus arrhizus* and *Aspergillus oryzae* demonstrated a promising efficacy compared to a conventional pancreatin formulation. Zentler-Munro et al. [96] demonstrated in an in vitro study that *Aspergillus niger* lipase has a wide pH range of 2.5 to 5.5. Some authors showed that the optimal pH of an *Aspergillus niger* lipase is approximately 4.5 and the inhibition by pH is completely reversible. In the in vitro study, they also showed that *Aspergillus niger* lipase mainly generates diglyceride instead of monoglyceride [96].

Enzyme supplements are sensitive proteins that may lose part of their activity during storage, during the processing of food or in the acid or proteolytic degradation of this in the animal intestine, so the results obtained by supplementing them in animal feed may depend on several factors. Providing enzymes with a physical barrier against external adversity is an approach of great interest today. There is a wide variety of materials for coverage that can be used to encapsulate food ingredients, which include hydrogenated oils, waxes, maltodextrins, starches, and gums.

KEYWORDS

- **fructooligosaccharides**
- **generally regarded as safe**
- **magnetic nanoparticles**
- **magnetite**
- **solid state**
- **submerged fermentation systems**

REFERENCES

1. Krishania, M., Sindhu, R., Parameswaran, B., Ahluwalia, V., Kumar, V., Sangwan, R. S., & Pandey, A. Design of bioreactors in solid-state fermentation. In: *Current Developments in Biotechnology and Bioengineering* (pp. 83–96). Elsevier, B.V., **2018**.
2. Peng, R., He, Z., Gou, T., Du, J., & Li, H. Detection of parameters in solid state fermentation of monascus by near infrared spectroscopy. *Infrared Physics and Technology*, **2019**, *96*, 244–250.
3. Pandey, A., Soccol, C. R., & Mitchell, D. A. New developments in solid-state fermentation, I: Bioprocesses and products. *Proc. Biochem.*, **2000**, *35*(10), 1153–1169.
4. Höfer, M., & Lenz, J. Biotechnological advantages of laboratory-scale solid-state fermentation with fungi. *Applied Microbiology and Biotechnology*, **2004**, *64*(2), 175–186.
5. Marín, M., Sánchez, A., & Artola, A. Production and recovery of cellulases through solid-state fermentation of selected lignocellulosic wastes. *Journal of Cleaner Production*, **2019**, *209*, 937–946.
6. Kim, D., & Han, G. D. Fermented rice bran attenuates oxidative stress. In: Watson, R. R., Preedy, V. R., Zibadi, S., & Zibadi, H., (eds.), *Wheat and Rice in Disease Prevention and Health* (pp. 467–480). San Diego: Academic Press, **2014**.
7. Fazenda, M. L., Seviour, R., McNeil, B., & Harvey, L. M. Submerged culture fermentation of 'Higer fungi': The macrofungi. *Adv. Appl. Microbiol.*, **2008**, *63*, 33–103.
8. Gutiérrez, C. M., & Villena, G. K. *Surface Adhesion Fermentation: A New Fermentation Category.* Facultad de Ciencias Biologicas, UNMM, **2003**.
9. Zune, Q., Delepierre, A., Gofflot, S., Bauwens, J., Twizere, J. C., Punt, P. J., & Francis, F. A fungal biofilm reactor based on metal structured packing improves the quality of a Gla: GFP fusion protein produced by *Aspergillus oryzae*. *Applied Microbiology and Biotechnology*, **2015**, *99*(15), 6241–6254.
10. Yung-Hua, L., & Xiaolin, T. Quorum sensing and bacterial social interactions in biofilms. *Sensors (Basel)*, **2012**, *12*(3), 2519–2538.
11. Donlan1, R. M., & Costerton, J. William. biofilms: Survival mechanisms of clinically relevant microorganisms. *Clin. Microbiol.*, **2002**, *15*(2), 167–193.

12. Kregiel, D., Berlowska, J., & Ambroziak, W. Adhesion of yeast cells to different porous supports, stability of cell-carrier systems and formation of volatile by-products. *World Journal of Microbiology and Biotechnology,* **2012**, *28*(12), 3399–3408.

13. Linder, M. B. Hydrophobins: Proteins that self assemble at interfaces. *Current Opinion in Colloid and Interface Science,* **2009**, *14*(5), 356–363.

14. Sunde, M., Pham, C. L. L., & Kwan, A. H. molecular characteristics and biological functions of surface-active and surfactant proteins. *Annual Review of Biochemistry,* **2017**, *86*(1), 585–608.

15. Gutiérrez-Correa, M., Ludeña, Y., Ramage, G., & Villena, G. K. Recent advances on filamentous fungal biofilms for industrial uses. *Applied Biochemistry and Biotechnology,* **2012**, *167*(5), 1235–1253.

16. Mohamad, N. R., Marzuki, N. H. C., Buang, N. A., Huyop, F., & Wahab, R. A. An overview of technologies for immobilization of enzymes and surface analysis techniques for immobilized enzymes. *Biotechnology and Biotechnological Equipment,* **2015**, *29*(2), 205–220.

17. Nedović, V., Gibson, B., Mantzouridou, T. F., Bugarski, B., Djordjević, V., Kalušević, A., Paraskevopoulou, A., Sandell, M., Šmogrovičová, D., & Yilmaztekin, M. Aroma formation by immobilized yeast cells in fermentation processes. *Yeast,* **2015**, *32*(1), 173–216.

18. Aranda, V., Martínez-Hernández, Y., & Anna, I. Production of enzymes of industrial interest through *Mucor gryseocyamus* using whey as a source of nutrients and *Omputia imbricate* as support. Master's Thesis. Autonomous University of Coahuila, **2008**.

19. Mata, G. M. A. Production of fungal enzymes and their application in the production of compounds of pharmaceutical interest. Master's Thesis. Autonomous University of Coahuila, **2009**.

20. Cano, J. C., Uresti N. D., Ilina, A., & Martínez, J. L. *Lipase Production by Surface-Adhesion Fermentation (SAF) Of Filamentous Fungi from Coahuila's Semi-Desert,* **2010**.

21. Abdullah, N., Iqbal, M., & Zafar S. I.. Potential of immobilized fungi as viable inoculum. *Mycologist,* **1995**, *4*(9), 168–171.

22. Akhtar, Z. N. Impact assessments of transgenic cry1ab rice on the population dynamics of five non-target thrips species and their general predatory flower bug in Bt and non-Bt rice fields using color sticky card traps. *Journal of Integrative Agriculture,* **2004**.

23. García, G. C. Design of lipase biocatalysts and their application in bioprocesses. PhD's Thesis. Autonomous University of Madrid, Spain, Universidad Autonoma de Madrid, España, **2014**.

24. Ooijkaas, L. P., Weber, F. J., Buitelaar, R. M., Tramper, J., & Rinzema, A., Defined media and inert supports: Their potential as solid-state fermentation production systems. *Trends Biotechnol.,* **2000**, *18*, 356–360.

25. Rocha-Pino, Z., Vigueras, G., Sepúlveda-Sánchez, J. D., Hernández-Guerrero, M., Campos-Terán, J., Fernández, F. J., & Shirai, K. The hydrophobicity of the support in solid state culture affected the production of hydrophobins from *Lecanicillium lecanii. Process Biochemistry,* **2015**, *50*(1), 14–19.

26. Ciappellano, S., Tedesco, E., Venturini, M., & Benetti, F. *In vitro* toxicity assessment of oral nanocarriers. *Advanced Drug Delivery Reviews,* **2016**, *106*, 381–401.

27. Astruc D., F. Lu, J. R., & Aranzaes, A. Nanoparticles as recyclable catalysts: The frontier between homogeneous and heterogeneous catalysis. *Chem. Int. Ed.,* **2005**, *44*, 7852–7872.

28. Luo, X. L., Morrin, A., Killard, A. J., & Smyth, M. R. Application of nanoparticles in electrochemical sensors and biosensors. *Electroanalysis,* **2006**, *18*, 319–326.
29. Shen, Y. C., Tang, Z., Gui, M., Cheng, J. Q., Wang, X., & Lu, Z. H. Nonlinear optical response of colloidal gold nanoparticles studied by hyper-Rayleigh scattering technique. *Chem. Lett.,* **2000**, 1140–1141.
30. Jazirehpour, M., & Seyyed, E., S. A. Effect of aspect ratio on dielectric, magnetic, percolative and microwave absorption properties of magnetite nanoparticles. *Journal of Alloys and Compounds,* **2015**, *638*, 188–196.
31. Seabra, A. B., Pelegrino, M. T., & Haddad, P. S. Antimicrobial applications of super-paramagnetic iron oxide nanoparticles: Perspectives and challenges. *Micro and Nano Technologies* (pp. 531–550). Elsevier, **2017**.
32. Li, G. Y., Jiang, Y. R., Huang, K. L., Ding, P., & Chen, J. Preparation and properties of magnetic Fe3O4-chitosan nanoparticles. *Journal of Alloys and Compounds,* **2008**, *466*(1/2), 451–456.
33. Hai, N. H., Luong, N. H., Chau, N., & Tai, N. Q. Preparation of magnetic nanoparticles embedded in polystyrene microspheres. *Journal of Physics: Conference Series,* **2009**, *187*(1), 1–6.
34. Xu, Y., Mahmood, M., Fejleh, A., Li, Z., Watanabe, F., Trigwell, S., et al. Carbon-covered magnetic nanomaterials and their application for the thermolysis of cancer cells. *International Journal of Nanomedicine,* **2010**, *5*, 167–176.
35. Shagholani, H., Ghoreishi, S. M., & Sharifi, S. H. Conversion of amine groups on chitosan-coated SPIONs into carbocyclic acid and investigation of its interaction with BSA in drug delivery systems. *Journal of Drug Delivery Science and Technology,* **2018**, *45*, 373–377.
36. Ali, A., & Ahmed, S. A review on chitosan and its nanocomposites in drug delivery. *International Journal of Biological Macromolecules,* **2018**, *109*, 273–286.
37. Mitra, S., Priyam, B., Ratan, G., & Joydeep, M. Cellulase and xylanase activity in relation to biofilm formation by two intertidal filamentous fungi in a novel polymethylmethacrylate conico-cylindrical flask. *Bioprocess and Biosystems Engineering,* **2011**, *34*(9), 1087–1101.
38. Rakchai, N., H-Kittikun, A., & Zimmermann, W. The production of immobilized whole-cell lipase from *Aspergillus nomius* ST57 and the enhancement of the synthesis of fatty acid methyl esters using a two-step reaction. *Journal of Molecular Catalysis B: Enzymatic,* **2016**, *133*, S128–S136.
39. Gricajeva, A., Kazlauskas, S., Kalėdienė, L., & Bendikienė, V. Analysis of *Aspergillus* sp. lipase immobilization for the application in organic synthesis. *International Journal of Biological Macromolecules,* **2018**, *108*, 1165–1175.
40. Khootama, A., Putri, D. N., & Hermansyah, H. Techno-economic analysis of lipase enzyme production from *Aspergillus niger* using agro-industrial waste by solid state fermentation. *Energy Procedia,* **2018**, *153*, 143–148.
41. Prabaningtyas, R. K., Putri, D. N., Utami, T. S., & Hermansyah, H. Production of immobilized extracellular lipase from *Aspergillus niger* by solid state fermentation method using palm kernel cake, soybean meal, and coir pith as the substrate. *Energy Procedia,* **2018**, *153*, 242–247.
42. Tacin, M. V., Massi, F. P., Fungaro, M. H. P., Teixeira, M. F. S., De Paula, A. V., & De Carvalho, S. E. V. Biotechnological valorization of oils from agro-industrial wastes to produce lipase using *Aspergillus* sp. from Amazon. *Biocatalysis and Agricultural Biotechnology,* **2019**, *17*, 369–378.

43. Pabai, F., Kermasha, S., & Morin, A. Interesterification of butter fat by partially purified extracellular lipases from Pseudomonas putida, *Aspergillus niger* and *Rhizopus oryzae*. *World J Microbiol Biotechnol,* **1995,** *11*, 669–677.
44. Nadeem, H., Rashid, M. H., Siddique, M. H., Azeem, F., Muzammil, S., Javed, M. R., Ali, M. A., Rasul, I., & Riaz, M. Microbial invertases: A review on kinetics, thermodynamics, physiochemical properties. *Process Biochemistry,* **2015,** *50*(8), 1202–1210.
45. De Vries, R. P., & Visser, J. *Aspergillus* enzyme involved in degradation of plant cell wall polysaccharides. *Microbiology and Molecular Biology Reviews,* **2001,** *65*, 497–522.
46. Neumann, N. P., & Lampen, J. O. Purification and properties of yeast invertase. *Biochemistry,* **1967,** *6*(2), 468–475.
47. Guo, P. C., Wang, Q., Wang, Z., Dong, Z., He, H., & Zhao, P. Biochemical characterization and functional analysis of invertase Bmsuc1 from silkworm, *Bombyx mori*. *International Journal of Biological Macromolecules,* **2018,** *107*, 2334–2341.
48. Chand B. T., Bansuli, T. N., & Savitri, T. N. Invertase of *Saccharomyces cerevisiae* SAA-612: Production, characterization and application in synthesis of fructo-oligosaccharides. *Food Science and Technology,* **2017,** *77*, 178–185.
49. Oyedeji, O., Bakare, M. K., Adewale, I. O., Olutiola, P. O., & Omoboye, O. O. Optimized production and characterization of thermostable invertase from *Aspergillus niger* IBK1, using pineapple peel as alternate substrate. *Biocatalysis and Agricultural Biotechnology,* **2017,** *9*, 218–223.
50. Mirela, M. G., Materi, S., Pele, M., Dumitrescu, F., & Matei, A. Invertase production by fungi, characterization of enzyme activity and kinetic parameters. *Revista de Chimie.,* **2017,** *68*(10), 2205–2208.
51. Kulshrestha, S., Tyagi, P., Sindhi, V., & Yadavilli, K. S. Invertase and its applications: A brief review. *Journal of Pharmacy Research,* **2013,** *7*(9), 792–797.
52. Belcarz, A., Ginalska, G., & Lobarzewskij, A. The novel non-glycosylated invertase from *Candida utilis* (the properties and the conditions of production and purification). *Biochim. Biophys. Acta,* **2002,** *1594*, 40–53.
53. Gascon, S., Neumann, N. P., & Lampen, J. O. Comparative study of the properties of the purified internal and external invertases. *Journal of Biological Chemistry,* **1968,** *243*, 1573.
54. Carlson, C., & Botstein, D. Two differentially regulated MRNA with different 5' ends encode secreted and intracellular forms of yeast invertase. *Cell,* **1982,** *28*, 145–154.
55. Vargas, L., Pião, A. C. S., Domingos, R. N., & Carmona, E. C. Ultrasound effect on invertase from *Aspergillus niger*. *World J. Microbiol. Biotechnol.,* **2004,** *20*, 137–142.
56. Boddy, L. M., Bergés, T., Barreau, C., Vainstein, M. H., Dobson, M. J., Balance, D. J., & Peberdy, J. F. Purification and characterization of an *Aspergillus niger* invertase and its DNA sequence. *Curr. Genet.,* **1993,** *24*, 60–66.
57. Chen, Ch. Liu. Production of β-fructofuranoside by *Aspergillus japonicus*. *Enzyme Microb. Technol.,* **1996,** *18*, 153–160.
58. Mukherjee, S., & Suman, K. Regulation of cellobiase secretion in termitomyces clypeatus by co-aggregation with sucrase. *Current Microbiology,* **2002,** *45*(1), 70–73.
59. Mitchell, D. A., Berovic, M., & Krieger, N. Overview of solid state bioprocessing. *Biotechnol. Annu. Rev.,* **2002,** *8*, 183–225.
60. Romero, G. Production of invertase by *Aspergillus niger* in liquid fermentation and solid fermentation. PhD's Thesis. Autonomous Metropolitan University. México, D.F, **2001.**
61. Dinarvand, M., Rezaee, M., & Foroughi, M. Optimizing culture conditions for production of intra and extracellular inulinase and invertase from *Aspergillus niger* ATCC 20611 by

response surface methodology (RSM). *Brazilian Journal of Microbiology*, **2017**, *48*(3), 427–441.

62. Marsh, A. J., Mitchell, D. A., Stuart, D. M., & Howes, T. O_2 uptake during solid- state fermentation in a rotating drum bioreactor. *Biotechnology Letters*, **1998**, *20*(6), 67–61.

63. Salihu, A., Bala, M., & Alam, M. Z. Lipase production by *Aspergillus niger* using sheanut cake: An optimization study. *Journal of Taibah University for Science*, **2016**, *10*(6), 850–859.

64. Sethi, B. K., Nanda, P. K., & Sahoo, S. Characterization of biotechnologically relevant extracellular lipase produced by *Aspergillus terreus* NCFT 4269.10. *Brazilian Journal of Microbiology*, **2016**, *47*(1), 143–149.

65. Eijkman, C. About enzymes in bacteria and molds. *Cbl. Bakt. Parasitenk Infektionskr.*, **1901**, *29*, 841–1901.

66. Lai, O. M., Lee, Y. Y., Phuah, E. T., & Akoh, C. C. Lipase/esterase: Properties and industrial applications. In: Melton, L., Shahidi, F., & Varelis. P. (eds.), *Module in Food Science Encyclopedia of Food Chemistry* (pp. 158–167). Oxford: Academic Press, **2019**.

67. Saraiva, R. N., Bandeira, P. B., Pessoa P. M., Mendes B. R., Dos Santos, J. C. S., & Barros, G. L. R. Biotechnological potential of lipases from pseudomonas: Sources, properties and applications. *Process Biochemistry*, **2018**, *75*, 99–120.

68. Jaeger, K. E. Lipases for biotechnology. *Current Opinion in Biotechnology*, **2002**, *13*, 390–397.

69. Fernández-Lorente, G., Palomo, J. M., Fuentes, M., Mateo, C., Guisán, J. M., & Fernández-Lafuente, R. Self-assembly of Pseudomonas fluorescens lipase into bimolecular aggregates dramatically affects functional properties. *Biotechnol. Bioeng*, **2003**, *82*, 232–237.

70. Palomo, J. M., Fuentes, M., Fernández-Lorente, G., Mateo, C., Guisán, J. M., & Fernández-Lafuente, R. General trend of lipase to self-assembly living bimolecular aggregates greatly modifies the enzyme functionality. *Biomacromol.*, **2003**, *4*, 1–6.

71. Camacho-Páez, Robles-Medina, Camacho, R., Esteban, C., & Molina-Grima, Y. Kinetics of lipase-catalyzed interesterification of triolein and caprylic acid to produce structured lipids. *Journal of Chemical Technology and Biotechnology*, **2003**, *78*, 461–470.

72. Neves, P. F., & Petersen, S. B. How do lipases and esterases work: The electrostatic contribution. *Journal of Biotechnology*, **2001**, *85*, 115–147.

73. Cheng, C., Jiang, T., Wu, Y., Cui, L., Qin, S., & He, B. Elucidation of lid open and orientation of lipase activated in interfacial activation by amphiphilic environment. *International Journal of Biological Macromolecules*, **2018**, *119*, 1211–1217.

74. Pascoal, A., Estevinho, L. M., Martins, I. M., & Choupina, A. B. Review: Novel sources and functions of microbial lipases and their role in the infection mechanisms. *Physiological and Molecular Plant Pathology*, **2018**, *104*, 119–126.

75. Malcata, F. X. Engineering of/ with lipases: Scope and strategies. In: Malcata, F. X., (ed.), *Engineering of/ with Lipases* (pp. 1–16). Netherlands: Kluwer Academic Publishers, **1996**.

76. Ransac, S., Carrière, F., Rogalska, E., Verger, R., Marguet, F., & Buono, G. The kinetics, specificities, and structural features of lipases. In: Malcata, F. X., (ed.), *Engineering of/ with Lipases* (pp. 143–182). Netherlands: Kluwer Academic Publishers, **1996**.

77. Petersen, S. B. Lipases and esterases: Some evolutionary and protein engineering aspects. In: Malcata, F. X., (ed.), *Engineering of with Lipases* (pp. 125–142). Netherlands: Kluwer Academic Publishers, **1996**.

78. Foresti, M. L., & Ferreira, M. L. Computational approach to solvent-free synthesis of ethyl oleate using *Candida rugosa* and *Candida antarctica* B Lipases. I. Interfacial activation and substrate (ethanol, oleic acid) adsorption. *Biomacromolecules*, **2004**, *5*(6), 2366–2375.

79. Mead, J. R., Irvine, S. A., & Ramji, D. P. Lipoprotein lipase: Structure, function, regulation, and role in disease. *J. Mol. Med.,* **2002,** *80,* 753–769.
80. Geoffry, K., & Achur, R. N. Screening and production of lipase from fungal organisms. *Biocatalysis and Agricultural Biotechnology,* **2018,** *14,* 241–253.
81. Hofvendahl, K., & Hahn-Hägerdal, B. Factors affecting the fermentative lactic acid production from renewable resources. *Enzyme. Microb. Technol.,* **2000,** *26,* 87–107.
82. Navvabi, A., Razzaghi, M., Fernandes, P., Karami, L., & Homaei, A. Novel lipases discovery specifically from marine organisms for industrial production and practical applications. *Process Biochemistry,* **2018,** *70,* 61–70.
83. Sahay, S., & Chouhan, D. Study on the potential of cold-active lipases from psychrotrophic fungi for detergent formulation. *Journal of Genetic Engineering and Biotechnology,* **2018,** *16,* 319–325.
84. Joshi, R., Sharma, R., & Kuila, A. Lipase production from *Fusarium incarnatum* KU377454 and its immobilization using Fe3O4 NPs for application in waste cooking oil degradation. *Bioresource Technology Reports,* **2019,** *5,* 134–140.
85. Undurraga, D., Markovits, A., & Erazo, S. Cocoa butter equivalent through enzymic interesterification of palm oil midfraction. *Process Biochem.,* **2001,** *36,* 933–939.
86. Eş, I., Vieira, J. D. G., & Amaral, A. C. Principles, techniques, and applications of biocatalyst immobilization for industrial application. *Applied Microbiology and Biotechnology,* **2015,** *99*(5), 2065–2082.
87. Tian, X., Dai, L., Liu, D., & Du, W. Improved lipase-catalyzed methanolysis for biodiesel production by combing in-situ removal of by-product glycerol. *Fuel,* **2018,** *232,* 45–50.
88. Wolf, I. V., Meinardi, C. A., & Zalazar, C. A. Production of flavor compounds from fat during cheese ripening by action of lipases and esterases. *Protein Pept. Lett.,* **2009,** *15,* 1235–1243.
89. Rubin, B., & Dennis, E. A. Lipases: Part B. In: *Enzyme Characterization and Utilization Methods in Enzymology* (Vol. 286, pp. 1–563). New York: Academic Press, **1997.**
90. Kazlauskas, R. J., & Bornscheuer, U. Biotransformations with lipases. In: Rehm, H. J., & Reeds, G., (eds.), *Biotechnology, 8a* (pp. 37–192). Wiley-VCH, New York, **1998.**
91. Klibanov, A. M. Immobilized enzymes and cells as practical catalysts. *Science,* **1983,** *219,* 722–727.
92. Arroyo, M., & Sinisterra, J. V. Influence of chiral corvones on selectivity of pure lipase-B from Candida Antarctica. *Biotechnol. Lett.,* **1995,** *17,* 525–530.
93. Biehl, R. R., & Baker, D. H. Efficacy of supplemental 1α-hydroxycholecalciferol and microbial phytase for young pigs fed phosphorus- or amino acid-deficient corn-soybean meal diets. *J. Anim. Sci.,* **1996,** *74,* 2960–2966.
94. Fan, M. Z., & Archbold, T. Novel and disruptive biological strategies for resolving gut health challenges in monogastric food animal production. *Animal Nutrition,* **2015,** *1*(3), 138–143.
95. Ojha, B. K., Singh, P. K., & Shrivastava, N. Enzymes in the Animal Feed Industry. In: Kuddus, M. (eds.), *Enzymes in Food Biotechnology* (pp. 93–109). Academic Press, **2019.**
96. Zentler-Munro, P., Assoufi, A., Balasubramanian, K., Cornell, D., Benoliel, D., Northfield, T., & Hodson M. Therapeutic potential and clinical efficacy of acid resistant fungal lipase in the treatment of pancreatic steatorrhoea due to cystic fibrosis. *Pancreas,* **1992,** *7,* 311–319.

CHAPTER 11

Recent Advances in the Bioconversion of 2-Phenylethanol Through Biotechnological Processes for Using as a Natural Food Additive

ITZA NALLELY CORDERO-SOTO,[1,2] SERGIO HUERTA-OCHOA,[2]
MARWEN MOUSSA,[3] LUZ ARACELI OCHOA-MARTÍNEZ,[1]
NICOLÁS OSCAR SOTO-CRUZ,[1] and
OLGA MIRIAM RUTIAGA-QUIÑONES[1]

[1]*National Technological Institute of Mexico, Technological Institute of Durango, Department of Chemical and Biochemical Engineering, 34080, Durango, Mexico, E-mail: omrutiaga@itdurango.edu.mx (O. M. Rutiaga-Quiñones)*

[2]*Autonomous Metropolitan University, Iztapalapa Campus, 09340 CDMX, Mexico, E-mail: sho@xanum.uam.mx (S. Huerta-Ochoa)*

[3]*UMR 782 Genetic and Microbiology of Food Processes (GMPA), AgroParisTech., INRA, Université Paris-Saclay, F-78850, Thiverval-Grignon, France*

ABSTRACT

2-Phenylethanol (2-PE) is an alcohol that emits an aroma of the fresh and delicate roses. Natural 2-PE is present in many plants, especially in the rose oil. Because of its delicate aroma and sweet taste, 2-PE is the second-most commonly used flavor after vanillin. 2-PE and its high valued derivatives can be used in food, tobacco, and cosmetics industries. The global annual output of 2-PE is nearly 10,000 tons in 2010. Chemical synthesis is the basic method used to produce 2-PE. However, benzene or styrene—the raw materials used to synthesize 2-PE—are regulated chemicals. Methods to produce natural

2-PE include plant extraction and microbial transformation. The production of 2-PE through biotransformation presents broad application prospects. Several microorganisms can produce 2-PE. However, yeasts with generally recognized as safe (GRAS) designation *Kluyveromyces marxianus* and *Saccharomyces cerevisiae* stand out among them. The drawback of the biocatalytic process is the inhibition of yeast growth by the accumulating product. To overcome this challenging limitation, production strategies have been developed, including identification of highly productive microorganism, biocompatible *in situ* product removal techniques (ISPR), use of industrial wastes as substrates, and more recently, genetic modification of microorganisms.

11.1   INTRODUCTION

Flavors have a high influence on foodstuff, beverages, cosmetics, cleaning products, and the pharmaceutical industry. Its economic contribution has increased in the last few decades. Thus, there is a growing interest in the industry and academia for developing the process of production of these numerous compounds [1]. The concept of flavor is used to explain the simultaneous reactions of taste and smell when food and beverages are placed in the oral cavity. Food is composed of a great variety of components—non-volatile or volatile—responsible for the sensation perceived by mouth and nasal cavity receptors [2, 3]. Volatile compounds pass through the nasal cavity and are absorbed by the olfactory receptors, reaching a stimulation that induces the odor. All components that cause these sensations are known as aromas, which are organic compounds at room temperature characterized by having a low molecular weight, and generally are present in beverages and foods at low concentrations [4, 5].

The aromatic compounds can be classified as chemical or natural aromas, according to the process of obtaining them. Currently, the chemical routes are the most common for the production of this type of compounds; however, in recent decades, studies have focused on the development and application of alternative technologies to produce aromatic compounds through natural routes. The reasons behind are that the consumer demand has a greater tendency towards the natural products, and the quality of the final product is affected by the chemical routes [4, 6–9].

According to the American Food and Drug Administration (FDA) and European regulation EEC 1334/2008, a natural compound is obtained physically or by enzymatic and microbiological processes [10]. Among the natural aromatic compounds synthesized by a variety of microorganisms, various molecules can be found, such as ethyl esters, acetate esters, aldehydes, ketone, higher alcohols, etc. [4, 7, 11, 12]. Natural isoamyl acetate, vanillin, and 2-phenyletanol are the most important molecules used as flavors and additives in the food industry [3, 5, 6].

11.2   2-PHENYLETHANOL: AROMA AND FLAVOR MOLECULE IN FOOD

The 2-Phenylethanol (2-PE) is a colorless liquid with a characteristic fresh rose aroma and sweet taste, which is considered as a high-value compound for the food, cosmetic, and fragrance industry [6]. The global demand for it in the market was approximately 10,000 tons in 2010 [10, 13]. Recently, it has started to have a major influence on the food industry, especially on the fermented food and beverages, where a significant number of volatile compounds are produced, by the microorganisms conferring flavor and aroma, and thus improving the quality of the product and meeting the consumer's interest. 2-PE has been used as an additive for the formulations of non-alcoholic beverages, candy, ice cream, and chewing gum [3, 6, 10, 14]. Table 11.1 presents some chemical and physical properties of 2-PE.

Naturally, 2-PE is found in essential oils (EOs) of flowers, especially in roses [15]. The presence of 2-PE has been identified in the significant volatile compounds of roasted coffee, honey, extra virgin olive oil (OO), the aromatic profile of alcoholic drinks, such as wine, beer, tequila, and mescal, conferring important sensory characteristics that influence the quality of these beverages [6, 14, 16–20].

2-PE is highly valued by consumers when it is produced by natural means, i.e., from rose petals or bioconversion through microorganisms. The reported cost of natural 2-PE is 1,000 dollars per kg, whereas 2-PE produced by chemical synthesis costs 3.5 dollars per kg. The difference in cost is because of the lower acceptance of chemical 2-PE than the natural one by consumers, caused by the less purified and lower quality final products by following the chemical pathways, such as catalytic reduction of styrene oxide and Friedel Craft reaction [6].

11.3   NATURAL PRODUCTION OF 2-PHENYLETHANOL BY MICROBIOLOGICAL WAYS

The Ehrlich pathway is the most used for the production of 2-PE by microbiological means [6, 11, 21–23]. It is carried out in three enzymatic reactions

TABLE 11.1 2-Penylethanol Physical and Chemical Properties

Formula	Molecular Structure	Molecular Weight	Boiling Point	$T_{Ignition}$	$T_{Melting}$	Density (20°C)	Solubility	Log P_{OW} Value	CAS Number
$C_8H_{10}O$	OH (phenylethanol structure)	122.2 g/mol	219.8°C	102°C	–27°C	1.02 g/cm^3	water 19 g/L (20°C), alcohols, esters, and aldehydes	1.36	60-12-8

for the bioconversion of 2-phenylethanol, from the amino acid L-phenylal-anine [21]. The first reaction involves a transamination reaction, in which ammonium is released, which is condensed with α-ketoglutarate, forming glutamate [24]. L-phenylalanine is transaminated to phenylpyruvate by a transaminase, then decarboxylated to phenyl acetaldehyde by phenylpyru-vate decarboxylase, which is finally reduced to 2-PE by a dehydrogenase [6] (Figure 11.1).

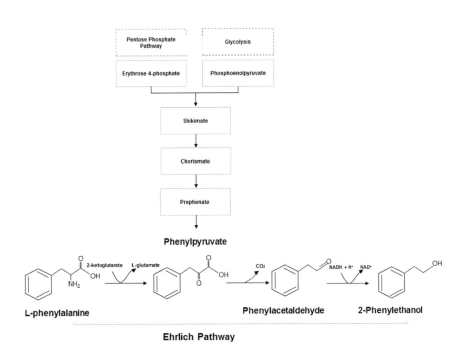

FIGURE 11.1 Metabolic pathway for 2-penyelthanol production.

However, using L-phenylalanine as a substrate increases the produc-tion cost of 2-PE. One of the alternatives for the production of natural 2-PE without the need for the bioconversion of L-phenylalanine is via de *novo* synthesis through the Shikimate pathway. The erythrose-4-phosphate and the phosphoenolpyruvate, coming from the Pentose phosphate route and the Glycolysis, respectively, are condensed into chorismate and prephenate as intermediaries forming phenylpyruvate [6]. This stage is catalyzed by the 3-deoxy-D-arabino-heptulosonate-7-phosphate synthase enzyme. The shikimate undergoes a series of reactions catalyzed by chorismate synthase enzyme forming the chorismate, which is converted to prephenate. The prephenate undergoes dehydration by means of the prephenate dehydratase enzyme, forming phenylpyruvate. From phenylpyruvate, there may be a transamination with glutamate, which acts as a donor of amino groups, forming L-phenylalanine. In this way, phenylpyruvate usually follows the Ehrlich route and is decarboxylated and reduced to 2-PE [6].

11.4 MICROORGANISMS AND STRATEGIES FOR 2-PHENYLETHANOL PRODUCTION

Several strategies have been explored in order to implement robust biotechnological processes for the production of 2-PE, such as the identification of highly productive microorganisms, *in situ* product removal techniques (ISPR), use of industrial waste to reduce the production costs, and more recently, genetic modification of microorganisms. In this sense, according to literature, the first systematic efforts were focused on the aromatic profiles of several strains. *Microbacterium* sp. and *Brevibacterium linens* bacteria were reported as 2-PE producer strains, using the standard soy medium culture [25]. Fungi strains, such as *Geotrichum penicillium* and *Ischnoderma benzoinum* presented the ability to produce 2-PE, however, these strains were not very efficient and only produced concentrations below 0.3 gL^{-1} [6]. Besides, *Aspergillus niger* strain was proved using L-phenylalanine as a precursor for the 2-PE production, reaching a concentration of 1.37 gL^{-1}, after nine days of fermentation [26].

Yeasts, on the other hand, have shown interesting results and are the most used microorganisms for the 2-PE bioconversion. *Saccharomyces cerevisiae* has been widely reported reaching high productions of 2-PE, especially when extraction systems are coupled to the bioconversion process [12, 22, 24, 27–34]. *S. cerevisiae* has also been used for the production of high-quality 2-PE aroma using L-phenylalanine, glucose, and ascorbic acid to suppress the formation of by-products by the improvement of redox reactions.

Etschmann et al. [35] studied different yeasts named, *Kluyveromyces marxianus, Saccharomyces cerevisiae, Zygosaccharomyces rouxii, Clavispora lusitaniae, Pichia anomala, Pichia membranaefaciens,* and *Schizosaccharomyces pombe* in a molasses medium. In the study, they concluded that the strain with the highest 2-PE production capacity was *Kluyveromyces marxianus* CBS 600 y CBS 397. *K. marxianus* yeast has been considered as promising yeast for flavor production because it has been widely used for 2-PE production [23, 36, 37]. Several characteristics made this strain interesting such as thermotolerance, rapid growth, and versatility on the uptake of a carbon source like lactose, glucose, or inulin, and that is secure yeast to be used in the food industry [3].

However, other strains have also shown a promising behavior for 2-PE bioconversion, but have been poorly studied for this process. Chrepto-wicz et al. [27] isolated 28 yeast strains from the biological sources and fermented food to evaluate the capability of 2-PE production on waste-based media supplemented with L-phenylalanine. From these yeasts, eight strains were selected after evaluated 2-PE production on the shake batch cultures, including species like *Meyerozyma caribbica*, *Metschnikowia chrysoperlae*, *Meyerozyma guilliermondii*, *Pichia fermentans*, and *Saccharomyces cerevisiae*. *Metschnikowia pulcherrima* strain (NCYC 373) was also proved by Chantasuban et al. [38] in an optimized system in flasks, producing adequate yields under *de novo* synthesis conditions and reaching a final concentration of about 1.5 gL^{-1}, once the process was scaling to batch and continuous operation on the stirred bioreactor. Moreover, *Candida glycerinogenes* WL2002-5 reported as robust and stress-tolerant yeast was evaluated for 2-PE production in batch fermentation, achieving a production of 5 gL^{-1} with the supplementation of L-phenylalanine. This is the highest concentration reported at the moment, for the case of fermentations without the coupling of the *in situ* recovery techniques [8].

Other studies have focused on the genetically manipulated microorganisms, mainly because bioconversions have not been competitive with the chemical routes regarding the operating costs. In this sense, sustainable strategies of the engineering microorganisms have been sought, where the results showed a favorable tendency for the 2-PE production [39–44]. Celińska et al. [39] reported a novel strain, *Yarrowia lipolytica* NCYC3825, genetically designed to produce 2-PE. This yeast was evaluated on a non-optimized culture, reaching a final concentration of 2 gL^{-1}. On the other hand, *K. marxianus*, genetically manipulated to overproduce 2-PE from glucose, reached a concentration of 1.3 gL^{-1} of 2-PE from 20 gL^{-1} of glucose, without the addition of L-phenylalanine as a precursor [41].

Moreover, a designed strain *Enterobacter* sp. CGMCC 5087, able to use renewable sugar and NH_4Cl as carbon and nitrogen source, respectively, produced 2-PE by *de novo* synthesis, following the phenylpyruvate route, reaching a production of 0.33 gL^{-1} a concentration 3.26 times higher than the wild strain [9]. However, the use of the modified strains presents several disadvantages. The concentrations achieved for these types of microorganisms are still not remarkable for the 2-PE production, and the stability of the strain could be questioned. Also, those native strains are nowadays preferred by consumers because of a growing tendency for natural products.

Recently, some strains have been proved on waste-based cultures. Non-conventional yeasts, such as *Metschnikowia chrysoperlae* WUT25 and *Pichia fermentans* WUT36 were used to produce the 2-PE through the hydrolyzed corn stover, reaching a final concentration of 3.67 gL^{-1} with *Pichia fermentans* yeast [45]. On the other hand, *K. marxianus* ITD00262 was evaluated on a cheese waste by-product, composed by lactose and proteins for the 2-PE production. Maximal production was found on the sweet whey, reaching a low concentration of 0.96 gL^{-1}. However, other strategies can be used to favor this process with further investigation of whey as a substrate for the aroma bioconversions [46] (see Table 11.2).

TABLE 11.2 2-Penylethanol Producer Microorganisms Reported on the Last Five Years

Strain	References
Y. lipolytica NCYC3825[i]	[39]
S. cerevisiae[i]	[40]
K. marxianus[i]	[41]
K. marxianus LOCK0024	[47]
Enterobacter sp. CGMCC 5087[i]	[9]
E. coli[i]	[42]
S. cerevisiae AS 2.1182	[48]
S. cerevisiae[i]	[43]
S. cerevisiae JM2014	[49]
C. glycerinogenes WL2002-5	[8]
K. marxianus ITD00262	[46]
E. coli[i]	[50]
S. cerevisiae strains	[51]
K. marxianus ATCC 10022	[52, 53]
M. chrysoperlae WUT25	[45]
P. fermentans WUT36	
M. caribbica, M. chrysoperlae, M. guilliermondii, P. fermentans, S. cerevisiae	[27]
M. pulcherrima NCYC 373	[38]
S. cerevisiae[i]	[44]
S. cerevisiae BCRC 21812	[54]

[i]Genetic engineered strain.

11.5 INTENSIFICATION OF THE NATURAL 2-PE BY COUPLING EXTRACTION TECHNIQUES

The most limiting factor for the 2-PE bioconversion is 2-PE toxicity causing by-product inhibition. The cell membrane is the main site where the effect of the toxicity of this compound is reflected because the transport systems of sugars and amino acids are affected along with the membrane permeability, which increases and accelerates the passive diffusion of ions and small metabolites [6]. Most of the microorganisms proved for the 2-PE inhibition are very sensitive to the inhibitory effect. However, some strains have shown high tolerance to 2-PE, such as *S. cerevisiae* and *Candida glycerinogenes* WL2002-5 tolerated around 4 gL^{-1} of 2-PE [8, 24]. The inhibition effect of the 2-PE has allowed a continuous investigation to overcome this problem, mainly developing *in situ* extraction techniques and integrated processes for the 2-PE production.

Several *in situ* product recovery techniques (ISPR) have been developed for the natural 2-PE recovery over the last decades, trying to reach a high-quality compound that can be used as a flavor and an aroma on the food [22, 23, 36]. One of the most recent and promising phenomena for the recovery of 2-PE is the immobilization of the solvent by membrane contactor. This technique has several advantages, such as easier operation, high affinity to the interest compound, and no toxic effect on the cells because of the contact with the solvent. Nevertheless, some of the main disadvantages of this technique are that in a system with high oxygen supply, the extraction process can be affected because of the bubbles formed inside the solvent transfer pipes. In addition, the final recovery of the product from the solvent is complicated because of the evaporation temperature [33, 34, 55]. Mihal et al. [33] evaluated an integrated process of microfiltration and membrane for the intensification of 2-PE produced by *S. cerevisiae*. The final concentration of 2-PE reached in a fed-batch bioreactor was 4 g L^{-1}. On the other hand, Adler et al. [55] studied an integrated bioconversion using a membrane extraction process, for the production of 2-PE by the yeast *Kluyveromyces marxianus* CBS 600. They developed a model based on data from the fermentation and separation processes and found that the overall product concentration was increased and was higher than the predicted by the model.

Pervaporation also has been proved for 2-PE recovery by Etschmann et al. [36]. An organophilic pervaporation unit was coupled to a bioreactor using the yeast *Kluyveromyces marxianus* CBS 600, reaching a final 2-PE concentration of 3.47 gL^{-1}. On the other hand, at 45°C, they found that 2-phenylethylacetate production was favored, producing 4 gL^{-1}. However, pervaporation has been poorly studied for 2-PE production; this may be due to it is necessary to use high temperatures to carry out the extraction of the aroma, affecting the viability of the microorganisms. The microencapsulation, on the other hand, was studied for 2-PE production in order to remove 2-PE and reduce the inhibition effect on the cells of *S. cerevisiae* using capsules that were formulated with a hydrophobic core of dibutylsebacate and an alginate-based wall. This process presented several advantages such as fast mass transfer, reduction of by-product inhibition, etc. However, the final overall concentration of 2-PE in a fed-batch culture was only increased from 3.8 to 5.6 gL^{-1}. This because a part of the inhibitory product dissolved in the dibutylsebacate core [29].

The liquid-liquid extraction has been widely reported for the 2-PE recovery presenting favorable results [22, 27, 36, 49]. This extractive technique has several advantages, such as low cost, simplicity, and mass transport. However, the main challenge for this technique is the extractant selection, because solvents are generally difficult to evaporate and confer residual aromas to the final product, affecting the final quality. Moreover, it must be biocompatible [23, 56]. Log $P_{O/W}$ is often used as a criterion in order to predict the biocompatibility of a given solvent. Furthermore, Laane et al. [57] established a correlation between solvent toxicity and its log $P_{O/W}$ value. Increased toxicity is associated with the increased polarity of the organic solvent [56, 58].

Stark et al. [22] carried out fermentations for the 2-PE production in a fed-batch process with the strain *Saccharomyces cerevisiae* GIV 2009, reaching a final concentration of 3.8 gL^{-1} of 2-PE. Later they used an *in situ* recovery technique by liquid-liquid extraction using oleic acid (OA) as organic phase and reported a concentration of 12.6 gL^{-1} of 2-PE. On the other hand, studies with ionic liquids using the yeast *S. cerevisiae* Ye9-612 found that 2-PE was increased, obtaining the concentration of 5.0 gL^{-1} [31]. Other authors proved rapeseed oil as the organic phase, with *S. cerevisiae* JM2014, reaching 9.7 gL^{-1} of 2-PE after 72 h. Rapeseed oil was also used as a 2-PE matrix, so that it can be directly used on the food, without a further separation [59]. Etschmann et al. [37] studied the strain *K. marxianus* CBS 600, in a fed-batch system using polypropylene glycol 1200 as the aqueous two-phase system. At the end of the process, they obtained a 2-PE concentration of 10.2 gL^{-1} and 24.0 gL^{-1} in the extraction phase.

The adsorption technique has been explored in the last decade for the 2-PE recovery. Several studies have focused on the adsorbent screening to select the one that presents adequate characteristics for the bioprocess, or even in the standard solution, as a promising material for further application on bioconversions [7, 16, 23, 28, 30, 60]. The adsorption processes have been considered as promising alternatives to recover 2-PE, because of several advantages, such as low cost of the adsorbent materials, regeneration, and lack of residual aromas on the final product. Moreover, adsorbents do not affect the viability of the cells and recovery of the interest compound is less complicated than the liquid extraction [23, 61]. Nevertheless, the main efforts for the adsorptive processes and their scaling are related to the selection of the adsorbent, where high selectivity and biocompatibility are needed [28, 30]. Gao and Daugulis [23] proved several materials for the 2-PE adsorption, including polyurethane and nylon. However, the most effective was the Hytrel® resin, because it showed compatibility with the cells in a coupled system and it was possible to produce 20.4 gL^{-1} of 2-PE.

On the other hand, Mei et al. [28] studied a non-polar macroporous resin D101 to enhance the biotransformation of L-phenylalanine to 2-PE, reaching a final concentration of 6.17 gL^{-1} using *S. cerevisiae* BD strain. Other approaches reported the use of macroporous resin FD0816 to avoid the toxic effects of 2-PE for *Saccharomyces cerevisiae* sp. (strain R-UV3), reaching a final concentration of 3.5 gL^{-1} of 2-PE [30]. The HZ818 resin was used as an adsorbent to reduce the toxicity of the culture, and the production of 2-PE was favored, yielding 6.6 gL^{-1} from 12 gL^{-1} of L-phenylalanine [32]. Table 11.3 summarizes the principal studies of 2-PE recovery with several extraction techniques.

11.6 ENVIRONMENTAL FACTORS INVOLVED IN THE MICROBIOLOGICAL PRODUCTION OF 2-PE

The 2-PE is a secondary metabolite that has been related to the microorganism growth. For this reason, the culture medium is a key factor for the production of this compound. Important nutrients, such as nitrogen, phosphorus, and sulfur are included in the formulation and optimization of the culture medium. On the other hand, studies have shown that the amino acids influence the production of higher alcohols depending on the balance established between their anabolic and catabolic routes, which are controlled by the amount of available nitrogen [6]. In the 2-PE production, the addition of the amino acid L-phenylalanine is important for the activation of the Ehrlich pathway to obtain higher concentrations [20]. However, recent studies have shown that with engineered microorganisms, 2-PE can be produced without L-phenylalanine, reducing the costs.

TABLE 11.3 2-Phenylethanol Production Through Biotechnological Processes Using an Extraction Technique

Bioproduction System	$C_{[2\text{-PE}]}$ (gL^{-1})	References
Extraction with oleic acid in a two-phase fed-batch system. *S. cerevisiae* Giv 2009	12.6	[22]
Oleyl alcohol as extractant in a two-phase system. *K. marxianus* CBS 600	3.0	[35]
Microencapsulation, capsules of 2.2 mm diameter, dibutyl sebacate, and an alginate-based wall in a fed-batch culture. S. cerevisiae Giv 2009	3.8 to 5.6	[29]
Pervaporation with an organophilic unit. K. marxianus CBS 600	3.47	[36]
PPG 1200 as organic phase in a two-phase fed-batch system. *K. marxianus* CBS 600	26.5	[37]
D101 nonpolar macroporous resin as adsorbent. S. cerevisiae BD	6.2	[28]
Hytrel as adsorbent on a two-phase partitioning system. *K. marxianus* CBS 600	20.4	[23]
Biphasic system with ionic liquids. MPPyr [Tf2N], OMA [Tf2N] or BMIM [Tf2N]. *S. cerevisiae* Ye9-612	~ 5.0	[31]
HZ818 resin as adsorbent in a liquid-solid extraction. *S. cerevisiae P-3*	6.6	[32]
FD0816 macroporous resin as adsorbent on a continuous system. *S. cerevisiae R-UV3*	3.5	[30]
Membrane contactor in a fed-batch culture. S. cerevisiae	4.0	[33]
PPG 1200 as organic phase *S. cerevisiae*	6.1	[40]
Ethyl acetate used as extractant from broth *S. cerevisiae* JM2014	3.60	[49]
Rapeseed oil as organic phase in a two-phase system. S. cerevisiae JM2014	18.50	[59]
Oleyl alcohol as *in situ* extraction Solid-phase extraction with activated carbon *M. pulcherrima* NCYC 373	3.0 14.0	[38]

Besides the culture media, the operation conditions of the system are also involved in 2-PE production, because they directly influence the production of biomass. The temperature may affect the growth of microorganisms—high temperatures can denature proteins that act as catalysts or transporters and may cause severe damage to the cell membrane [6]. The pH can also affect microbial growth, influencing the metabolism of the microorganisms and consequently, the products generated in the production system [13]. Several factors cause pH changes in a bioprocess, such as the formation of acid products, the entry of nutrients, and reduced agitation. Drastic changes of pH in the medium can damage microorganisms affecting the cell membrane or inhibiting the activity of enzymes and transport proteins in the membrane. In general, higher alcohols are formed at slightly acidic pH [6]. On the other hand, the oxygen supply also has a significant influence on the metabolism of microorganisms; when the culture medium is aerated, it increases cell growth and viability. The limited oxygen supply during the bioconversions may cause qualitative and quantitative changes in the lipid composition of the cell membrane [62].

11.7 CONCLUSION

The molecules of flavor and aroma have a significant effect on the quality standard of food and beverages. There is a growing demand for natural and friendly environment products nowadays. For this reason, in recent years, several efforts have been made to increase the production of these compounds by natural routes. Microbiological processes, by bioconversion of L-phenylalanine to 2-PE, have great potentials for producing this compound at the industrial scale; however, it is necessary to develop new strategies for the recovery and intensification of this high-value metabolite from the fermentation media.

KEYWORDS

- **bioproduction**
- **extraction**
- **generally recognized as safe**
- ***in situ* product removal techniques**
- **natural fragrance**
- **yeasts**

REFERENCES

1. Carlquist, M., Gibson, B., Karagul, Y. Y., Paraskevopoulou, A., Sandell, M., Angelov, A. I., et al. Process engineering for bioflavor production with metabolically active yeasts: A mini-review. *Yeast,* **2015,** *32,* 123–143.

2. Delwiche, J. The impact of perceptual interactions on perceived flavor. *Food Qual. Prefer.,* **2004**, *15,* 137–146.

3. Morrissey, J. P., Etschmann, M. M. W., Schrader, J., & De Billerbeck, G. M. Cell factory applications of the yeast *Kluyveromyces marxianus* for the biotechnological production of natural flavor and fragrance molecules. *Yeast,* **2015**, *32,* 3–16.

4. Cheetham, P. S. J. The use of biotransformations for the production of flavors and fragrances. *Trends Biotechnol.,* **1993**, *11,* 478–488.

5. Krings, U., & Berger R. G. Biotechnological production of flavors and fragrances. *Appl. Biochem. Biotechnol.,* **1998**, 1–8.

6. Etschmann, M. M., Bluemke, W., Sell, D., & Schrader, J. Biotechnological production of 2-phenylethanol. *Appl. Microbiol. Biotechnol.,* **2002**, *59,* 1–8.

7. Šimko, I., Roriz, E., Gramblička, M., Illeová, V., & Polakovič, M. Adsorption separation of 2-phenylethanol and L-phenylalanine on polymeric resins: Adsorbent screening, single-component and binary equilibria. *Food Bioprod. Process,* **2015**, *95,* 254–263.

8. Lu, X., Wang, Y., Zong, H., Ji, H., Zhuge, B., & Dong, Z. Bioconversion of L-phenyl-alanine to 2-phenylethanol by the novel stress-tolerant yeast *Candida glycerinogenes* WL2002-5. *Bioengineered,* **2016**, *7,* 418–423.

9. Zhang, H., Cao, M., Jiang, X., Zou, H., Wang, C., Xu, X., & Xian, M. De-Novo Synthesis of 2-phenylethanol by *Enterobacter* sp. CGMCC 5087. *BMC Biotechnol.,* **2014**, *14,* 30.

10. Hua, D., & Xu, P. Recent advances in biotechnological production of 2-phenylethanol. *Biotechnol. Adv.,* **2011**, *29,* 654–660.

11. Cordente, A. G., Curtin, C. D., Varela, C., & Pretorius, I. S. Flavor-active wine yeasts. *Appl. Microbiol. Biotechnol.,* **2012**, *96,* 601–618.

12. Eshkol, N., Sendovski, M., Bahalul, M., Katz-Ezov, T., Kashi, Y., & Fishman, A. Production of 2-phenylethanol from L-phenylalanine by a stress tolerant *Saccharomyces cerevisiae* Strain. *J. Appl. Microbiol.,* **2009**, *106,* 534–542.

13. Huang, C. J., Lee, S. L., & Chou, C. C. Production of 2-phenylethanol, a flavor ingredient, by *Pichia fermentans* L-5 under various culture conditions. *Food Res. Int.,* **2001**, *34,* 277–282.

14. León-Rodríguez, A., González-Hernández, L., Barba de la Rosa, A. P., Escalante-Minakata, P., & López, M. G. Characterization of volatile compounds of mezcal, an ethnic alcoholic beverage obtained from *Agave salmiana. J. Agric. Food Chem.,* **2006**, *54,* 1337–1341.

15. Hwang, J. Y., Park, J., Seo, J. H., Cha, M., Cho, B. K., Kim, J., & Kim, B. G. Simul-taneous synthesis of 2-phenylethanol and L-Homophenylalanine using aromatic trans-aminase with yeast Ehrlich pathway. *Biotechnol. Bioeng.,* **2009**, *102,* 1323–1329.

16. Carpiné, D., Dagostin, J. L. A., Da Silva, V. R., Igarashi-Mafra, L., & Mafra, M. R. Adsorption of volatile aroma compound 2-phenylethanol from synthetic solution onto granular activated carbon in batch and continuous modes. *J. Food Eng.,* **2013**, *117,* 370–377.

17. Piasenzotto, L., Gracco, L., & Conte, L. Solid phase microextraction (SPME) applied to honey quality control. *J. Sci. Food Agric.,* **2003**, *83,* 1037–1044.

18. Cajka, T., Riddellova, K., Tomaniova, M., & Hajslova, J. Recognition of beer brand based on multivariate analysis of volatile fingerprint. *J. Chromatogr. A.,* **2010**, *1217,* 4195–4203.

19. Lambrechts, M. G., & Pretorius, I. S. Yeast and its importance to wine aroma: A review. *South African J. Enol. Vitic.,* **2000**, *21,* 97–129.

20. Wittmann, C., Hans, M., & Bluemke, W. Metabolic physiology of aroma-producing *Kluyveromyces marxianus*. *Yeast*, **2002**, *19*, 1351–1363.

21. Hazelwood, L. A., Daran, J. M., Van Maris, A. J., Pronk, J. T., & Dickinson, J. R. The Ehrlich pathway for fusel alcohol production: A century of research on *Saccharomyces cerevisiae* metabolism. *Appl. Environ. Microbiol.*, **2008**, *74*, 2259–2266.

22. Stark, D., Münch, T., Sonnleitner, B., Marison, I. W., & Von Stockar, U. Extractive bioconversion of 2 phenylethanol from L phenylalanine by *Saccharomyces cerevisiae*. *Biotechnol. Prog.*, **2002**, *18*, 514–523.

23. Gao, F., & Daugulis, A. J. Bioproduction of the aroma compound 2-phenylethanol in a solid-liquid two-phase partitioning bioreactor system by *Kluyveromyces marxianus*. *Biotechnol. Bioeng.*, **2009**, *104*, 332–339.

24. Stark, D., Zala, D., Münch, T., Sonnleitner, B., Marison, I. W., & Von Stockar, U. Inhibition Aspects of the bioconversion of l-phenylalanine to 2-phenylethanol by *Saccharomyces cerevisiae*. *Enzyme Microb. Technol.*, **2003**, *32*, 212–223.

25. Jollivet, N., Bézenger, M. C., Vayssier, Y., & Belin, J. M. Production of volatile compounds in liquid cultures by six strains of coryneform bacteria. *Appl. Microbiol. Biotechnol.*, **1992**, *36*, 790–794.

26. Lomascolo, A., Lesage-Meessen, L., Haon, M., Navarro, D., Antona, C., Faulds, C., & Marcel, A. Evaluation of the potential of *Aspergillus niger* species for the bioconversion of L-phenylalanine into 2-phenylethanol. *World J. Microbiol. Biotechnol.*, **2001**, *17*, 99–102.

27. Chreptowicz, K., Sternicka, M. K., Kowalska, P. D., & Mierzejewska, J. Screening of yeasts for the production of 2-phenylethanol (rose aroma) in organic waste-based media. *Lett. Appl. Microbiol.*, **2018**, *66*, 153–160.

28. Mei, J., Min, H., & Lü, Z. Enhanced biotransformation of l-phenylalanine to 2-phenyl-ethanol using an *in situ* product adsorption technique. *Process Biochem.*, **2009**, *44*, 886–890.

29. Stark, D., Kornmann, H., Münch, T., Sonnleitner, B., Marison, I. W., & Von Stockar, U. Novel type of *in situ* extraction: Use of solvent containing microcapsules for the bioconversion of 2-phenylethanol from L-phenylalanine by *Saccharomyces cerevisiae*. *Biotechnol. Bioeng.*, **2003**, *83*, 376–385.

30. 30. Wang, H., Dong, Q., Meng, C., Shi, X. A., & Guo, Y. A continuous and adsorptive bioprocess for efficient production of the natural aroma chemical 2-phenylethanol with yeast. *Enzyme Microb. Technol.*, **2011**, *48*, 404–407.

31. Sendovski, M., Nir, N., & Fishman, A. Bioproduction of 2-phenylethanol in a biphasic ionic liquid aqueous system. *J. Agricul. Food Chem.*, **2010**, *58*(4), 2260–2265.

32. Hua, D., Lin, S., Li, Y., Chen, H., Zhang, Z., Du, Y., Zhang, X., & Xu, P. Enhanced 2-phenylethanol production from L-phenylalanine via *in situ* product adsorption. *Biocatal. Biotransformation*, **2010**, *28*, 259–266.

33. Mihal, M., Vereš, R., & Markoš, J. Investigation of 2-phenylethanol production in fed-batch hybrid bioreactor: Membrane extraction and microfiltration. *Sep. Purif. Technol.*, **2012**, *95*, 126–135.

34. Mihal', M., Vereš, R., Markoš, J., & Štefuca, V. Intensification of 2-phenylethanol production in fed-batch hybrid bioreactor: Biotransformations and simulations. *Chem. Eng. Process. Process Intensif.*, **2012**, *57–58*, 75–85.

35. Etschmann, M. M. W., Sell, D., & Schrader, J. Screening of yeasts for the production of the aroma compound 2-phenylethanol in a molasses-based medium. *Biotechnol. Lett.*, **2003**, *25*(7), 531–536.

36. Etschmann, M. M., Sell, D., & Schrader, J. Production of 2-phenylethanol and 2-phenylethylacetate from L-phenylalanine by coupling whole-cell biocatalysis with organophilic pervaporation. *Biotechnol. Bioeng.,* **2005,** *92,* 624–634.
37. Etschmann, M. M., & Schrader, J. An aqueous-organic two-phase bioprocess for efficient production of the natural aroma chemicals 2-phenylethanol and 2-phenylethylacetate with yeast. *Appl. Microbiol. Biotechnol.,* **2006,** *71,* 440–443.
38. Chantasuban, T., Santomauro, F., Gore-Lloyd, D., Parsons, S., Henk, D., Scott, R. J., & Chuck, C. Elevated production of the aromatic fragrance molecule, 2-phenylethanol, using *Metschnikowia Pulcherrima* through both de novo and ex novo conversion in batch and continuous modes. *J. Chem. Technol. Biotechnol.,* **2018,** *93,* 2118–2130.
39. Celińska, E., Kubiak, P., Białas, W., Dziadas, M., & Grajek, W. *Yarrowia Lipolytica*: The novel and promising 2-phenylethanol producer. *J. Ind. Microbiol. Biotechnol.,* **2013,** *40,* 389–392.
40. Kim, B., Cho, B. R., & Hahn, J. S. Metabolic engineering of *Saccharomyces cerevisiae* for the production of 2-phenylethanol via Ehrlich pathway. *Biotechnol. Bioeng.,* **2014,** *111,* 115–124.
41. Kim, T. Y., Lee, S. W., & Oh, M. K. Biosynthesis of 2-phenylethanol from glucose with genetically engineered *Kluyveromyces marxianus. Enzyme Microb. Technol.,* **2014,** *61–62,* 44–47.
42. Kang, Z., Zhang, C., Du, G., & Chen, J. Metabolic engineering of *Escherichia coli* for production of 2-phenylethanol from renewable glucose. *Appl. Biochem. Biotechnol.,* **2014,** *172,* 2012–2021.
43. Yin, S., Zhou, H., Xiao, X., Lang, T., Liang, J., & Wang, C. Improving 2-phenylethanol Production via Ehrlich pathway using genetic engineered *Saccharomyces cerevisiae* strains. *Curr. Microbiol.,* **2015,** *70,* 762–767.
44. Li, W., Chen, S. J., Wang, J. H., Zhang, C. Y., Shi, Y., Guo, X. W., Chen, Y. F., & Xiao, D. G. Genetic engineering to alter carbon flux for various higher alcohol productions by *Saccharomyces cerevisiae* for Chinese Baijiu fermentation. *Appl. Microbiol. Biotechnol.,* **2018,** *102,* 1783–1795.
45. Mierzejewska, J., Dąbkowska, K., Chreptowicz, K., & Sokołowska, A. Hydrolyzed corn stover as a promising feedstock for 2-phenylethanol production by non-conventional yeast. *J. Chem. Technol. Biotechnol.,* **2019,** *94,* 777–784.
46. Conde-Báez, L., Castro-Rosas, J., Villagómez-Ibarra, J. R., Páez-Lerma, J. B., & Gómez-Aldapa, C. Evaluation of waste of the cheese industry for the production of aroma of roses (Phenylethyl Alcohol). *Waste Biomass Valor.,* **2017,** *8,* 1343–1350.
47. Wilkowska, A., Kregiel, D., Guneser, O., & Karagul, Y. Y. Growth and by-product profiles of *Kluyveromyces marxianus* cells immobilized in foamed alginate. *Yeast,* **2014,** *32,* 217–225.
48. Tian, X., Ye, R., Wang, J., Chen, Y., Cai, B., Guan, S., Rong, S., & Li, Q. Effects of aroma quality on the biotransformation of natural 2-phenylethanol produced using ascorbic acid. *Electron. J. Biotechnol.,* **2015,** *18,* 286–290.
49. Chreptowicz, K., Wielechowska, M., Główczyk-Zubek, J., Rybak, E., & Mierzejewska, J. Production of natural 2-phenylethanol: from biotransformation to purified product. *Food Bioprod. Process,* **2016,** *100,* 275–281.
50. Wang, P., Yang, X., Lin, B., Huang, J., & Tao, Y. Cofactor self-sufficient whole-cell biocatalysts for the production of 2-phenylethanol. *Metab. Eng.,* **2017,** *44,* 143–149.

51. Mierzejewska, J., Tymoszewska, A., Chreptowicz, K., & Krol, K. Mating of 2 laboratory *Saccharomyces cerevisiae* strains resulted in enhanced production of 2-phenylethanol by biotransformation of L -phenylalanine. *J. Mol. Microbiol. Biotechnol.,* **2017**, *27,* 81–90.

52. Martínez, O., Sánchez, A., Font, X., & Barrena, R. Valorization of sugarcane bagasse and sugar beet molasses using *Kluyveromyces marxianus* for producing value-added aroma compounds via solid-state fermentation. *J. Clean. Prod.,* **2017**, *158,* 8–17.

53. Martínez, O., Sánchez, A., Font, X., & Barrena, R., Bioproduction of 2-phenylethanol and 2-phenethyl acetate by *Kluyveromyces marxianus* through the solid-state fermentation of sugarcane bagasse. *Appl. Microbiol. Biotechnol.,* **2018**, *102,* 4703–4716.

54. Shu, C. H., Chen, Y. J., Nirwana, C., & Cahyani, C. Enhanced bioconversion of L-phenylalanine into 2-phenylethanol via an oxygen control strategy and *in situ* product recovery. *J. Chem. Technol. Biotechnol.,* **2018**, *93,* 3035–3043.

55. Adler, P., Hugen, T., Wiewiora, M., & Kunz, B. Modeling of an integrated fermentation/membrane extraction process for the production of 2-phenylethanol and 2-phenylethylacetate. *Enzyme Microb. Technol.,* **2011**, *48,* 285–292.

56. Bruce, L. J., & Daugulis, A. J. Solvent selection-strategies for extractive biocatalysis. *Biotechnol. Prog.,* **1991**, *7,* 116–124.

57. Laane, C., Boeren, S., Vos, K., & Veeger, C. Rules for optimization of biocatalysis in organic solvents. *Biotechnol. Bioeng.,* **1987**, *30,* 81–87.

58. Barton, W. E., & Daugulis, A. J. Evaluation of solvents for extractive butanol fermentation with *Clostridium-Acetobutylicum* and the use of poly (propylene glycol) 1200. *Appl. Microbiol. Biotechnol.,* **1992**, *36,* 632–639.

59. Chreptowicz, K., & Mierzejewska, J. Enhanced bioproduction of 2-phenylethanol in a biphasic system with rapeseed oil. *N. Biotechnol.,* **2018**, *42,* 56–61.

60. Cordero-Soto, I. N., Rutiaga-Quiñones, O. M., Huerta-Ochoa, S., Saucedo-Rivalcoba, V., & Gallegos-Infante, A. On the understanding of the adsorption of 2-phenylethanol on polyurethane-keratin based membranes. *Int. J. Chem. Eng.,* **2017**, *15,* 1–22.

61. Kulkarni, S., & Kaware, J. Regeneration and recovery in adsorption: A review. *Int. J. Innov. Sci. Eng. Technol.,* **2014**, *1,* 61–64.

62. Lafon-Lafourcade, S., Geneix, C., & Ribereau-Gayon, P. Inhibition of alcoholic fermentation of grape must by fatty acids produced by yeasts and their elimination by yeast ghosts. *Appl. Environ. Microbiol.,* **1984**, *47,* 1246–1249.

CHAPTER 12

Natural Antimicrobials from Vegetable By-Products: Extraction, Bioactivity, and Stability

RICARDO GÓMEZ-GARCÍA, D. A. CAMPOS, A. VILAS-BOAS,
A. R. MADUREIRA, and M. PINTADO

*Universidade Católica Portuguesa, CBQF–Centro de Biotecnologia e
Química Fina–Laboratório Associado, Escola Superior de Biotecnologia,
Rua Arquiteto Lobão Vital 172, 4200-374 Porto, Portugal*

ABSTRACT

Nowadays, food plant agro-industrial processes are the main principal sources of a wide range of vegetable by-products, which have been considered as residues and despite its nutritional value are still discharged by industries, and their accumulation represents a severe environmental problem. The use of these plant by-products adds value to the industry by obtaining bioactive compounds (BCs) and essential oils (EOs), which currently could have many potential applications in medicinal and food fields. These compounds are well-recognized with some health potential benefits due to their bioactive properties like antioxidant activity and as antimicrobial agents against different microorganisms. A significant part of these BCs and EOs are not directly accessible in the vegetable by-products, and some studies have been advanced in order to develop new and efficient methodologies for their extraction through an environmentally friendly process and sustainable chemistry. Hence, this chapter reviews the compounds that can be extracted from plant sources, with antioxidant and antimicrobial activities and their potential applications.

12.1 INTRODUCTION

Over the last 10 years, considerable effort has been extended between many researchers' groups around the world to improve extraction of bioactive compounds (BCs) and essential oils (EOLs) from different sources, such as vegetable by-products from processing of beer, wine, sugar cane, coffee dregs, fruits, and vegetables among others [1]. This is the result of an increasing demand for natural and functional ingredients to replace the synthetic ones currently used owing to the proven some undesirable negative effects on the human body [2]. These agro-industrial matrices include a wide variety of by-products comprising pulp remnants, bagasse, peels, husks, stems and seeds [3]. Briefly, vegetable plants are well-documented for their high content of BCs and EOLs [4, 5] which play important roles as plants protectors from insects and microorganisms. These bioactive extracts may be used in the food industry with several applications as food fortifiers, coloring agents, and food preservatives [6]. Additionally, these compounds have been demonstrated effective bioactive properties such as anti-inflammatory, antioxidant, antiproliferative, and antimicrobial effects against different microorganisms such as bacteria and fungi [7]. Moreover, vegetable by-products also still hold a considerable number of BCs and EOLs, but not directly available making it challenging to extract them, and some research studies have been carried out in order to develop new and efficient methods for their recovery. These are environmentally friendly processes which under suitable chemistry context provide the valorization of these renewable organic materials reducing the pollution generated by their disposal [8].

The extraction process of these valuable molecules from raw matrices represents the most relevant stage in the recovery and downstream process, because plant by-products have a complex composition in polysaccharides, proteins, and secondary metabolites among others. Also, there are some different physiochemical properties between BCs and EOLs making their extraction the most challenging task. Conventionally, extractive methodologies such as water distillation, steam distillation, cohobation, cold, and hot maceration often imply complex and time-consuming multi-step process, thus leading to low selectivity, massive utilization of organic solvents, water pollution, and low recovery yields [9, 10]. In the present, BCs are obtained using methanolic solutions and chromatography (gel and ion-exchange); however, these conventional methods have some limitations, including high costs, low scale-up for industries purposes, and low yield of recovery and several cases losses of stability. While, EOLs have been extracted by traditional methods such as Soxhlet system, cold pressing, and distillation through the exposure to boiling water or steam [9]. These methods also have disadvantages, related to high-energy costs and long process time consuming.

Currently, several new methods have been tested in order to extract these important molecules from vegetable by-products focused on fast, efficient, and less cost extraction process with the quality enhancement of BCs and EOLs [9] such as microwave-assisted extraction (MAE), supercritical fluid extraction (SFE), ultrasound extraction, sub-critical water extraction and controlled pressure drop process [10, 11]. By using these extractive methodologies, the natural compounds obtained from agro-food by-products either as isolated compounds or extracts could allow a wide range of applications in food and pharmacological industries, since they are categorized as safe or GRAS grade [12], as well as, preserve their chemical stability exhibiting their good effective bioactive properties [13]. The specific aim of this chapter is to gather and critically review relevant information regarding extraction of BCs and EOLs that can be obtained through safe methodologies from the raw vegetable by-products discarded by agro-industries and envisage their potential applications in the modern market industries.

12.2 BRIEFLY HISTORY OF ANTIMICROBIAL AGENT

The first time in science history that a paper was published mentioning antimicrobial agents was in 1933, but only in 1950, the term natural antimicrobial agents was introduced by Kane et al. [14]. In such review, the use of natural sources (plants, microorganisms, and animals) was described and discussed which compounds could be considered natural antimicrobials. Until 1980, an increase in publication was observed concerning this subject, and more than 400 documents were published. Nevertheless, concerning the antimicrobials of natural sources only 8 published papers were found. In the next years several authors started to publish about the antimicrobial capacity of several compounds present in nature, such as polypeptides, phenols, and lipids. As an example, Watson, and Bloom [15] were the first authors to describe the antimicrobial properties of a tissue polypeptide and to compare the activity of a synthetic lysine polypeptide and the extracted one. In the same context, Jurd et al. [16] described for the first time the antimicrobial properties of natural phenols that were identified and extracted from *Dalbergia* a wood-related *Machaerium* species. In such species, several *C*-cinnamylated phenols were identified being the first report about the antimicrobial activity of obtusastyrene (*trans*-4-cinnamylphenol). Later, a synthesized variety of nuclear alkylated analogues of *C*-cinnamylated phenols, these compounds were tested against 22 microorganisms, Gram-positive and Gram-negative bacteria, but also yeast and molds [17]. The results showed that several di-*C*-cinnamylphenols, as well as various derivatives (acetates, benzoates, and methyl ethers) of the mono-*C*-cinnamylphenols showed no activity at high concentrations. Other derivatives demonstrated antimicrobial activity, such as 2-cinnamylphenol, dihydro derivative, obtusastyrene and dihydro-obtusastyrene that showed strong inhibition of Gram-positive bacteria (*Bacillus cereus, Sarcina lutea, Staphylococcus aureus*, and *Streptococcus lactis*) at concentration of 25 µg/mL, 11 species of yeasts and molds at concentrations until of 50 µg/mL inhibition of *Aspergillus* species at 50–100 µg/mL. Nevertheless, they were not effective against Gram-negative bacteria species. The study concluded that the natural extracted compounds showed antimicrobial activity against Gram-positive strains and low against fungi, owing to the nuclear alkylation of the cinnamylphenols, which introduces a methyl, ethyl, and propyl group in the active compounds [17].

The same authors preceded with the evaluation of other phenols such as umbelliferone and coumarins derivates [18–20]. But in this study, Jurd et al. [20] went further and evaluated the difference between natural extracted compounds and the synthetic ones. They reported that the natural umbelliferone derivatives were ineffective against the wide variety of yeasts and molds and associated the inhibitory effect of these derivatives with the 7-hydroxyl group and the antimicrobial activity may increase by the simple alkylation of this functional group. Concluding, the derivatives presented antimicrobial activity, and the synthetic derivatives presented higher antimicrobial activity but were not statistically significant from the natural compounds [20], Kabara et al. [21] described the antimicrobial properties of lipids and compared the activity between naturals and synthetic fatty-acids and monoglycerides, providing the first relationship between structure-function of lipids. As an example of the study, brassylic acid and its corresponding diesters were all inactive against the tested microorganisms, confirming that long-chain dicarboxylic acids are not antimicrobial. As monoglycerides, C_{11} and C_{13} fatty acids (FA) as showed to be less active than monolaurin (mono-ester formed from glycerol and lauric acid (LaA)). Hence, lower chain FA ($<_{16}$) are biologically more active when mono esterified to glycerol.

These examples also showed a representation of the antimicrobial activity of FA and its derivatives mainly against Gram-positive bacteria and yeasts, and open the possibility to pursuit the studies of antimicrobial activity of other FA [21]. In the 90s reports appeared about natural products containing secondary metabolites with potential antimicrobial activity. With this, Silver, and Bostian [22] applied for the need for innovation regarding antimicrobial research of discover and development of novel antibacterial and antifungal agents and his desire was accomplished owing to need to search for new chemotherapeutic agents. New publications appeared reporting the search of antimicrobial compounds, identification, and extraction, application in food, medicine, and finally incorporation in vehicles that increase the resistance to external factors, bioavailability, and bio-accessibility increasing.

Until know more than 14,000 manuscripts were published concerning natural antimicrobial agents, the most recent studies described the applications of extracts or EOs, as well as, the measure of the antimicrobial activity spectrum Ng et al. [23] described the antimicrobial and antioxidant activities of flavonoids and compared the activity of flavonoid-producing yeast with the natural flavonoids. Results showed that the phenolic compounds engineered exhibited good, broad-spectrum antibacterial activity surpassing all other pure flavonoids and phenolics tested in this study. Also, the flavonoids extract demonstrated to moderate to strong antimicrobial activities against common food pathogens, such as *Campylobacter* sp., *Salmonella* sp., *Staphylococcus aureus, Escherichia coli, Listeria monocytogenes,* and *Clostridium botulinum*. Thus, the authors conclude that the microbial phenolic compounds presented higher antioxidant and antimicrobial activity while compared than pure compounds, which could provide a cost-effective and sustainable source of BCs in the future [23].

Other recent work [24], described the encapsulation of two EOLs for food preservatives applications. Two EOLs (*Origanum vulgare* and *Thymus capitans*) with antimicrobial activity were studied and these extracts presented higher antimicrobial activity when compared with pure compounds, since they act synergistically with major components, such as phenolic compounds thymol and carvacrol, and other minor components, including monoterpene hydrocarbons such as *p*-cymene and γ-terpinene. Also, Granata et al. [24] described the interaction of these compounds with the food matrix and reported the loss of activity of these phenolic compounds. Therefore, the development of novel carriers to incorporate such BCs and understand the maintenance of bioactivity and bio-accessibility was also shown. Thus, the antimicrobial activity was tested against Gram-positive and Gram-negative bacteria, *Staphylococcus aureus, Escherichia coli, Pseudomonas aeruginosa,* and *Listeria monocytogenes*, and the authors concluded that the produced nanosystems could be applied to the EOLs as a natural alternative to prevent food-borne diseases. Also, the nanosystems demonstrated a high percentage of loading of EOLs and the encapsulation potentiated the antimicrobial activity of the EOLs against the tested bacteria [24].

Furthermore, several chapters and research papers are still being written by current the scientists of today, according to their experiences and results-focused on the improvement of availability, applicability, as well as, the discovery of new antimicrobial sources.

12.3 BIOACTIVE COMPOUNDS (BCS): PHENOLIC COMPOUNDS AND CAROTENOIDS

Since the beginning of civilization, vegetable plants have shown some health beneficial bioactivities in the treatment of common diseases such as analgesic, purgative painful discharges, dysuria, and anti-inflammatory properties attributed to the high content of BCs, through fresh consumption or aqueous extracts of the edible part [25]. Subsequently, many research studies have been carried out in order to investigate the content of these molecules in the different parts of vegetable tissues as pulp, seeds, leaf, and peels and to try to correlate with the different bioactive activities such as antioxidant and antimicrobial activities [26, 27]. Moreover, some studies have demonstrated that seeds and peels, which are mostly considered as by-products, have a higher amount of BCs and for those better biological activities than the edible part [28]. These BCs, mainly polyphenols, and carotenoids are the two types of molecules present on these by-products which are frequently responsible for the well-effective biological activities. Polyphenols are widely distributed in the plant kingdom and are the most copious group in the vegetable plant which is well-documented for their biological properties [29]. Nowadays, different classifications of polyphenols are available based on their chemical structure as tannins, lignans, flavonoids, and phenolic acids among others, which carry out different functions such as pigments and plant protective agents [7].

Many vegetable plants are an excellent source of carotenoids which are other class of phytochemicals responsible for the common pigments, such as lutein (yellow), lycopene (red) and b-carotene (orange) that occur in various fruits [30]. These compounds are tetraterpenoid pigments (C_{40}) and also can be classified in two groups based on the functional groups, such as xantho-phylls (lutein and zeaxanthin) that contain oxygen as a functional group or only a parent hydrocarbon chain without any functional group (β-carotene and lycopene) [31, 32]. All these natural BCs play an important role in the human diet due to their major bioactive activities, e.g., polyphenols could prevent some neurodegenerative diseases like Alzheimer and Parkinson [33] and carotenoids promote photoprotection in the eye with pro-vitamin A activity, enhancing immune function and prevention of chronic diseases

[34]. By these reasons and the existence of relevant reports on BCs extraction from raw sources, there is a potential to open a new field of investigation for researchers, nutritionists, and consumers in order to develop new functional products towards health promotion.

12.4 ESSENTIAL OILS (EOS) FROM LIPIDS FRACTION

In addition to the phenolic compounds, lipids are also considered as antimicrobial agents. The lipids are biomolecules that are soluble in nonpolar solvents and therefore hydrocarbons. Thus, these molecules could be present in the EOLs extracted from the vegetable by-products, unwrapping roles as antimicrobial agents. The lipids can be fractionated in several molecules, such as FA, waxes, sterols, fat-soluble vitamins (such as vitamins A, D, E, and K), monoglycerides, diglycerides, triglycerides, and phospholipids (PL). In the last decade, the valorization of vegetable by-products has been performed for the recovery of EOLs, which have been the target molecules due to their potential applications in food and pharmacological uses supported by their biological activities namely control of cholesterol metabolism (associated with cardiovascular illness) among others [3]. Ripe oils are an organic mixture of esters derived from glycerol with a chain of FA.

Physical and chemical characteristics of oils are greatly influenced by the type and proportion of the FA on the triacylglycerol [35]. Fatty acid can be classified depending of their length and number of carbons (C) into short-chain (2–8 C), medium (8–12 C) and long-chain (13–24 C) or into saturated (SFA), monounsaturated (MUFA) and polyunsaturated fatty acids (PUFA) associated to the presence or absence of double bonds [36]. Long-chain fatty acids (LCFA) are meanly composed of oleic acid (OA) (C18:1), linoleic acid (LA) (C18:2) and palmitoleic acid (C16:0) [37]. The FAs can be released from lipids, typically by enzyme action, becoming a free fatty acid (FFAs). The function of the lipids, FAs, and FFAs includes storing energy and can act as structural components of cell membranes. Their action as an antimicrobial agent, have been proved by the capacity to disrupt the bacterial cell membranes, hindering DNA replication or inhibited other intracellular targets. From all the lipids described in bibliography, the most powerful inhibitory structure are the long-chain FAs, when compared with the short chain, while the unsaturated are more effective than the saturated ones, thus the long-chain unsaturated fatty-acids (LC-PUFAs) have a bigger broad spectrum of activity and therefore lack of classical resistance mechanisms against the actions of these compounds.

12.5 BIOACTIVITIES OF NATURAL ANTIMICROBIALS

In the past, our antecedents used vegetable plants as treatment agents of several common diseases such as anti-inflammatory, analgesic, purgative, and dysuria. Along time all these properties were attributed to the high content of natural metabolites (BCs and EOLs) that at that time were achieved through normal fresh consumption of vegetable mesocarp (e.g., salads, jam or juices) which is the most studied part in terms of the determination of their nutritional composition [25, 38]. Moreover, many studies showed that these metabolites possess different beneficial biological activities such as anti-cancer, antioxidant, pro-vitamin A and antimicrobial activity. Currently, some other studies have been carried out in order to extract BCs and EOLs from vegetable by-products to explore *in vitro* their biological activities and some applicability for industrial purposes.

12.5.1 ANTIOXIDANT ACTIVITY

Antioxidant agents could be defined as a molecule or substance that could inhibit the oxidation of lipids or other biomolecules and thus, prevent or repair the damage of the body cell that is caused by free radical actions [38, 39]. Antioxidant activity of BCs, principally polyphenols, is due to their redox properties and is important for the absorption of neutralization of free radicals [40]. Regarding carotenoids, protective effects are referred by their antioxidant, redox-sensitive cell signaling and provitamin A properties [41]. The characteristic shape of alternating single and double bonds in the backbone of carotenoids allow them to absorb excess energy from other molecules, which may be related from their antioxidant property. The antioxidant activity provided by the BCs from natural sources is already well documented since they promote human health among other effects [31]. Also, in food industries are regularly used for food formulation to increase the oxidative stability and prolong the shelf life of food products. Several BCs used are chemically synthesized, such as, butylated hydroxyl anisole (BHA) and butylated hydroxyl toluene (BHT) which is categorized with some toxicity effects [42]. Actually, in Europe, the use of these compounds was prohibited, and the finding for new and natural ones has is being promoted. Tocopherols or vitamin E (E306) are examples of natural antioxidants that were suggested to be used as a replacer of the synthetic ones. Nevertheless, they showed good effect on the prevention of the lipidic oxidation on food and the prices could be 10 times higher than BHA or BHT (400 US$/Kg).

The high content of polyphenols and carotenoids present in vegetables and their by-products could be more explored as natural antioxidant agents for foods avoiding the negative impact of the synthetic ones.

12.5.2 ANTIMICROBIAL ACTIVITY

Antimicrobial agents are defined as a group of molecules that suppress multiplication and growth, can also promote the death of microorganisms such as bacteria, fungi, or viruses [43]. Usually, synthetic pesticides are used for control of postharvest microorganisms in vegetable plants; however, these agents still have some limitations because of the residues that remain in the products and that have been shown to be carcinogenic [44]. For these reasons, the focus has been given to identification and extraction of new natural antimicrobial agents from natural sources through eco-friendly extractive methodologies to surpass these constraints. Traditionally, some vegetables were found with anthelmintic effect and were used as an anti-parasitic agent [45]. This kind of effects is attributed to the concentration of polyphenols, which also are recognized with antimicrobial activity by the inhibition of enzymatic reactions which promote microorganisms' growth [7]. In several reports, the antimicrobial activity of EOLs was studied, and they have demonstrated high activity against several microorganisms, but in a general manner, the EOLs are more effective against to Gram-positive strains than Gram-negative. A study concerning grape seed and pine bark demonstrated inhibitory effect against several Gram-negative strains, with high virulence factors, such as *Escherichia coli* 0157:H7 and *Salmonella typhimurium*, but also demonstrate inhibition against *Listeria monocytogenes* a Gram-positive [46]. Also, a water extract of olive leaf at 0.6% (w/v) demonstrated inhibition after 3 h of contact, but when testing higher concentrations of extract a total inhibition was an exhibit for the two tested yeasts (*Dermatophytes* and *Candida albicans*) [47].

Other authors such as Pereira et al. [48] have linked the antimicrobial activity with several phenolic compounds present in a water extract of olive leaf such as 4-hydroxybenzoic, 3,4-dihydroxy-phenylacetic acids, and cinnamic acid derivatives (p-coumaric, ferulic, and caffeic acids). Besides, catechol, 4-methyl-catechol, tyrosol, hydroxytyrosol, verbascoside, and relatively high concentrations of flavonoids (luteolin-7-glucoside, apigenin-7-glucoside, rutin, and quercetin) are also found in olive by-products [49, 50]. The individual study of several identified compounds was performed in order to understand the extent of the antimicrobial activity of each phenolic compound. Thus, Fernandez-Bolanos et al. [50]) studied the antimicrobial activity of two main phenolic compounds present in olive water extract, hydroxytyrosol, and oleuropein. The results showed a large board of action against several strains for hydroxytyrosol, while oleuropein demonstrates antimicrobial activity but in a much lesser extent. However, the common factor between the reported activity and EOLs it was always the phenolic compounds composition present in each EOLs. All reported activity due to the presence of several derivates, such as hydroxytyrosol, carvacrol, eugenol, cinnamic, gallic acid and flavonoids between other derivatives [38].

12.5.3 GALLIC ACID AND FLAVONOIDS AS ANTIMICROBIAL COMPOUNDS

Gallic acid is also known as 3, 4, 5-trihydroxybenzoic acid is an important phenolic compound, and it is present in several plants, vegetables, tea leaves, and fruits, and therefore is present in their by-products derived from their processing. Beyond the pure compound of gallic acid, several derivates also where identified [51]. The literature reports reveal that this compound and its derivatives possess a wide spectrum of biological activities besides the antimicrobial potential [52]. Concerning the antimicrobial activity of gallic acid, Khatkar et al. [53] synthesized 33 derivatives of gallic acid and tested against several microorganisms, Gram-positive, and Gram-negative bacteria, yeast, and fungi. The relationship between structure and activity was compared and in was concluded that the gallic acid with ester and amides were more active as an antimicrobial agent than the anilides. Also, when the ester contained an aromatic group, these were more effective against Gram-negative bacteria (*Escherichia coli*) and yeasts (*Candida albicans*). On the other hand, when the gallic acid contained a bicyclic aromatic ring was more potent against Gram-positive bacteria (*Staphylococcus aureus*). Thus, the authors concluded that differences in structure promoted differences in broad-spectrum activity and different structures different targets [53], Chanwitheesuk et al. [52] studied the antimicrobial activity of this phenolic against several microorganisms (eight bacterial strains, one yeast, and five fungi), using different extraction solvents (water, ethanol, chloroform, and acetone). Results showed that all extracts possessed antimicrobial activity. The aqueous extracts tested at 250 mg/mL showed high activity against all bacteria and against dermatophytic fungi. On the other hand, the ethanolic extracts did not shown the same antimicrobial activity, because the extract it was not able to inhibit two Gram-negative strains (*Klebsiella pneumoniae* and *Pseudomonas aeruginosa*) and it was found that this was the only extract that contained gallic acid. Nevertheless, exhibited the activity against the bacteria (*Salmonella typhi* and *Staphylococcus aureus)* with the MIC values of 2500 and 1250 µg/ml, respectively [52]. Through several years, research works have described the antimicrobial activity of chemical made gallic acid presented notably good results against microorganisms; however, this synthetic molecule represents a costly inconvenience. Since this evidence, scientific efforts are focused in the extraction of gallic acid and its derivatives from different natural and cheap sources as vegetable by-products trying to solve worldwide issues like pollution and food supplies.

Flavonoids are phenolic structures which contain one carbonyl group synthesized by plants in response to microbial infections and are effective antimicrobial components to an extensive list of microorganisms [54]. Extracts from plant-based sources rich in flavonoids showed antimicrobial activity against Gram-negative bacteria (*Escherichia coli* and *Klebsiella pneumoniae*) and Gram-positive bacteria (*Staphylococcus aureus* and *Enterococcus faecalis*) [55–57]. Numerous research reports describe the antimicrobial activity of flavonoids, Wong et al. [58] described the total flavonoid content of palm kernel, a by-product of palm oil (PO) extraction and demonstrated a high level of total flavonoids, M'hiri et al. [59] described the bioactivities of lemon industrial by-product and where able to demonstrate the richness of this waste as a source of BCs, such as flavonoids (4.35 g quercetin/100 g dry basis).

On the other hand, some authors studied the antimicrobial activity of flavonoids, used leaves from *Trianthema decandra* L. to produce aqueous bioactive extracts, and after performed the isolation of flavonoids through chloroform extraction. The obtained crude extracts were tested against several Gram-positive and Gram-negative bacteria and the results showed high activity against *Enterococcus faecalis, Escherichia coli,* and *Pseudomonas vulgaris* and low inhibition for *Salmonella typhi* [54], Sati et al. [60] investigated the effect of the extraction method on the recovery of flavonoids glycosides and studied the antimicrobial and antioxidant activity of *Ginkgo biloba* leaves. The four methods of extraction tested were maceration, reflux, shaker, and Soxhlet and the antimicrobial activity was evaluated against Gram-positive and Gram-negative bacteria and fungi. The authors identified in all extracts three flavonoids, quercetin, kaempferol, and isorhamnetin and at different, proportions; regarding the antimicrobial activity, these had the highest antimicrobial activity against the Gram-positive bacteria (*Bacillus subtilis*) followed by Gram-negative bacteria (*Pseudomonasputida*) and fungi. In conclusion, the reflux method was the most efficient extraction method for recovery of flavonoid glycosides and also for obtaining the higher antimicrobial and antioxidant activities [60]. Until now, various research works evaluated the flavonoids from plant-based materials, but only in the last years, the researchers were focused in the industrial vegetable by-products which are categorized as rich sources of flavonoids with bioactive properties, opening high potential to valorize these raw materials into new industrial fields.

12.6   MECHANISM OF ACTION OF NATURAL ANTIMICROBIALS

The antimicrobial properties of BCs and EOLs have been attributed to some different action modes, in particular to their ability to interact with proteins and inhibit enzyme reactions. Hydrolysable tannins were shown to damage lipid bilayer membranes allowing to release of intracellular components, the mode of action by which they inhibit as an example the fungal growth is related is the complexation of metal ions.

The antimicrobial mechanism of action of natural agents can interact with the microorganisms in four different ways. The antimicrobial mechanism of action includes: (i) membrane rupture with ATP-ase activity inhibition, (ii) leakage of essential biomolecules from the cell, (iii) disruption of the proton motive force, and (iv) enzyme inactivation. In general, the phenolic compounds act by the change of the microbial cell permeability, allowing the loss of biomolecules from intercellular content; therefore, these compounds interfere with the membrane functionality, by the interaction of membrane proteins inducing changes in the structure and function of the cells. Overall, the affected functions are electron transport, nutrient uptake, synthesis of proteins and nucleic acid, and enzyme activity.

12.6.1   *DISRUPTION OF THE ELECTRON TRANSPORT CHAIN*

Several are the carriers in the electron transport chain, which are embedded within the membrane, pass electrons from one carrier to another until two electrons combine with the final acceptor, usually oxygen, and two protons to form water. The protons are exported from the inside of the cell while the concentration of electrons in the cytosol increases; this generates a proton gradient and membrane potential which are crucial for ATP production [61]. Therefore, the medium and long-chain saturated and unsaturated FFAs can bind to the carriers of the membrane causing the electron carriers to move apart or be displaced from the membrane entirely [62, 63], this result in a reduction in ATP production and the bacterium becomes deprived of an essential source of energy [64]. On the other hand, the detergent effect of FFAs could account for the loss of components through the membrane [63].

12.6.2 UNCOUPLING OF OXIDATIVE PHOSPHORYLATION

ATP synthase uses the energy from the proton motive force, which results from the proton gradient and membrane potential, created by electron transport chain, to convert ADP to ATP, the FFAs can interact with the inner part of the cell membrane, affecting this process, and reducing the ATP production [62]. The interaction of FFAs could be directed into the ATP synthase, preventing the well-function of the enzyme, therefore can interfere with the proton gradient and membrane potential, weakening the proton motive force upon which ATP synthesis relies [63].

12.6.3 CELL LYSIS

As explained before, the unsaturated FFAs onto the bacterial inner membrane causes it to become more fluid and permeable, thus the connection of these structures to the membrane leads to internal content to leak from the cell, which causes growth inhibition. This effect could lead to cell death [62], alternatively if the FFAs increase excessively the membrane can become unstable and the cell will ultimately lyse [65], these behaviors were noticed in several kinds of cells, bacteria, or even enveloped viruses [66]. Another part of the mechanism of action and when the FFAs have detergent action, these are able to solubilize large sections of the cell membrane, leading to cell lysis [63].

12.6.4 INHIBITION OF ENZYME ACTIVITY

The most important molecules being responsible for inhibition of enzyme activity, crucial for cell growth and multiplication, are the unsaturated FFAs [67], but also, high molecular weight phenolic compounds [68]. The mechanisms are based on the inhibition of enzymes in the membrane or cytosol that is crucial for bacterial survival and growth could account for some of the antibacterial effects. Also, this connection could cause altered and inappropriate cell membrane fluidity and permeability leading to the membrane-related problems described above [63].

12.6.5   IMPAIRMENT OF NUTRIENT UPTAKE

The FFAs can inhibit the ability of bacteria to take up nutrients, such as amino acids, thereby effectively starving the bacterium of the nutrients it requires to remain viable [62, 69]. As explained before, several are the BCs from the same family and with similar structures that show different grades of antimicrobial activity, for example, carvacrol, and thymol differ on the location of a hydroxyl group, the action intensity of antimicrobial activity is different. However, the mechanism of action is the same; these compounds increase membrane permeability, by dissolution of binding between the PL and the FAs chains of the cells. These molecules improve the penetration of other antimicrobial agents into the cell because FAs can increase cell membrane permeability [63], but also, the molecules can affect the expression of bacterial virulence factors of some bacteria, which are important or essential for the establishment of a contamination process, probably by disrupting cell-to-cell signaling. Thus, FFAs can prevent initial bacterial adhesion and subsequent biofilm formation [70]. Still, the action mechanism of FFAs are unclear, the only knowledge is about the prime target, the bacterial cell membrane and the various essential processes that occur within and at the membrane. The FFAs present detergent properties because of its amphipathic structures and therefore can create a transient or permanent pore, that can vary in pore size, changing the membrane permeability, this property can lead at the end to cell lysis, inhibition of enzymatic activity, impairment of nutrient uptake and the generation of toxic peroxidation and auto-oxidation products.

12.7   VEGETABLE BY-PRODUCTS AS SOURCES OF NATURAL ANTIMICROBIALS

As already, mentioned, vegetable plants and their by-products generated during agro-industrial activities are well-recognized for their high content of BCs and EOLs [71]. Currently, characterization, identification, and quantification of these molecules from fruits and vegetables by-products have been carried out in order to categorize them as an excellent source of these beneficial molecules which constitute the main class of natural antioxidants and antimicrobials present in plants-based matrixes. Indeed, phenolic compounds are higher accumulated in the peels and seeds, as opposed to the most pulp of the fruit and vegetables. By this reason, besides their interesting properties, extraction of these biomolecules from processed by-products has been a challenging task between researchers. For example, Cevallos-Casals et al. [72] observed a 3–4 fold higher phenolic concentration in the skin than in the flesh among plum cultivar, ranging from 292–672 mg chlorogenic acid/100 g fresh weight (FW) and also has 3–9 fold higher anthocyanin content compared to plum flesh.

Similarly, Valavanidis et al. [73] reported that apple peels contain 1.5–2.5 fold higher phenolic compounds compared to the apple mesocarp. Also, Pastrana-Bonilla et al. [74] informed that total phenolic content among muscadine grapes is about 80 times more in the seeds compared to a pulp. Melon pulp also showed less concentration of phenolics (168 mg gallic acid/100 g extract) than peels (470 mg gallic acid/100 g extracts) and seeds (285 mg/100 g extracts) [38]. Currently [75] reported the highest concentration of polyphenols in melon seeds (22.91 mg/100 g extract) followed by peels (14.1 mg/100 g extract) and the lowest concentration in pulp (9.54 mg/100 g extract). Recently, melon seeds have been gained great interest aimed at their rich content of oil around 30.60% on a dry basis [76] and its FAs and BCs profile. Three predominant FAs are identified in the oil of the seeds, such as LA (68.98%), OA (15.84%), and palmitic acid (8.71%) among others in less concentration such as steric, myristic, and auric acid as well as phenolic acids and flavones [77]. Citrus fruits are used in food industries for producing fresh juice or citrus-based drinks, through these daily processes enormous amounts of by-product are generated, mainly peels [78]. Orange, lemon, and tangerine by-products represent a rich source of EOLs and BCs with bioactive properties [79]. Moreover, while flavonoids are abundant elsewhere in the plant kingdom, there are several compounds (e.g., flavanones, flavanone glycosides, and polymethoxylated flavones) unique to citrus, which are relatively rare in other plants [80, 81].

Presently, there are different numerous of vegetable by-products with a significant content of BCs and EOLs, which are categorized as natural antimicrobial agents with several other bioactive properties. Owing to the richness in such compounds, there is an increase in research of new eco-green friendly methodologies from the raw tissues which are discarded as residues and their disposal/accumulation cause environmental pollution and poorly valorized [7]. Regarding these facts, there is a growing interest in the recovery of natural compounds from these organic materials under the valorization context [71, 82–84].

12.8 EXTRACTIVE METHODOLOGIES OF NATURAL ANTIMICROBIALS FROM VEGETABLE BY-PRODUCTS

BCs and EOLs have been widely studied and are attributed with several important biological activities; however, part of them are not they are usually not free in the vegetable tissues, and are embedded within the plant cellular matrix, so it is necessary to apply extractive methodologies for their recovery [85]. The aim of extraction, besides extracting the added-value molecules of interest, is to maximize the yields, while minimizing the extraction of undesirable compounds. In the past, fresh plant material was used for the extraction, however, nowadays the extraction of BCs and EOLs from plants tissues are mostly done using dried plant as the starting material in order to inhibit the metabolic processes which can cause degradation of the active compounds.

The drying process decreases the water activity (A_w), which in turn decrease the enzyme activity which can promote the degradation of the material and BCs present. Also, the drying facilitates the milling process which is very important to decrease the particle size and increase the extraction yields owing to increase of solvent reaction surface [86]. Typically, extraction of BCs and EOLs from natural sources has been obtained through conventional methods such as water distillation, steam distillation, combined water, and steam distillation, cohobation, maceration with alcohol-water mixtures or hot fats and enfleurage. Steam and water distillation are the techniques most often used for extraction from plant tissues [2, 87] together with Soxhlet extraction which has been used for many years in the extraction of lipophilic compounds from vegetable matrixes with organic solvents as mobile phase. Soxhlet is often used as a reference for evaluating other solid-liquid extraction methods or new non-conventional extraction methods [88].

Moreover, in the performance of carrying out extraction of BCs and EOLs, there are a lot of factors that affect the efficiency of the extraction that are important to know and keep in mind before to develop the recovery process such as, extraction method, solvent type, and polarity, ratio solvent/ plant material, temperature, time, pH, particle size as well the matrix and target compounds [2, 89]. Depending on the polarity of the target molecules, the extraction process may need to be afforded by mixtures of methanol and water [90]; so the water can extract the more hydrophilic compounds and the methanol the more hydrophobic compounds although. As example, Morais et al. [91] employed pure methanol to extract polyphenols from different processed vegetable peels such as avocado [181.17 mg of gallic acid equivalents (GAE)/100 g dry weight (DW)], pineapple (428.13 mg GAE/100 g

DW), banana (425.24 mg GAE/100 g DW), papaya (756.96 mg GAE/100 g DW), passion fruit (504.06 mg GAE/100 g DW), watermelon (75.10 mg GAE/100 g DW) and melon (357.80 mg GAE/100 g DW).

Additionally, Table 12.1 shows a general spectrum of diverse studies focuses on the extraction of BCs from different origins of vegetable by-products. Methanol has been shown good extraction yields, still has some toxicity characteristics with a negative impact in human health. This fact has promoted using another extractor agent such as ethanol which is considered as a non-toxic solvent and "generally recognized as safe" (GRAS) and presents many uses in pharmaceutical and food industries as well as shown higher global extraction yields than methanol [86]. The pH of the extractor solvent can also influence the extraction yield of certain BCs for example, anthocyanins are unstable at neutral or alkaline pH and because these acidic aqueous solvents are often used for the extraction of these target compound [92].

Nevertheless, the conventional techniques still maintain some drawbacks which include low yields of recovery, selectivity, and stability, thermal decomposition of thermolabile molecules, as well as, pollution issues involving long extraction time that typically implies large quantities of organic solvents, heat, and agitation systems [93]. To overcome these limitations, non-conventional extractive methodologies (eco-friendly and innovative methodologies) have already developed for improvement these unfavorable facts and it has led to the development of more innovative approaches, with higher yields, low processing times and minimal consumption of extract solvents, water, and energy [2, 94]. These eco-friendly methodologies include ultrasound-assisted extraction (UAE), MAE, pulsed electric fields (PEF), SFE, enzymes assisted extraction and microbial processes. In addition, today still are some limitations of innovative technologies concerning the very high investment costs, full control of the variables during the process operation, lack of regulatory approval and well-defined industrial applications.

12.9 ULTRASOUND-ASSISTED EXTRACTION (UAE)

Ultrasound treatment is considered as an accessible, practical, and reproducible method for the extraction of value molecules. Recently, a considerable number of articles have been published focused on pointing out all of the benefits uses of UAE over conventional extraction concerning extraction efficiency, use of non-toxic and GRAS solvents and less energy and time consumption aimed at the employment of various vegetable by-product sources [95].

TABLE 12.1 General Spectrum of Some Studies Focused in the Extraction of BCs and EOLs from Different Vegetable by-Products

a. Conventional Extraction

Agro-Food	By-Products	Extraction Method	Type of Compounds	Yield	Bioactivity	References
Plumb	Skin	Methanol	Polyphenols	292–672 mg chlorogenic acid/100 g FM	Antioxidant/DPPH 1254 to 3244 µg/g FW Antimicrobial activity *E. coli* 0157:H7 *S. enteritidis*	[72]
Watermelon	Rinds	Ultrasonic water bath Methanol (70%v/v, phosphoric acid)	4-Hydroxybenzoic acid	958.3 µg/g FW	Antioxidant/DPPH 39.7% inhibition	[90]
			Vanillin	851.8 µg/g FW		
			Chlorogenic acid	196.3 µg/g FW		
			Caffeic acid	41.4 µg/g FW		
			Cumaric acid	8.8 µg/g FW		
Lemon	Peel	Shoxlet system	EOLs	0.95%	nd	[9]
Mango Passion fruit	Seeds	Cold extraction with chloroform, methanol, and water	Phenolic compounds	130.7 to 271.5 mg EAG/kg	Antioxidant/ DPPH	[3]
Papaya Melon			Fatty acids Carotenoids	38.8 to 79.4% 4.4 to 11.9 µg β-carotene/g	ABTS FRAP	
Mango	Peels	Water 121C for 10 min	Polyphenols	72.61 mg/g DM	Antioxidant/DPPH 50% inhibition	[26]

TABLE 12.1 *(Continued)*

Agro-Food	By-Products	Extraction Method	Type of Compounds	Yield	Bioactivity	References
					Antimicrobial	
					Colletotrichum gloeosporioides, Sclerotinia sclerotiorum, Mucor sp.	
					Fusarium oxysporum	
Melon	Seeds	Ethanol (60%v/v) Shaking at 30°C for 24 h	Linoleic acid	68.98%	nd	[76]
			Oleic acid	15.84%		
			Palmitic acid	8.71%		
			Amentoflavone	32.80 µg/g		
			Luteolin-7-O-glycoside	9.60 µg/g		
			Gallic acid	7.26 µg/g		
			b. Eco-Friendly Extraction			
Grape	Pomace	SSC	Gallic acid	9.0 mg/g DM	DPPH 81.4% inhibition	[28]
					ABTS 87% inhibition	
					LOI 85% inhibition	
Grape	Pomace	EAE	Gallic acid	nd	DPPH	[35]
			Resorcinol		80 to 90% inhibition	
			O-cumaric acid			
Orange	Peel	SFE	EOLs	5 to 6%	nd	[10]
Tomato	Peels	MAE	Lycopene	13.592 mg/100 g DM	nd	[132]

TABLE 12.1   (*Continued*)

Agro-Food	By-Products	Extraction Method	Type of Compounds	Yield	Bioactivity	References
Olive	Pomace	Multi-frequency Multi-mode Modulated (MMM) ultrasonic technique	Oleic acid	75%	Antioxidant activity FRAP 7 µmol FSE/ mL	[4]
			Palmitic	10%		
			Linoleic acid	9%		
			Steric acid	3%	DPPH 1200 mg TE/mL	
			Hydroxytyrosol	83.6 mg/100 g DM		
			Tyrosol	3.4 mg/100 g DM		

DM: dry matter; FW: fresh matter; LOI: Lipid oxidation inhibition; nd: non-defined.

Ultrasound is a special type of sound wave beyond human hearing. Usually, the lowest frequency applied is 20 kHz and the higher 100 mHz [96]. In this technique, the ultrasound waves pass through a medium by creating compression and expansion: this phenomenon is called cavitation which means production, growth, and collapse of bubbles [85]. The passage of ultrasound waves through a liquid matrix leads to the generation of mechanical homogenization, and can create a strong vibration between sample molecules, facilitating the mass transfer between solute and solvent, which facilitates organic and inorganic compounds leaching from plant matrix into the solvent [97].

The UAE mechanism requires two main types of physical phenomena, first the diffusion through the cell wall and second, rinsing the content from the cell after membrane and wall severe damage [96]. The main factors influencing UAE are frequency and power, solvent (solvent properties, including vapor pressure, surface tension, viscosity, and density) [98], temperature, pressure, time, sample moisture content and sample texture [96]. Khan et al. [99] evaluated several parameters that could potentially affect the efficiency of the UAE of polyphenols (especially flavanone glycosides) from orange peels. Observed that smaller particles decreased extraction rates since at smaller sizes particles started floating, thereby limiting their exposition to ultrasonic waves. An optimal concentration of 70.3 mg naringin and 205.2 mg hesperidin/100 g fresh peels was achieved at a temperature of 40°C, ethanol: water ratio a 4:1 (%v/v) and a sonication power of 150 W. It is believed that this technique can increase the total phenolic yield due to cavitation and mechanical effects on cell walls, leading to membrane damage and the consequent increase of contact surface area between residue containing the compounds of interest and appropriate solvent phases. As such, the greater the caused damage, the greater will be the obtained extraction yield, that is why is necessary to optimize the UAE method [97].

The UAE has been widely used to extract BCs from vegetable and vegetable by-products. Anthocyanins from black chokeberry wastes [100], grape by-products [101]; carotenoids from pomegranate [102], tomato wastes [103] and citrus peels [104]; lycopene from tomato wastes [105]. In a general manner, phenolic compounds can be extracted from several by-products plant-based sources as, citrus, apple, grapefruit, mandarin, grape, pomegranate, plum, lemon, and mango, on the other hand, the EOLs can also be obtained from kiwi, orange, pomegranate, and papaya. Therefore, the UAE can be effectively used to improved total phenolic compounds extraction, as well as, of specific phenolic compounds extraction.

12.10   MICROWAVE-ASSISTED EXTRACTION (MAE)

MAE is an advanced approach that uses microwave energy to release intracellular content along with the solvent, to heat and extract target molecules from plant sources. It was shown that MAE improved polyphenol extraction in shorter periods at lower temperatures and with less expenditure of solvent than conventional extraction [89]. The microwave is an electromagnetic radiation with a wavelength from 0.001 m to 1 m (i.e., with a frequency from 3×10^{11} to 3×10^{8} Hz), which can be transmitted as the wave [106]. When microwave passes across the medium, their energy may be absorbed and converted into thermal energy.

The energy is efficiently absorbed by some substances inside vegetable materials, especially the polar molecules like water causing the water inside the cell to transform into vapor. Consequently, the internal temperature of the plant cells increases drastically and increases the pressure on the cell wall with a modification of the physical properties of the biological tissues, which improving the porosity of the biological matrix (rupture the cell walls and/ or plasma membranes) and resulting in a better penetration of the extracting solvent and improved yield of the BCs [107]. For example, Liazid et al. [108] evaluated the MAE of anthocyanins from grape skins obtained 1857.9 of malvidin/Kg FW and allowed to identify three different anthocyanin derivatives such as malvidin 3-coumaroylglucoside (18.8 mg/Kg FW), malvidin 3-caffeoylglucoside (51.3 mg/Kg FW) and petunidin 3-*p*-coumaroylglucoside (17.5 mg/Kg FW) at 40% methanol, 100°C and 500 W for 5 min. Results of total malvidin are higher than that obtained by solid-liquid maceration (1545.9 mg of mavidin/Kg FW) using ultraturrax at 24,000 rpm with methanol and formic acid (95% v/v), while the anthocyanin derivatives were not detected.

On the other hand, since the lycopene is extremely nonpolar and susceptible to degradation, MAE potentially can offer efficient lycopene recovery from tomato peels. Ho et al. [109] observed that the optimal MAE conditions were determined with 10% ethyl acetate solvent at 400 W obtained 13.592 mg/100 g of the extracted all-trans-lycopene. Despite worthy extraction of BCs, some scientists prefer other extractive methods that MAE because microwave apparatus is expensive compared with others equipment and some cases still require some toxic solvents.

12.11   PULSED ELECTRIC FIELD (PEF)

The application of electrical currents has stimulated intensive research as a non-thermal process since the beginning of the past century; the first applications of PEFs were about the disintegration of biological material, as well as, for microbial inactivation as an alternative to pasteurization [110]. Due to these peculiar properties, PEF could have different potential applications; some studies have recently concentrated on the permeabilization of cell membranes, with the aim of enhancing mass transfer from the inner part of the cells. In fact, in plant cells permeabilization of the membrane is easier to attain, usually requiring lower electric field intensities, which is reflected in lower energy consumption [111]. This is thanks to the reduction of the resistance to mass transfer due to the induced permeabilization of plant cells induced by the electroporation phenomenon [112]. Pulse electric field technology can be used as a pre-treatment to increase the yield to enhance the recovery of valuable intracellular compounds such as antioxidants from plant food materials, food wastes, and by-products. Recently, these tissues have shown be a promise green extraction technology for obtaining these compounds from different fruit and vegetable tissues [111, 113, 114] and an economic and sustainable point of view, mainly due to its ability to soften and disrupt the cell membrane, thus facilitating the release of intracellular compounds.

The PEF technology consists of placing the biological material between two electrodes and apply an electrical treatment of short time (from several nanoseconds to several milliseconds) with pulse electric field strength from 100–300 V/cm to 20–80 kV/cm which leads to cell membrane disruption, thus increasing cell permeability [115]. The charge accumulation on the membrane surface of biological material induces an increasing of transmembrane potential on both sides of the cell membrane, which is electrically pierced, causing lethal or sublethal damage to the cells, this rupture can release more intracellular high-added value compounds which are linked to the cell membranes [116,117]. Assisted extraction of polyphenols by PEF has been investigated on different vegetal matrices e.g., Corrales et al. [101] observed that the application of PEF on *Dornfelder* grape skins, using 3 kV/cm, 30 pulses, 10 kJ/kg, with a mixture of ethanol and water (50:50 v/v) and temperature (70 °C) increased the total polyphenolic content recovery approximately two-fold higher than the untreated control sample, consequently presented the highest antioxidant activity values (data obtained through ABTS· method). Also, they reported an increase on the total anthocyanin content with 17% up of extraction yields compared to the untreated control sample. Furthermore, this technology represents one of the most convenient emerging non-thermal technologies for BCs recovery from by-products, due to its high versatility, short treatment times and reduced heating effects [115], as well as, advantage related with its easy scale-up for pilot or industrial application [118].

12.12 SUPERCRITICAL FLUID EXTRACTION (SFE)

Every earthly substance has three basic physical states namely: solid, liquid, and gas. Supercritical state is a distinctive state and can only be attained when temperature and pressure exceed the critical values [93]. Supercritical fluid possesses gas-like properties of diffusion, viscosity, and surface tension, and liquid-like density and solvation power. These properties make it suitable for extracting compounds in a short time with higher yields [119]. In fact, SFE has significant advantages over conventional solvent methods: it is an environmentally friendly technique; it enhances extraction efficiency and selectivity, and avoids the expensive post-processing of the extracts for solvent elimination.

The extraction of compounds from natural sources is the most widely studied application of supercritical fluids [87]. Carbon dioxide (CO_2) is the most widely used supercritical fluid. This gas becomes a supercritical fluid at temperatures above 31.1°C and close to room temperature and 7380 kPa. The interest in supercritical-CO_2 (SC-CO_2) extraction increased due to the excellent solvent for nonpolar analytes and also CO_2 is readily available at low cost and has low toxicity. Even though SC-CO_2 has a poor solubility for polar compounds, modification such as adding a small amount of ethanol and methanol enables the extraction of polar compounds. For example, Paes et al. [120] employed SC-CO_2 for extraction phenolic compounds from fresh blueberry pomace. The authors achieved the highest recovery of phenolics (134 mg GAE/g) and anthocyanins (808 mg/100 g) at 20 MPa of pressure with 90% CO_2, 5% H_2O, and 5% ethanol.

12.13 ENZYMATIC ASSISTED EXTRACTION (EAE)

Fruits and vegetables have most of their polyphenols in the free or soluble conjugate forms but apart from these soluble compounds within the plant cell wall, there are bound phenolics to the cell wall linked to various polysaccharides such as hemicellulose, cellulose, and pectin [121], which are linked by hydrophobic interactions and hydrogen bonds. Other phenolic acids from either linkage with lignin through their hydroxyl groups in the aromatic ring or ester linkages with structural carbohydrates and proteins through their carboxylic groups [122]. In the case of flavonoids, they are covalently linked by a glycosidic bond with sugar moieties through hydroxyl (-OH) group (*O*-glycosides) or carbon-carbon bonds (*C*-glycosides) [122]. These links reduce the extraction efficiency of conventional extraction and of some green extraction methods [123, 124]. Bound polyphenols comprise an average of 24% of the total phenolics content present in these food matrices [123]. On the other hand, the basic principle of EAE is the disruption of plant cell wall by hydrolyzing through enzymes as a bound catalyst, to release the bound phenolics and increase the extraction yield.

Carbohydrate hydrolyzing enzymes such as pectinases, cellulases, amylases, hemicellulases, and glucanases had been effectively used to release bound polyphenols, because they can disintegrate the plant cell wall matrix and consequently facilitating the polyphenol extraction [125–127]. Cellulase, hemicellulase, pectinase, and protease cocktail form have been employed to extract oil from grape seed [128], pancreatin has been used to extract lycopene from tomatoes [129] and xylanase, amylase, papain are also used as a pre-treatment step before extraction in order to maximize the extraction yield [130]. The application of these enzymes promotes the release of bound phenolic compounds. In this instance, it is possible decrease the effect of solvent pre-treatment and either reduces the amount of solvent needed for extraction or increases the yield of extractable compounds [131]. The recent studies [132] with EAE application have shown faster extraction, higher recovery, reduced solvent usage and lower energy consumption when compared to non-enzymatic methods. Additionally, the efficiency of extraction by EAE depends on various parameters involving temperature, time of reaction, pH of system, solvent, and enzyme concentration, mode of actionenzymes and also particle size of the substrate [133]. In this scenario, some studies related the successful application of enzymes for the BCs extraction, including carotenoids from marigold flower [134], lycopene from tomato peels [135, 136] and corilagin (ellagitannin) and geraniin (dehydroellagitannin) from a plant [124]. Also, Zheng et al. [125] analyzed polyphenol extraction from unripe apples by an enzymatic commercial complex (Viscozyme L), the extraction was performed using pH 3.7, 50°C for 12 h (enzyme reaction) and ethanol (70%v/v). These parameters allowed the released phenolic compounds increasing total phenolics content and the extraction yield by about 3 and 2 times, respectively, together with the contents of *p*-coumaric, ferulic, and caffeic acids 8, 4, and 32 times, respectively. Also, the contents of phloretin, quercetin-3-glycoside, and quercetin increased about 25, 45, and 5 times, respectively, compared with the control treatment. Cuccolini et al. [137] extracted lycopene from tomato peels using hydrolytic enzymes and pH changes avoiding organic solvent. The final product shows a lycopene content around 8–10% w/w on a dry basis, which represents a 30-fold increase with respect to the lycopene concentration of the untreated tomato peels.

Similarly, in an attempt to valorize grape (*Vitis vinifera* L.) by-products, Gómez-García et al. [138] evaluated three different types of enzymatic complex (Celluclast®, 1.5 L Pectinex® Ultra and Novoferm®) for release of phenolic compounds increasing the free radical-scavenging activity (DPPH) in each treatment. These interesting studies reveal the reproducibility of phenolics extraction by enzymes from vegetable by-products, representing a potential, sustainable, and eco-friendly alternative against conventional methodologies.

12.14 MICROBIAL PROCESSES

During the last decade, a microbial process such as liquid culture (LC) and solid-state culture (SSC) have been employed under the biotransformation and valorization context of vegetable by-products in order to produce/extract value-added bioproducts. Both processes can achieve by microorganisms (fungi, bacteria or yeast), LC by fungi was the most common bioprocess applied for enzymes production [139], and therefore, it was used in the extraction of BCs, however, there is a lack of information regarding this fact which are reported with low production yields [140]. On the other hand, SSC also is a well-documented bioprocess for enzymes and polyphenols compounds production and extraction [8, 141]. This method presents some advantages against LC involving low energy consumption, less cost, more efficiency, better control parameters, less water content and high productivity of the target molecules.

The SSC has defined by various authors as a biotechnological method that develops not only in the surface but also within a solid porous matrix with the absence or nearby absence of free water but with enough moisture for microbial growth [142, 143]. Fungi are the microorganisms mostly used for their advantages related to better adaptation for the high requirement of oxygen and low content of free water promoting their development [144]. In SSC, fungi produce enzymes which are excreted into the medium, this is other important advantages during degradation of complex plant tissues. In this instance, microbial enzymes play an important role as in the same as in the EAE method. They can release the bound phenolics which are covalently linked to the structural components of cell-wall from the plant substrates [145]. In this microbial process, also there are some factors aimed at the obtainment of target molecules. For example, if it is necessary aeration flux (ratio), bioreactor characteristics; shape (columns, flask, trays, bags) and material (glass, plastic, aluminum), type of fungi, inoculum concentration, the particle size of substrate, pH, and temperature. These are the most considered factors that it is important to establish before to carry out the SSC.

Currently, there exist some different studies involving plant vegetable by-products employment as a solid support/substrate matrix on a dry basis, because they represent a natural attractive source of BCs with antioxidant and antimicrobial properties [146]. In this instance, many researchers prefer the application of SSC as an eco-friendly extractive methodology against EAE method due to, it employs costly commercial enzymes and on the other hand, SSC allows employing vegetable by-products in major quantity for BCs obtainment. Moreover, SSC is considered as multi-product production process in which it is possible to obtain more than one bioproduct such as microbial biomass, enzymes, and BCs for different purposes. Besides, utilization of these organic raw materials can solve quickly the pollution generated by their disposal.

12.15 STABILITY OF NATURAL ANTIMICROBIALS

The stability of BCs and EOLs in the vegetable plant is normally affected by many factors when they are added to the food preparation. Some factors related with (i) conditions of storage such as packaging, freezing, and aeration, (ii) enzymatic oxidation during slicing, peeling pulping or juicing, (iii) chemical factors as pH variation, oxidation of lipid and antioxidant constituents. All these factors also can contribute to the stability of these natural important molecules. Due to all of those unfavorable effects, nowadays, many studies of research have been focused in the utilization of vegetable plant-based derivatives considering as safety molecules, as well as to have both antioxidant and antimicrobial activity for food industrial applications in order to improve nutritional quality and shelf life. On the other hand, also to replace the synthetic ones used as preservatives which have been reported with potential toxicity to human health [147]. Natural vegetable extracts rich in BCs and EOLs are becoming important ingredients to accomplish these facts. Nevertheless, their incorporation to food matrixes is a challenging task in innovation and development terms due to their low stability when they are subject to processing and packing [148]. Drying and storage temperature, oxygen availability, and light conditions are important factors in determining the stability of carotenoids compounds: however, these factors have been only recently tested for maize with increased levels of vitamin A carotenoids. Many factors may determine carotenoids synthesis and accumulation in vegetable plants, but just as important to the food manufacturer and consumer are the influences on carotenoids stability after harvest.

Not only industrial food processing, cooking, and storage can have marked effect on content, but also form carotenoids acids, heat treatment and exposure to light and oxygen can promote degradation and isomerization of the all-trans-carotenoids to the cis isomers. The same effects are applied to some polyphenols, for instance, the direct incorporation of EOLs carvacrol as an active antimicrobial agent into foods is a challenge, because of the relative low water-solubility. This insolubility tends to provoke phase separation when mixed with water, which reduces product quality but also can lead to loss of activity because of the pH and composition of the food matrix [149]. This is a clear example of one of the problems associated with incorporating EOLs into commercial products, and novel strategies have been developed to overcome this problematic [149].

In this context, Chang et al. [150] encapsulated carvacrol within oil-in-water nanoemulsions, involved titrating an essential oil/ripening inhibitor/surfactant mixture into water. When measuring the antimicrobial activity of the carvacrol nanoemulsions, the results showed to be effective inhibition against Salmonella enteriditis and Escherichia coli O157:H7 in contaminated foods [151]. Also, Ryu et al. [149] produced carvacrol nanoemulsions, using another type of oil, and demonstrate a higher antimicrobial activity against the same microorganisms. Procopio et al. [152] described the application of cinnamon bark oleoresin as an antimicrobial agent, being delivered through microparticles, since oleoresins are susceptible to degradation by light, oxygen, and temperature, making difficult to maintain their stability. The extracts showed a high content of several phenolic compounds, such as cinnamic ester, eugenol, and coumarin, all susceptible to external factors. The antimicrobial activity was performed against several microorganisms and studied the oleoresin microparticles, as well as the extract alone; results showed that the process of microencapsulation leads to loss of the volatile antimicrobial compounds, decreasing the general antimicrobial activity.

The vegetable extracts and EOs have a high variety of compounds, and therefore different stability behaviors, the loss of these constituents can result in bioactivity decreases. Some of such compounds also have low bioavailability and are poorly absorbed either due to their large molecular size and poor lipid solubility and cannot be absorb by passive diffusion [153, 154]. But, other difficulties are encountered when there is a possibility of developing novel strategies to increase the stability and bioavailability of the BCs in the industrial manufacturing due to the poor long-term physical and chemical stability [155, 156]. The maintenance of stability of such compounds it is difficult, in addition, to find sustainable solutions that can be applied easily and at low cost have been the new challenging tasks for researchers.

12.16 CONCLUDING COMMENTS

By all the exposed in this chapter, the principal efforts of biotechnologists are focused in the approach to understand and learn the synergy between vegetable by-products and the better way to use them for further re-incorporation and valorization to solve the actual issues, such as human health, pollution, and food preservation, in order to develop effective and eco-friendly methodologies. This will permit to save saving operating outlays, time, and resources as well as to improve the extraction yields of BCs and EOLs maintaining and enhancing their stability regarding on the bioactive properties, mainly antioxidant and antimicrobial activity. Thereby, all the natural biomolecules obtained could increase the added value of the agro-food by-products considered as residues, since they can be potentially employed in many different industries.

KEYWORDS

- **antimicrobial properties**
- **antioxidant**
- **bioactive compounds**
- **extraction processes**
- **polyunsaturated fatty acids**
- **vegetable by-products**

REFERENCES

1. Yusuf, M. "Agro-industrial waste materials and their recycled value-added applications." *Handbook of Ecomaterials*, **2017**, pp. 1–11.
2. Giacometti, J., et al. "Extraction of bioactive compounds and essential oils from Mediterranean herbs by conventional and green innovative techniques: A review." *Food Research International*, **2018**.
3. Da Silva, A. C., & Jorge, N. "Bioactive compounds of the lipid fractions of agro-industrial waste." *Food Research International*, **2014**, *66*, 493–500.

4. Nunes, M. A., et al. "Olive pomace as a valuable source of bioactive compounds: A study regarding its lipid-and water-soluble components." *Science of the Total Environment,* **2018**, *644,* 229–236.

5. Rashid, U., Rehman, H. A., Hussain, I., Ibrahim, M., & Haider, M. S. "Muskmelon (*Cucumis melo*) seed oil: A potential non-food oil source for biodiesel production." *Energy, 36*(9), **2011**, 5632–5639.

6. Hąc-Wydro, K., Flasiński, M., & Romańczuk, K. "Essential oils as food eco-preservatives: Model system studies on the effect of temperature on limonene antibacterial activity." *Food Chemistry,* **2017**, *235*, 127–135.

7. Martins, S., Mussatto, S. I., Martínez-Avila, G., Montañez-Saenz, J., Aguilar, C. N., & Teixeira, J. A. "Bioactive phenolic compounds: Production and extraction by solid-state fermentation: A review." *Biotechnology Advances,* **2011**, *29*(3), 365–373.

8. Larios-Cruz, Ramón, et al. "Valorization of Grapefruit By-Products as Solid Support for Solid-State Fermentation to Produce Antioxidant Bioactive Extracts." *Waste and Biomass Valorization,* **2019**, *10*(4), 763–769.

9. Lopresto, C. G., Petrillo, F., Casazza, A. A., Aliakbarian, B., Perego, P., & Calabrò, V. "A non-conventional method to extract D-limonene from waste lemon peels and comparison with traditional soxhlet extraction." *Separation and Purification Technology,* **2014**, *137*, 13–20.

10. Hong, Y. S., & Kim, K. S. "Determination of the volatile flavor components of orange and grapefruit by simultaneous distillation-extraction." *Korean Journal of Food Preservation,* **2016**, *23*(1), 63–73.

11. Li, Y., Qi, H., Jin, Y., Tian, X., Sui, L., & Qiu, Y. "Role of ethylene in biosynthetic pathway of related-aroma volatiles derived from amino acids in oriental sweet melons (*Cucumis melo* var. *Makuwa Makino*)," *Scientia Horticulturae,* **2016**, *201*, 24–35.

12. Negi, P. S. "Plant extracts for the control of bacterial growth: Efficacy, stability, and safety issues for food application." *International Journal of Food Microbiology,* **2012**, *156*(1), 7–17.

13. Sepúlveda, L., Aguilera-Carbó, A., Ascacio-Valdés, J., Rodríguez-Herrera, R., Martínez-Hernández, J., & Aguilar, C. "Optimization of ellagic acid accumulation by *Aspergillus niger* GH1 in solid state culture using pomegranate shell powder as a support." *Process Biochemistry,* **2012**, *47*(12), 2199–2203.

14. Kane, J. H., Finlay, A., & Sobin, B. "Antimicrobial agents from natural sources." *Annals of the New York Academy of Sciences,* **1950**, *53*(2), 226–228.

15. Watson, D. W., & Bloom, W. L., "Antimicrobial activity of a natural and a synthetic polypeptide." *Proceedings of the Society for Experimental Biology and Medicine,* **1952**, *81*(1), 29–33.

16. Jurd, L., King, A., Mihara, K., & Stanley, W. "Antimicrobial properties of natural phenols and related compounds obtusastyrene I." *Applied Microbiology,* **1971**, *21*(3), 507–510.

17. Jurd, L., Stevens, K., King, A., & Mihara, K. "Antimicrobial properties of natural phenols and related compounds II: Cinnamylated phenols and their hydrogenation products." *Journal of Pharmaceutical Sciences, 60*(11), **1971**, 1753–1755.

18. Jurd, L., Corse, J., King, A. Jr., Bayne, H., & Mihara, K. "Antimicrobial properties of 6, 7-dihydroxy-, 7, 8-dihydroxy-, 6-hydroxy-and 8-hydroxycoumarins." *Phytochemistry,* **1971**, *10*(12), 2971–2974.

19. King, A., Bayne, H., Jurd, L., & Case, C. "Antimicrobial properties of natural phenols and related compounds: Obtusastyrene and dihydro-obtusastyrene." *Antimicrobial agents and Chemotherapy,* **1972**, *1*(3), 263–267.

20. Jurd, L., King, A. Jr., & Mihara, K. "Antimicrobial properties of umbelliferone derivatives." *Phytochemistry*, **1971**, *10*(12), 2965–2970.
21. Kabara, J., Vrable, R., & Lie, K. J. M. "Antimicrobial lipids: Natural and synthetic fatty acids and monoglycerides." *Lipids*, **1977**, *12*(9), 753–759.
22. Silver, L., & Bostian, K. "Screening of natural products for antimicrobial agents." *European Journal of Clinical Microbiology and Infectious Diseases*, **1990**, *9*(7), 455–461.
23. Ng, K. R., Lyu, X., Mark, R., & Chen, W. N. "Antimicrobial and antioxidant activities of phenolic metabolites from flavonoid-producing yeast: Potential as natural food preservatives." *Food Chemistry*, **2019**, *270*, 123–129.
24. Granata, G., et al. "Essential oils encapsulated in polymer-based nanocapsules as potential candidates for application in food preservation." *Food Chemistry*, **2018**, *269*, 286–292.
25. Shofian, N. M., et al., "Effect of freeze-drying on the antioxidant compounds and antioxidant activity of selected tropical fruits." *International Journal of Molecular Sciences*, **2011**, *12*(7), 4678–4692.
26. Rojas, R., et al. "Valorization of mango peels: Extraction of pectin and antioxidant and antifungal polyphenols." *Waste and Biomass Valorization*, **2018**, pp. 1–10.
27. Goulas, V., & Manganaris, G. A. "Exploring the phytochemical content and the antioxidant potential of citrus fruits grown in cyprus," *Food Chemistry*, **2012**, *131*(1), 39–47.
28. Martínez-Ávila, G. C., Aguilera-Carbó, A. F., Rodríguez-Herrera, R., & Aguilar, C. N. "Fungal enhancement of the antioxidant properties of grape waste." *Annals of Microbiology*, **2012**, *62*(3), 923–930.
29. Amaro, A. L., Oliveira, A., & Almeida, D. P. "Biologically active compounds in melon: Modulation by preharvest, post-harvest, and processing factors." In: *Processing and Impact on Active Components in Food* (pp. 165–171). Elsevier, **2015**.
30. Namitha, K., & Negi, P. "Chemistry and biotechnology of carotenoids." *Critical Reviews in Food Science and Nutrition*, **2010**, *50*(8), 728–760.
31. Saini, R. K., Nile, S. H., & Park, S. W. "Carotenoids from fruits and vegetables: Chemistry, analysis, occurrence, bioavailability, and biological activities." *Food Research International*, **2015**, *76*, 735–750.
32. Poiroux-Gonord, F., Bidel, L. P., Fanciullino, A. L., Gautier, H. L. N., Lauri-Lopez, F. L., & Urban, L. "Health benefits of vitamins and secondary metabolites of fruits and vegetables and prospects to increase their concentrations by agronomic approaches." *Journal of Agricultural and Food Chemistry*, **2010**, *58*(23), 12065–12082.
33. Rodriguez-Amaya, D. B. "Natural food pigments and colorants." In: *Bioactive Molecules in Food* (pp. 1–35). Springer, **2018**.
34. Dorni, C., Sharma, P., Saikia, G., & Longvah, T. "Fatty acid profile of edible oils and fats consumed in India." *Food Chemistry*, **2018**, *238*, 9–15.
35. Kostik, V., Memeti, S., & Bauer, B. "Fatty acid composition of edible oils and fats." *Journal of Hygienic Engineering and Design*, **2013**, *4*, 112–116.
36. Zhang, C., Su, H., Baeyens, J., & Tan, T. "Reviewing the anaerobic digestion of food waste for biogas production." *Renewable and Sustainable Energy Reviews*, **2014**, *38*, 383–392.
37. Tadmor, Y., et al. "Genetics of flavonoid, carotenoid, and chlorophyll pigments in melon fruit rinds." *Journal of Agricultural and Food Chemistry*, *58*(19), **2010**, 10722–10728.
38. Ismail, H. I., Chan, K. W., Mariod, A. A., & Ismail, M. "Phenolic content and antioxidant activity of cantaloupe (*Cucumis melo*) methanolic extracts." *Food Chemistry*, **2010**, *119*(2), 643–647.
39. Ascacio-Valdés, J. A., et al. "Fungal biodegradation of pomegranate ellagitannins." *Journal of Basic Microbiology*, **2014**, *54*(1), 28–34.

40. Cervantes-Paz, B., et al. "Antioxidant activity and content of chlorophylls and carotenoids in raw and heat-processed Jalapeño peppers at intermediate stages of ripening." *Food Chemistry*, **2014**, *146*, 188–196.
41. Lutterodt, H., Slavin, M., Whent, M., Turner, E., & Yu, L. L., "Fatty acid composition, oxidative stability, antioxidant and antiproliferative properties of selected cold-pressed grape seed oils and flours." *Food Chemistry*, **2011**, *128*(2), 391–399.
42. Musumeci, T., & Puglisi, G. "Antimicrobial agents." In: *Drug-Biomembrane Interaction Studies* (pp. 305–333), Elsevier, **2013**.
43. Cespedes, C. L., et al. "New environmentally-friendly antimicrobials and biocides from Andean and Mexican biodiversity," *Environmental Research*, **2015**, *142*, 549–562.
44. Vishwakarma, V. K., Gupta, J. K., & Upadhyay, P. K. "Pharmacological importance of *Cucumis melo* L.: An overview." *Asian Journal of Pharmaceutical and Clinical Research*, **2017**, *10*(3), 8–12.
45. Ruiz, B., & Flotats, X. "Citrus essential oils and their influence on the anaerobic digestion process: An overview." *Waste Management*, **2014**, *34*(11), 2063–2079.
46. Gadang, V., Hettiarachchy, N., Johnson, M., & Owens, C. "Evaluation of antibacterial activity of whey protein isolate coating incorporated with nisin, grape seed extract, malic acid, and EDTA on a turkey frankfurter system." *Journal of Food Science*, **2008**, *73*(8), M389–M394.
47. Markin, D., Duek, L. & I. Berdicevsky. "In vitro antimicrobial activity of olive leaves. Antimicrobial effectiveness of olive leaves in vitro." *Mycoses*, **2003**, *46*(3–4), 132–136.
48. Pereira, A. P., et al. "Phenolic compounds and antimicrobial activity of olive (*Olea europaea* L. *Cv. Cobrançosa*) leaves." *Molecules*, **2007**, *12*(5), 1153–1162.
49. Sousa, A., et al. "Phenolics and antimicrobial activity of traditional stoned table olives 'alcaparra.'" *Bioorganic and Medicinal Chemistry*, **2006**, *14*(24), 8533–8538.
50. Fernandez-Bolanos, J. G., Lopez, O., Fernandez-Bolanos, J., & Rodriguez-Gutierrez, G., "Hydroxytyrosol and derivatives: Isolation, synthesis, and biological properties." *Current Organic Chemistry*, **2008**, *12*(6), 442–463.
51. Luzi, F., et al. "Effect of gallic acid and umbelliferone on thermal, mechanical, antioxidant and antimicrobial properties of poly(vinyl alcohol-co-ethylene) films." *Polymer Degradation and Stability*, **2018**, *152*, 162–176.
52. Chanwitheesuk, A., Teerawutgulrag, A., Kilburn, J. D., & Rakariyatham, N. "Antimicrobial gallic acid from *Caesalpinia mimosoides* Lamk," *Food Chemistry*, **2007**, *100*(3), 1044–1048.
53. Khatkar, A., Nanda, A., Kumar, P., & Narasimhan, B. "Synthesis, antimicrobial evaluation and QSAR studies of gallic acid derivatives." *Arabian Journal of Chemistry*, **2017**, *10*, S2870–S2880.
54. Geethalakshmi, R., & Sarada, V. D. "*In vitro* and *in silico* antimicrobial activity of sterol and flavonoid isolated from *Trianthema decandra* L." *Microbial Pathogenesis*, **2018**, *121*, 77–86.
55. El-Abyad, M. S., Morsi, N. M., Zaki, D. A., & Shaaban, M. T. "Preliminary screening of some Egyptian weeds for antimicrobial activity." *Microbios*, **1990**, *62*(250), 47–57.
56. Dall'Agnol, R., et al. "Antimicrobial activity of some hypericum species." *Phytomedicine*, **2003**, *10*(6–7), 511–516.
57. Moon, A., Khan, A., & Wadher, B. "*Antibacterial Potential of Thespesia populnea (Linn.) Sol. ex corr. Leaves and its Corresponding Callus Against Drug Resistant Isolates,*" **2010**.

58. Wong, W. H., et al. "Two level half factorial design for the extraction of phenolics, flavonoids and antioxidants recovery from palm kernel by-product." *Industrial Crops and Products,* **2015**, *63*, 238–248.

59. M'hiri, N., Ghali, R., Nasr, I. B., & Boudhrioua, N. "Effect of different drying processes on functional properties of industrial lemon byproduct." *Process Safety and Environmental Protection,* **2018**, *116*, 450–460.

60. Sati, P., Dhyani, P., Bhatt, I. D., & Pandey, A. "Ginkgo biloba flavonoid glycosides in antimicrobial perspective with reference to extraction method." *Journal of Traditional and Complementary Medicine,* **2018**.

61. Mitchell, P. "Coupling of phosphorylation to electron and hydrogen transfer by a chemi-osmotic type of mechanism." *Nature,* **1961**, *191*(4784), 144–148.

62. Galbraith, H., Miller, T., Paton, A., & Thompson, J. "Antibacterial activity of long chain fatty acids and the reversal with calcium, magnesium, ergocalciferol and cholesterol." *Journal of Applied Bacteriology,* **1971**, *34*(4), 803–813.

63. Desbois, A. P., & Smith, V. J. "Antibacterial free fatty acids: Activities, mechanisms of action and biotechnological potential." *Applied Microbiology and Biotechnology,* **2010**, *85*(6), 1629–1642.

64. Desbois, A. P., & Lawlor, K. C. "Antibacterial activity of long-chain polyunsaturated fatty acids against Propionibacterium acnes and *Staphylococcus aureus*." *Marine Drugs,* **2013**, *11*(11), 4544–4557.

65. Carson, D. D., & Daneo-Moore, L. "Effects of fatty acids on lysis of Streptococcus faecalis." *Journal of Bacteriology,* **1980**, *141*(3), 1122–1126.

66. Thormar, H., Isaacs, C. E., Brown, H. R., Barshatzky, M. R., & Pessolano, T. "Inactivation of enveloped viruses and killing of cells by fatty acids and monoglycerides." *Antimicrobial Agents and Chemotherapy,* **1987**, *31*(1), 27–31.

67. Zheng, C. J., Yoo, J. S., Lee, T. G., Cho, H. Y., Kim, Y. H., & Kim, W. G. "Fatty acid synthesis is a target for antibacterial activity of unsaturated fatty acids." *FEBS Letters,* **2005**, *579*(23), 5157–5162.

68. Dunn, C., & Freeman, C. "The role of molecular weight in the enzyme-inhibiting effect of phenolics: The significance in peatland carbon sequestration." *Ecological Engineering,* **2018**, *114*, 162–166.

69. Galbraith, H., & Miller, T. "Effect of long chain fatty acids on bacterial respiration and amino acid uptake." *Journal of Applied Bacteriology,* **1973**, *36*(4), 659–675.

70. Davies, D. G., & Marques, C. N. "A fatty acid messenger is responsible for inducing dispersion in microbial biofilms." *Journal of Bacteriology,* **2009**, *191*(5), 1393–1403.

71. Kabir, F., Tow, W. W., Hamauzu, Y., Katayama, S., Tanaka, S., & Nakamura, S. "Antioxidant and cytoprotective activities of extracts prepared from fruit and vegetable wastes and by-products." *Food Chemistry,* **2015**, *167*, 358–362.

72. Cevallos-Casals, B. A., Byrne, D., Okie, W. R., & Cisneros-Zevallos, L. "Selecting new peach and plum genotypes rich in phenolic compounds and enhanced functional properties." *Food Chemistry,* **2006**, *96*(2), 273–280.

73. Valavanidis, A., Vlachogianni, T., Psomas, A., Zovoili, A., & Siatis, V. "Polyphenolic profile and antioxidant activity of five apple cultivars grown under organic and conventional agricultural practices." *International Journal of Food Science and Technology,* **2009**, *44*(6), 1167–1175.

74. Pastrana-Bonilla, E., Akoh, C. C., Sellappan, S., & Krewer, G. "Phenolic content and antioxidant capacity of muscadine grapes." *Journal of Agricultural and Food Chemistry,* **2003**, *51*(18), 5497–5503.

75. Fundo, J. F., Miller, F. A., Garcia, E., Santos, J. R., Silva, C. L., & Brandão, T. R. "Physicochemical characteristics, bioactive compounds and antioxidant activity in juice, pulp, peel and seeds of Cantaloupe melon." *Journal of Food Measurement and Characterization,* **2018,** *12*(1), 292–300.

76. Da Silva, L. M. R., et al. "Quantification of bioactive compounds in pulps and by-products of tropical fruits from Brazil." *Food Chemistry,* **2014,** *143,* 398–404.

77. Mallek-Ayadi, S., Bahloul, N., & Kechaou, N. "Chemical composition and bioactive compounds of *Cucumis melo* L. seeds: Potential source for new trends of plant oils." *Process Safety and Environmental Protection,* **2018,** *113,* 68–77.

78. Li, B., Smith, B., & Hossain, M. M. "Extraction of phenolics from citrus peels: II. Enzyme-assisted extraction method." *Separation and Purification Technology,* **2006,** *48*(2), 189–196.

79. Rouseff, R. L., Martin, S. F., & Youtsey, C. O. "Quantitative survey of narirutin, naringin, hesperidin, and neohesperidin in citrus." *Journal of Agricultural and Food Chemistry,* **1987,** *35*(6), 1027–1030.

80. Manthey, J. A., & Grohmann, K. "Phenols in citrus peel byproducts. Concentrations of hydroxycinnamates and polymethoxylated flavones in citrus peel molasses." *Journal of Agricultural and Food Chemistry,* **2001,** *49*(7), 3268–3273.

81. Moyer, R. A., Hummer, K. E., Finn, C. E., Frei, B., & Wrolstad, R. E. "Anthocyanins, phenolics, and antioxidant capacity in diverse small fruits: Vaccinium, rubus, and ribes." *Journal of Agricultural and Food Chemistry,* **2002,** *50*(3), 519–525.

82. Goula, A. M., & Lazarides, H. N. "Integrated processes can turn industrial food waste into valuable food by-products and/or ingredients: The cases of olive mill and pomegranate wastes." *Journal of Food Engineering,* **2015,** *167,* 45–50.

83. Ayala-Zavala, J., Rosas-Domínguez, C., Vega-Vega, V., & González-Aguilar, G. "Antioxidant enrichment and antimicrobial protection of fresh-cut fruits using their own byproducts: Looking for integral exploitation." *Journal of Food Science,* **2010,** *75*(8), R175–R181.

84. Schieber, A., Stintzing, F. C., & Carle, R. "By-products of plant food processing as a source of functional compounds—recent developments." *Trends in Food Science and Technology,* **2001,** *12*(11), 401–413.

85. Wijngaard, H. H., Trifunovic, O., & Bongers, P. "Novel extraction techniques for phytochemicals." *Handbook of Plant Food Phytochemicals: Sources, Stability and Extraction,* **2013,** 412–433.

86. Harbourne, N., Marete, E., Jacquier, J. C., & O'Riordan, D. "Conventional extraction techniques for phytochemicals." *Handbook of Plant Food Phytochemicals: Sources, Stability and Extraction,* **2013,** 397–411.

87. Do Carmo, C. S., Serra, A., & Duarte, C. "Recovery technologies for lipophilic bioactives." In: *Engineering Foods for Bioactives Stability and Delivery* (pp. 1–49), Springer, **2017.**

88. Wang, L., & Weller, C. L. "Recent advances in extraction of nutraceuticals from plants." *Trends in Food Science and Technology,* **2006,** *17*(6), 300–312.

89. Putnik, P., Kovačević, D. B., Penić, M., Fegeš, M., & Dragović-Uzelac, V. "Microwave-assisted extraction (MAE) of Dalmatian sage leaves for the optimal yield of polyphenols: HPLC-DAD identification and quantification." *Food Analytical Methods,* **2016,** *9*(8), 2385–2394.

90. Oliveira, A., & Pintado, M. "Stability of polyphenols and carotenoids in strawberry and peach yoghurt throughout *in vitro* gastrointestinal digestion." *Food and Function,* **2015,** *6*(5), 1611–1619.

91. Morais, D. R., et al. "Antioxidant activity, phenolics and UPLC-ESI (–)–MS of extracts from different tropical fruits parts and processed peels." *Food Research International,* **2015**, *77,* 392–399.

92. Mateus, N., & De Freitas, V. "Anthocyanins as food colorants." In: *Anthocyanins* (pp. 284–304), Springer, **2008**.

93. Azmir, J., et al. "Techniques for extraction of bioactive compounds from plant materials: A review." *Journal of Food Engineering,* **2013**, *117*(4) 426–436.

94. Chemat, F., & Khan, M. K. "Applications of ultrasound in food technology: Processing, preservation and extraction." *Ultrasonics Sonochemistry,* **2011**, *18*(4), 813–835.

95. Roselló-Soto, E., et al. "Application of non-conventional extraction methods: Toward a sustainable and green production of valuable compounds from mushrooms." *Food Engineering Reviews,* **2016**, *8*(2), 214–234.

96. Talmaciu, A. I., Volf, I., & Popa, V. I. "A comparative analysis of the 'green' techniques applied for polyphenols extraction from bioresources." *Chemistry and Biodiversity,* **2015**, *12*(11), 1635–1651.

97. Alexandre, E. M., Moreira, S. A., Castro, L. M., Pintado, M., & Saraiva, J. A. "Emerging technologies to extract high added value compounds from fruit residues: Sub/supercritical, ultrasound-, and enzyme-assisted extractions." *Food Reviews International,* **2018**, *34*(6), 581–612.

98. Huaneng, X., Zhang, Y., & Chaohong, H. "Ultrasonically assisted extraction of isoflavones from stem of *Pueraria lobata* (Willd.) Ohwi and its mathematical model." *Chinese Journal of Chemical Engineering,* **2007**, *15*(6), 861–867.

99. Khan, M. K., Abert-Vian, M., Fabiano-Tixier, A. S., Dangles, O., & Chemat, F. "Ultrasound-assisted extraction of polyphenols (flavanone glycosides) from orange (*Citrus sinensis* L.) peel." *Food Chemistry,* **2010**, *119*(2), 851–858.

100. D'Alessandro, L. G., Dimitrov, K., Vauchel, P., & Nikov, I. "Kinetics of ultrasound assisted extraction of anthocyanins from *Aronia melanocarpa* (black chokeberry) wastes." *Chemical Engineering Research and Design,* **2014**, *92*(10), 1818–1826.

101. Corrales, M., Toepfl, S., Butz, P., Knorr, D., & Tauscher, B. "Extraction of anthocyanins from grape by-products assisted by ultrasonics, high hydrostatic pressure or pulsed electric fields: A comparison." *Innovative Food Science and Emerging Technologies,* **2008**, *9*(1), 85–91.

102. Goula, A. M., Ververi, M., Adamopoulou, A., & Kaderides, K. "Green ultrasound-assisted extraction of carotenoids from pomegranate wastes using vegetable oils." *Ultrasonics Sonochemistry,* **2017**, *34,* 821–830.

103. Luengo, E., Condón-Abanto, S., Condón, S., Álvarez, I., & Raso, J. "Improving the extraction of carotenoids from tomato waste by application of ultrasound under pressure." *Separation and Purification Technology,* **2014**, *136,* 130–136.

104. Sun, Y., Liu, D., Chen, J., Ye, X., & Yu, D. "Effects of different factors of ultrasound treatment on the extraction yield of the all-trans-β-carotene from citrus peels." *Ultrasonics Sonochemistry,* **2011**, *18*(1), 243–249.

105. Kumcuoglu, S., Yilmaz, T., & Tavman, S. "Ultrasound assisted extraction of lycopene from tomato processing wastes." *Journal of Food Science and Technology,* **2014**, *51*(12), 4102–4107.

106. Zhang, H. F., Yang, X. H., & Wang, Y. "Microwave assisted extraction of secondary metabolites from plants: Current status and future directions." *Trends in Food Science and Technology,* **2011**, *22*(12), 672–688,.

107. Rostagno, M., Villares, A., Guillamón, E., García-Lafuente, A., & Martínez, J. "Sample preparation for the analysis of isoflavones from soybeans and soy foods." *Journal of Chromatography A,* **2009,** *1216*(1), 2–29.

108. Liazid, A., Guerrero, R., Cantos, E., Palma, M., & Barroso, C. "Microwave assisted extraction of anthocyanins from grape skins." *Food Chemistry,* **2011,** *124*(3), 1238–1243.

109. Ho, K., Ferruzzi, M., Liceaga, A., & San Martín-González, M. "Microwave-assisted extraction of lycopene in tomato peels: Effect of extraction conditions on all-trans and cis-isomer yields." *LWT-Food Science and Technology,* **2015,** *62*(1), 160–168.

110. Töpfl, S. Pulsed electric fields (PEF) for permeabilization of cell membranes in food-and bioprocessing-applications. *Process and Equipment Design and Cost Analysis,* **2006.**

111. Donsì, F., Ferrari, G., & Pataro, G. "Applications of pulsed electric field treatments for the enhancement of mass transfer from vegetable tissue." *Food Engineering Reviews,* **2010,** *2*(2), 109–130.

112. Kotnik, T., Kramar, P., Pucihar, G., Miklavcic, D., & Tarek, M. "Cell membrane electroporation-Part 1: The phenomenon." *IEEE Electrical Insulation Magazine,* **2012,** *28*(5), 14–23.

113. Abenoza, M., Benito, M., Saldaña, G., Álvarez, I., Raso, J., & Sánchez-Gimeno, A. C. "Effects of pulsed electric field on yield extraction and quality of olive oil," *Food and Bioprocess Technology,* **2013,** *6*(6), 1367–1373.

114. Luengo, E., Álvarez, I., & Raso, J. "Improving the pressing extraction of polyphenols of orange peel by pulsed electric fields." *Innovative Food Science and Emerging Technologies,* **2013,** *17*, 79–84.

115. Barba, F. J., et al. "Current applications and new opportunities for the use of pulsed electric fields in food science and industry." *Food Research International,* **2015,** *77*, 773–798.

116. Barba, F. J., Grimi, N., & Vorobiev, E. "New approaches for the use of non-conventional cell disruption technologies to extract potential food additives and nutraceuticals from microalgae." *Food Engineering Reviews,* **2015,** *7*(1), 45–62.

117. Barbosa-Canovas, G. V., Pothakamury, U. R., Gongora-Nieto, M. M., & Swanson, B. G. *Preservation of Foods with Pulsed Electric Fields.* Elsevier, **1999.**

118. Toepfl, S., & Heinz, V. "Pulsed Electric Field assisted extraction-A case study." *Nonthermal Processing Technologies for Food,* **2011,** 190–200.

119. Sihvonen, M., Järvenpää, E., Hietaniemi, V., & Huopalahti, R. "Advances in supercritical carbon dioxide technologies." *Trends in Food Science & Technology,* **1999,** *10*(6-7), 217–222.

120. Paes, J., Dotta, R., Barbero, G. F., & Martínez, J. "Extraction of phenolic compounds and anthocyanins from blueberry (Vaccinium myrtillus L.) residues using supercritical CO2 and pressurized liquids." *The Journal of Supercritical Fluids,* **2014,** *95*, 8–16.

121. Wong, D. W. "Feruloyl esterase." *Applied Biochemistry and Biotechnology,* **2006,** *133*(2), 87–112.

122. Nadar, S. S., Rao, P., & Rathod, V. K. "Enzyme assisted extraction of biomolecules as an approach to novel extraction technology: A review." *Food Research International,* **2018.**

123. Acosta-Estrada, B. A., Gutiérrez-Uribe, J. A., & Serna-Saldívar, S. O. "Bound phenolics in foods, a review." *Food Chemistry,* **2014,** *152*, 46–55.

124. Yang, Y. C., et al., "Optimisation of microwave-assisted enzymatic extraction of corilagin and geraniin from Geranium sibiricum Linne and evaluation of antioxidant activity." *Food Chemistry,* **2010,** *122*(1), 373–380.

125. Zheng, H. Z., Hwang, I. W., & Chung, S. K. "Enhancing polyphenol extraction from unripe apples by carbohydrate-hydrolyzing enzymes." *Journal of Zhejiang University Science B*, **2009**, *10*(12), 912.

126. Landbo, A. K., & Meyer, A. S. "Enzyme-assisted extraction of antioxidative phenols from black currant juice press residues (Ribes nigrum)." *Journal of Agricultural and Food Chemistry*, **2001**, *49*(7), 3169–3177.

127. Stalikas, C. D. "Extraction, separation, and detection methods for phenolic acids and flavonoids." *Journal of Separation Science*, **2007**, *30*(18), 3268–3295.

128. Passos, C. P., Yilmaz, S., Silva, C. M., & Coimbra, M. A. "Enhancement of grape seed oil extraction using a cell wall degrading enzyme cocktail." *Food Chemistry*, **2009**, *115*(1), 48–53.

129. Dehghan-Shoar, Z., Hardacre, A. K., Meerdink, G., & Brennan, C. S. "Lycopene extraction from extruded products containing tomato skin." *International Journal of Food Science and Technology*, **2011**, *46*(2), 365–371.

130. Sowbhagya, H. B., Srinivas, P., & Krishnamurthy, N. "Effect of enzymes on extraction of volatiles from celery seeds." *Food Chemistry*, **2010**, *120*(1), 230–234.

131. Puri, M., Sharma, D., & Barrow, C. J. "Enzyme-assisted extraction of bioactives from plants." *Trends in Biotechnology*, **2012**, *30*(1),37–44.

132. Jiao, J., et al. "Microwave-assisted aqueous enzymatic extraction of oil from pumpkin seeds and evaluation of its physicochemical properties, fatty acid compositions and antioxidant activities." *Food Chemistry*, **2014**, *147*, 17–24.

133. M'hiri, N., Ioannou, I., Ghoul, M., & Boudhrioua, N. M. "Extraction methods of citrus peel phenolic compounds." *Food Reviews International*, **2014**, *30*(4), 265–290.

134. Barzana, E., et al. "Enzyme-mediated solvent extraction of carotenoids from marigold flower (Tagetes erecta)." *Journal of Agricultural and Food Chemistry*, **2002**, *50*(16), 4491–4496.

135. Lavecchia, R., & Zuorro, A. "Improved lycopene extraction from tomato peels using cell-wall degrading enzymes." *European Food Research and Technology*, **2008**, *228*(1), 153.

136. Zuorro, A., Fidaleo, M., & Lavecchia, R. "Enzyme-assisted extraction of lycopene from tomato processing waste." *Enzyme and Microbial Technology*, **2011**, *49*(6), 567–573.

137. Cuccolini, S., Aldini, A., Visai, L., Daglia, M., & Ferrari, D. "Environmentally friendly lycopene purification from tomato peel waste: Enzymatic assisted aqueous extraction." *Journal of Agricultural and Food Chemistry*, **2012**, *61*(8), 1646–1651.

138. Gómez-García, R., Martínez-Ávila, G. C. G., & Aguilar, C. N. "Enzyme-assisted extraction of antioxidant phenolics from grape (Vitis vinifera L.) residues." *3 Biotech*, **2012**, *2*(4), 297–300.

139. Subramaniyam, R., & Vimala, R. "Solid state and submerged fermentation for the production of bioactive substances: A comparative study." *Int. J. Sci Nat.*, **2012**, *3*(3), 480–486.

140. Medina-Morales, M., Martínez-Hernández, J. L., De La Garza, H., & Aguilar, C. N. "Cellulolytic enzymes production by solid state culture using pecan nut shell as substrate and support." *AJABS*, **2011**, *6*(2), 196–200.

141. Ravindran, R., & Jaiswal, A. K. "Exploitation of food industry waste for high-value products." *Trends in Biotechnology*, **2016**, *34*(1), 58–69.

142. Soccol, C. R., da Costa, E. S. F., Letti, L. A. J., Karp, S. G., Woiciechowski, A. L., & de Souza Vandenberghe, L. P. "Recent developments and innovations in solid state fermentation." *Biotechnology Research and Innovation*, **2017**.

143. Pirota, R. D., Delabona, P. S., & Farinas, C. S. "Simplification of the biomass to ethanol conversion process by using the whole medium of filamentous fungi cultivated under solid-state fermentation." *BioEnergy Research,* **2014**, *7*(2), 744–752.

144. Xie, P. J., Huang, L. X., Zhang, C. H., & Zhang, Y. L. "Phenolic compositions, and antioxidant performance of olive leaf and fruit (Olea europaea L.) extracts and their structure–activity relationships." *Journal of Functional Foods,* **2015**, *16*, 460–471.

145. Manan, M., & Webb, C., "Design aspects of solid state fermentation as applied to microbial bioprocessing." *J. Appl Biotechnol Bioeng,* **2017**, *4*(1), 00091.

146. Arora, S., Rani, R., & Ghosh, S., "Bioreactors in solid state fermentation technology: Design, applications and engineering aspects." *Journal of Biotechnology,* **2018**, *269*, 16–34.

147. Cheng, J. R., Liu, X. M., Zhang, W., Chen, Z. Y., & Wang, X. P. "Stability of phenolic compounds and antioxidant capacity of concentrated mulberry juice-enriched dried-minced pork slices during preparation and storage." *Food Control,* **2018**, *89*, 187–195.

148. Madureira, A. R., et al. "Characterization of solid lipid nanoparticles produced with carnauba wax for rosmarinic acid oral delivery." *RSC Advances,* **2015**, *5*(29), 22665–22673.

149. Ryu, V., McClements, D. J., Corradini, M. G., Yang, J. S., & McLandsborough, L. "Natural antimicrobial delivery systems: Formulation, antimicrobial activity, and mechanism of action of quillaja saponin-stabilized carvacrol nanoemulsions." *Food Hydrocolloids,* **2018**, *82*, 442–450.

150. Chang, Y., McLandsborough, L., & McClements, D. J. "Physicochemical properties and antimicrobial efficacy of carvacrol nanoemulsions formed by spontaneous emulsification." *Journal of Agricultural and Food Chemistry,* **2013**, *61*(37), 8906–8913.

151. Landry, K. S., Micheli, S., McClements, D. J., & McLandsborough, L. "Effectiveness of a spontaneous carvacrol nanoemulsion against Salmonella enterica Enteritidis and Escherichia coli O157: H7 on contaminated broccoli and radish seeds." *Food Microbiology,* **2015**, *51*, 10–17.

152. Procopio, F. R., et al. "Solid lipid microparticles loaded with cinnamon oleoresin: Characterization, stability and antimicrobial activity." *Food Research International,* **2018**, *113*, 351–361.

153. Manach, C., Williamson, G., Morand, C., Scalbert, A., & Rémésy, C. "Bioavailability and bioefficacy of polyphenols in humans. I. Review of 97 bioavailability studies–." *The American Journal of Clinical Nutrition,* **2005**, *81*(1), 230S–242S.

154. Campos, D. A., Madureira, A. R., Sarmento, B., Gomes, A. M., & Pintado, M. M. "Stability of bioactive solid lipid nanoparticles loaded with herbal extracts when exposed to simulated gastrointestinal tract conditions." *Food Research International,* **2015**, *78*, 131–140.

155. Blasi, P., Giovagnoli, S., Schoubben, A., Ricci, M., & Rossi, C. "Solid lipid nanoparticles for targeted brain drug delivery." *Advanced drug Delivery Reviews,* **2007**, *59*(6), 454–477.

156. Campos, D. A., Madureira, A. R., Sarmento, B., Pintado, M. M., & Gomes, A. M. "Technological stability of solid lipid nanoparticles loaded with phenolic compounds: Drying process and stability along storage." *Journal of Food Engineering,* **2017**, *196*, 1–10.

CHAPTER 13

Fermentative Bioprocesses for Detoxification of Agri-Food Wastes for Production of Bioactive Compounds

LILIANA LONDOÑO-HERNANDEZ,[1] MÓNICA L. CHÁVEZ-GONZÁLEZ,[1]
JUAN ALBERTO ASCACIO-VALDÉS,[1] HÉCTOR A. RUIZ,[2,3]
CRISTINA RAMÍREZ TORO,[4] and CRISTÓBAL NOÉ AGUILAR[1]

[1]*Bioprocesses and Bioproducts Group, Food Research Department, School of Chemistry, Universidad Autónoma de Coahuila, Saltillo, 25280, Coahuila, Mexico, E-mail: cristobal.aguilar@uadec.edu.mx (C. N. Aguilar)*

[2]*Biorefinery Group, Food Research Department, Faculty of Chemistry Sciences, Autonomous University of Coahuila, 25280, Saltillo, Coahuila, Mexico*

[3]*Cluster of Bioalcoholes, Mexican Centre for Innovation in Bioenergy (Cemie-Bio), Mexico*

[4]*School of Food Engineering, Universidad Del Valle, Cali, 25360, Valle del Cauca, Colombia*

ABSTRACT

One of the industrial sectors with the highest production in the world is agriculture, and therefore it is also one of the sectors that generate the greatest amount of waste. This waste is not used in its entirety and the remnants are disposed of in the correct way generating serious environmental problems. Agro-industrial waste includes a wide range of products of heterogeneous nature, which include seeds, husks, roots, stems, among others. These residues are characterized by a high content of proteins, carbohydrates or minerals so they could be used for the design of new human and animal foods, however, the content of toxic compounds limits their use. These residues are mostly characterized by having a high content of saponins, phytates, oxalates, alkaloids, among others, which have a negative effect on health. Therefore, it is necessary to apply treatments to be able to use these products. Biotechnological processes are a viable alternative to take advantage of these residues, given that during these toxic compounds are hydrolyzed and compounds are generated with biological activities such as antioxidants, antimicrobials, anticancer, antihypertensive, among others, which could have application in other industries such as pharmaceuticals. Nowadays, research to obtain biocompounds with these characteristics is growing since the beneficial effects of their consumption have been demonstrated. Therefore, fermentation processes are a profitable strategy to obtain high value-added biocompounds from agro-industrial residues. The objective of this chapter is to know the main toxic compounds present in agro-industrial waste and the biocompounds that can be generated from such waste according to the microorganism (bacteria, yeasts, fungi) used for fermentation.

13.1   INTRODUCTION

Today, cereals, and pulses remain one of the world's most important sources of food. The world's most widely cultivated staple cereals include wheat and rice, and coarse grains include maize, sorghum, millet, rice, barley, oats, and regional cereals such as teff (Ethiopia) and quinoa (Bolivia and Ecuador) [1]. For the year 2016, the annual production of cereals was 2,849 billion tons. In the same year, around 38 million tons were cultivated in Mexico, with corn being the most cultivated cereal, followed by wheat [2].

In any productive system that generates goods and services, some waste is produced. Therefore, food production systems also generate this type of by-products. It is considered that the production of residues in the cereals and legumes industry is between 10 and 30%, for which reason, only in Mexico around 11 million tons are produced annually and in total for the agro-industrial sector around 76 million tons of residues are produced annually [3]. The inadequate handling of these products produces, among others, excessive accumulation in different zones, decomposition of materials generating bad odors and toxic compounds that can reach water sources and economic losses to the farmer, so it is necessary to seek effective treatments to use these materials. Residues from the food industry generally composed of proteins, carbohydrates, minerals, and other compounds of interest, could be converted into value-added products, achieving a potential for use, thus reducing the environmental pollution associated with the poor disposal of these materials and achieving sustainable production. However, many of these materials present antinutritional and toxic compounds, such as saponins, condensed tannins, and physic acid, among others (See Table 13.1), which limit their use.

TABLE 13.1 Toxic Compounds in Different Plant Products

Raw Material	Toxic Compounds	References
Jojoba seeds	Phytic acid, phenolic	[118]
Leaves of E. cyclocarpum	Tannins, saponins, phytic acid, oxalates	[119]
Rapeseed cake	Phytic acid, tannins, glucosinolates	[120]
Ginkgo biloba	Ginkgolic acid	[121]
Bean seeds	Phytates, tannins, alkaloids, lecithins, trypsin inhibitors, oxalates	[122]
Castor	Phytates, tannins	[111]
Jatropha cakes	Tannins, phytates, oxalates, trypsin inhibitors	[123]
	Phorbol esters, phytates, trypsin inhibitors	[124]
	Lecithin, saponins, phytates, phorbol ester	[125]
Soybean meal	Phytic acid, Trypsin inhibitors	[126]
Cotton seed	Gossypol	[127]

The chemical nature of these compounds is very varied, as well as the concentration in plants and the biological properties they may have. To understand the role of these compounds in plants, in recent years they have been isolated, finding that these compounds present biological activities of great interest. It has been found that these compounds act as prebiotics, protectors of the circulatory system, and regulators of glycemia. They have also been found to have antioxidant, anti-cancer, and anti-inflammatory properties, among others [4–7]. Therefore, its harmful effect is being defined according to the concentration or amount consumed in the diet. Similarly, research has been directed to determine various techniques of extraction of these compounds and thus take advantage of the byproducts of the food industry by giving them an added value.

Various physical, chemical, thermal, and biological processes have been used to detoxify the materials and extract these compounds. In recent years, biological processes such as fermentation have gained particular attention, due to the various beneficial effects that the process can have on the treated materials. The fermentation is a biological process carried out by microorganisms, which catalyze nutrients, synthesize nutrients, secondary metabolites and other compounds under aerobic or anaerobic conditions [8]. Bacteria, fungi, and yeasts have been used for the fermentation processes, obtaining in each case different compound according to the metabolic route followed by the microorganism. Fermentation is an ancient technology used to produce different compounds of industrial interest such as enzymes, organic acids, pigments, phenolic compounds, aromas, among others. It has also been used for the reduction of toxic and antinutritional compounds [9]. Fermentation processes vary according to the microorganisms used and the operating conditions of the process such as temperature, pH, moisture content, aeration, nutrient concentrations, and nature of the substrate [4].

Considering the beneficial effects that fermentation processes can have on the detoxification of agro-industrial by-products and the production of bioactive compounds (BCs), the objective of this chapter is to review the main toxic compounds present in food, and according to the microorganisms used (bacteria, fungi, yeasts), to describe the types of BCs that can be obtained in a fermentation process.

13.2   TOXIC COMPOUNDS IN AGRO-FOOD WASTES

To avoid the attacks by herbivores, insects, pathogens, or to survive to the adverse environmental conditions, the plants synthesize different compounds, some of which are considered toxic and antinutritional, because they diminish the digestibility of different nutrients of the diet like proteins, carbohydrates, and minerals, causing alterations in the nutrition [10, 11]. These compounds limit the use of many agricultural products and by-products, so it is necessary to apply some treatment to reduce their concentration. For this, it is necessary to know the nature of the compounds present and thus know the effectiveness of the treatment applied. Chemically, antinutritional compounds are different; however, they can be classified in some groups. According to their physical and chemical properties they can be classified as cyanogenic glycosides (phaseolunatin, dhurrin, linamarin, lutaustralin), enzyme inhibitors (alkaloids, protease inhibitors, cynogens, G-6-PD, cholinesterase inhibitors, amylose inhibitors) and physiological disorganizers (lectines, saponins, lathrogens, oxalates, nitrate, and nitrite) [12]. From the biochemical point of view, they can be proteins (protease inhibitors, inhibitors of α-amylases, lectins), glucides (α-galactósidos, vicina, convicina, saponins), non-protein amino acids (L-DOPA, β-ODAP), polyphenols (condensed tannins) or alkaloids [13, 14].

And according to their thermal resistance in thermolabiles (protease inhibitors, amylase inhibitors, antívitamins D, E, and B12) and thermostable (saponins, cyanogens, phytates, alkaloids, oligosaccharides, and tannins) [15].

13.2.1 SAPONIN

Saponins are high molecular weight compounds, generally recognized as secondary plant metabolites. The name saponin is derived from the Latin word "sapon" which means soap because its molecules form foam when agitated with water [16]. Some of the plants that have been reported to contain triterpenoid and steroid saponins are: Black oats (*Avenastrigosa*), Oats (*Avena sativa*), Switchgrass (*Panicum virgatum*), Kleingrass (*Panicum coloratum*), Beetroot (*Beta vulgaris*), Quinoa (*Chenopodium quinoa*), Common pea (*Pisum sativum*), Soybean (*Glycine max*), Alfalfa (*Medicago sativa*), Barrel medic (*Medicago truncatula*), Common bean (*Phaseoulus vulgaris*), Tea (*Camellia sinensi*), Chilipepper (*Capsicum trutescens*), Tomato (*Solanum lycopersicum*), Potato (*Solanum tuberosum*), Garlic (*Allium sativum*), Blue chives (*Allium nutans*), Leek (*Allium porrum*), Onion (*Allium cepa*) and Chives (*Allium schoenoprasum*).

Saponin molecules usually contain a triterpene or steroid aglycone (sapogenin) and one or more sugar chains attached to it. This combination of polar and non-polar structural elements explains its foaming behavior in aqueous solutions [17]. Saponins have a variety of properties including sweetness and bitterness, foaming, and emulsifying properties, pharmacological, and medicinal properties [18]. It has been reported that some of the most important biological properties of saponins are: antimicrobial activity, anticancer activity, anti-cardiovascular activity, anti-inflammatory activity. Due to these properties, they are significant compounds in the pharmaceutical, food, and cosmetics industries [17].

However, some saponins have a hemolytic activity. This activity is dependent on the type of aglycone and nature and number of sugar chains attached to it, on the concentration, composition, and source of the saponins. Intake of high amounts of saponins may irritate the intestinal epithelium and hydrolyze the intestinal mucosal cells, interfering with the absorption of nutrients. The saponins could be interacting with the sterols in the erythrocyte membrane increase in its permeability and the consequent loss of hemoglobin. Also, they inhibit metabolic and digestive enzymes such as proteases, amylases, or lipases [16, 17].

13.2.2   TANNINS

Polyphenolic compounds vary widely in structure, from the simplest (monomers and oligomers) to complex high molecular weight polymers (tannins). More than 4000 individual polyphenolic compounds have been identified, which have been divided into two large groups: flavonoids and nonflavonoids. Both flavonoid and nonflavonoid groups can be found forming compounds of very high molecular weight (>500 AMU), both called tannins. However, each group produces a specific type of tannin: nonflavonoids polymerize to form hydrolyzable tannins, while certain flavonoids when polymerized, and form condensed tannins [19].

Traditionally, tannins have been classified into the two groups mentioned above: condensed tannins and hydrolyzable tannins, but today the most widely accepted classification groups them into four groups: gallotannins, ellagitannins, condensed tannins and complex tannins [20]. Tannins are phenolic compounds of complex nature, soluble in water with a relative molecular weight between 500–3 000 Dalton (Da), which have the property of precipitating alkaloids, albuminoids, and gelatin, apart from their phenolic reactions [21, 22].

Gallotannins are characterized by the presence of several organic acid molecules, such as gallic, digallic, and chebulic acid. On the other hand, ellagitannins are composed of units of ellagic acid bound to glycosides. Molecules with a nucleus of quinic acid are also considered as ellagitannins. To maintain bonds, gallotannins, and ellagitannins must have more than two-unit components of esterified acids bound to the central glucose. Gallotannins can be easily hydrolyzed in the presence of weak acids or alkaline conditions, either in hot water or by enzymatic treatments. The ellagitannins are more stable than gallotannins [19, 23].

Condensed tannins or proanthocyanidins comprise a group of oligomers and polymers of polyhydroxy-flavan-3-ol bound by the carbon-carbon bond between the flavanol subunits; they are so-called probably because they present reddish pigments of the anthocyanidin class such as cyanidine and delphinidine [20, 24, 25]. Tannins are compounds widely distributed in nature. They are present in plant bark, wood, fruits, fruit pods, leaves, roots, and plant galls. It has been found that tannins have various biological properties of great interest, act as antioxidants and antimicrobials, also have astringent and anti-inflammatory activity. However, in high concentrations, they are considered antinutritional compounds because they bind to proteins, carbohydrates, minerals, and other macromolecules decreasing their absorption in the organism.

Among the adverse effects of tannins are the formation of complexes with proteins, carbohydrates, and minerals due to the multiple hydroxyl groups that the molecule has to bind, as well as their pro-oxidant effect at high concentrations. Tannins have the ability to precipitate proteins present in saliva, which causes astringency sensation. They bind strongly to proteins rich in amino acids such as proline, glycine, glutamic acid and peptides through two interactions: (i) hydrogen bridges (between the carbonyl group of peptides and the hydrogens of the hydroxyl group of polyphenols); and (ii) hydrophobic interaction (between the neutral amino acids and the aromatic rings of the tannins). The interactions depend on the preference of each tannin molecule for three-dimensional arrangement, and on its colloidal state; inhibit digestive enzymes, compromising the digestion of proteins and other macronutrients; and interact with divalent minerals such as non-hematic iron, inhibiting the absorption of metals [19].

13.2.3   *PHYTIC ACID*

Phytic acid is the hexaphosphoric ester of the hexahydric cyclic alcohol meso-inositol. Structurally, this is a ring of myo-inositol bound to 6 phosphate molecules (myo-inositol hexakisphosphate). Phytic acid is the principal storage form of phosphorus in many plant tissues. Any salt of phytic acid is called phytate and can be soluble or insoluble [27, 28]. Inositols with 4, 5 or 6 phosphate groups are common in the legumes and cereals [12].

Phytic acid is a compound that has a strong ability to chelate multivalent metallic ions, especially zinc, calcium, and iron; this union can lead to very insoluble salts with low bioavailability of minerals, called phytates [27]. Phytates are compounds that can represent 1.0% of the dry weight (DW) of a typical seed and is 75.0% of the total phosphorus of the seed [29]. Some studies have proved that monogastric animals are not capable of absorbing these nutrients, which could cause malnutrition in populations where most of their diet is cereals such as corn and cause the presence of anemia, iron deficiency, or decreased growth rate due to zinc deficiency [30, 31]. Although their main effect is the decrease in the bioavailability of minerals, phytates also interact with basic protein residues, participating in the inhibition of digestive enzymes such as pepsin, pancreatin, and α-amylase; possibly by the chelation of Ca ions of enzymes (essential for the activity of trypsin and α-amylase) or by an interaction with their substrates [27, 32].

13.2.4 GOSSYPOL

Gossypol ($C_{30}H_{30}O_8$) or 2,20-bis(1,6,7-trihydroxy-3-methyl-5-isopropyl-8-aldehydonaphthalene) is a yellow phenolic pigment and non-volatile produced naturally in the seeds, roots, and stems of the cotton plant of the genus *Gossypium* (fam. Malvaceae) as a defense agent against pathogens and herbivorous insects [33, 34]. This compound has been studied for its biological activities such as anti-cancer activity and antiviral activity [35]. However, this compound is highly toxic, so the use of cotton byproducts for human or animal consumption is limited. One of the main harmful effects is on the reproductive system. In the male reproductive system, infertility occurs through the inhibition of sperm motility, reduction in sperm count, damage to mitochondrial DNA and injury to the germinal epithelium. In the female's reproductive system, the compound produces alterations in oestrus, pregnancy, and embryo development [36, 37]. Another of the negative effects of gossypol is in nutrition. Gossypol is a highly reactive compound, due to the presence of two aldehyde groups and six phenolic hydroxyl groups, whereby the free gossypol binds to the proteins present in the food forming complexes, which are insoluble in the animal digestive tract [33, 38].

13.2.5 PHORBOL

Phorbol compounds are a group of compounds known as tiglaine, with polycyclic structure, belonging to the family of diterpenes [39]. Pholbol esters are 20 carbon tetracyclic diterpenoids made up of four isoprene units, in which tiglaine is the fundamental alcohol [40]. Hydroxylation of this basic tiglaine structure at different positions and the union of these two different acids allow the formation of different compounds of phorbol esters, which have different properties, including toxicity [39, 40]. Phorbol esters is found in different plants, reported mainly in fruits and aerial parts of Thymelaeceae and Euphorbiaceae family [41], and between the species of these family the one that presents greater concentration of these compounds is *Jatropha curcas* a small shrub or tree originating in America but widely cultivated in countries of Asia and Africa. *Jatropha* by-products are rich in protein and other compounds, but the content of these compounds limits their use in the food industry. The main toxic effect of phorbol esters is that it promotes the growth of tumor cells. Phorbol esters mimic a biochemical effect of diacylglycerol and cause prolonged stimulation of protein kinase C (PKC). Activation of PKC by phorbol esters induces many biological processes, such as cell proliferation, skin irritation, and tumor promotion [42]. The consumption of phorbol esters causes burning/pain in the mouth and throat, vomiting, delirium, muscle shock, a decrease of visual capacity, high pulse rate, among others [40].

13.2.6 CYANIDE AND CYANOGENIC GLYCOSIDES

Cyanogenic glycosides are natural bioactive plant secondary metabolites derived from amino acids, which generally accumulate in the vacuoles of plant cells. These compounds are intermediately polar, water-soluble, and composed of a α-hydroxynitrile type aglycone and a sugar moiety (mostly d-glucose) [43–45]. The cyanogenic compounds play important roles in plant growth, development, and resistance against abiotic/biotic stresses [45], however, during the hydrolysis of these compounds, by the action of enzymes β-glycosidases (EC 3.2.1) and hydroxynitrile lyases, the hydrogen cyanide (HCN) is release. The HCN is a compound that can become highly toxic, producing cyanogenesis [14].

The distribution of the cyanogenic compounds in the plants is relatively wide; more than 3000 species containing these compounds have been reported. Some of plant families that contain cyanogenic compounds are Rosaceae, Leguminosae, Gramineae, Araceae, Poaceae, Compositae, Euphorbiaceae, and Passifloraceae. Of these families, some of the main plants that contain these compounds are Cassava (*Manihot esculenta*), Sorghum (*Sorghum* spp.), Black Cherry (*Prunus* spp.), Bamboo (*Bambusa* spp.), Apricots (*Prunus* spp.), Apple (seeds) (*Malus* spp.), Eucalyptus (*Eucalyptus* spp.), Corn (*Zea mays*), and Almonds (*Prunusdulcis*) among others [43, 45, 46].

Toxic levels of cyanogenic glycosides are estimated based on the amount of free cyanide. For human consumption, a concentration of cyanogens between 10–20 mg/100 g is considered acceptable [14]. Higher levels of consumption of these compounds can lead to intoxication or rapid death due to neurological damage caused by cyanide. Generally, the content of cyanogenic compounds in plants for human consumption does not exceed the level of risk, however, in tropical countries where the consumption of products such as cassava or sorghum is high, the presence of diseases associated with these compounds are frequent. Patients with diseases known as "Tropical Ataxic Neuropathy" present paralysis of the lower extremities resulting in ataxia and other neurological effects of the central nervous system involving the optic and auditory nerves [46].

13.2.7 OXALATES

Oxalic acid ($C_2H_2O_4$) and its salts are generally the end products of metabolism in some plant tissues [47]. Oxalic acid readily donates protons thus forming oxalate ions ($C_2O_4^{2-}$) and oxalate salts; this is a chelating agent of metallic cations, forming water-soluble salts with metallic ions such as lithium, sodium, potassium, and ferrous salts and insoluble salts with calcium. These oxalates are generally insoluble at neutral or alkaline pH [48, 49]. Oxalates are toxic to mammals due to their low solubility in the intestine which limits calcium absorption. Oxalates bind with calcium forming complexes, in the form of crystals, avoiding the absorption and use of calcium by the body, causing among other diseases rickets, osteomalacia, and renal stones [14, 50].

Free-form oxalate ions, soluble oxalate salts, insoluble oxalates, and mineralized calcium oxalate crystals occur naturally in plants; these are even produced in the body humans as a product of metabolism. Oxalates display no beneficial health and must be excreted in the urine. Therefore, a diet rich in oxalate increases urinary oxalate excretion and causes the toxic effects already mentioned [47–49].

13.2.8 GLUCOSINOLATES

Glucosinolates or thioglucosides are water-soluble organic compounds of alkyl aldoxime-O-sulfate esters with a b-D-thioglucopyranosyl at the aldoxime carbon Z-configuration to the ester group that contains sulfur and nitrogen. These compounds are the product of the secondary metabolism of plants, formed with any of the eight amino acids: alanine, valine, leucine, isoleucine, phenylalanine, methionine, tyrosine, and tryptophan [51]. The plants use glucosinolates as a defense mechanism mainly against pathogens. The glucosinolates are found in all crucifer plants (Brassicaceae family) including horticultural crops, oilseed crops, condiments, and herbage, such as mustard, broccoli, Brussels sprouts, cabbage, cauliflower, horseradish, turnip, kale. These compounds are also found in other families such as Capparaceae, Limnanthaceae, and Resedacea, and in the species *Carica papaya* [52]. About 200 different types of glucosinolates have been identified and classified according to the different molecular structures of the side chains as aliphatic, aromatic, and heterocyclic (indole) compounds [51, 53]. Among these compounds, some of the most studied are sinigrin, progoitrin, gluconapin, epiprogoitrin, glucoraphanin, and glucoiberin.

The glucosinolates have been found to have interesting biological activities such as anti-cancer, antioxidant, anti-inflammatory and antimicrobial activity, also protect against cardiovascular, neurodegenerative, diabetes-related or *Helicobacter pylori* diseases. These biological properties depend on the levels and classes of glucosinolates present. However, when the plant is consumed by humans and animals or transformed by other processes (cutting, chopping, mixing), the cell walls are disrupted and the endogenous myrosinase enzyme (β-thioglucosidase) converts the glucosinolates in different compounds. The compound formed depends on the nature of the side chain (R) of glucosinolate, and the physicochemical conditions of the medium, such as glucose, sulfate, thiocyanates, cyanides, isothiocyanates, nitriles, and epithionitrile [7, 51, 54, 55] some of which may be toxic at high concentrations as well as acting as antinutritional factors as they bind to some amino acids decreasing their assimilation by the body. To take advantage of the biological potential of glucosinolates, in recent years different processes of extraction of these compounds have been studied as well as the mechanism of inactivation of the myrosinase enzyme to avoid catabolism of these compounds before they reach the digestive tract where they can exert their positive effect [51, 55].

13.2.9 ALKALOIDS

Alkaloids are a group of compounds that present diverse structures, however, generally they are derived from amino acids or from the transamination process, able to form salts with an acid. Therefore, the alkaloids are classified according to the amino acids that provide their nitrogen atom and part of their skeleton. Most alkaloids are compounds of the secondary metabolism of plants and act as a mechanism of defense against pathogens. Probably, alkaloids are the most widely distributed natural toxins in the world. More than 12,000 alkaloids have been identified and some of the most studied alkaloids are xanthine, atropine, caffeine, nicotine, sparteine, and cocaine, which are found in some plants families such as Boraginaceae (all genera), Asteraceae (tribes Senecioneae and Eupatorieae), Amaryllidaceae, Compositae, Leguminosae, Liliaceae, Papaveraceae, and Solanaceae Fabaceae (genus Crotalaria) [11, 56]. Alkaloids are widely recognized for their beneficial effects on health; these compounds are recognized biological, pharmacological or physiological and chemical activity. It has also been recognized that excessive consumption can cause toxicity [12]. High concentrations of alkaloids cause physiological alterations. In the nervous system, inappropriately alter the electrochemical transmission, which can cause increased heartbeat, paralysis, among others. It has also been shown that some alkaloids can cause intestinal disorders, infertility, among others.

Due to the high consumption of coffee and tea worldwide, one of the alkaloids of particular interest is caffeine. Caffeine is a methylxanthine with bitter characteristics. It is a heat-stable alkaloid, and its concentration changes in coffee, according to the environmental conditions where the plant is cultivated. Caffeine stimulates the central nervous system as an adenosine receptor antagonist. Although caffeine is the most widely consumed and studied psychoactive substance in history, its effects on health are controversial. While caffeine intake has been associated with high blood cholesterol levels, coronary heart disease, and cancer, other studies suggest that its consumption may decrease the incidence of suicide and liver cirrhosis. A low to moderate intake of caffeine is associated with increased alertness, learning ability, exercise performance, and perhaps better mood, but high doses can have negative effects on some individuals [57].

The physiological effect of this methylated purine alkaloid may cause an increase in motor activity in ruminants and rats. The result of this abnormal activity could be an increase in energy use that would ultimately result in a decrease in weight gain and conversion efficiency. Among the effects caused by the high levels of caffeine, in general, we can mention the increase in the animal's thirst, as well as the increase in urinary evacuation, which results in nitrogen excretion [58].

13.2.10   PROTEASE INHIBITORS

Protease inhibitors are a large and important group of molecules that inhibit the action of proteases enzymes. The best known are those that react with serine proteases (SPs), such as trypsin and chymotrypsin. These compounds are found widely in plants, mainly in legumes, although they have also been found in cereals and tubers. Together with other compounds of the secondary metabolism of plants, they are part of the defense system of these against pathogens [59].

Protease inhibitors have been shown to have both positive and negative health effects. Among the positive effects is that it helps prevent cancer, present anti-inflammatory activities, helps to prevent multiple sclerosis and Duchenne muscular dystrophy (DMD), as well as others benefits such as protection against radiation-induced birth defect, hair, and weight loss in cancer patients, improved efficacy of certain chemotherapeutic drugs and lifespan extension [59, 60]. However, these compounds have been associated with the low nutritional quality of some foods. They are attributed to a growth depressant effect due to a negative feedback mechanism, which is activated by the presence of dietary proteins in the digestive tract. In a simultaneous way, there is an inactivation of trypsin that results in the release of cholecystokinin (CCK), an intestinal mucosal hormone that stimulates pancreatic acinar cells to release more trypsin and other enzymes such as chymotrypsin, elastase, and amylase. Thus, in addition to the underutilization of dietary protein, the net result is the loss of endogenous protein rich in sulfur amino acids and the consequent growth depression [12, 59]. Although they are inactivated at high temperatures, some thermal resistance has been observed due to their rigid structure and their high content of disulfide bonds. They have also shown resistance to pepsin treatments [61].

13.3 MICROORGANISM AND BIOPROCESS FOR PRODUCTION OF BIOACTIVE COMPOUNDS (BCS) FROM AGRO-FOOD WASTES

There is currently a high interest in the production and recovery of compounds of natural origin with an important biological potential or with high bioactive potential. These compounds have applications in different areas such as pharmaceuticals, cosmetics, and of course food, as they contribute to maintaining a good state of health and prevent degenerative diseases. The inclusion of BCs that promote the benefits mentioned above in everyday life is an activity of great relevance today.

Within the group of compounds considered as bioactive are polyphenolic compounds, which possess the necessary characteristics to meet the requirements of disease prevention, among others. Phenolic monomers (derived from polyphenolic compounds) have similar characteristics to their predecessors and to obtain them it is necessary to use hydrolytic pathways, which of the most studied are biotechnological pathways or bioprocesses that involve the interaction of agricultural plant materials or agro-food residues with specific groups of microorganisms.

It is known that microorganisms play an important role in the development of bioprocesses, since they can produce various enzymes that are responsible for the release or bioavailability of polymeric and monomeric BCs. There are abundantly distributed plant materials that contain high levels of polyphenolic compounds that can be used to recover them using a specific bioprocess, such as a fungal, bacterial culture, etc. Therefore, in this section, we will talk about the use of different microorganisms in bioprocesses for the production and recovery of BCs from agricultural materials and agro-food residues.

BCs in most cases are obtained from plant sources and using conventional extraction methods, for example, maceration, infusion, and reflux [62], however, these conventional methods have operational disadvantages in terms of low yields, low concentrations of the compounds obtained, among others [63]. It has been reported that a suitable method of production of BCs not only requires an adequate recovery of them but also facilitates their identification and characterization [64]. Interest in the recovery of BCs has increased because they have biological properties of great relevance such as antitumoral, antiviral, anti-cancer, antioxidant, among others [58–65].

Alternative methods for recovery of BCs have been developed, and one of the most interesting alternatives is the use of bioprocesses involving a microorganism (generally fungi, yeasts, and bacteria) [69]. For example, Yoshida et al. [70] reported the production and recovery of BCs (polyphenols, particularly ellagitannins) from the *Multiflora thiouchin* plant using a filamentous fungus and a bioprocess (solid-state fermentation). The authors reported that the production of these BCs was carried out by the enzymes produced by the microorganism during the bioprocess. Huang et al. [72] achieved the extraction of polyphenolic compounds using a fungus and fermentation in solid-state as well, and the obtained compounds were recovered in high percentages of yield.

Despite the above, the mentioned authors cited the need to develop new bioprocesses using new microorganisms capable of producing enzymes that help the bioavailability of the compounds. Other studies have reported the development of bioprocesses using filamentous fungi to produce very particular BCs such as ellagic acid from agroindustrial residues such as pomegranate peel [73]. Also, partial purification and recovery (at high yields) of ellagic acid, which has excellent industrial and biological importance due to its high antioxidant activity [74], was achieved. In other reports, the biotransformation mechanism of precursor compounds of bioactive substances has been studied; this has allowed for a better understanding of the biochemical behavior of a microorganism when it is used in the bioprocess [75]. Once the biotransformation mechanism for the recovery of BCs by bioprocesses has been described, it is possible to carry out the optimization of the bioprocess to obtain the bioactive compound at high yield; the main parameters that are evaluated in an optimization experiment can be the microbial inoculum concentration, the composition of the culture medium, temperatures, pH, etc. [76]. Finally, if a microbial bioprocess is known in its biochemical aspects, it is an optimized bioprocess; it is possible to identify the intermediary compounds generated during the bioprocess as well. The identification of intermediary compounds is very important for the development of methodologies for the recovery of BCs [77].

Other substances considered bioactive are enzymes due to their catalytic activity, which can be produced by microorganisms by a bioprocess (e.g., solid fermentation). Microbial enzyme production has been developed using several methods; however, the most important method in recent years is the use of bioprocesses. For example, microbial production of the enzyme β-glucosidase has been reported by a fungal strain using a solid bioprocess [78], the same enzyme has been produced by bioprocesses using bacterial strains and yeasts [79, 80]. This production strategy is remarkable because it involves work using residual, renewable (in some cases) and low-cost materials [81]. One of the advantages of using bioprocesses (such as solid-state fermentation) is that they can be carried out under simple microbial growth conditions that favor the development of microorganisms such as filamentous fungi (simulates the environmental growth conditions of these fungi). Another advantage is low energy consumption [80] and perhaps most importantly, they offer high levels of productivity, low catabolic repression, and increased stability of biologically active enzymes produced [82].

As mentioned above, microorganisms (such as bacteria, fungi, and yeasts) are capable of fermenting or biotransforming the components of different substrates (such as agro-industrial or agricultural residues) to generate ATP molecules and thus ensure their growth [83]. Ecologically speaking, this is important, as biotransformation allows better carbon fixation in the microbial biomass and reduces carbon losses in the form of carbon dioxide or methane [84]. It has been shown that fungi such as *Aspergillus niger* are capable of biotransforming and allowing the recovery of hydrolyzable polyphenols [75]. The biotransformation carried out during a bioprocess gives rise to different molecules of BCs that can be recovered and detected by spectrophotometric techniques. It has been reported that tannic acid and polyphenols [85] that are present in different agroindustrial residues when biotransformed by microorganisms during the bioprocess give rise to compounds such as gallic acid and ellagic acid, which are smaller but biologically more active molecules. In conclusion, the development of microbial bioprocesses is a fundamental aspect at present, because the interest in the recovery of BCs has increased and bioprocesses also offer operational advantages by increasing recovery yields and reducing environmental pollution. The role of microorganisms in the development of bioprocesses for the production and recovery of BCs are described below.

13.3.1 BACTERIA

Due to population growth, economic growth and industrialization, today the generation of by-products and waste is high, and therefore the problems associated with these as well as environmental pollution has also increased. Therefore, much research has been dedicated to establishing appropriate methodologies for handling products, mitigating the damage that may occur due to mishandling. The food industry also generates residues that have an interesting content of macro and micronutrients that can be used. Among these, the most abundant are cellulosic and lignocellulosic materials, which are also characterized by having different compounds considered toxic [86, 87]. To take advantage of these materials, different treatments have been used, among which are the fermentation processes. One of the most common fermentations is fermentation with lactic acid bacteria (LAB), in which various compounds are generated such as bioactive peptides, short-chain fatty acids (SCFAs), among others. In addition, the content of anti-nutritional compounds is reduced, converting them into molecules with interesting biological activities such as antioxidants, anti-cancer, anti-inflammatory, among others. Among the bacteria most used to carry out the fermentation processes are *Lactobacillus plantarum*, *Lb. brevis*, *Lb. rhamnosus*, *Lb. acidophilus*, *Leuconostoc mesenteroides*, *Lc. citreum*, *Lc. fallax*, *Lc. kimchi*, *Pediococcus pentosaceus*, *P. acidilactici*, *Weissellaconfusa*, *W. cibaria* [88–90]. Therefore, the application of the fermentation process with bacteria is a viable alternative to manage and potentiate the use of by-products generated in the food industry.

Different investigations have been carried out in recent years around the biocomposites produced by bacteria during fermentation. Moreover, among the most studied agro-industrial products and by-products is the soybean, which is a legume, characterized by its high nutritional value and by its high content of toxic compounds. Yang et al. [91] studied the fermentation process with *Bacillus amyloliquefaciens* SWJS22 of soy flour, for the nutritional improvement of this by-product. After the fermentation process, they found that there was an increase in the concentration of daidzein, glycitein, genistein, protocatechuic, and p-hydroxybenzoic acids. These compounds are annealed by their antioxidant activity, which increased the fermented soybean meal. Therefore, the fermentation process with this microorganism is proposed as a potential for the nutritional improvement of agro-industrial by-products. Dai et al. [92] improved through the fermentation process the characteristics of soybean meal, a by-product generated from the soybean oil (SO) extraction process. They evaluated the solid-state fermentation process with *Bacillus subtilis* on the content of nutritional and antinutritional compounds present in the flour. After 24 hours of fermentation, they found an increase in the content of total phenols and the antioxidant activity associated with these compounds. They also found a decrease of approximately 50% in the content of trypsin inhibitors, an anti-nutritional compound. As an additional result, they found that fermented soybean meal could inhibit angiotensin I-converting enzyme (ACE) compound associated with blood pressure problems. Therefore, the authors suggest that the fermentation process significantly improves the nutritional characteristics of soy byproducts.

Sang-Hyun and Seong-Jun [93] evaluated the changes in soy protein during the fermentation process with *Bacillus subtilis* to know the biochemical changes that occurred during the process. They found that after 24 h of fermentation some compounds recognized as antinutritional such as trypsin inhibitors and beta-conglycin decreased considerably, thus proving the benefits of fermentation in biomodified products. Ming-Yen and Cheng-Chun [94] evaluated the production of some bioactive computations in black soybean through the fermentation process with *Bacillus subtilis* BCRC 14715. After 18 h of fermentation they found that the content of total phenols increased 4-fold, as well as the antioxidant activity increased, attributing this fact to the phenolic compounds produced or released during the process.

Among other studies, Fritsch et al. [95] evaluated the fermentation process with the strains *Lactobacillus plantarum, Pediococcus pentosaceus, Lactobacillus casei, and Bifidobacterium animalis* subsp. *lactis* to decrease the content of chlorogenic acid in sunflower by-products. Of the four bacteria studied, *Lb. plantarum* and *P. pentosaceus* presented the best behaviors, decreasing up to 19% the acid content. They also found that the caffeic acid content increased due to the hydrolysis of chlorogenic acid. Caffeic acid is a biocompound recognized for its antioxidant and anticarcinogenic activity, for which the fermentation process improved the nutritional characteristics of the sunflower byproducts.

In addition to phenolic compounds obtained with activities recognized as antioxidants, anti-cancer, antihypertensive, among others, through bacterial fermentation of agro-industrial residues, other compounds such as bioactive peptides have also been isolated. Bioactive peptides are derived from proteins, which may contain between 3 and 20 amino acid units. In the last years, the studies of these compounds have increased since it has been found that they present interesting biological activities such as immunomodulatory, antimicrobial, antioxidative, antithrombotic, hypocholesterolemic, and antihypertensive [96]. Through fermentation, soy, and its by-products have been obtained, among other peptides, Glycinin, and β-conglycinin, which exhibit activities such as anti-hypertensive, antimicrobial, antioxidant, anti-diabetic, and anticancer [97, 98].

13.3.2   YEAST

Another of the microorganisms used in fermentation processes to potentiate the use of agro-industrial waste are yeasts. The use of yeasts use in fermentation processes has been known since ancient times in products popularly known as beer, wine, soy sauce, and bread. Yeasts are microorganisms with great biotechnological potential; in fact, the products obtained with yeasts are currently the most commercialized in the world. Yeasts are used to produce enzymes, pigments, amino acids, organic acids, flavors, among others. Some of the characteristics that make these microorganisms special are their resistance to extreme conditions, for example, high concentrations of sugar, their rapid growth, and the varied production of aromatic compounds. One of the most researched and best-known yeasts is *Saccharomyces cerevisiae*, due to its unique physiology and important participation in different fermented foods. However, there are other species of biotechnological importance that are being exploited, among these are: *Schizosaccharomyces pombe, Kluyveromyces lactis, Kluyveromyces marxianus, Schwanniomyces occidentalis, Lipomyces* spp., *Saccharomycopsis* spp., *Debaryomyces hansenii, Ogataea polymorpha, Komagataella pastoris, Scheffersomyces stipites, Pichia* spp., *Rhodotorula* spp., *Rhodosporidium* spp., *Yarrowia lipolytica, Candida* spp., *Trichosporon* spp., *Blastobotrys adeninivorans,* and *Xanthophyllomyces dendrorhous* [99].

Fermentation processes have treated several agro-industrial residues with yeasts. One of these is Okara, a byproduct of soy processing which a high content of isoflavones compound has recognized for its biological functions as estrogenic, antioxidant, and anticarcinogenic activities. However, there are also compounds that are considered antinutritional that must be reduced to potentiate their use. In this sense, Queiroz Santos et al. [100] studied the solid-state fermentation process with *S. cerevisiae* using Okara as a substrate. During fermentation, they evaluated the content of phenols, the activity β-glucosidase, the antioxidant activity, and the biotransformation of the β-glucoside isoflavones to isoflavones aglycones, which are the compounds that present the bioactivity. They found that at 72 h of processing phenol content and antioxidant activity increased. They also found that fermentation promoted the bioconversion of isoflavones β-glucoside (daidzin, glycitin, and genistin) to aglycones (daidzein, glycitein, and genistein). Therefore, the study demonstrated that the biotechnological process is an adequate strategy to improve the nutritional quality of by-products of the food industry.

Using this same by-product, Vong et al. [101] studied the fermentation process using *Yarrowia lipolytica* yeast. They evaluated changes in proximal composition, antioxidant capacity, and production of other compounds. They found that Okara is a byproduct suitable to be biotransformed by the yeast used without the need to add an extra source of carbon or nitrogen, recording the highest biomass production at 48 hours of processing. After 5 days of fermentation, they found an increase in the concentration of amino acids, which was attributed to the yeast's ability to produce protease enzymes. Also, there was an increase in antioxidant activity due to the increase of bioactive peptides and the increase of phenolic compounds p-hydroxy-phenylethanol and p-hydroxyphenyacetic acids, which are recognized for having this biological activity. During the study, it was found that there was an increase in aglycone isoflavones possibly due to the action of the enzyme β-glucosidase of *Y. lipolytica*, which are more bioavailable. Therefore, the use of *Y. lipolytica* yeast allows the production of high value-added biocompounds using wasted agro-industrial residues.

Another strategy to use agro-industrial residues by biotechnology has been the use of microorganisms in co-culture, thus taking advantage of the capacities and metabolites that each microorganism can generate.

Vong et al. [102, 103] evaluated the nutritional characteristics of Okara after fermentation with *Rhizopus oligosporus* and *Yarrowia lipolytica*. At the end of the process, they found that using the co-culture, the content of phytic acid (phytates) was reduced in greater percentage and the content of amino acids, organic acids (acetic, citric, oxalic, and succinic), and the antioxidant capacity was increased. Therefore, it was shown that the use of these microorganisms presents positive synergistic effects, improving the nutritional characteristics of this agroindustrial residue and obtaining biomolecules with biological activities of interest. Zhao et al. [103] evaluated the effect of solid-state fermentation with a commercial yeast strain and two lactic acid strains, *Lactobacillus bulgaricus,* and *Streptococcus thermophiles*, on the physical characteristics of wheat bran. After fermentation, the concentration of phytic acid was reduced, and the content of phenolic compounds was increased, due to the hydrolysis carried out by the enzymes released by the microorganisms used. Also, other evaluated characteristics showed improvement, so it is proposed the fermentation using co-cultures as a viable technological alternative to produce biocompounds and the improvement of products in the food industry.

Taking advantage of other non-vegetable residues from the food industry, but whose inadequate disposal can generate problems of environmental contamination, Magdouli et al. [104] used the yeast *Yarrowia lipolytica* for the production of lipases and lipids from crustacean waste. Comparing the production of lipases with traditional means, the researchers found a considerable increase in the use of crustacean residues, so they propose that the use of agro-industrial residues could be a biotechnological strategy to reduce costs and potentiate the use of these by-products. Among other biocompounds of interest produced by yeasts such as *Rhodotorula spp., Candida utilis,* and *Pichia pastoris*, using means supplemented with agro-industrial residues from the production of rice, wheat, among others, is lycopene. Lycopene is a carotenoid compound recognized for its cardioprotective, anticarcinogenic, and antioxidant properties. Therefore, its biotechnological production using agro-industrial residues is a potential alternative [105].

13.3.3   *FILAMENTOUS FUNGI*

Cellulosic and lignocellulosic materials are the most abundant materials in agro-industrial wastes [106, 107]. Various strategies have been implemented to degrade these materials and subsequently be used in other bioprocesses such as the generation of microbial biomass, cellulose, proteins like enzymes, antioxidants, vitamins, lipids, pigments, among others.

Numerous genera of fungus have been used to detoxify agro-industrial wastes by the growth of these strains on these residues. For the growth and development of these organisms, it is necessary that their metabolism produces specific enzymes to grow on these complex materials. As part of the production and action of these enzymes, there is a release of BCs resulting from catalysis, so it is a promising alternative. Table 13.2 shows a list of fungi that have been used for the detoxification of agro-industrial waste as well as the main compounds recovered during the bioprocess.

An example is the use of fungi such as *P. ostreatus* and *Trametes versicolor* to produce ligno and cellulolytic enzymes using apple processing residues as a substrate, the enzymes obtained were used in the cleaning of wastewater from the textile industry [108]. Other works have been focused on the detoxification of toxic waste to be converted into animal feed material [109].

TABLE 13.2 Agro-Industrial Wastes Detoxicated by Fungi and Main Compounds Recovered During the Bioprocess

Wastes	Toxic Compounds	Fungus	Bioactive Compound Product	References
Hemicellulose-rich liquor	Phenolics	*Irpexlacteus, Ganoderma lucidum*	Ligninolytic enzymes	[128]
Dry olive-mill	Phenolic compounds	*Cyclocybe aegerita, Chondrostereum purpureum*	Peroxidase, Laccase, Peroxygenase, Manganese peroxidase	[113]
Rice husks, Straw, Sunflower seed hulls		*Ganoderma lucidum*	Laccase	[129]
Sorghum		*Fusarium oxysporum*	Lipids (oleic, linoleic, and palmitic acid)	[130]
Jatropha seed cake	Tannins, lectins, saponins, phytate, among others	*Aspergillus versicolor* CJS-98	Lipase, Protease	[116]
Castor bean	Ricin	*Paecilomyces variotti*	Tannase, Phytase	[111]
Apple		*Pleurotusostreatus, Trametes versicolor*	Laccases, xylanases	[107]
Orange peel		*A. niger, Fusarium oxysporum, Neurosporacrassa, Penicillium decumbens*	Polygalacturonase, Pectate-lyase, Endoclucanase, Xylanase, β-glucosidase, Invertase	[131]
Grape		*Monascus purpureus*	β-glucosidase	[132]
Pineapple		*Rhizopus oligosporus*	β-glucosidase, Antioxidants	[133]

Some of these works reported that it is possible to use filamentous fungus to detoxify substrates with a high content of antinutritional compounds [110].

Castor bean residue is an excellent source of high protein content (around 35%) but this residue also has an important content of ricin that is a lethal toxin [111]. Madeira et al. [112] profit castor bean residue to produce tannases and phytases by *Paecilomyces variotii*, they demonstrated that ricin was reduced during fermentation time. Also, *Penicillium simplicissmum* has been used to eliminate ricin from castor bean residue [109].

Wastes from oil production are one of the most studied to be used as fungi substrates. For example, oil palm wastes has been used to obtain several by-products like as structural carbohydrate such as a cellulase, monosaccharide xylose, enzymes, production of hydrogen, organic fertilizer, among others. In most of the processes, filamentous fungi have been used to obtain these products thanks to their ability to grow on complex substrates [113]. Also, olive-oil production generates a significant amount of wastes, this by-product is called dry olive residue. The dry olive residue has a high polyphenol content and aromatic compounds which makes it a phytotoxic residue, some works have been explored the possibility of detoxifying this wastes by the use of various fungus strains [114, 115]. Some other fungi such as *Coriolopsis rigida, Ganoderma applantaum, Poria subvermispora, Pleurotus pulmonaris* have been used to reduce polyphenol content and detoxify residues from vegetable oil production [116]. *Jatropha curcas* is a plant that is cultivated to obtain its seeds due oil is obtained from them. From process is generated a waste called seed cake that have a high protein content an also antinutrients like tannins, lectins, saponins, phytate, among others. It was reported that thanks to the action of the fungus *Aspergillus versicolor* on this residue is possible to produce lipases and proteases [117].

Another waste of great importance for the recovery of high added value compounds is coffee. Coffee waste is rich in polyphenols, proteins, carbohydrates, pectic substances. The wastes are produced in huge amount, and the implementation of processes that help to take advantage of the rich content of this waste is desirable [110]. Biological detoxification has been carried out using fungus strains like *Aspergillus, Rhizopus, and Phanerochaete, Pleurotus ostreatus, Neurospora crassa* [118]. These residues have been treated by solid-state fermentation and are an interesting alternative to detoxify this type of material and at the same time produce industrially important enzymes [8].

The use of fungi for the detoxification of residues is widely used through solid-state fermentation (SSF) thanks to the advantages offered by this type of bioprocess. The detoxification of waste products can result in numerous compounds with mainly antioxidant biological activities, as well as many hydrolytic enzymes with various industrial applications.

13.4 PERSPECTIVES

Although fermentation processes offer an interesting opportunity for the proper management of agro-industrial waste, where even waste that is considered highly toxic can be used for the recovery of high value-added compounds using microorganisms capable of developing in them. To solve the problem of agro-industrial waste management through biotechnological processes, it is necessary to link different scientific areas that allow the growth in the knowledge of fermentation processes in this kind of material.

One of the alternatives to improve the productivity of the processes is the improvement of the most used microorganisms from genetic engineering using genes from other microorganisms with different characteristics and whose use in food is not safe or adequate. Another strategy is the use of different microorganisms during the fermentation process, taking advantage of the characteristics and metabolites generated by each one. In this sense, it is necessary to define the microorganisms and the stages in which they must be used, according to the treated residues, so that there is a real synergic effect and bio-transformation of these residues effectively is achieved. By-products can also have a synergistic effect, so mixing residues can also be considered as another strategy to increase the effectiveness of the process. Considering that the biocompounds generated during the process may be different, it is necessary to implement techniques of identification, quantification, and separation of compounds. The implementation of technically and economically feasible extraction and recovery methodologies is necessary to ensure that detoxification of these by-products is a plausible alternative.

Most studies focus on the use of agro-industrial residues with bacteria or fungi, while studies with yeasts are scarce. Yeasts are commonly used in the production of alcohol and aromatic compounds, so it is necessary to delve deeper into the type of compounds that these microorganisms can release using different types of agro-industrial waste.

Finally, it is necessary to evaluate other agro-industrial wastes considering the most produced in the region and the environmental, social, and economic impacts that may have the implementation of this type of bioprocess for treatment.

KEYWORDS

- **agro-industrial wastes**
- **biocompounds**
- **cholecystokinin**
- **Duchenne muscular dystrophy**
- **hydrogen cyanide**
- **protein kinase C**

REFERENCES

1. FAO. Agricultura Mundial 2015–2030: Perspectivas Por Sectores Principales. Agricultura Mundialhacialosaños 2015/2030.World Agriculture 2015–2030: Perspectives by main sectors. *World Agriculture Towards 2015/2030*, (pp. 32–74). Summary report **2002.**

2. Organización de las Naciones Unidas para la Agricultura y la Alimentación. Producción de cereales (toneladas métricas). Food and Agriculture Organization of the United Nations. Cereal production (metric tons). https://datos.bancomundial.org/indicador/AG.PRD.CREL.MT?locations=MX&view=chart. (Accessed on 6 January 2020).

3. González-Sánchez, M. E., Pérez-Fabiel, S., Wong-Villarreal, A., Bello-Mendoza, R., & Yãnez-Ocampo, G. Residuos Agroindustriales con potencial para la producción de metano mediante la digestión anaerobia. Agroindustrial waste with potential for methane production through anaerobic digestion. *Rev. Argent. Microbiol.,* **2015**, *47*(3), 229–235.

4. Handa, C. L., de Lima, F. S., Guelfi, M. F. G., Fernandes, M. da S., Georgetti, S. R., & Ida, E. I. Parameters of the fermentation of soybean flour by *Monascus purpureus* or *Aspergillus oryzae* on the production of bioactive compounds and antioxidant activity. *Food Chem.,* **2019**, *271*(July 2018), 274–283.

5. Arellano-González, M. A., Ramírez-Coronel, M. A., Torres-Mancera, M. T., Pérez-Morales, G. G., & Saucedo-Castañeda, G. Antioxidant activity of fermented and nonfermented Coffee (*Coffea Arabica*) pulp extracts. *Food Technol. Biotechnol.,* **2011**, *49*(3), 374–378.

6. Faisal, M., & Prasad, L. A Potential source of methyl-eugenol from secondary metabolite of *Rhizopus oryzae* 6975. *Int. J. Appl. Biol. Pharm. Technol.,* **2016**, *7*(4), 187–193.

7. Mazumder, A., Dwivedi, A., & Plessis, J. Du. Sinigrin and Its Therapeutic Benefits. *Molecules,* **2016**, *21*(4), 1–11.

8. Soccol, C. R., Costa, E. S. F. da, Letti, L. A. J., Karp, S. G., Woiciechowski, A. L., & Vandenberghe, L. P. de S. Recent developments and innovations in solid state fermentation. *Biotechnol. Res. Innov.,* **2017**, *1*(1), 52–71.

9. Ghoshal, G., Basu, S., & Shivhare, U. Solid state fermentation in food processing. *Int. J. Food Eng.,* **2012**, *8*(3).

10. Bora, P. Anti-nutritional factors in foods and their effects. *J. Acad. Ind. Res.,* **2014**, *3*(6), 285–290.
11. Kathirvel, P. *Secondary Metabolites.* In: Kathirvel, P., (ed.), Darshan Publishers, **2016**.
12. Sinha, K., & Khare, V. *Review on: Antinutritional Factors in Vegetable Crops.* **2017**, *6*(12), 353–358.
13. Fekadu, G. H. Antinutritional factors in plant foods: Potential health benefits and adverse effects. *Int. J. Nutr. Food Sci.,* **2014**, *3*(4), 284.
14. Mohan, V. R., Tresina, P. S., & Daffodil, E. D. *Antinutritional Factors in Legume Seeds: Characteristics and Determination,* (1st edn.), Elsevier Ltd., **2015**.
15. Juárez, F. B. Cambios bioquímicos en semillas de *Lupinusmontanus* y *Lupinusexaltatus* asociados a tratamientos físicos, químicos y germinativos. Biochemical changes in seeds of *Lupinusmontanus* and *Lupinusexaltatus* associated with physical, chemical and germinative treatments. *Colegio de Postgraduados en Ciencias Agrícolas,* **2010**, *34*.
16. Moghimipour, E., & Handali, S. Saponin: Properties, methods of evaluation and applications. *Annu. Res. Rev. Biol.,* **2015**, *5*(3), 207–220.
17. Netala, V. R., Ghosh, S. B., Bobbu, P., Anitha, D., & Tartte, V. Triterpenoid saponins: A Review on biosynthesis, applications and mechanism of their action. *Int. J. Pharm. Pharm. Sci.,* **2015**, *7*(1), *24–28*.
18. Vincken, J. P., Heng, L., de Groot, A., Gruppen, H. Saponins, classification and occurrence in the plant kingdom. *Phytochemistry,* **2007**, *68*(3), 275–297.
19. Vázquez-Flores, A. A., Álvarez-Parrilla, E., López-Días, J. A., Walle-Medrano, A., &De la Rosa, L. A. Taninos hidrolizables y condensados: Naturaleza Química, Ventajas y Desventajas de Su Consumo. Hydrolyzable and Condensed Tannins: Chemical nature, advantages and disadvantages of their consumption. *Tecnociencia Chihuahua,* **2012**, *6*(2), *84–93*.
20. Chávez-González, M., Rodríguez-Durán, L. V., Balagurusamy, N., Prado-Barragán, A., Rodríguez, R., Contreras, J. C., & Aguilar, C. N. Biotechnological advances and challenges of tannase: An overview. *Food Bioprocess Technol.,* **2012**, *5*(2), 445–459.
21. Coello, C. L. Los Taninos en la alimentación de las aves comerciales. Producción animal. Tannins in the feeding of commercial birds. Animal production. *Ciência Animal Brasileira,* **2000**, *1*(1), *5–22*.
22. Lara, D. M., & Londoño, Á. S. El uso de taninos condensados como alternativa nutricional y sanitaria en rumiantes. The use of condensed tannins as a nutritional and sanitary alternative in ruminants. *Rev. Med. Vet., (Bogota).* **2008**, *87–109*.
23. Aguilar, C. N., Rodríguez, R., Gutiérrez-Sánchez, G., Augur, C., Favela-Torres, E., Prado-Barragan, L. A., Ramírez-Coronel, A., & Contreras-Esquivel, J. C. Microbial tannases: Advances and perspectives. *Appl. Microbiol. Biotechnol.,* **2007**, *76*(1), *47–59*.
24. Monteiro, J. M., De Albuquerque, P. U., & De Lima Araújo, E. Taninos: Uma abordagem da química à ecologia. Tannins: An approach from chemistry to ecology. *Quim. Nov.,* **2005**, *28*(5), 892–896.
25. Schofield, P., Mbugua, D. M., & Pell, A. N. Analysis of condensed tannins: A review. *Anim. Feed Sci. Technol.* **2001**, *91*(1-2), 21–40.
26. Kumar, V., Sinha, A. K., Makkar, H. P. S., & Becker, K. Dietary roles of phytate and phytase in human nutrition: A review. *Food Chem.,* **2010**, *120*(4), 945–959.
27. Hurrell, R. Phytic acid degradation as a means of improving iron absorption. *Int. J. Vitam. Nutr. Res.,* **2004**, *74*, 445–452.
28. Raboy, V. Myo-Inositol-1,2,3,4,5,6-Hexakisphosphate. *Phytochemistry,* **2003**, *64*(6), 1033–1043.

29. García, D. E. Principales factores antinutricionales de las leguminosas forrajeras y sus formas de cuantificación. Main anti-nutritional factors of forage legumes and their forms of quantification. *Pastos y Forrajes,* **2004**, *27*(2),

30. García, D. E., Medina, M. G., Soca, M., & Montejo, I. L. Toxicidad de las leguminosas forrajeras en la alimentación de los animales monogástricos. Toxicity of forage legumes in monogastric animal feeding. *Pastos y Forrajes,* **2005**, *28*(4), 279–289.

31. Bing L. L., Rafiq, A., Tzeng, Y. M., & Rob, A. The induction and characterization of phytase and beyond. *Enzyme Microb. Technol.,* **1998**, *22*(5), 415–424.

32. Zhang, W. J., Xu, Z. R., Zhao, S. H., Sun, J. Y., & Yang, X. Development of a microbial fermentation process for detoxification of gossypol in cottonseed meal. *Anim. Feed Sci. Technol.,* **2007**, *135 (1–2),* 176–186.

33. de Peyster, A. *Gossypol. Encycl. Toxicol.* (3ʳᵈ edn). **2014**, *2,* 782–785.

34. Pelitire, S. M., Dowd, M. K., & Cheng, H. N. Acidic solvent extraction of gossypol from cottonseed meal. *Anim. Feed Sci. Technol.,* **2014**, *195,* 120–128.

35. Luz, V. B., Gadelha, I. C. N., Cordeiro, L. A. V., Melo, M. M., & Soto-Blanco, B. In vitro study of gossypol's ovarian toxicity to rodents and goats. *Toxicon,* **2018**, *145,* 56–60.

36. Hatamoto-zervoudakis, L. K., Júnior, M. F. D., Zervoudakis, J. T., Motheo, T. F., Silva-marques, R. P., Tsuneda, P. P., Nichi, M., Santo, B. S. E., & Almeida, R. D. Theriogenology free gossypol supplementation frequency and reproductive toxicity in young bulls. *Theriogenology,* **2018**, *110,* 153–157.

37. Krempl, C., Heidel-Fischer, H. M., Jiménez-Alemán, G. H., Reichelt, M., Menezes, R. C., Boland, W., Vogel, H., Heckel, D. G., & Joußen, N. Gossypol toxicity and detoxification in helicoverpaarmigera and heliothisvirescens. *Insect. Biochem. Mol. Biol.,* **2016**, *78,* 69–77.

38. Gogoi, R., Niyogi, U. K., & Tyagi, A. K. Reduction of phorbol ester content in jatropha cake using high energy gamma radiation. *J. Radiat. Res. Appl. Sci.,* **2014**, *7*(3), 305–309.

39. Joshi, C., Mathur, P., & Khare, S. K. Degradation of phorbol esters by pseudomonas aeruginosa psea during solid-state fermentation of deoiled jatropha curcas seed cake. *Bioresour. Technol.,* **2011**, *102*(7), 4815–4819.

40. Prinsloo, G., Nogemane, N., & Street, R. The use of plants containing genotoxic carcinogens as foods and medicine. *Food Chem. Toxicol.,* **2018**, *116,* 27–39.

41. Nakao, M., Hasegawa, G., Yasuhara, T., & Ishihara, Y. Degradation of jatropha curcasphorbol esters derived from jatropha oil cake and their tumor-promoting activity. *Ecotoxicol. Environ. Saf.,* **2015**, *114,* 357–364.

42. Vetter, J. Plant cyanogenic glycosides. *Toxicon,* **2000**, *38*(1), 11–36.

43. Rosenthal, G. A., & Berenbaum, M. R. *Herbivores: Their Interactions with Secondary Plant Metabolites* (Vol. I). The Chemical Participants, **1991**.

44. Sun, Z., Zhang, K., Chen, C., Wu, Y., & Tang, Y. Biosynthesis and regulation of cyanogenic glycoside production in forage plants. *Appl. Microbiol. Biotechnol.,* **2017**, *102*(1), *9–16.*

45. Panter, K. E. *Cyanogenic Glycoside-Containing Plants,* (3ʳᵈ edn.). Elsevier Inc., **2018**.

46. Morrison, S. C., & Savage, G. P. Oxalates. In: Caballero, B., (ed.), *Encyclopedia of Food Sciences and Nutrition* (2ⁿᵈ edn., pp. 4282–4287), Academic Press, Oxford, **2003**.

47. Norton, S. K. Lost seasonality and overconsumption of plants: Risking oxalate toxicity. *J. Evol. Heal,* **2018**, *2*(3), 4.

48. Shi, L., Arntfield, S. D., & Nickerson, M. Changes in levels of phytic acid, lectins and oxalates during soaking and cooking of canadian pulses. *Food Res. Int.,* **2018**, *107,* 660–668.

49. Sadaf, H., Raza, S. I., & Hassan, S. W. Role of gut microbiota against calcium oxalate. *Microb. Pathog.,* **2017**, *109,* 287–291.
50. Deng, Q., Zinoviadou, K. G., Galanakis, C. M., Orlien, V., Grimi, N., Vorobiev, E., Lebovka, N., & Barba, F. J. The effects of conventional and non-conventional processing on glucosinolates and its derived forms, isothiocyanates: Extraction, degradation, and applications. *Food Eng. Rev.,* **2015**, *7*(3), 357–381.
51. Rechcigl, M. *Handbook of Naturally Occurring Food Toxicants,* CRC press, **2018**.
52. Jeschke, V., Kearney, E. E., Schramm, K., Kunert, G., Shekhov, A., Gershenzon, J., & Vassão, D. G. How glucosinolates affect generalist lepidopteran larvae: Growth, development and glucosinolate metabolism. *Front. Plant Sci.,* **2017**, *8,* 1–12.
53. Latimer, I., Collett, M., Matthews, Z., Tapper, B., & Cridge, B. The in vitro toxicity of nitrile and epithionitrile derivatives of glucosinolates from rutabaga in human and bovine liver cells. *Bio Rxiv,* **2018**.
54. Barba, F. J., Nikmaram, N., Roohinejad, S., Khelfa, A., Zhu, Z., & Koubaa, M. Bioavailability of glucosinolates and their breakdown products: Impact of processing. *Front. Nutr.,* **2016**, *3,* 1–12.
55. Jank, B., & Rath, J. The risk of pyrrolizidine alkaloids in human food and animal feed. *Trends Plant Sci.,* **2017**, *22*(3), 191–193.
56. Farah, A. Coffee Constituents. In: Chu., Y. F., (ed.) *Coffee: Emerging Health Effects and Disease Prevention,* (pp. 21–58), John Wiley & Sons, Inc., **2012**.
57. Gutiérrez-Sánchez, G., Roussos, S., & Augur, C. Effect of caffeine concentration on biomass production, caffeine degradation, and morphology of *Aspergillus tamarii*. *Folia Microbiol. (Praha),* **2013**, *58*(3), 195–200.
58. Zhao, J., & Yaw Ee, K. *Protease Inhibitors* (vol. 1). Elsevier, **2015**.
59. Kostekli, M., & Karakaya, S. Protease Inhibitors in various flours and breads: effect of fermentation, baking and in vitro digestion on trypsin and chymotrypsin inhibitory activities. *Food Chem.,* **2017**, *224,* 62–68.
60. Vagadia, B. H., Vanga, S. K., & Raghavan, V. Inactivation methods of soybean trypsin inhibitor – A Review. *Trends Food Sci. Technol.,* **2017**, *64,* 115–125.
61. Paz, J. E. W., Márquez, D. B. M., Ávila, G. C. G. M., Cerda, R. E. B., & Aguilar, C. N. Ultrasound-Assisted extraction of polyphenols from native plants in the mexican desert. *Ultrason. Sonochem.,* **2015**, *22,* 474–481.
62. Patil, D. M., & Akamanchi, K. G. Ultrasound-assisted rapid extraction and kinetic modeling of influential factors: Extraction of camptothecin from nothapodytes nimmoniana plant. *Ultrason. Sonochem.,* **2017**, *37,* 582–591.
63. Ribeiro de Souza, E. B., da Silva, R. R., Afonso, S., & Scarminio, I. S. Enhanced extraction yields and mobile phase separations by solvent mixtures for the analysis of metabolites in Annona Muricata L. leaves. *J. Sep. Sci.,* **2009**, *32*(23–24), 4176–4185.
64. Khaizil, E. Z., Nik Aina, S. N. Z., & Mohd, D. S. Preliminary study on anti-proliferative activity of methanolic extract of nepheliumlappaceum peels towards breast (MDA-MB-231), Cervical (HeLa) and Osteosarcoma (MG-63) cancer cell lines. *Health (Irvine. Calif).* **2013**, *4*(2), 66–79.
65. Fang, E. F., & Ng, T. B. A Trypsin inhibitor from rambutan seeds with antitumor, Anti-HIV-1 reverse transcriptase, and nitric oxide-inducing properties. *Appl. Biochem. Biotechnol.,* **2015**, *175*(8), 3828–3839.
66. Yuvakkumar, R., Suresh, J., Saravanakumar, B., Nathanael, A. J., Hong, S. I., & Rajendran, V. Rambutan peels promoted biomimetic synthesis of bioinspired zinc

oxide nanochains for biomedical applications. *Spectrochim. Acta part A Mol. Biomol. Spectrosc.,* **2015,** *137,* 250–258.

67. Mistriyani, Riyanto, S., & Rohman. Antioxidant activities of rambutan (*Nephelium Lappaceum* L) peel in vitro. *Food Res.,* **2018,** *2*(1), 119–123.

68. Yang, Y. C., Li, J., Zu, Y. G., Fu, Y. J., Luo, M., Wu, N., & Liu, X. L. Optimisation of microwave-assisted enzymatic extraction of corilagin and geraniin from geranium sibiricumlinne and evaluation of antioxidant activity. *Food Chem.,* **2010,** *122*(1), 373–380.

69. Yoshida, T., Amakura, Y., Yokura, N., Ito, H., Isaza, J. H., Ramirez, S., Pelaez, D. P., & Renner, S. S. Oligomeric hydrolysable tannins from tibouchina multiflora. *Phytochemistry,* **1999,** *52*(8), 1661–1666.

70. Huang, W., Ni, J., & Borthwick, A. G. L. Biosynthesis of valonia tannin hydrolase and hydrolysis of valonia tannin to ellagic acid by *Aspergillus* SHL 6. *Process Biochem.,* **2005,** *40*(3), 1245–1249.

71. Huang, L. P., Jin, B., & Lant, P. Direct fermentation of potato starch wastewater to lactic acid by *Rhizopus oryzae* and *Rhizopus arrhizus. Bioprocess Biosyst. Eng.,* **2005,** *27*(4), 229–238.

72. Aguilera-Carbo, A., Hernandez-Rivera, J. S., Prado-Barragán, L. A., Augur, C., Favela-Torres, E., & Aguilar, C. N. Ellagic acid production by solid state culture using a *punicagranatum* husk aqueous extract as culture broth. In: *Proceedings of the 5th International Congress of Food Technology,* Thessaloniki, Greece, **2007.**

73. Hernández-Rivera, J. *Production, Purification and Characterization of Aspergillus Niger GH1 Enzyme Responsible of the Ellagitannins Hydrolysis.* Universidad autónoma de coahuila, México, **2008.**

74. Ascacio-Valdés, J. A., Buenrostro, J. J., De la Cruz, R., Sepúlveda, L., Aguilera, A. F., Prado, A., Contreras, J. C., Rodríguez, R., & Aguilar, C. N. Fungal biodegradation of pomegranate ellagitannins. *J. Basic Microbiol.,* **2014,** *54*(1), 28–34.

75. de la Cruz, R., Ascacio, J. A., Buenrostro, J., Sepúlveda, L., Rodríguez, R., Prado-Barragán, A., Contreras, J. C., Aguilera, A., & Aguilar, C. N. Optimization of ellagitannase production by *Aspergillus niger* GH1 by solid-state fermentation. *Prep. Biochem. Biotechnol.,* **2015,** *45*(7), 617–631.

76. Ascacio-Valdés, J. A., Aguilera-Carbó, A. F., Buenrostro, J. J., Prado-Barragán, A., Rodríguez-Herrera, R., & Aguilar, C. N. The complete biodegradation pathway of ellagitannins by *Aspergillus niger* in solid-state fermentation. *J. Basic Microbiol.,* **2016,** *56*(4), 329–336.

77. Almeida, J. M., Lima, V. A., Giloni-Lima, P. C., & Knob, A. Passion fruit peel as novel substrate for enhanced β-glucosidases production by penicilliumverruculosum: Potential of the crude extract for biomass hydrolysis. *Biomass and Bioenergy,* **2015,** *72,* 216–226.

78. Ng, I. S., Li, C. W., Chan, S. P., Chir, J. L., Chen, P. T., Tong, C. G., Yu, S. M., & Ho, T. H. D. High-level production of a thermoacidophilic β-glucosidase from penicillium citrinum YS40-5 by solid-state fermentation with rice bran. *Bioresour. Technol.,* **2010,** *101*(4), 1310–1317.

79. Garcia, N. F. L., Santos, F. R. da S., Gonçalves, F. A., Paz, M. F. da, Fonseca, G. G., & Leite, R. S. R. Production of β-glucosidase on solid-state fermentation by *Lichtheimia Ramosa* in agroindustrial residues: Characterization and catalytic properties of the enzymatic extract. *Electron. J. Biotechnol.,* **2015,** *18*(4), 314–319.

80. de Andrade Silva, C. A., Lacerda, M. P. F., Leite, R. S. R., & Fonseca, G. G. Physiology of *Lichtheimia Ramosa* Obtained by solid-state bioprocess using fruit wastes as substrate. *Bioprocess Biosyst. Eng.,* **2014**, *37*(4), 727–734.

81. Pandey, A. Solid-state fermentation. *Biochem. Eng. J.,* **2003**, *13*(2), 81–84.

82. Aguilar, C. N., Gutiérrez-Sánchez, & Sources, G., Properties, applications and potential uses of tannin acyl hydrolase. *Food Sci. Technol. Int.,* **2001**, *7*(5), 373–382.

83. Blummel, M., Moss, A., Givens, I., Makkar, H. P. S., & Becker, K. Preliminary studies on the relationship of microbial efficiencies of roughages in vitro and methane production in vivo. In: *Proceedings of the Society of Nutrition Physiology* (Germany), **1999**.

84. Feldman, K. S., Iyer, M. R., & Liu, Y. Ellagitannin chemistry. Studies on the stability and reactivity of 2, 4-HHDP-containing glucopyranose systems. *J. Org. Chem.,* **2003**, *68*(19), *7433–7438.*

85. Sarma, S. J., Dhillon, G. S., Hegde, K., Brar, S. K., & Verma, M. Biotransformation of waste biomass into high value biochemicals. In: Brar, S. K., Dhillon, G. S., & Soccol, C. R., (eds.) *Biotransformation of Waste Biomass into High Value Biochemicals,* (pp. 99–115). Springer New York, **2014**.

86. Anwar, Z., Gulfraz, M., & Irshad, M. Agro-industrial lignocellulosic biomass a key to unlock the future bio-energy: A brief review. *J. Radiat. Res. Appl. Sci.,* **2014**, *7*(2), 163–173.

87. Septembre-Malaterre, A., Remize, F., & Poucheret, P. Fruits and vegetables, as a source of nutritional compounds and phytochemicals: Changes in bioactive compounds during lactic fermentation. *Food Res. Int.,* **2018**, *104 (April 2017),* 86–99.

88. Wang, Y., Liu, X. T., Wang, H. L., Li, D. F., Piao, X. S., & Lu, W. Q. Optimization of processing conditions for solid-state fermented soybean meal and its effects on growth performance and nutrient digestibility of weanling pigs. *Livest. Sci.,* **2014**, *170*, 91–99.

89. Pedersen, M. B., Gaudu, P., Lechardeur, D., Petit, M. A., & Gruss, A. Aerobic respiration metabolism in lactic acid bacteria and uses in biotechnology. *Annu. Rev. Food Sci. Technol.,* **2012**, *3*(1), 37–58.

90. Yang, J., Wu, X. bin, Chen, H. lei, Sun-waterhouse, D., Zhong, H. B., & Cui, C. A value-added approach to improve the nutritional quality of soybean meal byproduct: Enhancing its antioxidant activity through fermentation by *Bacillus amyloliquefaciens* SWJS22. *Food Chem.,* **2019**, *272* (April 2018), 396–403.

91. Dai, C., Ma, H., He, R., Huang, L., Zhu, S., Ding, Q., & Luo, L. Improvement of nutritional value and bioactivity of soybean meal by solid-state fermentation with *Bacillus subtilis. LWT – Food Sci. Technol.,* **2017**, *86,* 1–7.

92. Sang-Hyun, S., & Seong-Jun, C. Changes in allergenic and antinutritional protein profiles of soybean meal during solid-state fermentation with *Bacillus subtilis. LWT – Food Sci. Technol.,* **2016**, *70,* 208–212.

93. Ming-Yen, J., & Cheng-Chun, C. Enhancement of antioxidant activity, total phenolic and flavonoid content of black soybeans by solid state fermentation with *Bacillus subtilis* BCRC 14715. *Food Microbiol.,* **2010**, *27*(5), 586–591.

94. Fritsch, C., Heinrich, V., Vogel, R. F., & Toelstede, S. Phenolic acid degradation potential and growth behavior of lactic acid bacteria in sunflower substrates. *Food Microbiol.,* **2016**, *57,* 178–186.

95. Rizzello, C. G., Tagliazucchi, D., Babini, E., SeforaRutella, G., TaneyoSaa, D. L., & Gianotti, A. Bioactive peptides from vegetable food matrices: Research trends and novel biotechnologies for synthesis and recovery. *J. Funct. Foods,* **2016**, *27,* 549–569.

96. Singh, B. P., Vij, S., & Hati, S. Functional significance of bioactive peptides derived from soybean. *Peptides, 2014, 54,* 171–179.

97. Sanjukta, S., & Rai, A. K. Production of bioactive peptides during soybean fermentation and their potential health benefits. *Trends Food Sci. Technol., 2016, 50,* 1–10.

98. Johnson, E. A., & Echavarri-Erasun, C. *Yeast Biotechnology* (vol. 1). Elsevier B.V., **2011.**

99. Queiroz Santos, V. A., Nascimento, C. G., Schimidt, C. A. P., Mantovani, D., Dekker, R. F. H., & da Cunha, M. A. A. Solid-State fermentation of soybean okara: Isoflavones biotransformation, Antioxidant activity and enhancement of nutritional quality. *Lwt* **2018,** *92,* 509–515.

100. Vong, W. C., Au Yang, K. L. C., & Liu, S. Q. Okara (Soybean Residue) biotransformation by Yeast *Yarrowia lipolytica. Int. J. Food Microbiol., 2016, 235,* 1–9.

101. Vong, W. C., Hua, X. Y., & Liu, S. Q. Solid-state fermentation with rhizopus oligosporus and yarrowia lipolytica improved nutritional and flavor properties of okara. *LWT – Food Sci. Technol.,* **2018,** *90(August 2017),* 316–322.

102. Zhao, H. M., Guo, X. N., & Zhu, K. X. Impact of solid-state fermentation on nutritional, physical and flavor properties of wheat bran. *Food Chem., 2017, 217,* 28–36.

103. Magdouli, S., Guedri, T., Tarek, R., Brar, S. K., & Blais, J. F. Valorization of raw glycerol and crustacean waste into value added products by *Yarrowia lipolytica. Bioresour. Technol., 2017, 243,* 57–68.

104. Hernández-Almanza, A., Montañez, J., Martínez, G., Aguilar-Jiménez, A., Contreras-Esquivel, J. C., & Aguilar, C. N. Lycopene: Progress in microbial production. *Trends Food Sci. Technol., 2016, 56,* 142–148.

105. Zabed, H., Sahu, J. N., Boyce, A. N., & Faruq, G. Fuel Ethanol production from lignocellulosic biomass: An overview on feedstocks and technological approaches. *Renew. Sustain. Energy Rev.* 2016, *66,* 751–774.

106. Jahnavi, G., Prashanthi, G. S., Sravanthi, K., & Rao, L. V. Status of Availability of lignocellulosic feed stocks in India: Biotechnological strategies involved in the production of bioethanol. *Renew. Sustain. Energy Rev.,* 2017, *73 (November 2016),* 798–820.

107. Iandolo, D., Amore, A., Birolo, L., Leo, G., Olivieri, G., & Faraco, V. Fungal solid-state fermentation on agro-industrial wastes for acid wastewater decolorization in a continuous flow packed-bed bioreactor. *Bioresour. Technol., 2011, 102*(16), 7603–7607.

108. Godoy, M. G., Gutarra, M. L. E., Castro, A. M., Machado, O. L. T., & Freire, D. M. G. Adding value to a toxic residue from the biodiesel industry: Production of two distinct pool of lipases from *Penicillium simplicissimum* in castor bean waste. *J. Ind. Microbiol. Biotechnol., 2011, 38*(8), 945–953.

109. Murthy, P. S., & Madhava, N. M. Sustainable management of coffee industry by-products and value addition – A review. *Resour. Conserv. Recycl., 2012, 66,* 45–58.

110. Merkley, E. D., Jenson, S. C., Arce, J. S., Melville, A. M., Leiser, O. P., Wunschel, D. S., & Wahl, K. L. Ricin-like proteins from the castor plant do not influence liquid chromatography-mass spectrometry detection of ricin in forensically relevant samples. *Toxicon,* 2017, *140,* 18–31.

111. Madeira, J. V., Macedo, J. A., & Macedo, G. A. Detoxification of castor bean residues and the simultaneous production of tannase and phytase by solid-state fermentation using *Paecilomyces variotii. Bioresour. Technol., 2011, 102*(15), 7343–7348.

112. Mohammad, N., Alam, M. Z., Kabbashi, N. A., & Ahsan, A. Effective composting of oil palm industrial waste by filamentous fungi: A review. *Resour. Conserv. Recycl., 2012, 58,* 69–78.

113. Reina, R., Liers, C., García-Romera, I., & Aranda, E. Enzymatic mechanisms and detoxification of dry olive-mill residue by *Cyclocybe aegerita, Mycetinis alliaceus* and *Chondrostereum purpureum. Int. Biodeterior. Biodegrad.,* **2017**, *117*, e1.

114. Sampedro, I., Cajthaml, T., Marinari, S., Petruccioli, M., Grego, S., & D'Annibale, A. Organic matter transformation and detoxification in dry olive mill residue by the saprophytic fungus *Paecilomyces farinosus. Process Biochem.,* **2009**, *44*(2), 216–225.

115. Saparrat, M. C. N., Jurado, M., Díaz, R., Romera, I. G., & Martínez, M. J. Transformation of the water soluble fraction from "alpeorujo" by coriolopsis rigida: The role of laccase in the process and its impact on azospirillum brasiliense survival. *Chemosphere,* **2010**, *78*(1), 72–76.

116. Veerabhadrappa, M. B., Shivakumar, S. B., & Devappa, S. Solid-state fermentation of jatropha seed cake for optimization of lipase, protease and detoxification of anti-nutrients in jatropha seed cake using *Aspergillus versicolor* CJS-98. *J. Biosci. Bioeng.,* **2014**, *117*(2), 208–214.

117. Kourmentza, C., Economou, C. N., Tsafrakidou, P., & Kornaros, M. Spent coffee grounds make much more than waste: Exploring recent advances and future exploitation strategies for the valorization of an emerging food waste stream. *J. Clean. Prod.,* **2018**, *172*, 980–992.

118. Abu-Salem, F. M., Mohamed, R. K., Gibriel, A. Y., & Rasmy, N. M. H. Levels of some antinutritional factors in tempeh produced from some legumes and jojobas seeds. **2014**, *12622*(3), 288–293.

119. Ayuk, A. A., Iyayi, E. A., Okon, B. I., Ayuk, J. O., & Jang, E. Biodegradation of antinutritional factors in whole leaves of enterolobium cyclocarpum by *Aspergillus Niger* using solid state fermentation. *J. Agric. Sci.,* **2014**, *6*(10), 188–196.

120. Shi, C., He, J., Yu, J., Yu, B., Huang, Z., Mao, X., Zheng, P., & Chen, D. Solid state fermentation of rapeseed cake with *Aspergillus niger* for degrading glucosinolates and upgrading nutritional value. *J. Anim. Sci. Biotechnol.,* **2015**, *6*(1), *13*.

121. Zhou, H., Wang, C. Z., Ye, J. Z., Chen, H. X., Tao, R., & Zhang, Y. S. Solid-state fermentation of Ginkgo biloba L. residue for optimal production of cellulase, protease and the simultaneous detoxification of *Ginkgo biloba* L. Residue using *candida tropicalis* and *Aspergillus oryzae*. Eur. *Food Res. Technol.,* **2014**, *240*(2), 379–388.

122. Harouna, D. V. *Detoxification of Common Bean (Phaseolus vulgaris) Flour through Open and Controlled Fermentation Methods,* **2014**, *3*(9), 819–825.

123. Oseni, O. A., & Akindahunsi, A. A. Some phytochemical properties and effect of fermentation on the seed of Jatropha curcas L. *Am. J. Food Technol.,* **2011**, *6*(2), 158–165.

124. Phengnuam, T., & Suntornsuk, W. Detoxification and anti-nutrients reduction of *Jatropha curcas* seed cake by Bacillus fermentation. *J. Biosci. Bioeng.,* **2013**, *115*(2), 168–172.

125. Belewu, M. A., & Sam, R. Solid state fermentation of *Jatropha curcas* kernel cake: Proximate composition and antinutritional components. *J. Yeast Fungal Res.,* **2010**, *1*, 44–46.

126. Gao, Y. L., Wang, C. S., Zhu, Q. H., & Qian, G. Y. Optimization of solid-state fermentation with lacto *Bacillus brevis* and *Aspergillus oryzae* for trypsin inhibitor degradation in soybean meal. *J. Integr. Agric.,* **2013**, *12*(5), 869–876.

127. Zhang, W. J., Xu, Z. R., Zhao, S. H., Jiang, J. F., Wang, Y. B., & Yan, X. H. Optimization of process parameters for reduction of gossypol levels in cottonseed meal by *Candida tropicalis* ZD-3 during solid substrate fermentation. *Toxicon,* **2006**, *48*(2), 221–226.

128. García-Torreiro, M., Martínez-Patiño, J. C., Gullón, B., Lú-Chau, T. A., Moreira, M. T., Lema, J. M., & Eibes, G. Simultaneous valorization and detoxification of the

hemicellulose rich liquor from the organosolv fractionation. *Int. Biodeterior. Biodegrad.*, **2018**, *126* (October 2017), 112–118.

129. Postemsky, P. D., Bidegain, M. A., Gonzalez-Matute, R., Figlas, N. D. &, Cubitto, M. A. Pilot-scale bioconversion of rice and sunflower agro-residues into medicinal mushrooms and laccase enzymes through solid-state fermentation with *Ganoderma lucidum*. *Bioresour. Technol.*, **2017**, *231*, 85–93.

130. Matsakas, L., Giannakou, M., & Vörös, D. Effect of synthetic and natural media on lipid production from *Fusarium oxysporum* electron. *J. Biotechnol.*, **2017**, *30*, 95–102.

131. Mamma, D., Kourtoglou, E., & Christakopoulos, P. Fungal multienzyme production on industrial by-products of the citrus-processing industry. *Bioresour. Technol.*, **2008**, *99*(7), 2373–2383.

132. Daroit, D. J., Silveira, S. T., Hertz, P. F., & Brandelli, A. Production of extracellular β-glucosidase by *Monascus purpureus* on different growth substrates. *Process Biochem.*, **2007**, *42*(5), 904–908.

133. Correia, R. T. P., McCue, P., Magalhães, M. M. A., Macêdo, G. R., & Shetty, K. Production of phenolic antioxidants by the solid-state bioconversion of pineapple waste mixed with soy flour using *Rhizopus oligosporus*. *Process Biochem.*, **2004**, *39*(12), 2167–2172.

CHAPTER 14

Lipids as Components for Formulation of Functional Foods: Recent Trends

LETICIA XOCHITL LOPEZ-MARTINEZ,[1]
JOSÉ JUAN BUENROSTRO-FIGUEROA,[2] EDWIN ROJO-GUTIÉRREZ,[2]
HUGO SERGIO GARCÍA-GALINDO,[3] and RAMIRO BAEZA-JIMÉNEZ[2]

[1]CONACYT–Food and Development Research Center, 80110, Culiacán, Sinaloa, Mexico

[2]Food and Development Research Center, 33089, Delicias, Chihuahua, Mexico, E-mail: ramiro.baeza@ciad.mx (R. Baeza-Jiménez)

[3]UNIDA, Technological Institute of Veracruz, National Technological Institute of Mexico, 91897, Veracruz, Mexico

ABSTRACT

Nutritionally beneficial components are found in nature. These include glycosides, oligosaccharides, terpenoids, proteins, and peptides, as well as different kinds of lipids. These so-called bioactive compounds (BCs) are known to improve health and prevent diseases. For years, lipids have been considered as villains; however, several biochemical and clinical studies have documented their benefits as health promoters. In this chapter, the applications of lipids for the preparation of functional foods via enzyme-catalyzed reactions are reviewed.

14.1 INTRODUCTION

Lipids are a diverse and ubiquitous group of organic compounds insoluble in water that can be extracted from cells and tissues with organic solvents (e.g., chloroform, ether, benzene). They have many key biological functions, such as acting as structural components of cell membranes, serving as energy storage sources and participating in signaling pathways [1, 2]. Other lipids, although present in relatively small quantities, play crucial roles as enzyme cofactors, electron carriers, and light-absorbing pigments, hydrophobic anchors for proteins, emulsifying agents, hormones, and intracellular messengers [3]. However, it is not uncommon to hear among people that the consumption of oils and fats must be reduced in order to reduce cholesterol-related health problems such as obesity, diabetes, heart attack, and other diseases.

During the past few years, much research has been focused on the modification of fats and oils for the production of functional and health-promoting lipids, also displaying technological properties that make their use in processing food feasible [4, 5]. Such modifications are known as structured lipids (SL) that enhance the role of fats and oils in food, nutrition, and health applications [6–8].

SL is generally constituents of functional foods also known as nutraceuticals [9], and they are lipids with different FA composition or positional distribution. Nowadays, lipids are a critical component in food, pharmaceutical, cosmetics, and other industries, either as natural or SL. In the present chapter, the applications of lipids for the preparation of functional foods via enzyme-catalyzed reactions are reviewed.

14.2 BIOCHEMISTRY OF LIPIDS

The wide varieties of natural products we can find in lipids include fatty acids (FAs) and their derivatives, steroids, terpenes, carotenoids, and bile acids [10]. The International Lipid Classification and Nomenclature Committee (ILCNC) under the sponsor of LIPID MAPS developed a comprehensive classification system that was published in 2005. The LIPID MAPS classification system is based on the concept of two fundamental "building blocks": ketoacyl groups and isoprene groups. Therefore, lipids are defined as hydrophobic or amphipathic molecules that may originate entirely or in part by carbanion-based condensations of thioesters (fatty acyls, glycerolipids, glycerophospholipids, sphingolipids, saccharolipids, polyketides) and/or by carbocation-based condensations of isoprene units (sterol and prenol lipids) [11].

14.2.1 *FATTY ACIDS (FAS)*

A fatty acid (FA) is a hydrocarbon chain, saturated or unsaturated, with a methyl group at one end (n), and a carboxylic functional group at the other (delta). A saturated fatty acid (SFA) possesses an alkane-like structure with a fully saturated hydrocarbon chain, while a monounsaturated fatty acid (MUFA) has one double bond, and a polyunsaturated (PUFA) contains several double bonds, being naturally in the *cis* configuration [12]. Over 1000 FA are known with different structures and compositions, but only around 20 FA occurs widely in nature; of these, palmitic, oleic, and linoleic acids (LAs) make up approximately 80% of commodity oils and fats [13]. To describe precisely the structure of an FA molecule, it is important to know the carbon length, the number of double bonds, and their exact position, in order to define the biological reactivity of the FA molecule and even the lipid-containing the FA studied [14]. Figure 14.1 describes the classification of FA, as well as some examples and physiological implications.

14.2.1.1 SHORT-CHAIN FATTY ACIDS (SCFA)

SFA are the 2, 3 and 4 carbon saturated monocarboxylic acids produced in the medium colon by bacterial enzymatic breakdown of dietary fiber, undigested starch, and non-absorbed simple carbohydrates. They provide an important energy source for the colon and are metabolized both locally and systematically [15].

Some examples of SCFA are ethanoic acid (C2:0; which is the biosynthetic precursor of FA and other metabolites), propanoic acid (C3:0; which is a biosynthetic precursor of some amino acids) and butyric acid (C4:0; which is found in ruminant milk fat).

14.2.1.2 MEDIUM CHAIN FATTY ACIDS (MCFA)

This type of FA has backbones of C_6 to C_{12}, and is quickly metabolized, generating energy. MCFA have immense medicinal and nutritional importance [16, 17]. Other applications include detergents, as in the case of lauric acid (LaA, C12:0), which is used in soaps, shampoos, and the widely employed SDS (sodium dodecyl sulfate).

14.2.1.3 LONG-CHAIN FATTY ACIDS (LCFA)

Carbon chains of 14 and longer are defined as LCFA. In this category both MUFA and PUFA are formed. Oleic acid (OA, C18:1 n–9) is a MUFA

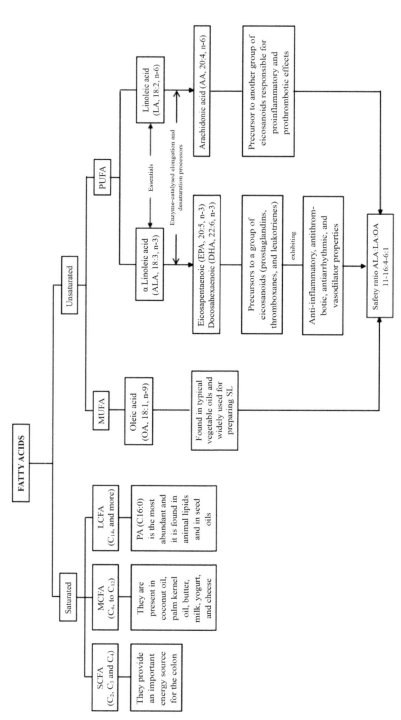

FIGURE 14.1 Biochemical aspects of fatty acids.

SCFA: short-chain fatty acids; MCFA: medium-chain fatty acids; LCFA: long-chain fatty acids; MUFA: monounsaturated fatty acids; PUFA: polyunsaturated fatty acids.

Source: Adapted from Baeza-Jiménez et al. [14].

found in common vegetable oils and is useful in SL for fulfilling the long-chain triacylglycerol requirements [18]. PUFA can be divided into omega-3 (n–3) and omega-6 (n–6) FA. Both groups are considered "essential" because they cannot be synthesized by humans and must be obtained through diet or supplementation. a-Linolenic acid (ALA, 18:3 n–3) is the precursor of this group and is found in certain plant oils, green leafy vegetables, beans, and nuts. Through an enzymatic process of desaturation and elongation, ALA is converted into eicosapentaenoic (EPA, 20:5 n–3) and docosahexaenoic (DHA, 22:6 n–3) acids, which are further metabolized to eicosanoids. On the other hand, LA (18:2 n–6) is the precursor of this group and is obtained from grains, meats, and the seeds of most plants. Under the same enzymatic cascade, LA is converted into arachidonic acid (AA, 20:4 n–6), which is further metabolized to another group of eicosanoids (see Figure 14.1).

14.2.2 GLYCEROLIPIDS

Glycerolipids are composed mainly of mono-, di-, and trisubstituted glycerol. This simple lipid class constitutes the bulk of storage fat in mammalian tissues and is the most abundant structure in oils and fats of animal and plant origin of commercial importance. They represent the most common lipid class present in foods. From natural sources, mono-(MAG) and diacylglycerols (DAG) are usually present as precursors for the formation of triacylglycerols (TAG) or phospholipids (PL). In foods, their presence may also indicate the degradation of TAG components by chemical or enzymatic deacylation. All animal and vegetable oils and fats are TAG, and thus, they are the main components of oil in the food industry. Designed TAG can be produced using the chemical or enzymatic process, namely, hydrogenation, fractionation, blending, interesterification, esterification, and recently via genetically-modified plants.

The reason why TAG is modified is to regio-design their FA content, and then, the benefit of the incorporated FA can be useful when they are hydrolyzed in the gastrointestinal tract (GIT). If the FA residues incorporated are SCFA or MCFA, the calorie intake will be lower, controlling body weight and moreover, they function as a treatment for lipid malabsorption. Palmitic acid (PA, 16:0) is also used in the rearrange of TAG molecules in infant formula for energy supply [19]. At the same time, MAG, and DAG including n–3 FA have been widely studied since their use as additives or carries in food, medicinal, and cosmetic industries were noted. On the other hand, DAG help to prevent obesity-related disorders as it prevents the accumulation of body fat [20].

14.2.3   *GLYCEROPHOSPHOLIPIDS*

In general, glycerophospholipids are composed of glycerol, FA, phosphate, and (usually) an organic base or polyhydroxy compound. They are also known as PL and can be divided into a variety of subclasses defined by the nature of the head group. This end group also defines the properties and biological functions of the PL [21]. PL are fundamental for life because are essential constituents of cell membranes and structure [22]. Because of their physical properties, biocompatibility, and nutritional functions, PL is useful in food, cosmetics, and pharmaceuticals industries, as emulsifiers, components of cosmetics, medical formulations and for liposome preparations.

14.2.4   *SPHINGOLIPIDS*

Sphingolipids can be divided into several major classes, all sharing a common structural feature: a sphingoid-base backbone synthesized *de novo* from serine and a long-chain fatty acyl-CoA, then converted into ceramides, phosphosphingolipids, glycosphingolipids, and protein adducts [23]. A wide variety of polar head groups may replace the simple proton attached to the -OH at *sn*–1 resulting in the different major classes of sphingolipids. They are important membrane components in both plant and animal cells. They are present in especially large amounts in brain and nerve tissue. Many biologically active sphingolipids have become commercially available because of their health benefits. Therefore, the situation for the technological and engineering study of sphingolipid modifications is changing rapidly. Ceramides, for example, have significant commercial potential in cosmetics and pharmaceuticals due to their major role in maintaining the water-retaining properties of the epidermis.

14.2.5   *STEROLS*

Sterol lipids are ringed lipids that play a role in the membrane integrity of eukaryotes [24]. The major sterol in animals, fungi, and plants are cholesterol, ergosterol, and b-sitosterol, respectively. Membrane sterols can be divided into three distinct parts: (1) the polar head; (2) the rigid sterol rings which are responsible of controlling membrane fluidity and packing a breathing space for protein function; (3) a floppy tail. Serum cholesterol in particularity has been the cause of controversies because of its strong relation to high blood pressure and heart disease in humans. However, cholesterol is a major structural component of the plasmatic membrane, where it provides mechanical strength, control phase behavior, and support membrane lateral organization and stability. In addition, it is a precursor for all steroid hormones, bile salts, and vitamin D. Therefore, cholesterol is an essential component of membranes and precursors for the synthesis of other vital biochemicals [25].

14.2.6 PRENOL LIPIDS

Phenol lipids are synthesized from the 5-carbon precursor's isopentenyl diphosphate and dimethylallyl diphosphate. The simple isoprenoids (linear alcohols, diphosphates, etc.) are formed by the successive addition of C_5 units and are classified depending on the number of the terpene units attached. When the prenol lipids contain more than 40 carbon atoms, they are defined as polyterpenes. Important isoprenoids include carotenoids, which are precursors of vitamin A and also possess antioxidant effects. Another biologically important class of molecules is exemplified by the quinones and the ubiquinones, as well as the hydroquinones, like vitamin K and vitamin E [23]. Polyprenols and their phosphorylated derivatives play important roles in the transportation of oligosaccharides across membranes [11].

14.3 PREPARATION OF STRUCTURED LIPIDS (SL)

During the past few decades, people have become more concerned about what they eat. Consumers now realized that a balanced diet contributes directly to their health [26]. Today, consuming carbohydrates, lipids, and proteins, is not only relevant for their nutritional value but also for their health implications. The latter is related to the bioactive compounds (BCs) present in foods and the ways through which they can be extracted define recent trends.

The production of functional foods or nutraceuticals has been explored. This type of novel foods is important for their nutritional value and their biological function such as antioxidants, antimicrobial, anticancer, cardiovascular implications, fat absorption, anti-inflammatory, neuroprotection, and others.

14.3.1   *FUNCTIONAL AND NUTRACEUTICAL FOODS*

A global or unified definition for functional or nutraceutical foods is not found in the technical literature; additionally, there is no legislative definition of the term in many countries [27–29], and sometimes the terms are used interchangeably. Because of that, several national authorities, academic, and industrial organizations, have proposed definitions for functional foods. For example, some authors define them as *"those foods that can provide health benefits beyond basic nutrition"* or *"foods that look similar to conventional foods that were designed to be consumed as part of a normal diet but that have been modified to favor physiological functions beyond simple nutritional requirements"* [30, 31].

More precisely, the term nutraceutical was coined in 1989 by Stephen Defelice, founder and chairman of foundation for innovation in medicine, who stated that *"a nutraceutical is any substance that is a food or a part of food and provides medical or health benefits, including the prevention and treatment of disease"* [32, 33]. The food products used as nutraceuticals comprise: probiotic, prebiotic, dietary fiber, n–3 FA and antioxidants.

Figure 14.2 shows that depending on the strategy or the combination of strategies performed, several types of functional and nutraceutical foods can be produced: fortified, enriched, altered, and enhanced foods [34]. As it can be seen in Figure 14.2, SL is considered as functional/nutraceutical foods, provided the incorporation of FA residues that were not naturally present or were linked in a different position of glycerol structure. Next, the operational parameters for the enzymatic synthesis of SL will be described.

14.3.2   *ENZYME-CATALYZED SYNTHESIS OF SL*

A structured lipid (SL) can be defined as a re-structured or modified TAG in order to change their FA residues and/or their positional distribution in the glycerol molecule (Figure 14.3). The newly incorporated FA enhances melting behavior, digestion, absorption, and metabolism, some with lower energetic value and health benefits [35].

SL can be obtained by chemical or enzymatic routes. Concerning chemical processes, the use of high temperatures and pressures, toxic catalyst, low specificity, and selectivity, lack of food-grade status and high-energy consumption, are inconvenient. On the other hand, enzymatic reactions are more specific, require mild reaction conditions and by-products are reduced or prevented. Phospholipases (EC 3.1.1.32) and lipases (EC 3.1.1.3) are the main groups of enzymes currently employed for producing SL.

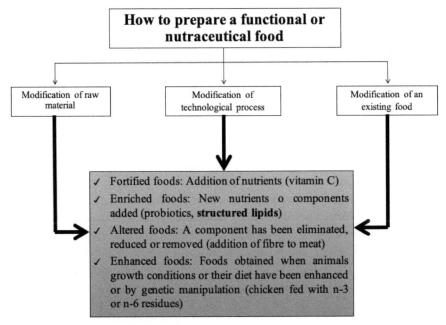

FIGURE 14.2 Preparation of functional and nutraceutical foods.
Source: Adapted from Siró et al. [34].

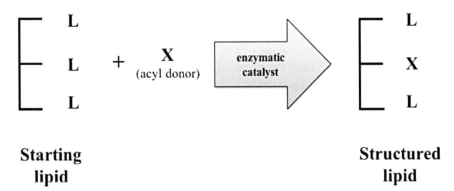

FIGURE 14.3 General schematic illustration of the enzymatic synthesis of structured lipid.

According to the works carried out in our group, we refer to some relevant aspects to take into consideration when an SL is about to be synthesized.

14.3.2.1   WHAT FA ARE INTENDED TO BE INCORPORATED?

Based on their biochemical and biological functions, both n–3 and n–6 are the main FA residues employed for preparing SL. FA of the n–3 family is needed for brain and eye development of the growing fetus during pregnancy and for maintaining and promoting health throughout life. Their potential health-promoting benefits include: reduce the inflammation in heart disease, inflammatory bowel disease (IBD), and rheumatoid arthritis, help prevent blood vessels and heart attacks, prevent hardening of arteries, decrease the risk of sudden death and abnormal heart rates, decrease TAG levels and lower blood pressure. N-3 is found in fish oil (FO), canola or soybean oils (SO), walnuts, and ground flaxseed or flaxseed oil.

FA of the n–6 family plays an important role in brain and heart function, and normal growth and development. Their potential health-promoting benefits include: neutral or lower levels of inflammatory markers, replacing saturated and *trans* fat with n–6 FA associated with decreasing risk of heart disease, improve insulin resistance and reduce the incidence of diabetes, lower blood pressure and lower cholesterol levels. N-6 is commonly found in vegetable oils. Table 14.1 is depicted as the FA contents of the common oils employed for preparing SL.

14.3.2.2   WHAT ENZYMATIC REACTIONS HAVE BEEN CONDUCTED?

The major enzymatic reactions for the modification of lipids are:

- **Acidolysis:** This reaction occurs between and ester and an acid resulting in an exchange of acyl groups.
- **Alcoholysis:** This reaction takes place between an ester and an alcohol, producing an ester with a different alkyl group.
- **Hydrolysis:** This reaction is carried out to prepare the partly hydrolyzed form of PL, lysophospholipids, as well as for glycerolipids.
- **Esterification:** It is the reverse reaction to hydrolysis and conducts to the production of an ester and water.

14.3.2.3   WHAT OPERATIONAL CONDITIONS AFFECT THE EXTENT OF THE ENZYMATIC REACTIONS?

Despite the occurring reaction, the operational conditions to be explored are:

TABLE 14.1 FA Content of the Common Oils Employed for Preparing SL

FA Content (wt.%) *	Coconut Oil	Palm Oil	Olive Oil	Canola Oil	Sunflower Oil	Soybean Oil	Camelina Oil	Fish Oil
SFA	92.3	50.4	15.3	5.9	12.8	12.6	9.7	4.12
MUFA	5.9	38.4	73.8	64.9	21.5	21.2	32.81	9.28
PUFA	1.7	10.5	10	28.6	66.1	65.5	54.1	83.9
Main FA residue	LaA (C12:0)	PA (C16:0)	OA (C18:1, n-9)	OA (C18:1, n-9)	LA (C18:2, n-6)	LA (C18:2, n-6)	ALA (C18:3, n-3)	DHA (C22:6, n-3)

*The fatty acid contents listed are referring to averages of different oils used in our different works.

Abbreviations: SFA: saturated fatty acids, MUFA: monounsaturated fatty acids, PUFA: polyunsaturated fatty acids, LA: linoleic acid, ALA: alpha-Linolenic acid, DHA: docosahexaenoic acid.

- **Reactivity of the Substrates:** The reactivity of different FA is influenced by enzyme specificity and by some inhibition effects.
- **Substrates Molar Ratio:** This parameter can increase yield; however, changes in polarity or viscosity must be overseen.
- **Temperature:** Increasing temperature allows attaining better yields, but fit is high enough, it can irreversibly cause enzyme denaturation.
- **Enzyme Load:** Sufficient enzyme load is needed to incorporate the desired FA into the TAG structure; however, high loads can involve agitation problems and hindered mass transfer, and the most important, a very expensive process.
- **pH:** Depending on the origin and the ionization state of residues in their active sites, enzyme can be positively affected or negatively damaged by pH.
- **Water Content and Water Activity (A$_w$):** Water is responsible for maintaining the active conformation of proteins, facilitating reagent diffusion, and maintaining enzyme dynamics.

14.4   RECENT TRENDS FOR SL APPLICATIONS IN FUNCTIONAL FOODS

As it is described in technical literature, the properties of fats (functional, nutritional, and sensorial) depend on three main factors: (1) FA composition (SFA and PUFA), (2) carbon chain length, and (3) position of the different FA residues in TAG (*sn*–1, *sn*–2 or *sn*–3). Therefore, those factors allow the production of new and novel SL to be used in the absorption of lipids, reduction of cholesterol and TAG, enteral, and parental nutrition, and to prevent some responses associated to their metabolism.

For the above mentioned, it has been referred to how SL are prepared, and now the question is: where are they applied? These newly designed lipids will inherit new properties to meet specials purposes for nutrition, food, and pharmaceutical applications.

14.4.1   ENTERAL AND PARENTAL FEED

Patients having difficulties or being unable to eat due to surgery and/or medical conditions, are undernutrition support, since it's critical to maintain their homeostasis and to deliver part or all of their nutrition and caloric requirements to preserve and enhance organ function and healing processes, as well as to promote their physical well-being. These conditions can refer to patients mechanically-ventilated, undernourished, and obese, among others, together with premature infants to the elderly [36]. Nutrition support can be offered by enteral or parenteral feeding. Enteral nutrition (EN) feeding uses the GIT to deliver a normal oral diet or specialized liquid nutrition through a nasogastric tube to meet the daily requirements [37]. On the other hand, parenteral nutrition (PN) is provided through the intravenous administration of nutrients and calories [38]. PN is commonly used when EN is contraindicated or if patients cannot tolerate EN [39].

Lipids have become essential components for both EN and PN because they can provide a concentrated source of metabolic energy, essential FA and act as carriers of pre-formed fat-soluble vitamins (A, D, E, and K) [40], maintaining the important metabolic and functional properties of lipids previously described. Thus, employing lipid emulsions (LE) in EN and PN has minimized the dependence of carbohydrates (e.g., dextrose) for the obtaining of no protein calories and the daily recommended ingest of FA. TAG is the primary component of EN and PN lipids emulsions, which are conformed by MCFA, LCFA, or very LCFA.

TAG-rich in MCFA are known as medium-chain triacylglycerols or MCTs. Therefore, TAG rich in long-chain are termed long-chain triacylglycerols or LCTs. The different compositions of FA attached to the glycerol backbone of TAG, can widely influence physiological processes like metabolism, inflammation, blood coagulation, oxidative stress, among others [41]. Therefore, current LE is unique in their FA profile, having different biological advantages and disadvantages. Standard EN formulas consist of 15 to 30% lipids and are specifically designed for a variety of diseases, while some selected commercial formulas of LEs available for use in PN are detailed in Table 14.2. The first generation of PN emulsions is SO based. They have been used widely for an extensive amount of time. SO-based lipids are rich in PUFA (mainly OA) and phytosterols. Unfortunately, their use increases the risk of intrahepatic biliary cholestasis, and sepsis. Other undesirable effects are the promoting of inflammation and suppressing immune function caused by its high LA content [42]. This evidence caused that the German Society of Nutritional Medicine banned the use of pure SO-LE in PN support of critically ill patients [43].

A second generation of intravenous LE includes MCTs (mainly from coconut and tropical nut oils) and SL. MCTs base LE are blended 50/50 with SO to provide the necessary essential FA. Studies support the use of SO/MCTs LE instead of pure SO, since SO/MCTs are resistant to peroxidation, have less pro-inflammatory effects and can improve the recovery of patients in the same extent than pure SO LE [44]. However, MCTs oils in high doses may provoke acidemia.

TABLE 14.2   FA Content of Some LE Commercially Available for PN

	Source	MUFA	PUFA	SA	n–3: n–6
Clin Oleic[a]	80% olive oil + 20% soybean oil	64	22	14	1:6
Intralipid[a]	100% soybean oil	24	61	15	1:7
Lipofundin[a]	50% MCT + 50% soybean oil	11	31	58	1:7
Structolipid[a]	36% MCT + 64% soybean oil	14	40	46	1:7
Omegaven[a]	100% fish oil	23	56	21	10:1

Source: Modified from Miles and Calder [47].

A third generation of LE incorporated olive oil (OO), prepared with a mixture of 80:20 (OO:SO). As MUFA predominate in the composition of these emulsions, they are less susceptible to peroxidation and produce little or no inflammation compared to MCTs/SO or pure SO emulsions [45].

The fourth-generation consists of any mixture containing FO. LE with FO formulas provides high concentrations of EPA and DHA, which have demonstrated beneficial effects on cell membranes and inflammatory processes. Besides, emulsions enriched with FO are probably responsible for a reduction in the period of recovery of critically ill patients [39].

Emulsions derived from SO are the formulas most extensively studied (preclinical and clinical), and the results demonstrated that vital nutrients are delivered efficiently and safely to all patients in the need of support nutrition. New generations of LE formulas emerge as alternative lipids such as SO/MCTs, or SO/MCTs/OO/FO blend combinations, where evaluations have indicated potential benefits because they exert reduced impacts on oxidative stress and inflammatory properties. They are immune modulating and have higher antioxidant content, decrease of the risk of cholestasis, and presented lower morbidity and mortality in specific groups of patients [41, 46]. When PN is considered, the n–3/n–6 FA ratio is an important factor because it plays an important role in immunity and inflammation, influencing patient outcomes [47].

In general, newer LE components seem to be safer and better tolerated than pure SO, where FO-enriched EN and PN have great potential due to their high anti-inflammatory and immune-modulating effects [48]. Despite the studies conducted, there is still limited data of newer LE on their biologic effects and clinical benefits. The lack of data is mainly attributed to the inconsistent amounts, different types of lipids, and heterogeneous patient populations [49]. Nevertheless, actual data consistently favor and encourage the use of standard enteral and parenteral lipid formulations enriched with different mixes of n–3, n–6, and MCTs.

14.4.2 HUMAN MILK ANALOGS

Maternal milk is regarded as the optimal food for newborn infants, providing essential micro and macronutrients as well as BCs (cells, anti-infectious compounds, grow factors, anti-inflammatory agents and prebiotics) to support healthy growth and development of infants. Lactose is the most abundant macronutrient in human milk (6.7 to 7.8 g/dL), followed by fat (3.2 to 3.6 g/dL) and protein (0.9 to 1.2 g/dL) [50]. Lipids are the primary energy source in human milk and infant formulas since they provide approximately 50% of total calories. Different factors can alter human milk fat (HMF) composition, namely, lactation stages, dietary habits, seasons, genetics, and individual conditions (e.g., maternal protein intake) [51]. More than 98% of HMF is composed by TAG, containing SFA, MUFA aw well as n–3 and n–6 PUFA (mainly DHA, EPA, and AA) [52]. MCFA are the main source of rapid energy, while DHA and AA are essential for retinal and brain development [53]. It is important to maintain a balance in DHA, AA, and EPA contents, and also of the corresponding precursors, LA, and ALA, in order to achieve optimal nutrition [54]. Essential FA and PUFA allow the brain to grow and function, and develop cognitive and motor skills, and neurological reflexes [55].

PA is the predominant FA and is mainly located at *sn*–2 position of glycerol backbone, while *sn*–1,3 positions are occupied by unsaturated FA. This specific configuration enhances the efficiency of nutrients and the energy obtained by infants. It is known that formulas containing high levels of PA at the *sn*–1,3 positions exhibit a lower absorption of both PA and Ca^{++}. This occurs because PA at those positions is release as free PA in the alkaline intestinal medium by pancreatic lipases. Free PA reacts with divalent cations such as Ca^{++} and Mg^{++} producing insoluble soaps which are excreted as hard stools [56]. A bioavailable source of calcium is important because during the first year of life of an infant, his/her weight is triplicated and height increases 50%. Therefore, the structure of TAG in fed-formulas for infants' determines the availability of those components [57].

The starting point to define an infant formula is the average values of human milk components, which are known as the "gold standard" [58].

Continuous analysis of the interactions between HMF and other components in human breast milk are leading the pursuit of an innovative analog product that fulfills the requirements of infant daily nutrition, as human milk does [59]. Remarkable advances have been achieved in lipid technology for the development of human milk analogs. FA found in human milk is typically mimicked through vegetable oil blends (coconut, palm olein, sunflower, canola, and others.). DHA and AA are directly supplemented since infant metabolism is not fully capable of processing parental n–6 and n–3 FA. Another tendency is the use of SL to meet TAG composition as well as the FA distribution in human milk. These SL are synthesized by enzymatic reactions (esterification, acidolysis, alcoholysis, and interesterification), and allows the incorporation of ca. 60% of PA in the *sn*–2 position, and unsaturated FA in the outer positions. HMF containing SL is termed sometimes b-palmitate [60]. Lipids in human milk are emulsified in a structure called milk-fat globules (MFG), organized with TAG as core, and surrounded by a membrane (milk-fat globules membrane (MFGM)). An important aim is to successfully recreate this MFGM by a lipid matrix to obtain similar physiological properties [61], which provide benefits in the adulthood like fat accumulation and an improvement in the metabolic profile [62].

Production of HMF analogs can be obtained typically by two reaction routes. The first one is performed in a single step with three possible options: (1) acidolysis of a TAG with FA; (2) interesterification between two TAG; (3) interesterification between a TAG and ethyl or methyl ester of FA [60]. The second route is conducted by a two-step process, where the first step involves the alcoholysis of a TAG by the *sn*–1,3-lipase to obtain a 2-MAG, and secondly, a direct esterification of purified 2-MAG with unsaturated FA [63]. The first route is easily performed and do not need additional hydrolysis steps, is faster and allows higher incorporation of acyl groups, such as AA and DHA, whereas in the second route the enzyme has better reusability, and the reaction presented high purity products and higher yield [60]. Many typical studies for obtaining of HMF analogs through enzyme-catalyzed synthesis and clinical trials are summarized by Zou et al. [60].

Since infant formulas serve as the principal providers of nutrients for infants besides human breast milk, the need to provide optimal nutrition and health benefits is extremely important. Besides, sometimes for different reasons infants cannot inherit human milk and thus HMF analogs are so important. Novel SL for the development of HMF analogs with high PA content at *sn*–2 position, enriched with important long-chain PUFA such as AA, DHA, and docosapentaenoic (DPA, 22:5 n–3), can provide the necessary nutritional requirements of an infant with potential benefits. However, more clinical trials are required for a better understanding of the physio-logical and health outcomes of HMF analogs, to take advantage of their potential fully.

14.4.3 COCOA BUTTER ALTERNATIVES (CBA)

Cocoa butter (CB) is an important ingredient in confectionery and also an indispensable component in manufacturing chocolate, because of its specific physical properties, as it forms the continuous phase of choco-late. It is responsible for the gloss, texture, and typical melting behavior of chocolate. CB conforms to the only continuous phase of fat found in chocolate, providing a homogeneous dispersion of the rest of the ingredi-ents [64]. However, its true significance lies in the fact that CB can crys-tallize into four polymorphic forms: a, g, b,' and b crystals; each one with different melting points (17, 23, 26, and 35–37°C, respectively). b crystals are the structure that confers superior sensory (excellent sheen, snap, and smooth texture) and physicochemical properties to chocolate because it only melts when the temperature is above 30–32°C since it has the highest melting point of the polymorphic forms. The unique characteristics that CB provides make this product highly demanded. However, the CB production is decreasing because 30% of the world's cocoa crops are destroyed by pests and has been subject to climate change [65]. The increasing demand for CB combined with its limited production; greatly raise its value in the market, because besides the decreased production issue, cocoa beans have less fat [66]. Therefore, food industries are keen to find alternatives to CB fat. CB is mainly composed of three different kinds of TAGs: (1) palmitic, oleic, palmitic (POP), (2) palmitic, oleic, stearic (POS), and (3) stearic, oleic, stearic (SOS). In general, TAG found in CB is rich in PA, stearic acid, OA, and LA, and small amounts of lauric and myristic acids. As a result of the alternative fats research, CBA was developed.

CBA are grouped in CB substitutes (CBS), CB replacers (CBR) and CB equivalents (CBE). CBE is non-lauric fats that share many physicochemical properties of CB fats and blended in varied proportions without altering the properties of the final product. CBA can be obtained through physical and/or chemical modifications of lipids maintaining their characteristic proper-ties [67]. CBE can be classified in two main groups: CB extenders (CBEX) which are fats that dilute CB to minimize costs [68] and CB improvers (CBI) that increase the solid fat and consequently melting point and hardness of chocolate caused by the high content of SOS. Because of this, CBI is used in the improvement of the resistance of CB against softness and fat bloom [68]. CBR, as well as CBE, are non-lauric fats but obtained through hydrogenation and fractionation of vegetable oils. CBR contains similar chain length of FA but different TAG compositions. Do not need tempering, can mix well with CB and moreover, they possess good sensorial properties. However, CBR contains high levels of trans-FA, who are related to cardiovascular diseases.

Contrary to CBE and CBR, CBS is rich in short-chain TAG like lauric and myristic acids (45–55% and 15–20%, respectively), followed by palmitic (9.2%) and stearic acid (8.75%) [69]. The presence of lauric and myristic acid confers a pleasant mouth feeling to CBS chocolates. CBS has similar physical properties to CB but are chemically incompatible derived from their different TAG composition [70]. Therefore, CBS are suitable to substitute completely CB. Unlike CB, CBS does not need tempering since they acquired a stable crystal form directly from melting.

The lipid sources for the production of all CBA are diverse and natural. Palm kernel oil (PKO), palm oil (PO), mango seed fat (MFK), kokum butter, salt fat, shea butter, and illipé fat are the fats and oils most commonly employed. The methods for the production of CBA entail hydrogenation, fractionation, interesterification, and blending of natural fats. The main purpose of these processes is to mimic the CB composition to obtain similar physicochemical properties, by reducing the degree of unsaturation of the acyl groups (hydrogenation), by redistributing FA chains (interesterification) or by a physical separation of TAG via selective crystallization and filtration (fractionation) [71]. Fractioning is applied when the content of SOS is too low for its effective use as a raw material for CBE formulation. By fractionation, the SOS levels can be increased similar to those found in CB [72]. Among CBA, CBE is the most important. The modification of fats through interesterification reactions catalyzed by enzymes has been widely studied given their advantages like low energy consumption, the absence of isomerization by-products, and better control of products, in comparison to chemical interesterification [73]. On the other hand, the use of blending is also used as a method to obtain CBE with the appropriate ratio of POP, POS, and SOS by mixing different vegetable fats [67].

PO is inexpensive and commonly available; PO mid fraction (PMF) is obtained by double fractionation of PO and is often used as a raw material to produce CBE since it resembles more closely to CB fat composition. PMF TAG is mainly constituted by POP (51.8%), which can be converted to POS and SOS by interesterification using a 1,3-specific lipase and a stearic acid-rich acyl donor, to obtain a similar composition to CB. Many reports have used PO and PMF to obtain CBE by fractioning and blending [66, 74, 77] and enzymatic interesterification [73, 78] who obtained physicochemical properties and composition similar to commercial CB.

MFK is mainly constituted by oleic, palmitic, and stearic acids [79], while its TAG present are POS (11%), SOS (40%), SOO (23%), POO (5%), SOA (4%) and OOO (5%) [9]. OA is the most abundant acid in the middle *sn*–2 position [80]. The extraction and fractionation of different varieties of MFK are extensively done nowadays since many groups have been conducted because of the physical and chemical properties similar to CB [81]. Among the studies conducted, the fractioning of MFK and blending with PMF have been studied by Sonwai et al. [77]. They developed a blending composed of 80/20 (MKF/PMF). Their blending showed that the crystallization and melting behavior of the blend was similar to that of CB, being pertinent for its use as CBE. Kaphueakngam et al. [82], on the other hand, made different blends of mango seed almond fat (MAF) and PMF. They reported that the blend containing 80% of MAF and 20% of PMF approached the composition, melting behavior and slip melting point found in commercial CB. Solís-Fuentes and Durán-de-Bazúa [79] studied the thermal behavior of MAF and mixtures of MAF and CB. They reported that FA contents and physicochemical properties of MAF are very similar to those of CB.

Kokum butter is composed by two major FA: stearic (50–60%) and oleic (36–40%) acids, where SOS is the predominant TAG (72%) [9]. Kokum butter is used in countries with hot climate, as a suitable raw material for the production of hard butter with high resistance to temperature because of the composition of FA. Studies have demonstrated that when CB is combined with small amounts of SOS or SOS-rich fats, its hardness is increased, the fat bloom is improved, and the tempering time is decreased [83, 84].

In salt fat, the SOS is also the main TAG present (42%), followed by SOO (14%) and SOA (13%). For that reason, the dominance of oleic and stearic acids gives to salt fat a FA profile similar to that of CB. Due to the high SOS concentration, the need of the fractioning of salt fat is essential to make it suitable as a raw material for CBEs [9].

Shea kernel shares similar composition and properties to CB. Gunstone [9] reported that SOS (42%), SOO (26%) are the main TAG contained in shea kernel fat. The FA profile is predominating by oleic (46.4%), stearic (41.5%), linoleic (6.6%) and palmitic (4%) [85]. The ratio of stearic and OAs produces differences in shea butter consistency. The high stearic acid content gives the shea butter its solid consistency, while a higher percentage of OA leads to soft and liquid-like texture [86]. Shea fats with melting points equal to or less than hard CB are designated CBE, while those with a higher melting point are known as CBI. The latter is used for improving soft CB [70]. Shea kernel fat is often used as a substitute for CB because it is sweet and oily; characteristics conferred by its TAG composition [87].

Illipé butter, unlike other raw materials, can be used as CBE without further processing since its TAG composition greatly matches that of CB. The major FA founded is stearic (39–46%), oleic (34–37%), and palmitic (18–21%). The main TAG is SOS (45%), POS (34%), and POP (7%) [9].

The distinctive properties of CB make it an extensively valuable and demanded product. As the production of CB does not meet the demand, its short supply increases significantly the price. In that matter, efforts have been made to find an alternative to CB and improve economic and techno-logical matters. Studies have shown that there is a wide variety of plant fats that with the help of fat modification methods can be transformed into CBA. However, there still exists the need of further investigations to prepare or improve CBAs in order to fulfill completely CB demand.

14.4.4 LIPID BASE DELIVERY SYSTEMS

Delivery systems of many bioactive agents specifically designed to promote or maintain human health within the GIT have been developed for diverse applications in food and pharmaceutical industries. The main application of delivery systems is for the encapsulation, protection, controlled release, and texture modification of BCs [88]. These BCs, when delivered effectively at specific locations (GIT, mouth, stomach, small intestine, colon), can be employed to control the release of lipophilic drugs, vitamins (A and D), minerals, or nutraceuticals (coenzyme Q) in supplements or food [89]. Traditional administration routes of lipid-base delivery systems are through oral and parenteral administration. However, effectiveness of delivery systems is still limited because of their poor oral bioavailability profile [90]. Other existent issues in delivery systems are the difficulty to incorporate the bioactive components in the carrier system caused by low water-solubility, poor chemical stability, and high melting point. Consequently, a variety of emulsion-based delivery systems have been developed, with the objective of potentiating the bioavailability of bioagent compounds [90, 92].

Among lipid-based delivery systems, we can find conventional emul-sions, nanoemulsions, multilayer emulsions, and solid lipid nanoparticles (SLN), among others [88, 93, 95]. Each one has specific properties that make them stand out from each other [96]. In order to achieve the optimal performance of delivery systems whatever their application (e.g., food, pharmaceutics), they have to fulfill some functional attributes, and technical characteristics as shown in Figure 14.4.

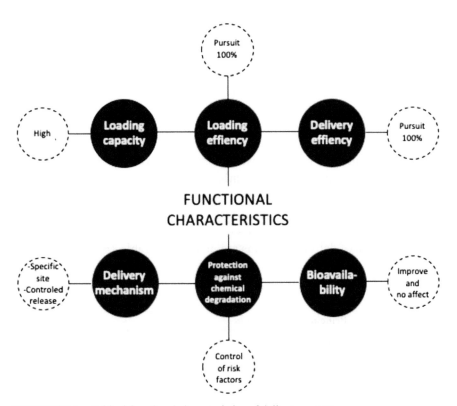

FIGURE 14.4 Critical functional characteristics of delivery systems.
Source: Adapted from McClements and Li [88].

14.4.4.1 CONVENTIONAL EMULSIONS AND NANOEMULSIONS

An emulsion can be defined as a substance that reduces the surface tension between a miscible (continuous phase) and immiscible phase (disperse phase), which favors the emulsification and increases its stability. Conventionally the miscible and immiscible substance is described as "water" and "oil." In that matter, the terms "water" and "oil" will be used through this chapter for better understanding. Most common emulsions are conformed by oil droplets disperse in water (oil-in-water, o/w) and water droplets contained in oil (water-in-oil, w/o); moreover, multiple emulsions can also be found which contain oil-in-water-in-oil (o/w/o) or water-in-oil-in-water (w/o/w) [97].

The main differences between emulsions and nanoemulsions are the size of the lipid droplets. The diameter of emulsions is generally between 0.1 and 100 mm [98], whereas in nanoemulsions are in the range of 10 nm to 100 nm. The physicochemical characteristics of emulsions and nanoemulsions like optical properties, stability, rheology, and release characteristics, are mainly determined by their composition, concentration, size, physical state, and interfacial properties [99]. Since emulsions with different droplet sizes can be made, their physicochemical and functional properties strongly depend on their particle size [88]. Conventional emulsions tend to have a cloudy or opaque appearance, while nanoemulsions tend to be transparent or with slight turbidity [100]. Conventional emulsions are the most common option as a delivery system since they are ease to prepare and scale-up. Besides, conventional emulsions can offer a wide variety of properties and functions because of their non-polar, polar, and amphiphilic regions where different active ingredients can interact [99]. Additionally, emulsions can be simply manufactured as food-grade for their use in the food industry. Common components used in the fabrication of delivery systems are included in Table 14.3.

TABLE 14.3 Example of Common Components That Can Be Used in the Production of Delivery Systems

Name	Principal Constituents	Examples
Major structural components	Lipids	Animal fats, fish, and plant oils, flavor, and essential oils and indigestible oils.
	Surfactants	Non-ionic, anionic, cationic, and zwitterionic
	Biopolymers	Globular proteins, flexible proteins, non-ionic polysaccharides, anionic polysaccharides, and cationic polysaccharides
Oil phase	Non-polar components	TAG, DAG, MAG, flavor oils, essential oils, mineral oils, fat substitutes, waxes, weighting agents, oil-soluble vitamins
Aqueous phase	Polar components	Water, co-solvents, acids, bases, minerals, carbohydrates, and proteins.
Stabilizers	Emulsifier, texture modifier, weighting agent and ripening retarder.	

Source: Modified from McClements [102] and McClements [103].

Depending on the droplet concentration contained in emulsions, their rheology can vary from low viscosity liquids to high viscosity liquids, to semi-solids, at low (<20%), mid (20–40%) and high (>40%) droplet concentrations, respectively [101]. However, emulsions are susceptible to physical instability when they are exposed to environmental stresses like heating, chilling, extreme pH values, among others. Emulsions are thermodynamically unfavorable systems that tend to break down over time due to a variety of physicochemical mechanisms that include coalescence, flocculation, Ostwald ripening, and gravitational separation. Emulsions in general often prone to gravitational separation and droplet aggregation because of the size of the droplets. On the other hand, their ability to protect and release the active agents is limited because of their micrometric size and nanometric interfacial layer. Also, the emulsions that can be used for the coating of oil drops (interfacial layers) are restricted [102]. These disadvantages enhance the need for more sophisticated delivery systems that can lead to more specialized applications.

Nanoemulsions, by having a smaller particle size, exhibit different physicochemical properties than conventional emulsions, which makes them more appropriate for some specific food applications [103]. Nanoemulsions are thermodynamically unstable too, but unlike emulsions, they are usually highly stable to gravitational separation because the relatively small droplet size means that Brownian motion effects dominate gravitational forces. In addition, nanoemulsions tend to have better stability against droplet aggregation because when the strength of the colloidal forces acting between droplets usually falls with decreasing droplet size, whereas steric forces are less dependent on droplet size [99]. The bioavailability offered by nanoemulsions of the active agents they carry and delivered is higher than the obtained from emulsions because by having a smaller droplet size, the active ingredients diffuse easily and quickly out of them [104]. Additionally, since nanoemulsions are transparent or slightly turbid, they can be used in food or beverages without been unpleasantly noted.

The methods for the production of conventional emulsions and nanoemulsions can be divided into two groups, high energy, and low energy. High energy methods involve homogenizers, which provide strong forces that separate and mix the miscible and immiscible phases, producing small drops of oil [105]. Some examples of high energy equipment's include high-pressure homogenizers, colloid mills, ultrasonic homogenizers, and membrane homogenizers [106]. Low energy relies on the spontaneous formation of small drops of oil in O/W systems with emulsifiers, where the solution to environmental conditions was altered [107]. Some of the methods of low energy are spontaneous emulsification, emulsion inversion points, and phase inversion temperature [102].

Emulsions (O/W) and nanoemulsions have been used to encapsulate, protect, and deliver lipophilic BCs, such as n–3FA, phytosterols, flavonoids, carotenoids, vitamins, coenzymes, and drugs [88, 108–113].

14.4.4.2   SOLID LIPID PARTICLES (SLPS)

Solid lipid particle (SLP) suspensions have similar compositions and structures to nanoemulsions or emulsions since they consist of solid lipids stabilized with an emulsifying layer in an aqueous dispersion, in other words, they resemble the nanoemulsions by replacing the inner liquid lipid with a solid lipid [114]. Different physicochemical properties of SLP can be obtained by modifying the composition, size, shape, concentration of the lipid phase, as well as the characteristics of the interfacial layers coating them. In addition, specific attributes of the lipid crystal matrix can be controlled (concentration, packing, and morphology) [115, 116]. The SLN are sub-micron colloidal carriers between 50 and 1000 nm, and the particles may have a spherical morphology, depending on the nature of the lipid crystal matrix. The lipid employed to fabricate SLN includes FA, steroids, waxes, MAG, DAG, and TAG. SLN provide a number of advantages such a great control and/or target release, improve stability of pharmaceuticals, can carry lipophilic and hydrophilic drugs, excellent biocompatibility, are easy to scale up and sterilize, more affordable, can avoid organic solvents, and are faster and easier to validate and gain regulatory approval [117].

As for disadvantages, low drug loading capacity (this is limited because of the formation of a perfect lipid crystal matrix), drug expulsion after polymeric transition during storage and high-water content of the dispersions (70–99.9%) [114, 116]. As emulsions and nanoemulsions, SLN is prone to destabilization mechanisms since they present an unpredictable gelation tendency and lipid particle grows [118, 119]. Three models of SLN for drug incorporation are currently developed: (1) a solid solution model, where there is a strong interaction between lipid and drug; (2) a core-shell model (drug-enriched shell), where the drug is concentrated in the surrounding layer that coats the lipid crystals; (3) a core-shell model (drug-enriched core), where there is a formation of a drug-enriched core (lipid matrix) [120–122]. The second generation of SLP known as nanostructured lipid carriers (NLC) was introduced to improve the difficulties of SLN [123, 124]. In NLC the lipidic phase is contained both solid and liquid lipids at ambient temperature [125]. Contrary to SLN, NLC show more loading capacity, some less water in the dispersion, and prevent or minimize the drug expulsion during storage [126].

Three forms of structure for NLC have been presented and are showed in Table 14.4.

TABLE 14.4   Structures of NLC

Name	Characteristics
Imperfect type	Solid and liquid lipids are mixed in various structures. The drug attach to the lipid crystal matrix in the gaps between FAs and TAGs of the different lipid structures conforming the matrix.
Formless Type	They lack a crystalline structure (by using certain lipid mixtures), preventing the expulsion of loaded drug.
Multiple Type	The drug solubilizes mainly in the liquid lipid, consequently, the solid lipid avoids the decomposition.

Source: Naseri et al. [126].

Commonly used methods for the preparation of SLN are high-pressure homogenization (hot homogenization and cold homogenization included), solvent emulsification, evaporation or diffusion, supercritical fluid, ultrasonic, or high-speed homogenization and spray drying [119, 127, 128]. SLN and NCL have shown great results in a wide range of applications such as oral drug delivery, pulmonary delivery [129–132], gene transfer applications [133], food [134–136], cosmetics, and dermatologists [137].

14.4.4.3   FILLED HYDROGEL PARTICLES

Filled hydrogel particles are typically classified as an oil-in-water-in-water $(o/w_1/w_2)$ emulsion, as oil droplets are dispersed throughout a hydrogel (w_1) in an aqueous medium (w_2) [103]. The size of the filled hydrogel particles can vary from 10 nm to 1000 μm, depending on the methods and materials used [138]. Filled hydrogel particles can be performed with a wide range of different physicochemical characteristics depending on the nature of the starting materials and the fabrication procedure [103]. Filled hydrogel particles have the potential to provide high bioavailability of BCs in foods, since they improve their intestinal absorption. In addition, they also allow for better control and release characteristics are modifying different parameters such as temperature, enzymes, dilution, and pH [138]. Filled hydrogel particles are susceptible to many of the same instability mechanisms as emulsions. For the formation of the filled hydrogel particles, an O/W emulsion or nano-emulsion is previously prepared for most of the methods by homogenizing an oil phase with an aqueous phase containing a water-soluble emulsifier. By combining this O/W emulsion with a biopolymer solution and proper environmental conditions the hydrogel particle formation can be promoted, and a filled hydrogel particle can be obtained [103]. Numerous methods to fabricate filled hydrogel particles have been discussed in different studies [88, 138], including simple coacervation, complex coacervation, injection, emulsion templating, spray-drying, and macroscopic gel disruption methods, among others. For the applications of filled hydrogel particles, there have been studies working with n–3 FA [139–141], flavor oils [142], FO [138, 143] and curcumin [144]. The application of filled hydrogel particles in food is still limited since in mostly of the studies use synthetics polymers, which not have been tested for safety [138].

Nowadays, there is a wide variety of options of delivery systems available for the protection, bioavailability, and accurate control and release of the active agents of interest. Conventional O/W emulsions are currently the most widely used method because of their ease preparation and also specific characteristics, as well as the many applications where they can be applied. However, conventional emulsions as their limited ability to protect and control release of certain types of delivery systems have advantages and disadvantages. The search for more food-grade delivery systems which are economically available is necessary. Moreover, it is essential to develop more *in vitro* and *in vivo* methods that can provide us with the certainty and confidence of the safety and performance of the actual and future delivery systems.

14.4.5   COMMERCIAL PRODUCTS BASED ON SL

The above-described applications for SL, have led to the commercialization of some products, as the ones listed next and well detailed in a review by Zam [35]:

- **Betapol:** For infant and food formulation (45% PA content).
- **Impact:** Pharmaceuticals for patients suffering from trauma or surgery, sepsis or cancer (interesterification of LaA and high LA oil).
- **Laurical:** Medical nutrition and confectionery coating, coffee whiteners, whipped toppings and filling fats (FA profile: 40% of LaA as well as OA, LA, and ALA).
- **Neobee:** Nutritional or medical beverages. (FA profile: C8:0, C10:0 and LCFA).
- **Captex:** Improved the absorption of LA when administered to cystic fibrosis patients (FA profile: C8:0, C10:0, LA).

- **Caprenin:** SL containing C8:0, C10:0, and C22:0 esterified to glycerol and it is manufactured from coconut, palm kernel and rapeseed oils by a chemical transesterification. It is recognized as GRAS by the FDA for its use in candy bars and confectionery coatings (nuts, cookies).
- **Salatrim:** SL in the form of TAG containing SCFA (mainly C2:0, C3:0 or C4:0) and at least one LCFA (mainly C18:0). It is recognized as GRAS by the FDA for its use in baking chips, chocolate-flavored coatings, baked, and dairy products, dressing, dips, and sauces, or as a CB substitute in foods. Now is marketed as Benefit.

14.5 CONCLUSIONS

Lipids, as well as their monomeric units, play important roles in human health. Due to little or null physical activities, sedentary life and carbohydrate-rich diets, people around the world are suffering from diabetes, obesity, and cardiovascular diseases. Then, recent concern about what we are eating had led to the preparation of functional or nutraceutical foods. SL are one of those novel foods, containing different FAs which provide specific nutritional properties and health benefits. As it was described previously, the potential of these SL is related to the new FAs incorporated and the specific location in triacylglycerol structure, leading to applications in the food industry and medicine.

Even when several efforts have been conducted, further research is still needed at the following levels: (1) reaction system, including type of reactor and the operational conditions, (2) omics tools for the complete metabolically analysis of the designed lipids, (3) stability and storage of the new lipids, and (4) large scale production by means of technological and economical feasible processes.

KEYWORDS

- **arachidonic acid**
- **fatty acids**
- **food applications**
- **structured lipids**

REFERENCES

1. Watson, A. D. Thematic review series: Systems biology approaches to metabolic and cardiovascular disorders. lipidomics: A global approach to lipid analysis in biological systems. *Journal of Lipid Research,* **2006**, *47*(10), 2101–2111.
2. Orešič, M., Hänninen, V. A., & Vidal-Puig, A. Lipidomics: A new window to biomedical frontiers. *Trends in Biotechnology,* **2008**, *26*(12), 647–652.
3. Nelson, D. L., Lehninger, A. L., & Cox, M. M. *Lehninger Principles of Biochemistry.* W. H. Freeman, **2008**.
4. Rosseau, D., & Marangoni, A. G. The effects of interesterification on the physical properties of fats. In: Marangoni, A. G., & Narine, S. S., (eds.) *Physical Properties of Lipids* (pp. 479–527). Marcel Decker: New York, **2002**.
5. Feltes, M. M. C., de Oliveira Pitol, L., Gomes Correia, J. F., Grimaldi, R., Block, J. M., & Ninow, J. L. *Incorporation of Medium Chain Fatty Acids Into Fish Oil Triglycerides by Chemical and Enzymatic Interesterification.* **2009**, *60*(2), 9.
6. Osborn, H. T., & Akoh, C. C. Structured lipids-novel fats with medical, nutraceutical, and food applications. *Comprehensive Reviews in Food Science and Food Safety,* **2002**, *1*(3), 110–120.
7. Lavers, B. Designer oils revisited. *Food Ingredients And Analysis International* **2003**, *25*(5), 10–12.
8. Lavers, B. Technical Focus – Oils and fats: The growth of "designer oils." *Food Ingredients, Health & Nutrition,* **2002**, *12*, 12–15.
9. Gunstone, F. D., & Hamilton, R. J. *Oleochemical Manufacture and Applications.* Sheffield Academic Press, **2001**.
10. Christie, W. W., & Han, X. *Lipid Analysis,* (p. 448). Pergamon Press: New York, **1982**.
11. Fahy, E., Subramaniam, S., Brown, H. A., Glass, C. K., Merrill, A. H. J., Murphy, R. C., Raetz, C. R., Russell, D. W., Seyama, Y., Shaw, W., Shimizu, T., Spener, F., van Meer, G., Van Nieuwenhze, M. S., White, S. H., Witztum, J. L., & Dennis, E. A. A comprehensive classification system for lipids. *Journal of Lipid Research,* **2005**, *46*(5), 839–861.
12. Naudet, N., Soulier, J., & Farines, M. Principaux constituants chimiques des corps gras. In: Karleskind, A. (ed.), *Manuel des Corps Gras.* Lavoisier Tec and Doc, Paris, **1992**, *1*, 65–115.
13. Scrimgeour, C. M., & Harwood, J. L., Fatty acid and lipid structure. In: Gunstone, F. D., Harwood, J. L., & Dijkstra, A. J. (eds.). *The Lipid Handbook* (pp. 1–16). CRC Press: Boca Raton, **2007**.
14. Baeza-Jiménez, R., López-Martínez, L. X., & García, H. S., Biocalytic modification of food lipids: *Reactions and Applications Revista Mexicana de Ingeniería Química,* **2014**, *13*(1), 29–47.
15. Rombeau, J. L. Investigations of short-chain fatty acids in humans. *Clinical Nutrition Supplements* **2004**, *1*(2), 19–23.
16. Trivedi, R., & Singh, R. P. Modification of oils and fats to produce structured lipids. *Journal of Oleo Science,* **2005**, *54*(8), 423–430.
17. Nunes, P. A., Pires-Cabral, P., & Ferreira-Dias, S. Production of olive oil enriched with medium chain fatty acids catalyzed by commercial immobilized lipases. *Food Chemistry,* **2011**, *127*(3), 993–998.
18. Akoh, C. C., & Min, D. B., *Food Lipids Chemistry, Nutrition and Biotechnology,* (p. 1005). Marcel Dekker: New York, **2002**.

19. Iwasaki, Y., & Yamane, T., Enzymatic synthesis of structured lipids. *Advances in Biochemical Engineering/Biotechnology*, **2004**, *90*, 151–171.
20. Lo, S. K., Tan, C. P., Long, K., Yusoff, M. S. A., & Lai, O. M. Diacylglycerol oil—Properties, processes and products: A review. *Food and Bioprocess Technology*, **2008**, *1*(3), 223.
21. Guo, Z., Vikbjerg, A. F., & Xu, X. Enzymatic modification of phospholipids for functional applications and human nutrition. *Biotechnology Advances*, **2005**, *23*, 203–259.
22. Szuhaj, B. F., & Nieuwenhuyzen, W. V. *Nutrition and Biochemistry of Phospholipids* (vol. 24, p. 250). AOCS Press Technovation: Illinois, **2003**.
23. Donato, P., Dugo, P., & Mondello, L. Separation of lipids. In: Fanali, S., Haddad, P. R., Poole, C. F., Schoenmakers, P., & Lloyd, D. (eds.) *Liquid Chromatography*, (pp. 203–248). Elsevier: Amsterdam, **2013**.
24. Leblond, J. D., Lasiter Ad Fau Li, C., Li C Fau Logares, R., Logares R Fau Rengefors, K., Rengefors K Fau Evens, T. J., & Evens, T. J. A data mining approach to dinoflagellate clustering according to sterol composition: Correlations with evolutionary history. *International Journal of Data Mining and Bioinformatics*, (pp. 431–451), **2010**, *4* (1748–5673 (Print)).
25. Stillwell, W., Chapter 5 – Membrane Polar Lipids. In Stillwell, W., (ed.) *An Introduction to Biological Membranes*, (2nd edn., pp. 63–87). Elsevier, **2016**.
26. Mollet, B., & Rowland, I. Functional foods: At the frontier between food and pharma. *Current Opinion in Biotechnology*, **2002**, *13*(5), 483–485.
27. Niva, M. 'All foods affect health': Understandings of functional foods and healthy eating among health-oriented finns. *Appetite*, **2007**, *48*(3), 384–393.
28. Mark-Herbert, C. Innovation of a new product category-functional foods. *Technovation*, **2004**, *24*(9), 713–719.
29. Alzamora, S. M., Salvatori, D., Tapia, M. S., López-Malo, A., Welti-Chanes, J., & Fito, P. Novel functional foods from vegetable matrices impregnated with biologically active compounds. *Journal of Food Engineering* **2005**, *67*(1), 205–214.
30. Bech-Larsen, T., & Grunert, K. G. The perceived healthiness of functional foods: A conjoint study of Danish, Finnish and American consumers' perception of functional foods. *Appetite*, **2003**, *40*(1), 9–14.
31. Spence, J. T. Challenges related to the composition of functional foods. *Journal of Food Composition and Analysis*, **2006**, *19*, S4–S6.
32. De Felice, S. L. The nutraceutical initiative: A recommendation for U.S. economic and regulatory reforms. *Genetic Engineering News*, **1992**, *12*, 13–15.
33. Pandey, M., Verma, R. K., & Saraf, S. A. Nutraceuticals: New era of medicine and health. *Asian Journal of Pharmaceutical and Clinical Research*, **2010**, *3*(1), 11–15.
34. Siró, I., Kápolna, E., Kápolna, B., & Lugasi, A. Functional food. Product development, marketing and consumer acceptance-A review. *Appetite*, **2008**, *51*(3), 456–467.
35. Zam, W. Structured lipids: Methods of production, commercial products and nutraceutical characteristics. *Progress in Nutrition*, **2015**, *17*, 198–213.
36. Singer, P., Hicsmayr, M., Biolo, G., Felbinger, T. W., Berger, M. M., Goeters, C., Kondrup, J., Wunder, C., & Pichard, C. Pragmatic approach to nutrition in the ICU: Expert opinion regarding which calorie protein target. *Clinical Nutrition*, **2014**, *33*(2), 246–251.
37. Alkhawaja, S., Martin, C., Butler, R. J., & Gwadry-Sridhar, F. Post-pyloric versus gastric tube feeding for preventing pneumonia and improving nutritional outcomes in critically ill adults. *Cochrane Database of Systematic Reviews*, **2015**, (8).

38. Cederholm, T., Barazzoni, R., Austin, P., Ballmer, P., Biolo, G., Bischoff, S. C., Compher, C., Correia, I., Higashiguchi, T., Holst, M., Jensen, G. L., Malone, A., Muscaritoli, M., Nyulasi, I., Pirlich, M., Rothenberg, E., Schindler, K., Schneider, S. M., de van der Schueren, M. A. E.; Sieber, C., Valentini, L., Yu, J. C., Van Gossum, A., & Singer, P. ESPEN guidelines on definitions and terminology of clinical nutrition. *Clinical Nutrition,* **2017**, *36*(1), 49–64.

39. Singer, P., Berger, M. M., Van den Berghe, G., Biolo, G., Calder, P., Forbes, A., Griffiths, R., Kreyman, G., Leverve, X., & Pichard, C. ESPEN guidelines on parenteral nutrition: Intensive care. *Clinical Nutrition,* **2009**, *28*(4), 387–400.

40. Richards, P. M. Lipid chemistry and biochemistry. In: Hui, Y. H., (ed.) *Handbook of Food Science, Technology and Engineering* (vol. 1, pp. 8–17). Taylor & Francis: New York, **2006**.

41. Calder, P. C., Jensen, G. L., Koletzko, B. V., Singer, P., & Wanten, G. J. A., Lipid emulsions in parenteral nutrition of intensive care patients: current thinking and future directions. *Intensive Care Medicine,* **2010**, *36*(5), 735–749.

42. Fell, G. L., Nandivada, P., Gura, K. M., & Puder, M., Intravenous lipid emulsions in parenteral nutrition. *Advances in Nutrition,* **2015**, *6*(5), 600–610.

43. Adolph, A., Heller, A., Koch, T., Koletzko, B., Kreymann, K. G., Krohn, K., Pscheidl, E., & Senkal, M. *6 Lipidemulsionen* (vol. 32). 2007.

44. Mateu-de Antonio, J., Grau, S., Luque, S., Marín-Casino, M., Albert, I., & Ribes, E. Comparative effects of olive oil-based and soyabean oil-based emulsions on infection rate and leucocyte count in critically ill patients receiving parenteral nutrition. *British Journal of Nutrition,* **2008**, *99*(4), 846–854.

45. Reimund, J. M., Rahmi, G., Escalin, G., Pinna, G., Finck, G., Muller, C. D., Duclos, B., & Baumann, R. Efficacy and safety of an olive oil-based intravenous fat emulsion in adult patients on home parenteral nutrition. *Alimentary Pharmacology & Therapeutics,* **2005**, *21*(4), 445–454.

46. Anez-Bustillos, L., Dao, D. T., Baker, M. A., Fell, G. L., Puder, M., & Gura, K. M. Intravenous fat emulsion formulations for the adult and pediatric patient. *Nutrition in Clinical Practice,* **2016**, *31*(5), 596–609.

47. Miles, E. A., & Calder, P. C. Fatty acids, lipid emulsion and the inmunne and inflammatory systems. *World Review of Nutrition and Dietetics,* **2015**, *112*, 17–30.

48. Calder, P. C. Functional roles of fatty acids and their effects on human health. *Journal of Parenteral and Enteral Nutrition,* **2015**, *39*(1S), 18S-32S.

49. Munroe, C., Frantz, D., Martindale, R. G., & McClave, S. A. The optimal lipid formulation in enteral feeding in critical illness: Clinical Update and review of the literature. *Current Gastroenterology Reports,* **2011**, *13*(4), 368–375.

50. Ballard, O., & Morrow, A. L. Human milk composition: Nutrients and bioactive factors. *Pediatric Clinics of North America,* **2013**, *60*(1), 49–74.

51. Jensen, R. G. Human milk lipids as a model for infant formula. *Lipid Technology,* **1998**, *10*, 34–38.

52. Bokor, S., Koletzko, B., & Decsi, T. Systematic review of fatty acid composition of human milk from 331 mothers of preterm compared to full-term infants. *Annals of Nutrition and Metabolism,* **2007**, *51*, 550–556.

53. Schuchardt, J. P., Huss, M., Stauss-Grabo, M., & Hahn, A. Significance of long-chain polyunsaturated fatty acids (PUFAs) for the development and behavior of children. *European Journal of Pediatrics,* **2010**, *169*(2), 149–164.

54. Michalski, M. C. Lipids and milk fat globule properties in human milk. In: Zibadi, S., Watson, R. R., & Preedy, V. R., (eds.), *Handbook of Dietary and Nutritional Aspects of Human Breast Milk,.* Wageningen Academic Publisher, **2013**.

55. Yehuda, S., Rabinovitz, S., & Mostofsky, D. I. Essential fatty acids and the brain: From infancy to aging. *Neurobiology of Aging,* **2005**, *26*(1), 98–102.

56. Kennedy, J. Structured lipids: Fats of the future. *Food Technology,* **1991**, *45*, 76–83.

57. Ostrom, K. M., Cordle, C. T., Schaller, J. P., Winship, T. R., Thomas, D. J., Jacobs, J. R., Blatter, M. M., Cho, S., Gooch, W. M., Granoff, D. M., Faden, H., & Pickering, L. K. Inmune status of infants fed soy-based formulas with or without added nucleotides for 1 year: pat 1: vaccines responses, and morbidity. *Journal of the American College of Nutrition,* **2002**, *21*(6), 564–569.

58. Stam, J., Sauer, P. J. J. & Boehm, G. Can we define an infant's need from the composition of human milk? *The American Journal of Clinical Nutrition,* **2013**, *98*(2), 521S–528S.

59. Long, A. C., Kaiser, J. L., & Katz, G. E. Lipids in infant formulas: Current and future innovations. *Lipid Technology,* **2013**, *25*(6), 127–129.

60. Zou, L., Pande, G., & Akoh, C. C. Infant formula fat analogs and human milk fat: New focus on infant developmental needs. *Annual Review of Food Science and Technology,* **2016**, *7*(1), 139–165.

61. Bar-on, Z., Ben, D. G., Pelled, D., & Shulman, A. Mimetic lipids and dietary supplements comprising the same. **2004**.

62. Oosting, A., Kegler, D., Wopereis, H. J., Teller, I. C., van de Heijning, B. J. M., Verkade, H. J., & van der Beek, E. M. Size and phospholipid coating of lipid droplets in the diet of young mice modify body fat accumulation in adulthood. *Pediatric Research,* **2012**, *72*, 362.

63. Soumanou, M. M., Pérignon, M., & Villeneuve, P. Lipase-catalyzed interesterification reactions for human milk fat substitutes production: A review. *European Journal of Lipid Science and Technology,* **2013**, *115*(3), 270–285.

64. Wang, H. X., Wu, H., Ho, C. T., & Weng, X. C. Cocoa butter equivalent from enzymatic interesterification of tea seed oil and fatty acid methyl esters. *Food Chemistry,* **2006**, *97*(4), 661–665.

65. Naik, B., & Kumar, D. V. *Cocoa Butter and Its Alternatives*(p 2:1–11)*, A Reveiw*. 2014.

66. Zaidul, I. S. M., Nik Norulaini, N. A., Mohd Omar, A. K., & Smith, R. L. Blending of supercritical carbon dioxide (SC-CO$_2$) extracted palm kernel oil fractions and palm oil to obtain cocoa butter replacers. *Journal of Food Engineering,* **2007**, *78*(4), 1397–1409.

67. Beckett, S. T. *The Science of Chocolate*. RSC Publishing: Cambridge, **2008**.

68. Timms, R. E. Cocoa butter, a unique vegetable fat. *Lipid Technology,* (pp. 101–107), **1999**.

69. Sabariah, S., Md. Ali, A. R., & Chong, C. L. Chemical and physical characteristics of cocoa butter substitutes, milk fat and Malaysian cocoa butter blends. *Journal of the American Oil Chemists' Society,* **1998**, *75*(8), 905–910.

70. Lipp, M., & Anklam, E. Review of cocoa butter and alternative fats for use in chocolate—Part A. compositional data. *Food Chemistry,* **1998**, *62*(1), 73–97.

71. Kellens, M., Gibon, V., Hendrix, M., & De Greyt, W. Palm oil fractionation. *European Journal of Lipid Science and Technology,* **2007**, *109*(4), 336–349.

72. Buchgraber, M., Ulberth, F., Anklam, E., *Method Validation For Detection And Quantification Of Cocoa Butter Equivalents In Cocoa Butter And Plain Chocolate,* **2004**, *87*, 1164–1172.

73. Undurraga, D., Markovits, A., & Erazo, S. Cocoa butter equivalent through enzymic interesterification of palm oil midfraction. *Process Biochemistry,* **2001**, *36*(10), 933–939.

74. Ali, A. R. M. Effect of co-fractionation technique in the preparation of palm oil and sal fat based cocoa butter equivalent. *International Journal of Food Sciences and Nutrition,* **1996**, *47*(1), 15–22.

75. Calliauw, G., Foubert, I., De Greyt, W., Dijckmans, P., Kellens, M., & Dewettinck, K. Production of cocoa butter substitutes via two-stage static fractionation of palm kernel oil. *Journal of the American Oil Chemists' Society,* **2005**, *82*(11), 783–789.

76. Hashimoto, S., Nezu, T., Arakawa, H., Ito, T., & Maruzeni, S. Preparation of sharp-melting hard palm midfraction and its use as hard butter in chocolate. *Journal of the American Oil Chemists' Society,* **2001**, *78*(5), 455–460.

77. Sonwai, S., Kaphueakngam, P., & Flood, A. Blending of mango kernel fat and palm oil mid-fraction to obtain cocoa butter equivalent. *Journal of Food Science and Technology,* **2014**, *51*(10), 2357–2369.

78. Bloomer, S., Adlercreutz, P., & Mattiasson, B. Triglyceride interesterification by lipases. 1. Cocoa butter equivalents from a fraction of palm oil. *Journal of the American Oil Chemists' Society,* **1990**, *67*(8), 519–524.

79. Solís-Fuentes, J. A., & Durán-de-Bazúa, M. C. Mango seed uses: thermal behaviour of mango seed almond fat and its mixtures with cocoa butter. *Bioresource Technology,* **2004**, *92*(1), 71–78.

80. Holcapek, M., Lisa, M., Jandera, P., & Kabatova, N. Quantitation of triacylglycerols in plant oils using HPLC with APCI-MS, evaporative light-scattering and UV detection. *Journal of Separation Science,* **2005**, *28*, 1315–1333.

81. Jahurul, M. H. A., Zaidul, I. S. M., Norulaini, N. A. N., Sahena, F., Jinap, S., Azmir, J., Sharif, K. M., & Omar, A. K. M. Cocoa butter fats and possibilities of substitution in food products concerning cocoa varieties, alternative sources, extraction methods, composition, and characteristics. *Journal of Food Engineering,* **2013**, *117*(4), 467–476.

82. Kaphueakngam, P. *Production of Cocoa Butter Equivalent From Mango Seed Almond Fat And Palm Oil Mid-Fraction.* **2018**.

83. Jeyarani, T., & Reddy, S. Y. Heat-resistant cocoa butter extenders from mahua (*Madhuca latifolia*) and kokum (*Garcinia indica*) fats. *Journal of the American Oil Chemists' Society,* **1999**, *76*(12), 1431–1436.

84. Maheshwari, B., & Yella Reddy, S. Application of kokum (*Garcinia indica*) fat as cocoa butter improver in chocolate. *Journal of the Science of Food and Agriculture,* **2005**, *85*(1), 135–140.

85. Akihisa, T., Kojima, N., Katoh, N., Ichimura, Y., Suzuki, H., Fukatsu, M., Maranz, S., & Masters, T., E. Triterpene alcohol and fatty acid composition of shea nuts from seven African Countries. *Journal of Oleo Science,* **2010**, *59*(7), 351–360.

86. Maranz, S., & Wiesman, Z. Evidence for indigenous selection and distribution of the shea tree, Vitellaria paradoxa, and its potential significance to prevailing parkland savanna tree patterns in sub-Saharan Africa north of the equator. *Journal of Biogeography,* **2003**, *30*(10), 1505–1516.

87. Olajide, J. O., Ade-Omowaye, B. I. O., & Otunola, E. T. Some physical properties of shea kernel. *Journal of Agricultural Engineering Research,* **2000**, *76*(4), 419–421.

88. McClements, D. J., & Li, Y. Structured emulsion-based delivery systems: Controlling the digestion and release of lipophilic food components. *Advances in Colloid and Interface Science,* **2010**, *159*(2), 213–228.

89. Kosaraju, S. L. Colon targeted delivery systems: Review of polysaccharides for encapsulation and delivery. *Critical Reviews in Food Science and Nutrition,* **2005,** *45*(4), 251–258.

90. Williams, H. D., Trevaskis, N. L., Charman, S. A., Shanker, R. M., Charman, W. N., Pouton, C. W., & Porter, C. J. H. Strategies to address low drug solubility in discovery and development. *Pharmacological Reviews,* **2013,** *65*(1), 315.

91. Kumar, S., Bhargava, D., Thakkar, A., & Arora, S. *Drug Carrier Systems for Solubility Enhancement of BCS Class II Drugs: A Critical Review.* **2013,** *30*(3), 217–256.

92. Wang, G., Wang, J., Wu, W., Tony To, S. S., Zhao, H., & Wang, J. Advances in lipid-based drug delivery: Enhancing efficiency for hydrophobic drugs. *Expert Opinion on Drug Delivery,* **2015,** *12*(9), 1475–1499.

93. Azhar Yaqoob, K., Sushama, T., Zeenat, I., Farhan Jalees, A., & Roop Krishan, K. Multiple Emulsions: An overview. *Current Drug Delivery,* **2006,** *3*(4), 429–443.

94. Ahmed, K., Li, Y., McClements, D. J., & Xiao, H. Nanoemulsion- and emulsion-based delivery systems for curcumin: Encapsulation and release properties. *Food Chemistry,* **2012,** *132*(2), 799–807.

95. Aditya, N. P., Macedo, A. S., Doktorovova, S., Souto, E. B., Kim, S., Chang, P. S., & Ko, S. Development and evaluation of lipid nanocarriers for quercetin delivery: A comparative study of solid lipid nanoparticles (SLN), nanostructured lipid carriers (NLC), and lipid nanoemulsions (LNE). *LWT – Food Science and Technology,* **2014,** *59*(1), 115–121.

96. McClements, D. J., Decker, E. A., & Park, Y. Controlling lipid bioavailability through physicochemical and structural approaches. *Critical Reviews in Food Science and Nutrition,* **2008,** *49*(1), 48–67.

97. Aulton, M. E. Properties of solutions. In: Aulton, M., & Taylor, K. M. G., (eds.) *Pharmaceutics: The Design And Manufacture Of Medicines,* (pp. 38–48). Elsevier: Edinburg, **2013.**

98. Singh, H., Ye, A., & Horne, D. Structuring food emulsions in the gastrointestinal tract to modify lipid digestion. *Progress in Lipid Research,* **2009,** *48*(2), 92–100.

99. McClements, D. J. *Food Emulsions: Principles, Practice and Techniques,* (p. 632). CRC Press: Boca Raton, **2005.**

100. Wooster, T. J., Golding, M., & Sanguansri, P. Impact of oil type on nanoemulsion formation and ostwald ripening stability. *Langmuir,* **2008,** *24*(22), 12758–12765.

101. McClements, D. J. Theoretical prediction of emulsion color. *Advances in Colloid and Interface Science,* **2002,** *97*(1), 63–89.

102. McClements, D. J. *Food Emulsions: Principles, Practice and Techniques,*(p. 690). CRC Press: Boca Raton, **2015.**

103. McClements, D. J., & Rao, J. Food-grade nanoemulsions: Formulation, fabrication, properties, performance, biological fate, and potential toxicity. *Critical Reviews in Food Science and Nutrition,* **2011,** *51*(4), 285–330.

104. Acosta, E. Bioavailability of nanoparticles in nutrient and nutraceutical delivery. *Current Opinion in Colloid and Interface Science,* **2009,** *14*(1), 3–15.

105. Solè, I., Maestro, A., González, C., Solans, C., & Gutiérrez, J. M. Optimization of Nano-emulsion preparation by low-energy methods in an ionic surfactant system. *Langmuir,* **2006,** *22*(20), 8326–8332.

106. Walstra, P. *Physical Chemistry Of Foods,* (p. 832). Marcel Decker: New York, **2003.**

107. Anton, N., & Vandamme, T. F. The universality of low-energy nano-emulsification. *International Journal of Pharmaceutics,* **2009,** *377*(1), 142–147.

108. Taylor, T. M., Weiss, J., Davidson, P. M., & Bruce, B. D. Liposomal nanocapsules in food science and agricu lture. *Critical Reviews in Food Science and Nutrition,* **2005,** *45* (7–8), 587–605.

109. Flanagan, J., Singh, H., Microemulsions: A potential delivery system for bioactives in food. *Critical Reviews in Food Science and Nutrition,* **2006,** *46*(3), 221–237.

110. Kesisoglou, F., Panmai, S., & Wu, Y. Application of nanoparticles in oral delivery of immediate release formulations. *Current Nanoscience,* **2007,** *3*(2), 183–190.

111. Tagne, J. B., Kakumanu, S., Ortiz, D., Shea, T., & Nicolosi, R. J. A Nanoemulsion formulation of tamoxifen increases its efficacy in a breast cancer cell line. *Molecular Pharmaceutics,* **2008,** *5*(2), 280–286.

112. Yang, Y., & McClements, D. J. Encapsulation of vitamin E in edible emulsions fabricated using a natural surfactant. *Food Hydrocolloids,* **2013,** *30*(2), 712–720.

113. Cho, H. T., Salvia-Trujillo, L., Kim, J., Park, Y., Xiao, H., & McClements, D. J. Droplet size and composition of nutraceutical nanoemulsions influences bioavailability of long chain fatty acids and Coenzyme Q10. *Food Chemistry,* **2014,** *156,* 117–122.

114. Martins, S., Sarmento, B., & Ferreira, D. C. Lipid-based colloidal carriers for peptide and protein delivery-liposomes versus lipid nanoparticles. *International Journal of Nanomedicine,* **2007,** *2*(4), 595–607.

115. Saupe, A., Wissing, S. A., Lenk, A., Schmidt, C., & Muller, R. H. Solid lipid nanoparticles (SLN) and nanostructured lipid carriers (NLC)-Structural investigations on two different carrier systems. *Bio-Medical Materials and Engineering,* **2005,** *15*(5), 393–402.

116. Wissing, S. A., Kayser, O., & Müller, R. H. Solid lipid nanoparticles for parenteral drug delivery. *Advanced Drug Delivery Reviews,* **2004,** *56*(9), 1257–1272.

117. Mukherjee, S., Ray, S., & Thakur, R. S. Solid lipid nanoparticles: A Modern formulation approach in drug delivery system. *Indian Journal of Pharmaceutical Sciences,* **2009,** *71*(4), 349–358.

118. Dash, M., Chiellini, F., Ottenbrite, R. M., & Chiellini, E. Chitosan—A versatile semi-synthetic polymer in biomedical applications. *Progress in Polymer Science,* **2011,** *36*(8), 981–1014.

119. Blasi, P., Giovagnoli, S., Schoubben, A., Ricci, M., & Rossi, C. Solid lipid nanoparticles for targeted brain drug delivery. *Advanced Drug Delivery Reviews,* **2007,** *59*(6), 454–477.

120. Müller, R. H., Radtke, M., & Wissing, S. A. Solid lipid nanoparticles (SLN) and nanostructured lipid carriers (NLC) in cosmetic and dermatological preparations. *Advanced Drug Delivery Reviews,* **2002,** *54,* S131–S155.

121. Üner, M., & Yener, G. Importance of solid lipid nanoparticles (SLN) in various administration routes and future perspectives. *International Journal of Nanomedicine,* **2007,** *2*(3), 289–300.

122. Pardeshi, C., Rajput, P., Belgamwar, V., Tekade, A., Patil, G., Chaudhary, K., & Sonje, A. Solid lipid based nanocarriers: An overview. *Acta Pharmaceutica,* **2012,** *62*(4), 170–184.

123. Mehnert, W., & Mäder, K. Solid lipid nanoparticles: Production, characterization and applications. *Advanced Drug Delivery Reviews,* **2001,** *47*(2), 165–196.

124. Guimarães, K. L., & Ré, M. I. Lipid nanoparticles as carriers for cosmetic ingredients: The first (SLN) and the second generation (NLC). In: Beck, R., Guterres, S., & Pohlmann, A., (eds.) *Nanocosmetics and Nanomedicines: New Approaches for Skin Care,* (pp. 101–122). Springer: Germany, **2011.**

125. Tamjidi, F., Shahedi, M., Varshosaz, J., Nasirpour, A., Nanostructured lipid carriers (NLC): A potential delivery system for bioactive food molecules. *Innovative Food Science & Emerging, Technologies,* **2013**, *19,* 29–43.

126. Naseri, N., Valizadeh, H., & Zakeri-Milani, P. Solid lipid nanoparticles and nanostructured lipid carriers: Structure, preparation and application. *Advanced Pharmaceutical Bulletin,* **2015**, *5*(3), 305–313.

127. Yuan, H., Huang, L. F., Du, Y. Z., Ying, X. Y., You, J., Hu, F. Q., & Zeng, S. Solid lipid nanoparticles prepared by solvent diffusion method in a nanoreactor system. *Colloids and Surfaces B: Biointerfaces,* **2008**, *61*(2), 132–137.

128. Yoon, G., Park, J. W., & Yoon, I. S. Solid lipid nanoparticles (SLNs) and nanostructured lipid carriers (NLCs): Recent advances in drug delivery. *Journal of Pharmaceutical Investigation,* **2013**, *43*(5), 353–362.

129. Souto, E. B., & Müller, R. H. Lipid nanoparticles: Effect on bioavailability and pharmacokinetic changes. In: Schäfer-Korting, M., (ed.), *Drug Delivery* (pp. 115–141). Springer Berlin Heidelberg: Berlin, Heidelberg, **2010**.

130. Jawahar, N., & Reddy, G. Nanoparticles: A novel pulmonary drug delivery system for tuberculosis. *Journal of Pharmaceutical Sciences and Research,* **2012**, *4,* 1901–1906.

131. Paranjpe, M., & Muller-Goymann, C. C. Nanoparticle-mediated pulmonary drug delivery: A review. *International Journal of Molecular Science,* **2014**, *15*(4), 5852–5873.

132. Weber, S., Zimmer, A.,& Pardeike, J. Solid Lipid Nanoparticles (SLN) and nanostructured lipid carriers (NLC) for pulmonary application: A review of the state of the art. *European Journal of Pharmaceutics and Biopharmaceutics,* **2014**, *86*(1), 7–22.

133. Han, Y., Li, Y., Zhang, P., Sun, J., Li, X., Sun, X., & Kong, F. Nanostructured lipid carriers as novel drug delivery system for lung cancer gene therapy. *Pharmaceutical Development and Technology,* **2016**, *21*(3), 277–281.

134. Lai, F., Wissing, S. A., Müller, R. H., & Fadda, A. M. *Artemisia arborescens* L essential oil-loaded solid lipid nanoparticles for potential agricultural application: Preparation and characterization. *AAPS Pharm. Sci. Tech.,* **2006**, *7*(1), E10.

135. Weiss, J., Decker, E. A., McClements, D. J., Kristbergsson, K., Helgason, T., & Awad, T. Solid lipid nanoparticles as delivery systems for bioactive food components. *Food Biophysics,* **2008**, *3*(2), 146–154.

136. Waghmare, A. S., Grampurohit, N. D., Gadhave, M. V., Gaikwad, D. D., & Jadhav, S. I. *Solid Lipid Nanoparticles: A Promising Drug Delivery System.* 2012, *3,* 100–107.

137. Pardeike, J., Hommoss, A., & Müller, R. H. Lipid nanoparticles (SLN, NLC) in cosmetic and pharmaceutical dermal products. *International Journal of Pharmaceutics,* **2009**, *366*(1), 170–184.

138. Zhang, Z., Decker, E. A., & McClements, D. J. Encapsulation, protection, and release of polyunsaturated lipids using biopolymer-based hydrogel particles. *Food Research International,* **2014**, *64,* 520–526.

139. Lamprecht, A., & U. S. C. M. L. Influences of process parameters on preparation of microparticle used as a carrier system for O – 3 unsaturated fatty acid ethyl esters used in supplementary nutrition. *Journal of Microencapsulation,* **2001**, *18*(3), 347–357.

140. Wu, K. G., & Xiao, Q. Microencapsulation of fish oil by simple coacervation of hydroxypropyl methylcellulose. *Chinese Journal of Chemistry,* **2005**, *23*(11), 1569–1572.

141. Salcedo-Sandoval, L., Cofrades, S., Ruiz-Capillas, C., Matalanis, A., McClements, D. J., Decker, E. A., & Jiménez-Colmenero, F. Oxidative stability of n–3 fatty acids

encapsulated in filled hydrogel particles and of pork meat systems containing them. *Food Chemistry,* **2015,** *184,* 207–213.

142. Weinbreck, F., Minor, M., & de Kruif, C. G. Microencapsulation of oils using whey protein/gum arabic coacervates. *Journal of Microencapsulation,* **2004,** *21*(6), 667–679.

143. Kim, H. J., Decker, E. A., & Julian McClements, D. Preparation of multiple emulsions based on thermodynamic incompatibility of heat-denatured whey protein and pectin solutions. *Food hydrocolloids,* **2006,** *20*(5), 586–595.

144. Zeeb, B., Saberi, A. H., Weiss, J., & McClements, D. J. Formation and characterization of filled hydrogel beads based on calcium alginate: Factors influencing nanoemulsion retention and release. *Food Hydrocolloids,* **2015,** *50,* 27–36.

CHAPTER 15

Milk-Clotting Enzymes: *S. elaeagnifolium* As An Alternative Source

NÉSTOR GUTIÉRREZ-MÉNDEZ,[1] JOSÉ ALBERTO LÓPEZ-DÍAZ,[2]
DELY RUBI CHÁVEZ-GARAY,[1] MARTHA YARELY LEAL-RAMOS,[1] and
ANTONIO GARCÍA-TRIANA[1]

[1]*School of Chemical Sciences, Autonomous University of
Chihuahua, 31125 Chihuahua, Chih, Mexico, Tel.: +52 (614) 236 6000,
E-mail: ngutierrez@uach.mx (N. Gutiérrez-Méndez)*

[2]*Institute of Biomedical Sciences, Autonomous University of Ciudad
Juarez, Ciudad Juárez, Chihuahua, México*

ABSTRACT

Proteases are one of the most used enzymes in the food industry, including the dairy industry. The coagulation of milk requires a protease capable of hydrolyzing caseins in such a way that electrostatic and steric repulsion be reduced. Animal-origin proteases like chymosin and pepsin have been used for centuries as milk coagulants. Nowadays, microbial proteases and recombinant calf rennet are commercially available. However, different factors such as religious or ethical concerns, diet, and price, have promoted the search for alternative milk coagulants. In this book chapter, it will be addressed and discussed the most recent information about milk-clotting proteases from the fruits of *S. elaeagnifolium,* an endemic plant from the northeast of Mexico, the southwest of the United States, and Argentina.

15.1 INTRODUCTION

Cheese is a fermented milk-based food product, which allows preserving the milk components for a longer time. Cheese production is one of the most widely distributed activities in the dairy industry. Nowadays, cheese is produced all over the world, and it is estimated there are more than 1000 varieties. The increasing global demand for dairy products like cheese has promoted the improvement of manufacturing processes, and the search for new technologies or new additives like starters and enzymes [1, 2]. The process of cheese manufacturing varies with the type of cheese, though some steps are common in most cheese varieties. For instance, the milk or curd fermentation (acidification), the milk clotting (coagulation), the separation of the curd from the whey (syneresis), and the application of pressure to control the moisture, and to give the cheese a particular form (pressing). The coagulation of milk in cheese making is achieved by adding proteases, and for centuries, the source of proteases for milk clotting was the liquid from the abomasums of calves [2, 3].

Proteases can be found in all living organisms (including mammals, plants, and microorganisms) since they are used to degrade nonfunctional proteins into amino acids [4, 5]. These enzymes cleave peptide bonds that can be internal (endopeptidases), N-terminal (aminopeptidases), or C-terminal (carboxypeptidases). All proteases stabilize the oxygen from the carbonyl group in an oxyanion hole, which polarizes the carbonyl group of the substrate peptide bond. As a result, the carbon atom is more vulnerable for attack by an activated nucleophile. Four major activated nucleophiles are involved in proteolysis: Cys-His (cysteine proteases (CPs)), Ser-His (serine proteases (SPs)), H_2O-Me^{2+} (metalloproteases), and H_2O-Asp (aspartic proteases (APs)) [5]. The main protease in the calf rennet is chymosin (EC 3.4.23.4), which is an aspartic protease secreted in the abomasal mucosa of newborn ruminants [3].

Contrary to other gastric APs, chymosin exhibits a low proteolytic activity, but such action is highly specific to κ-caseins. Hydrolysis of κ-caseins (Phe_{105}-Met_{106}) releases a glyco-macro-peptide and exposes the positive charges of *para*-κ-caseins that remain in the casein micelles. The exposition of positive charges decreases of the repulsive electric forces between casein micelles (primary enzymatic phase of coagulation). The subsequent aggregation of casein micelles depends on non-enzymatic factors such as the pH, the temperature, but chiefly by the presence of Ca^{2+} (secondary non-enzymatic phase) [6]. Only those enzymes with a high ratio of milk-clotting activity to proteolytic activity (MCA/PA) are suitable for cheese manufacture [3].

Historically, the calf rennet obtained from the stomachs of young ruminants (mainly calves) was the primary source of proteases for cheese making. In the '60s and '70s, the increase of cheese production raised the demand for coagulating enzymes markedly. Since then, the demand for milk coagulants started exceeding the supply, and today only 20–30% is covered by calf rennet. The most suitable substitutes of calf rennet include microbial proteases and recombinant enzymes. Extracellular proteases of microbial origin have been used in the commercial manufacture of cheese since the '60s. Fungal strains like *Rhizomucor miehei, Rhizomucor pusillus,* and *C. parasitica* are used in the large-scale production of chymosin-like proteases for cheese production. Lacto-vegetarians and consumers that belong to specific religions (e.g., Islam, Judaism, and Buddhism) will accept these types of coagulating enzymes. However, most microbial proteases have a higher proteolytic activity than chymosin, which may lead to adverse effects on cheese production.

On the other hand, there are recombinant enzymes metabolized by genetically modified microorganisms (GMOs). It is estimated that recombinant versions of chymosin comprise 70–80% of the global market for coagulants, headed by the recombinant *Bos taurus* chymosin expressed in *Aspergillus niger* var. *awamori.* Even though the GMOs might provide milk-clotting enzymes almost unlimited, there are some constraints. For instance, various European countries (France, Germany, and The Netherlands) have prohibited the involvement of GMOs in the manufacture of organic cheeses, and cheeses with protected designation of origin (PDO) [6–8]. Therefore, there is a growing interest in other sources of milk-clotting proteases like the plant-derived proteases. This book chapter discusses the current knowledge on *Solanum elaeagnifolium* and the use of proteases from the fruits of this plant as milk coagulant in the manufacture of cheese.

15.2 PLANT PROTEASES AS MILK-CLOTTING ENZYMES

All the plants contain proteases, and plant genomes encode hundreds of different proteases [5]. The wide distribution of proteases among plants demonstrates that they play a pivotal role in processes such as protein turnover, degradation of misfolded proteins, senescence, and the ubiquitin/proteasome pathway [9]. Proteases are also responsible for the post-translational modification of proteins by limited proteolysis at highly specific sites. They are involved in a great diversity of cellular processes, including photoinhibition in the chloroplast, defense mechanisms, programmed cell death, and photomorphogenesis in the developing seedling. Proteases are thus involved in all aspects of the plant life cycle ranging from mobilization of storage proteins during seed germination to the initiation of cell death and senescence programs [4]. Plants contain all the four catalytic forms of proteases. Therefore, aspartic, serine, and CPs have been isolated from several plant sources and studied for milk-clotting ability [7].

APs are characterized by having two aspartic residues at their catalytic site. They are most active at acidic pH, are specifically inhibited by pepstatin with preferential activity for peptide bonds breakage between hydrophobic amino acid residues responsible for the catalytic activity [7]. Most plant APs identified so far are synthesized with a pre-prodomain and subsequently converted to mature two-chain enzymes. Also, they are present in a wide variety of plant species and have been characterized and purified from a variety of tissues such as seeds, flowers, and leaves. It has been suggested that plant APs are involved in the digestion of insects in carnivorous plants, in the degradation of plant proteins in response to pathogens, during development processes, protein-storage processing mechanisms, stress responses and senescence [4]. CPs, also known as thiol proteases. The catalytic mechanism of these enzymes involves a cysteine group in the active site. CPs have great potential in the food, biotechnology, and pharmaceutical industries owing to their property of being active over a wide range of temperature and pH [7]. These enzymes comprise a family of enzymes, consisting of papain and related plant proteases such as chymopapain, caricain, bromelain, actinidin, ficin, aleurain, and others. In plants, CPs are involved in protein maturation, degradation, and protein rebuilding in response to various external stimuli and also play a housekeeping function to remove abnormal, misfolded proteins. CPs has been shown to be implicated in proteolysis during senescence, under drought and during programmed cell death [4]. SPs possess a serine residue in their active site and share some biochemical and physiological features. In plants, they are widespread among taxonomic groups, from trees and crops to legumes and herbs and present in almost all plant parts, but most abundant in fruits [7]. SPs are usually considered to act principally as degradative enzymes and be involved in a variety of physiological processes including symbiosis, hypersensitive response, the infection of plant cells, pathogenesis in virus-infected plants, germination, signaling, tissue differentiation, xylogenesis, senescence, programmed cell death, and protein degradation/processing [4].

Plant-derived proteases are obtained from the aqueous maceration of parts of plants such as leaf, flowers, and branches [10]. However, only a few plant proteases are suitable for their use in cheese production. Compared to animal chymosin, proteases from plants have stronger proteolytic activity and produce non-specific hydrolysis on milk caseins which leads to a low ratio of milk clotting activity to proteolytic activity [7, 8]. Furthermore, the high proteolytic nature of most plant-derived proteases not only produces bitter peptides but also reduce the cheese yield [7]. Therefore, commercially available proteases like papain (from papaya leaves), bromelain (from pineapple), and ficin (from fig) are used to produce milk protein hydrolysates but not cheese [11]. Hardly any plant-derived protease is used in the industrial manufacture of cheese, except for those obtained from *Cynara* sp. and certain species of *Solanum*. The APs from *Cynara* sp. (cardosins) are used to produce a large variety of cheeses in Mediterranean, West African, and southern European countries. In Portugal and Spain, the ewe- and goat-milk cheeses manufactured with an aqueous extract of *Cynara cardunculus* have achieved the PDO. Some examples of cheeses made with cardosins are the Mestico de Tolosa cheese, de Nisa, de Castelo Branco, deÉvora, Serpa, de Azeitao, Serra da Estrela, La Serna, Torta del Casar, Flor de Guía, Media flor de guía, and Guía cheese [12]. Plant-derived coagulants from *Solanum dubium* and *Solanum elaeagnofolium* are also used in the manufacture of commercial cheeses in Sudan [13] and Mexico [14] respectively. Cheeses made with *S. dubium* are known in Sudan as Gubbein cheese [15]. In Mexico, a filata-type cheese named Asadero is manufactured with the ripe yellow fruits of *S. elaeagnifolium* [16]. Besides these plant proteases, there are different reports in the literature of other plant-derived proteases that could be potentially used as milk coagulants. Jacob and Rohm [8], as well as Shah et al. [7], have extensively reviewed the plant proteases that could be used as milk-clotting enzymes in cheese making. Table 15.1 shows a brief description of plant-derived proteases with reported milk-clotting activity.

15.3 THE GENUS *SOLANUM* AND THE SPECIES *S. ELAEAGNIFOLIUM*

The genus *Solanum* includes some economically important species such as the potato (*S. tuberosum*), tomato (*S. lycopersicum*), and eggplant (*S. melongena*). This genus is the largest one in the *Solanaceae* family (order *Solanales*) and contains approximately 1,500 species. *Solanum* plants possess anthers which open by terminal pores, and flowers that lack the specialized calyx observed in the genus *Lycianthes* from the same family [38, 58]. The *Solanaceae* family (and sometimes the genus *Solanum*) is referred as *"nightshade"* since involving hallucinogenic or deadly plants such as *S. dulcamara* (woody nightshade) *S. nigrum* (black nightshade), and *Atropa belladonna* (deadly nightshade) [59]. The presence of toxic glycoalkaloids is a typical feature of *Solanum* plants. Glycoalkaloids have a steroidal structure (aglycone) and

TABLE 15.1 Plants Reported Possessing Milk-Clotting Proteases

Source	References	Source	References
Actinidia deliciosa (kiwi)	[17, 18]	*Ficus carica*	[10]
Albizzia lebbeck	[19]	*Ficus religiose* (latex)	[20]
Allium fistulosum Linn (spring onions)	[21]	*Foeniculum vulgare* (fennel)	[22]
Allium macranthum Baker (chives)	[21]	*Helianthus annuus* (sunflower seed)	[19]
Allium porrum Linn (garlic sprout)	[21]	*Hemerocallis citrina* Baroni *(day-lily buds)*	[21]
Allium tuberosum Rottl. ex Spreng (leek)	[21]	*Jacaratia corumbensis* O. kuntze	[23]
Ananas Sativa (pineapple)	[24]	*Lactuca sativa* L. cv Romana	[17]
Asparagus officinalis Linn *(asparagus)*	[21]	*Lupines luteus* (yellow lupin)	[25]
Balanites aegyptiaca	[21]	*Malva crispa* Linn *(malva)*	[21]
Brassica alba (white mustard)	[25]	*Moringa oleifera*	[26]
Brassica oleracea Linn (kale)	[21]	*Onopordum turcicum*	[27]
Brassica chinensis Linn (rape)	[21]	*Opuntia ficus-indica*	[28]
Brassica napus (rape),	[25]	*Pedilanthus tithymaloides*	[29]
Brassica nigra (black mustard)	[25]	*Polyporus badius* (mushroom)	[20]
Brassica oleracea Linn (broccoli)	[21]	*Raphanus sativus* (radish)	[25]
Bromelia pinguin	[30]	*Solanum aculeastrum*	[31]
Bromelia hieronymi	[32]	*Solanum aethiopicum*	[31]
Carica papaya (papaya)	[24]	*Solanum anomalum*	[31]
Calotropis gigantea	[33]	*Solanum cerasiferum*	[31]
Calotropis procera (Sodom apple)	[33, 34]	*Solanum dasyphyllum*	[31]
Centaurea calcitrapa	[35]	*Solanum dubium* Fresen	[13, 15, 36–38]

TABLE 15.1 *(Continued)*

Source	References	Source	References
Cicer arietinum (chick pea)	[39]	*Solanum elaeagnifolium Cavanilles*	[14, 16, 40–42]
Coprinus lagopides (mushroom)	[43]	*Solanum indicum*	[31]
Corchorus olitorius (corchorus)	[25]	*Solanum nigrum*	[31]
Coriandrum sativum (coriander)	[25]	*Solanum nodiflorum*	[31]
Cucumis melo var. Reticulatus (melon)	[44]	*Solanum terminale*	[31]
Cyphomandra betacea (tamarillo)	[45]	*Stachys geobombycis (cordysep)*	[21]
Cynara cardunculus L. (cardo)	[46–49]	*Streblus asper*	[50]
Cynara cardunculus var. Sylvestris	[51]	*Silybum marianum*	[52]
Cynara scolymus (globe artichoke)	[53, 54]	*Trigonella foenum*	[25]
Eruca sativa (arugula)	[25]	*Vigna aconitifolia* (moth bean)	[39]
Euphorbia microsciadia	[55]	*Vigna radiata* (green gram)	[39]
Euphorbia neriifolia Linn	[56]	*Vigna unguiculate* (cowpea)	[25]
Euphoria nivulia	[29]	*Whitania coagulans*	[57]
		Zingiber officinale	[21]

a carbohydrate side chain which frequently is a trisaccharide (Figure 15.1). The α-chaconine, α-solanine (potato), α-tomatine, dehydrotomatine (tomato), solamargine, and the solasonine (eggplant) are examples of glycoalkaloids in *Solanum* plants [59, 60]. However, poisoning out barks due to the ingestion of potatoes or tomatoes are rare since the concentration of glycoalkaloids in potato tubers, and tomato fruits are low ($\leq$ 9 mg/100 g potatoes, and $\leq$ 16 mg /100 g green tomatoes) [60]. Typically, it is necessary for an oral dose of 225 to 1000 mg of solanine per kg of body weight to produce pathological effects in animals [61].

The genus *Solanum* is divided into 13 major clades, of which the Leptostemonum clade that involves the spiny solanums is the largest one. The species *elaeagnifolium* is one of the 550 accepted species in the clade Leptostemonum or subgenus *Leptostemonum* [62]. *Solanum elaeagnifolium* is a perennial herb that has an aerial growth (up to one meter) with an extensive root system (up to two meters deep). The flowers of this plant are blue, violet, and rarely white. The leaves are oblong or lanceolate (1 to 10 cm long and 1 to 2.5 cm wide) with entire or lobed margins and a petiole of 0.5 to 2 cm long. The stem is acicular, cylindrical, and sparingly branched with few acicular spines. The unripe fruits are fleshy, spherical (1 to 1.5 cm diameter) and green with white patches, although some plants produce dark brownish fruits. During the fall, the greenish fruits become dry and turn around yellow or orange. The dry fruits are brittle and break easily releasing flat seeds with 2 to 4 mm in diameter (Figure 15.2). It is estimated that a single plant produces up to 60 fruits, and an only fruit may contain 60 to 120 seeds [62–64].

Solanum elaeagnifolium Cavanilles is native to northeast Mexico, southwest United States of America (USA), and Argentina, though nowadays it can be found in Australia, southern India, South America, South Africa, and the Mediterranean [63]. Some familiar names of *S. elaeagnifolium* are silver leaf nightshade, silver leaf nettle, white horse-nettle, Silverleaf bitter apple, trompillo, revienta caballo and meloncillo del campo (Spanish). This plant is considered a weed with a negative impact on crop and livestock production. Farms in the USA have been abandoned due to infestation with *S. elaeagnifolium*. In Morocco, the fields infested with this plant decreased their value by 25% [63, 64]. Silver leaf nightshade is rarely used as cattle feed since it is associated with adverse toxicity [63]. Moderate poisoning symptoms have been reported in cattle fed with ripe berries at 0.1 to 0.3% of their body weight, although, the green unripe berries are less toxic than the yellow ripe fruits [64]. At nightshade, the toxicity associated with *S. elaeagnifolium* is attributed to the presence of steroidal alkaloids. The whole plant, but mostly

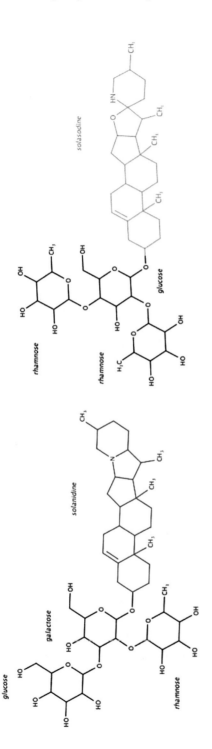

FIGURE 15.1 Structures of two glycoalkaloids reported in *Solanum* plants (solanine: Pubchem CID 9549171; solamargine: Pubchem CID 73611).

the fruits contain glycoalkaloids such as solamargine (Figure 15.1) [65] and solanine (Figure 15.1) [64]. This plant also has a large concentration of solasodine (aglycon of solamargine; Figure 15.1) with 1.0 to 2.15 mg per gram of dry weight (DW).

FIGURE 15.2   *Solanum elaeagnifolium* var. Cavanilles: A) flowers, B) leaves, C) stem, D) unripe fruits, E) dark brownish fruits, F) ripe fruits, G) seeds.

Beyond the negative impact of silverleaf nightshade on agriculture and cattle breeding, this plant has positive uses. Glycoalkaloids in *S. elaeagnifolium,* as well as their corresponding aglycons, have a steroid structure which can be used in the synthesis of commercial drugs. Similar *Solanum* plants like *S. laciniatum* and *S. khasianum* are used to extract solasodine for the synthesis of cortisone and progesterone [65]. Furthermore, both glycoalkaloids contained in *S. elaeagnifolium* (solasonine and solamargine) have been reported as inhibitors of human cancer cells from the colon, prostate, breast, and cervix [66]. The glycoalkaloids from *S. elaeagnifolium* could also be used as bio-herbicides or anti-microbial agents since these are effective against certain herbivores and pathogenic microorganisms [67]. The fruits of silverleaf nightshade also contain valuable compounds such as flavonoids, saponins, triterpenes, and tanines. Some of these compounds have anti-tumoral activity on breast tumor explants. According to Hernandez et al. [66] the methanolic extracts from the fruits of *S. elaeagnifolium* had quinic acid, chlorogenic acid, and dicaffeoylquinic acid which reduced the viability of tumoral cell lines Vero, HeLa, and MCF-7.

Similarly, Hawas et al. [67] have reported that the methanolic extract obtained from the flowering aerial parts of *S. elaeagnifolium* contains kaemp-ferol 8-C-β-galactoside and other similar glycosidic flavonoids. These flavo-noids showed a hepatoprotective effect against histopathological and histo-chemical damage induced in mice by paracetamol in the liver. On the other hand, the seeds in the fruits of *S. elaeagnifolium* have a rich variety of fatty acids (FAs) and sterols with potential commercial value. According to Feki et al. [68] the fatty acid composition of the seed oil from *S. elaeagnifolium* comprises linoleic acid (LA) (67.59%), oleic (16.70%), palimitic (8.52%), stearic (3.65%) and myristic (1.36%). Meanwhile, the sterol fraction consisted of β-sitosterol, campesterol, stigmasterol, and Δ_5-avenasterol.

Table 15.2 shows the proximate composition of dried yellow fruits of *S. elaeagnifolium*. When the fruits are harvested in the late fall, the mois-ture in the fruits decreases sharply. At this point, the fruits are mostly seeds protected by the dry and brittle peel (Figure 15.2); consequently, the fiber and lipid content are high (Table 15.2). The dried fruits also have an elevated content of proteins with 12 g per 100 g of dried fruits. Among the proteins enclosed in the fruits, there are some proteases with milk-clotting activity, which are used in the manufacture of cheese. Therefore, the fruits of *S. elaeagnifolium* can be a profitable source of proteases for the dairy industry as mentioned in the next section.

15.4 THE MILK-CLOTTING ENZYME FROM *S. ELAEAGNIFOLIUM*

In northern Mexico, native Pimas used the fruits of *Solanum elaeagnifolium* (commonly called trompillo) to make different types of cheese. The main cheese made with the fruits of *Solanum elaeagnifolium* is the Asadero cheese, which is a typical Mexican cheese similar to Oaxaca cheese. This cheese is a fresh, stretched curd cheese that is produced by artisans using raw cow's milk. In Chihuahua, Mexico, the Asadero cheese is produced and consumed throughout the state. The natives of the northern zone of Chihuahua use the ripe yellow fruits of *S. elaeagnifolium* as a milk coagulant [14, 16]. The fruits of *S. elaeagnifolium* have been used for cheese making, since the begin-ning of the last century. In 1916 and 1924, Bodansky [40, 71] reported the coagulant activity on milk and the conditions in which the enzymes present in the *Solanum* fruit act. Only the fruit of yellow color (mature) presents proteolytic activity, unlike the fruits of green color (not ripe) or brown (overripe). This is probably since, in the immature fruit, the proteases are inactive, and in the overripe fruits, they could be degraded [14].

TABLE 15.2 Proximate Composition of Dried (Yellow) Fruits from *Solanum Elaeagnifolium*

Component	Content	Mean Value	References
	(g/100 g Dried Fruits)	(g/100 g Dried Fruits)	
Moisture	26.34	10.54 ± 10.5	[41]
	4.9		[14]
	5.8		[69]
	5.14		[70]
Ash	4.12	5.63 ± 1.0	[41]
	6.7		[14]
	5.8		[69]
	5.9		[70]
Protein	4.5	12.0 ± 5.1	[41]
	16.2		[14]
	13.8		[69]
	13.5		[70]
Total fat	4.8	3.45 ± 2.0	[41]
	0.5		[14]
	5.0		[69]
	3.5		[70]
Total carbohydrates	60.2	68.25 ± 5.4	[41]
	71.5		[14]
	69.6		[69]
	71.7		[70]
Total fiber	51.7	48.4 ± 10.4	[14]
	56.8		[69]
	36.7		[70]
Soluble fiber	2.9	2.9	[69]
Insoluble fiber	53.8	53.8	[69]

S. elaeagnifolium is a wild plant that grows in the northern region of Mexico, and it is considered as a weed. This plant has in its fruits proteases that present a general proteolytic activity, which is useful in the traditional process of making Asadero cheese. Several authors have reported the use of plant-derived proteases, such as Cardosins from *Cynara* sp., in the elaboration of different types of cheeses and showed that the sensory attributes of the cheeses are similar or better than the cheeses made with commercial rennet [54, 72–75]. Cardosins from *Cynara* sp. share biochemical characteristics with chymosin. Both, the chymosin and the vegetable coagulant break the Phe_{105}–Met_{106} binding of κ-casein, site of action of chymosin; however, vegetable coagulants are more proteolytic and, because they do not have a specific cleavage site, they have higher activity on β- and κ-caseins [76, 77].

Typically, the aqueous extract of trompillo's fruits is used as a milk-clotting agent in the manufacture of Asadero cheese. Macerating the dried fruits with water (ratio 1:10) and a small portion of salt (~5%) obtain the crude aqueous extract which is used as a rennet substitute [14, 16]. According to Ramírez-Vigil and Gutiérrez-Méndez [69], the crude extract from the fruits of *S. elaeagnifolium* had a proteolytic activity of 1.52 *U-Tyr* mL^{-1} and specific proteolytic activity of 0.78 *U-Tyr* mg^{-1} of protein. In comparison, chymosin had lower proteolytic activity (0.29 *U-Tyr* mL^{-1}) but higher specific activity (2.51 *U-Tyr* mg^{-1} of protein). On the other hand, this plant-derived coagulant has an impressive ability to clot the milk (978.36 *U* mL^{-1}); though this milk-clotting activity (*U*) was not as good as observed in chymosin (2,285.70 *U* mL^{-1}). Overall, the aqueous crude extract from the fruits of *S. elaeagnifolium* has a high ratio (1,184.4 *U/U-Tyr*) of MCA/PA [69]. The higher the MCA/PA ratio, the better the use of coagulant for cheese making; since clotting time is shorter, and the volume of coagulant needed is lower [3, 78]. Even though the MCA/PA ratio of the plant-derived coagulant from *S. elaeagnifolium* is lower than observed in chymosin (8,278.5 *U/U-Tyr*), this ratio is higher than reported for other plant coagulants like Neriifolin S, papain, trypsin, ficin, and religiosin (433, 367, 3.6, 393 and 387 *U*/OD 660 nm) [78, 79]. Therefore, proteases from the berries of *S. elaeagnifolium* are suitable for cheese making.

In the electrophoretogram of Figure 15.3, the bands of the proteins present in the fruit of *S. elaeagnifolium* are shown. The molecular weights estimated are 67, 55, 45, 39, 36, 23, and 18 kDa; Up to 8 bands with molecular weights ranging from 22 to 64 kDa have also been reported [16]. Some of these identified bands with similar molecular weight as reported for aspartic proteinases. Greenberg and Winnick [80] identified the main proteins of *S. elaeagnifolium*, with milk-clotting properties, as *Solanain* in 1940. However, years later it was found that the *Solanain* shared structural homology with cucumisin, a serine protease extracted from *Cucumis melo* [81], having an estimated molecular mass of 66 kDa.

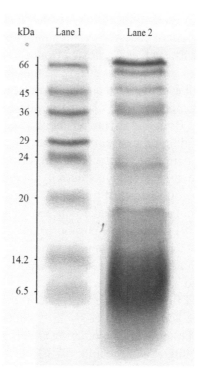

FIGURE 15.3 SDS-PAGE of proteins obtained from ripe yellow fruits of *S. elaeagnifolium*. Lane 1, molecular weight markers; lane 2 proteins extracted from the fruits of *S. elaeagnifolium*.

Vargas-Requena et al. [82] identified the milk-clotting protease in the fruits of *S. elaeagnifolium* as cucumisin type proteases (EC.3.4.21.25). Chavez-Garay et al. [14] have suggested that proteases from *S. elaeagnifolium* could be SPs with a molecular weight between 50 to 70 kDa. In contrast, Gutiérrez-Mendez et al. [16] have stated that milk-clotting proteases in the fruits of the trompillo plant could be an aspartic protease-like chymosin and those reported in the flowers of *Cynara cardunculus*. The APs present in their active site two residues of aspartic acid, which are responsible for the catalytic activity. It is needed an acidic pH to activate the enzyme, and it is inhibited especially by pepstatin. These enzymes show a specific preference for cleaving peptide bonds from hydrophobic amino acid residues [4]. The vegetable APs known up to now have been isolated and characterized from leaves, flowers, and seeds of different species [7].

The protein profiles of casein hydrolysates by *S. elaeagnifolium* are different from those produced by chymosin (Figure 15.4). Chymosin and *S. elaeagnifolium* protein extract show a proteolytic activity mainly on κ-casein, in addition to β-casein. The native proteins are hydrolyzed in fragments of lower molecular weight in both cases, but mainly with *S. elaeagnifolium* enzymes. However, *S. elaeagnifolium* has an extensive proteolytic capacity over κ-casein, since it eliminates it and is no longer present in cheese. Unlike the treatment with chymosin, that still has a small portion of κ-casein without degrading. The enzymatic extract of *S. elaeagnifolium* cleaves κ-casein in the same peptide bond as chymosin, since some of the polypeptides generated coincide in relative mobility, being different in concentration (optical density). Chazarra et al. [54] found something similar when comparing the effect of an artichoke flower extract (*Cynara scolymus, L.*) with calf stomach rennet.

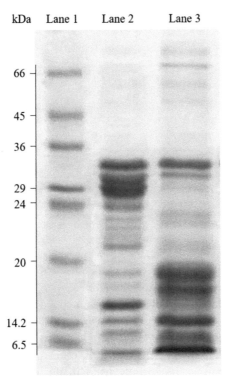

FIGURE 15.4 SDS PAGE of protein profiles of Asadero cheese proteins made with chymosin (lane 2) or enzymatic extract of *S. elaeagnifolium* (lane 3).

Regarding the proteolytic capacity on the β-casein, the *S. elaeagnifolium* enzymes degrade approximately half of that with chymosin; whereas on α-casein, the activity is not considered important. The main products of casein hydrolysis by proteases of *S. elaeagnifolium* are distributed below 20 kDa, while those of chymosin is less than 15 kDa. These polypeptides differ in number and size between the products of the two enzymes. When the effect of enzymes of *S. elaeagnifolium* on milk caseins is analyzed separately, it is observed that it has extensive hydrolysis on β-casein and on κ-casein. However, it has no activity on α-casein (Figure 15.5).

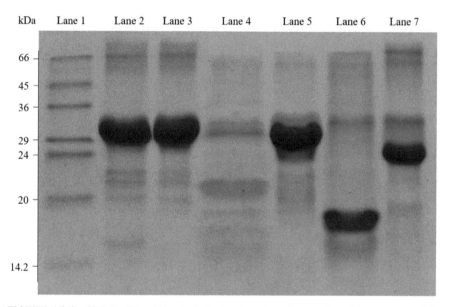

FIGURE 15.5   SDS-PAGE of the hydrolysis of α, β, and κ-casein with enzymes of *S. elaeagnifolium*, Lane 1, molecular weight markers; lane 2 hydrolyzed α-casein; lane 3: α-casein; lane 4: hydrolyzed β-casein; lane 5: β-casein; lane 6: hydrolyzed κ-casein; 7: κ-casein.

15.5   CONCLUDING REMARKS

Solanum elaeagnifolium is a plant considered as a weed by farmers and ranchers. Nevertheless, the fruits of this plant enclose a wide variety of compounds that can be potentially used in the pharmaceutical and dairy industry. Proteases from the ripe yellow fruits of *S. elaeagnifolium* have been used for almost a century in the manufacture of cheese in the northern region of Mexico. Unlike other plant-derived proteases, those obtained from *S. elaeagnifolium* have a high ratio of MCA/PA, making it ideal for cheese making. Notwithstanding the recent studies on proteases from the fruits of *S. elaeagnifolium*, the protease responsible for the milk-clotting activity has not been entirely identified and characterized. On the other hand, and as Figure 15.4 shows, proteases from *S. elaeagnifolium* hydrolyze caseins in a different way that chymosin does. Therefore, the plant-derived coagulant from *S. elaeagnifolium* could also be used to produce bioactive peptides from milk proteins and to modify the cheese texture. Furthermore, proteases from trompillo's fruits could also be used in the meat industry. According to Carballo-Carballo and Leal-Ramos [70], the proteases from *S. elaeagnifolium* have collagenolytic activity and can reduce the hardness of meat with high levels of collagen (e.g., the meat from cull cows).

KEYWORDS

- **milk-clotting enzymes**
- **plant-derived coagulants**
- **serine proteases**
- **Silverleaf**
- ***Solanum elaeagnifolium***
- **Trompillo**

REFERENCES

1. Walstra, P., Wouters, J. T. M., & Geurst, T. M. *Dairy Science and Technology*, (2nd edn.). CRC press: Boca Raton FL, **2006**.
2. Fox, P. F., McSweeney, P. L. H., Cogan, T. M., & Guinee, T. P. *Cheese: Chemistry, Physics and Microbiology: General Aspects*, (vol. 1). Elsevier Academic Press: San Diego, **2004**.
3. Horne, D. S., & Banks, J. M. Rennet-Induced coagulation of milk. In: Fox, P. F., McSweeney, P. L. H., Cogan, T. M., & Guinee, T. P. (eds.), *Cheese Chemistry, Physics And Microbiology* (vol. 1, pp. 47–70). Elsevier Academic Press: San Diego, **2004**, General.
4. González-Rábade, N., Badillo-Corona, J. A., Aranda-Barradas, J. S., & Oliver-Salvador, M. del C. Production of Plant Proteases in Vivo and in Vitro – A Review. *Biotechnol. Adv.*, **2011**, *29*(6), 983–996.
5. van der Hoorn, R. A. L. Plant Proteases: From phenotypes to molecular mechanisms. *Annu. Rev. Plant Biol.*, **2008**, *59*(1), 191–223.
6. Crabbe, M. J. C. Rennets: General and molecular aspects. In *Cheese: Chemistry, Physics and Microbiology* (vol. 1, pp. 19–45). Elsevier Academic Press: San Diego, **2004**.
7. Shah, M. A., Mir, S. A., & Paray, M. A. Plant proteases as milk-clotting enzymes in cheesemaking: A review. *Dairy Sci. Technol.*, **2014**, *94*(1), 5–16.

8. Jacob, M., Jaros, D., & Rohm, H. Recent advances in milk clotting enzymes. *Int. J. Dairy Technol.,* **2011,** *64*(1), 14–33.

9. Sun, Q., Zhang, B., Yan, Q. J., & Jiang, Z. Q. Comparative analysis on the distribution of protease activities among fruits and vegetable resources. *Food Chem.,* **2016,** *213*, 708–713.

10. Trani, A. Enzymes applications for the dairy industry. *Advances in Dairy Products,* **2017,** No. November.

11. Nongonierma, A. B., & FitzGerald, R. J. Enzymes exogenous to milk in dairy technology. proteinases. In: Fuquay, J. W. (ed.), *Encyclopedia of Dairy Sciences* (2nd edn., pp. 289–296), **2011,** San Diego: Academic Press.

12. Conceição, C., Martins, P., Alvarenga, N., Dias, J., Lamy, E., Garrido, L., Gomes, S., Freitas, S., Belo, A., Brás, T., et al. *Cynara cardunculus*: Use in cheese making and pharmaceutical applications. In: Koca, N., (ed.), *Technological Approaches for Novel Applications in Dairy Processing* (pp. 73–105). Intech Open, **2018.**

13. Ahmed, I. A. M., Morishina, I., Babiker, E. E., Mori, N., Morishima, I., Babiker, E. E., & Mori, N. Characterization of partially purified milk-clotting enzyme from *Solanum dubium* fresen seeds. *Food Chem.,* **2009,** *116*(2), 395–400.

14. Chávez-Garay, D. R., Gutiérrez-Méndez, N., Valenzuela-Soto, M. E., & García-Triana, A. Partial Characterization of a plant coagulant obtained from the berries of *Solanum elaeagnifolium. CYTA – J. Food,* **2016,** *14*(2).

15. Ahmed, I. A. M., Morishima, I., Babiker, E. E., Mori, & Dubiumin, N., A chymotrypsin-like serine protease from the seeds of *Solanum dubium* fresen. *Phytochemistry,* **2009,** *70*(4), 483–491.

16. Gutiérrez-Méndez, N., Chávez-Garay, D. R., & Jiménez-Campos, H. Exploring the milk-clotting properties of a plant coagulant from the berries of *S. elaeagni folium* var. *cavanilles. J. Food Sci.,* **2012,** *71*(1), 89–94.

17. Lo Piero, A. R., Puglisi, I., & Petrone, G. Characterization of "Lettucine": A Serine-like protease from lactuca sativa leaves, as a novel enzyme for milk clotting. *J. Agric. Food Chem.,* **2002,** *50*(8), 2439–2443.

18. Lo Piero, A. R., Puglisi, I., & Petrone, G. Characterization of the purified actinidin as a plant coagulant of bovine milk. *Eur. Food Res. Technol.,* **2011,** *233*(3), 517–524.

19. Egito, A. S., Girardet, J. M., Laguna, L. E., Poirson, C., Mollé, D., Miclo, L., Humbert, G., & Gaillard, J. L. Milk-clotting activity of enzyme extracts from sunflower and albizia seeds and specific hydrolysis of bovine κ-casein. *Int. Dairy J.,* **2007,** *17*(7), 816–825.

20. Shabani, R., Shahidi, S. A., & Rafe, A. Rheological and structural properties of enzyme-induced gelation of milk proteins by ficin and polyporus badius. *Food Sci. Nutr.,* **2018,** *6*(2), 287–294.

21. Sun, Q., Zhang, B., Yan, Q. J., & Jiang, Z. Q. Comparative analysis on the distribution of protease activities among fruits and vegetable resources. *Food Chem.,* **2016,** *213*, 708–713.

22. Bey, N., Debbebi, H., Abidi, F., Marzouki, M. N., & Ben Salah, A. The non-edible parts of Fennel (*Fœniculum vulgare*) as a new milk-clotting protease source. *Ind. Crops Prod.,* **2018,** *112*, 181–187.

23. Duarte, A. R., Duarte, D. M. R., Moreira, K. A., Cavalcanti, M. T. H.; de Lima-Filho, J. L., & Porto, A. L. F. Jacaratia Corumbensis O. kuntze a new vegetable source for milk-clotting enzymes. *Brazilian Arch. Biol. Technol.,* **2009,** *52*(1), 1–9.

24. Martin, H. Proteinase activities of kiwifruit, pineapple and papaya using ovalbumin, soy protein, casein and bovine serum albumin as substrates. *J. Food Nutr. Res.,* **2017,** *5*(4), 214–225.

25. Elmazar, M. M. E., El-Sayed, S. T., & Al-Azzouny, R. A. Screening some local Egyptian seeds extract for milk-clotting activity and physicochemical characterization of *Brassica napus* seed extract. *J. Agric. Food. Tech.,* **2012**, *2*(2), 28–34.

26. Pontual, E. V., Carvalho, B. E. A.; Bezerra, R. S., Coelho, L. C. B. B., Napoleão, T. H., & Paiva, P. M. G. Caseinolytic and milk-clotting activities from *Moringa oleifera* flowers. *Food Chem.,* **2012**, *135*(3), 1848–1854.

27. Tamer, I. M. Identification and partial purification of a novel milk clotting enzyme from *Onopordum turcicum. Biotechnol. Lett.,* **1993**, *15*(4), 427–432.

28. Teixeira, G., Santana, A. R., Salomé Pais, M., & Clemente, A. Enzymes of *Opuntia ficus-indica* (l.) miller with potential industrial applications-I. *Appl. Biochem. Biotechnol. – Part A Enzym. Eng. Biotechnol.,* **2000**, *88*(1-3), 299–312.

29. Mahajan, R. T., & Chaudhari, G. M. Plant latex as vegetable source for milk clotting enzymes and their use in cheese preparation. *Int. J. Adv. Res. Journalwww.journalijar.com Int. J. Adv. Res.,* **2014**, *2*(5), 1173–1181.

30. Moreno-Hernández, J. M., Bañuelos Pérez, M. J., Osuna-Ruiz, I., Salazar-Leyva, J. A., Ramirez-Suarez, J. C., & Mazorra-Manzano, M. Á. Exploring the milk-clotting properties of extracts from bromelia pinguin fruit. *J. Microbiol. Biotechnol. Food Sci.,* **2017**, *7*(1).

31. Guiama, V. D., Libouga, D. G., Ngah, E., & Mbofung, C. M. Milk-clotting activity of berries extracts from nine solanum plants. *African J. Biotechnol.,* **2010**, *9*(25), 3911–3918.

32. Bruno, M. A., Lazza, C. M., Errasti, M. E., López, L. M. I., Caffini, N. O., & Pardo, M. F. Milk clotting and proteolytic activity of an enzyme preparation from bromelia hieronymy fruits. *LWT – Food Sci. Technol.,* **2010**, *43*, 695–701.

33. Anusha, R., Singh, M. K., & Bindhu, O. S. Characterization of potential milk coagulants from *Calotropis gigan* tea plant parts and their hydrolytic pattern of bovine casein. *Eur. Food Res. Technol.,* **2014**.

34. Oseni, O. A., & Ekperigin, M. M. Distribution of Proteolytic and Milk Clotting Enzymes in the Plant of Sodom Apple Calotropis Procera (Ait.) R. Br. (*Asclepiadaceae*). *J. Biotechnol. Res.,* **2013**, *1*(2), 24–27.

35. Domingos, A., Cardoso, P. C., Xue, Z. T., Clemente, A., Brodelius, P. E., & Pais, M. S., Purification, Cloning and autoproteolytic processing of an aspartic proteinase from centaurea calcitrapa. *Eur. J. Biochem.,* **2000**, *267*(23), 6824–6831.

36. Yousif, B. H., McMahon, D. J., & Shammet, K. M. Milk-clotting enzyme from solanum dobium plant. *International Dairy Journal,* **1996**.

37. Kheir, S. E. O., El Owni, O. A. O., & Abdalla, M. O. M. Comparison of quality of sudanese white cheese (Gibna Bayda) manufactured with solanum dubium fruit extract and rennet. *Pakistan J. Nutr.,* **2011**, *10*(2), 106–111.

38. Roskov, Y., Abucay, L., Orell, T., Nicolson, D., Bailly, N., Kirk, P. M., Bourgoin, T., DeWalt, R. E., Decock, W., DeWever, A., et al. *Species 2000 & ITIS Catalogue of Life, 2018 Annual Checklist.* Leiden, the Netherlands, **2018**.

39. Bajulge, P. V., Bhutia, K. C., Awasarmol, A. V., Shaikh, F. K., Panche, A. N., & Harke, S. N. Evaluation of caseinolytic, gelatinolytic and milk-clotting activities from germinated seeds of green gram (*Vigna radiata*), Chick Pea (*Cicer arietinum*), Fenugreek (*Trigonellafoenum-graecum*) and Moth Bean (*Vigna aconitifolia*). *Int. Food Res. J.,* **2018**.

40. Bodansky, A. A study of a milk-coagulating enzyme of *Solanum elaeagnifolium. J. Biol. Chem.,* **1924**, *61*(2), 365–375.

41. Aquino-Favela, A., López-Díaz, J. A., Vargas-Requena, C. L., & Martínez-Ruiz, N. R. Manufacture of asadero cheese with a vegetable rennet from *Solanum elaeagnifolium*. In *3rd International congress Food Science and Food Biotechnology in Developing Countries,* (pp 76–80), Queretaro, **2008**.

42. Martínez-Ruiz, N. R., & López-Díaz, J. A. Optimización de la extracción y estandarización de un cuajo vegetal para la elaboración de queso asadero (Extraction Optimization and Standarization of a Plant Coagulant to Manufacture of Asadero Cheese). *Cienc. en la Front. Rev. Cienc. y Tecnol. la UACJ,* **2008**, *6*, 173–176.

43. Shamtsyan, M., Dmitriyeva, T., Kolesnikov, B., & Denisova, N. Novel milk-clotting enzyme produced by coprinus lagopides basidial mushroom. *LWT – Food Sci. Technol.,* **2014**, *58*(2), 343–347.

44. Gagaoua, M., Ziane, F., Nait Rabah, S., Boucherba, N., Ait Kaki El-Hadef El-Okki, A., Bouanane-Darenfed, A., & Hafid, K. Three phase partitioning, a scalable method for the purification and recovery of cucumisin, a milk-clotting enzyme, from the juice of *Cucumis melo* var. *reticulatus*. *Int. J. Biol. Macromol.,* **2017**, *102*, 515–525.

45. Li, Z., Scott, K., Hemar, Y., & Otter, D. Protease activity of enzyme extracts from tamarillo fruit and their specific hydrolysis of bovine caseins. *Food Res. Int.,* **2018**.

46. Agboola, S., Chen, S., & Zhao, J. Formation of bitter peptides during ripening of ovine milk cheese made with different coagulants. *Dairy Sci. Technol.,* **2004**, *84*(6), 567–578.

47. Pimentel, C., Van Der Straeten, D., Pires, E., Faro, C., & Rodrigues Pousada, C. Characterization and expression analysis of the aspartic protease gene family of *Cynara cardunculus* L. *FEBS J.,* **2007**, *274*(10), 2523–2539.

48. Almeida, C. M., & Simões, I. Cardoon-based rennets for cheese production. *Appl. Microbiol. Biotechnol.,* **2018**.

49. Ordiales, E., Martín, A., Benito, M. J., Hernández, A., Ruiz-moyano, S., Córdoba, M. D. G., & Ruiz-Moyan, S. Technological characterisation by free zone capillary electrophoresis (FCZE) of the vegetable rennet (*Cynara cardunculus*) used in 'Torta Del Casar' cheese-making. *Food Chem.,* **2012**, *133*(1), 227–235.

50. Ishak, R., Idris, Y. M. A., Mustafa, S., Sipat, A., Muhammad, S. K. S., & Manap, M. Y. A. Factors affecting milk coagulating activities of kesinai (*Streblus asper*) extract. *Int. J. Dairy Sci.,* **2006**, *1*(2), 131–135.

51. Ben Amira, A., Mokni, A., Yaich, H., Chaabouni, M., Besbes, S., Blecker, C., & Attia, H. Technological properties of milk gels produced by chymosin and wild cardoon rennet optimized by response surface methodology. *Food Chem.,* **2017**, *237*, 150–158.

52. Vairo-Cavalli, S., Claver, S., Priolo, N., & Natalucci, C. Extraction and partial characterization of a coagulant preparation from *Silybum marianum* flowers. Its action on bovine caseinate. *J. Dairy Res.,* **2005**, *72*(3), 271–275.

53. Llorente, B. E., Brutti, C. B., & Caffini, N. O. Purification and characterization of a milk-clotting aspartic proteinase from globe artichoke (*Cynara scolymus* L.). *J. Agric. Food Chem.,* **2004**, *52*(26), 8182–8189.

54. Chazarra, S., Sidrach, L., López-Molina, D., & Rodríguez-López, J. N. Characterization of the milk-clotting properties of extracts from artichoke (*Cynara scolymus* L.) flowers. *Int. Dairy J.,* **2007**, *17*(12), 1393–1400.

55. Rezanejad, H., Karbalaei-Heidari, H. R., Rezaei, S., & Yousefi, R. Microsciadin, a new milk-clotting cysteine protease from an endemic species, euphorbia microsciadia. *Biomacromol. J. Biomacromolecular J.,* **2015**, *1*(1), 93–103.

56. Yadav, R. P., Patel, A. K., Jagannadham, M. V. & Neriifolin, S. A dimeric serine protease from euphorbia *Neriifolia* linn.: Purification and biochemical characterisation. *Food Chem.*, **2012**, *132*(3), 1296–1304.

57. Nawaz, M. A., Masud, T., & Sammi, S. Quality evaluation of mozzarella cheese made from buffalo milk by using paneer booti (Withania Coagulans) and calf rennet. *Int. J. Dairy Technol.*, **2011**, *64*(2), 218–226.

58. Weese, T. L., & Bohs, L. A Three-gene phylogeny of the genus solanum (*Solanaceae*). *Systematic Botany,* **2017**, 445–463.

59. Müller, J. L. Love Potions and the ointment of witches: Historical aspects of the nightshade alkaloids. *J. Toxicol. Clin. Toxicol.,* **1998**, *36*(6), 617–627.

60. Barceloux, D. G. Potatoes, tomatoes, and solanine toxicity (Solanum Tuberosum L., Solanum Lycopersicum L.). *Disease-a-Month,* **2009**, *55*(6), 391–402.

61. Crocco, S. Potato sprouts and greening potatoes-potential toxic reaction. *Jama-Journal Am. Med. Assoc.* **1981**, *245*(6), 625.

62. Knapp, S., Sagona, E., Carbonell, A. K. Z., & Chiarini, F. A Revision of the solanum elaeagnifolium clade (*Elaeagnifolium Clade; Subgenus Leptostemonum, Solanaceae*). *Phyto Keys* **2017**, (84), 1–104.

63. OEPP. Solanum elaeagnifolium. *OEPP/EPPO Bull.* **2007**, *37*, 236–245.

64. Boyd, J. W., Murray, D. S., & Tyrl, R. J. Silverleaf nightshade, *Solanum elaeagnifolium*, origin, distribution, and relation to man. *Econ. Bot.,* **1984**, *38*(2), 210–217.

65. Nigra, H. M., Caso, O. H., & Giulietti, A. M. Production of solasodine by calli from different parts of *Solanum eleganifolium* cav. plants. *Plant Cell Rep.,* **1987**, *6*(2), 135–137.

66. Hernázndez, L., Carranza, P., Cobos, L., López, L., Ascasio, J., & Silva, Y. Bioguided fractionation from *Solanum elaeagnifolium* to evaluate toxicity on cellular lines and. *Rev. LA Fac. CIENCIAS Farm. Y Aliment.,* **2017**, *24*(2), 124–131.

67. Hawas Usama W; M, S. G., Abou, E. K. L. T., H, F. A. R., Khaled, M., & Francisco, L. A new flavonoid c-glycoside from solanum elaeagnifolium with hepatoprotective and curative activities against paracetamol- induced liver injury in mice. *Zeitschrift für Naturforschung C*, (p. 19), **2013**.

68. Feki, H., Koubaa, I., Jaber, H., Makni, J., & Damak, M. Characteristics and chemical composition of *Solanum elaeagnifolium* seed oil. *J. Eng. Appl. Sci.(Asian Res. Publ. Netw.),* **2013**, *8*(9), 708–712.

69. Ramírez-Vigil, P., & Gutiérrez-Méndez, N. *Aislamiento y Clasificación de Una Proteasa Extraída de Los Frutos de Solanum elaeagnifolium Cavanilles* (Isolation and Classification of the protease extracted from the fruits of *Solanum elaeagnifolium* Cavanilles), Universidad Autónoma de Chihuahua: Chihuahua, Chihuahua, México, **2016**.

70. Carballo-Carballo, D., & Leal-Ramos, M. Y. Evaluación de La Actividad Proteolítica de Las Bayas de Trompillo (*Solanum elaeagnifolium* Var. Cavanilles) Sobre La Terneza y Propiedades Físico-Químicas de La Carne de Vacas Viejas, (Assessment of Proteolytic Activity from Trompillo's Fruits (*Solanum elaeagnifolium* Cavanilles) on Hardness and Physico-Chemical Properies of Cull Cow Meat. *Universidad Autónoma de Chihuahua,* **2015**.

71. Bodansky, A. The Chymase of *Solanum elaeagnifolium* a preliminary note. *J. Biol. Chem.,* **1916**, *27*(1), 103–105.

72. Esteves, C. L. C., Lucey, J. A., Hyslop, D. B., & Pires, E. M. V. Effect of gelation temperature on the properties of skim milk gels made from plant coagulants and chymosin. *Int. Dairy J.,* **2003**.

73. Silva, S. V., Pihlanto, A., & Malcata, F. X. Bioactive peptides in ovine and caprine cheese-like systems prepared with proteases from cynara cardunculus. *J. Dairy Sci.,* **2006**.

74. Galán, E., Prados, F., Pino, A., Tejada, L., & Fernández-Salguero, J. Influence of different amounts of vegetable coagulant from cardoon *Cynara cardunculus* and calf rennet on the proteolysis and sensory characteristics of cheeses made with sheep milk. *Int. Dairy J.,* **2008**.

75. Martínez, R. N. D. R., Enriquez, S. F., Vázquez, N. R. E., & López, D. J. A. Microbiological quality of asadero cheese manufactured with a plant based coagulant from *Solanum elaeagnifolium. Food Nutr. Sci.,* **2013**.

76. Macedo, I. Q., Faro, C. J., & Pires, E. M. Specificity and kinetics of the milk-clotting enzyme from cardoon (*Cynara cardunculus* L.) toward Bovine κ-Casein. *J. Agric. Food Chem.,* **1993**.

77. Macedo, I. Q., Faro, C. J., & Pires, E. M. caseinolytic specificity of cardosin, an aspartic protease from the cardoon *Cynara cardunculus* l.: Action on bovine α_s – and β-casein and comparison with chymosin. *J. Agric. Food Chem.,* **1996**.

78. Yadav, R. P., Patel, A. K., & Jagannadham, M. V. Neriifolin S, a dimeric serine protease from Euphorbia *Neriifolia* Linn.: Purification and biochemical characterisation. *Food Chem.,* **2012**, *132*(3), 1296–1304.

79. Kumari, M., Sharma, A., & Jagannadham, M. V. Decolorization of crude latex by activated charcoal, purification and physico-chemical characterization of religiosin, a milk-clotting serine protease from the latex of *Ficus religiosa. J. Agric. Food Chem.,* **2010**, *58*(13), 8027–8034.

80. Greenberg, D. M., & Winnick, T. Plant proteases. I. activation-inhibition reactions. II. PH-activity curves. III. kinetic properties. *J. Biol. Chem.,* **1940**, *135*, 761–787.

81. Uchikoba, T., & Kaneda, M. Milk-clotting activity of cucumisin, a plant serine protease from melon fruit. *Appl. Biochem. Biotechnol. – Part A Enzym. Eng. Biotechnol.,* **1996**.

82. Vargas-Requena, C. L., Hernández-Santoyo, A., Jiménez-Vega, F., & López-Díaz, J. A. Purificación y Caracterización Parcial de Una Proteasa Tipo Cucumisina Presente En *Solanum elaeagnifolium* (Purification and Partial Caracterization of Cucumisin-like Protease from *Solanum elaeagnifolium*). *Cienc. en la Front. Rev. Cienc. y Tecnol. la UACJ,* **2008**, *6*, 161–164.

CHAPTER 16

Pomegranate (*Punica granatum* L.) Nutritional and Functional Properties

GERARDO MANUEL GONZÁLEZ-GONZÁLEZ,[1]
JESUS ANTONIO MORLETT-CHÁVEZ,[1]
ADRIANA CAROLINA FLORES-GALLEGOS,[1]
JUAN ALBERTO ASCACIO-VALDÉS,[1]
SANDRA CECILIA ESPARZA-GONZÁLEZ,[2] and
RAÚL RODRÍGUEZ-HERRERA[1]

*[1]School of Chemistry, Autonomous University of Coahuila,
25280 Saltillo, Coahuila, México,
E-mail: raul.rodriguez@uadec.edu.mx (R. Rodríguez-Herrera)*

*[2]School of Medicine, Autonomous University of Coahuila, 25000 Saltillo,
Coahuila, México*

ABSTRACT

Pomegranate (*Punica granatum*) is a plant known for its different beneficial health and therapeutic properties; also, this plant can grow on different types of soils and climates. Pomegranate fruit is considerate of high value, because of its antioxidant, anticancer, antimicrobial, antifungal, anti-parasitic properties, among others. The study of this plant is important because it contributes to the quality of life and prevention of diverse types of diseases such as the different effects mentioned above. Pomegranate has the same importance in the Middle East as well as in western countries, where the increase of cultivation area, utilization, and rational consumption of this fruit is recommended. Therefore, in this chapter, the nutritional and functional properties of pomegranate are discussed (*Punica granatum* l.).

16.1   INTRODUCTION

Pomegranate (*Punica granatum* L.) belongs to the Lythraceae family, since ancient time, it has been considered a plant with components that can confer benefits to human health [1]. This plant is mentioned since biblical times, and it was familiar to the Hebrews, who thought that this fruit was the apple of Eden's garden. One of the pillars of King Salomon's building had pomegranate as decoration. In Babylon, the pomegranate was an agent of resurrection, while the Persians believed that the pomegranate seeds conferred the battlefield's victory. In classical mythology, Persephone was forced to live with Hades because she ate the seed of pomegranate. Also, in Christianity, the pomegranate symbolizes longevity; while in China it also symbolizes longevity. Also, Turkish legends comment that a woman could predict the number of her children, depending on the number of seeds that fell on the ground [2].

16.2   POMEGRANATE ORIGIN

Pomegranate is native of northern India, the Himalayas, and the Mediterranean. However, this plant has been cultivated in regions like Asia, Europe, and North Africa [3]. It was introduced in Center, South, and Southwest of Asia and Indonesia in 1416, While in Latin America, Arizona, and California; it was introduced by the Spanish in 1796. Pomegranate is cultivated in some regions of Africa. The most important growing regions are Egypt, China, Afghanistan, Turkey, Syria Pakistan, Bangladesh, Iran, Iraq, and Saudi Arabia [2].

16.3   POMEGRANATE ADAPTATION

The pomegranate is a shrub or small tree that can reach 6 to 7 meters in height, with dark grayish bark. It grows under climatic conditions with no extreme or subtropical temperatures. In regions of cold winters and hot summers, trees can grow easily in regions of annual mean temperature between 20–24C, tolerating temperatures up to –11°C. It is capable of surviving semi desert's region with 500–1000 mm of rain [4]. Pomegranate leaves are coriaceous, green bright and oblong. Its flowers grow at the tips of the branches, are attractive large scarlet red or white flowers, bisexual, up to 4 cm (centimeter) wide. In addition, its flowers are characterized by crowned frets which have persistent reddish or violaceus chalice with six triangular and persistent lobes, six petals, widely obovate, wrinkles alternating with the sepal lobes, numerous stamens inserted into the flower tube, its center is editable with a bright red, and new layers. The flowers are terminals or axillaries and grow alone [2]. Pomegranate has adaptation to almost all types of soils which include types where other fruits trees are not able to grow, such as alkaline soils, with gravel and humid or acid locations. For commercial cultivation, well-drained, heavy, or light soils are recommended to support the plant under dry seasons. Irrigation is recommended to maintain high yields in dry areas [5].

16.4 USES AND EDIBLE PARTS

The fruit is known to have an astringent flavor, as well as a bittersweet taste. The pulp is refreshing, and it is sectioned in chambers that group grains with seeds surrounded by pulp, which are edible. The fruit is also consumed as a juice when placed in lemonades or drinks like wine [2]. In some regions of India, its juice is very popular in beverages, to give a sweet taste; also, it is used to produce jelly or grenadine syrup which is used on cocktails, drinks, and wines mixtures.

In Saudi Arabia, the juice is frozen or concentrated on using in different foods such as desserts, drinks, or meals. Also, it is processed to obtain a jelly by adding them pectin and sugar, the seeds, and the grains are used to spreads foods [4]. In India, it is used for desserts, and for softening different types of meat, because of its proteolytic enzymes. The dried pomegranate is used as a condiment, as well as, granola bars, ice creams, yogurt, and salad dressing. Also, pomegranate dried, and ground seeds are used as an acid condiment to prepare curries and avoid being trapped between the teeth [2].

In Turkey, pomegranate is used for salads and garnish of desserts such as the güllac, and in sauces for dressing and/or marinating meat. In Greece, it is used to prepare Kolliva, which is a mixture of seeds and sugar. In Iran, a traditional recipe "fesenjan" is made of a thick pomegranate sauce and ground nuts used to flavor duck and chicken, and flavor a popular ash-e nar sauce soup. In Azerbaiyán, is used for Narsarab sauce [2].

16.5 FRUIT NUTRIENTS

The competition of the fruit varies depending on the location of cultivation, crop management, water availability, and weather, among other things. The raw fruit is 44% of peel and membranes. A pomegranate fruit of 100 g of the Wonderful cultivar contains 77.93 g of water, 1.67 g of protein, 1.17 g of lipid acids, 4 g of fiber, 13.67 g of total sugars. Minerals content is 10 mg of calcium, 0.30 mg of iron, 12 mg of magnesium, 36 mg of phosphorus, 236 mg of potassium, 3 mg of sodium, 0.35 mg of zinc, 0.182 mg of copper, and 0.119 mg of manganese. Also contains 10.2 mg vitamin C (ascorbic acid), 0.067 mg of thiamine, 0.052 mg of riboflavin, 0.29 mg of niacin, 0.37 mg of pantothenic acid, 0.075 mg of vitamin B-6, 0.6 mg of vitamin E (Alpha-tocopherol), 16.4 mg of vitamin K and 0.120 g of fatty acids (FAs) within which are: palmitic acid (PA), lauric acid (LaA), stearic acid, etc. [6]

The pomegranate husk contains on average 14% of moisture, 6.8% of ash, 5.1% of protein, 1.5% of lipid acid, 30.5% of carbohydrate, and 42.61% of fiber [7]. The bottled pomegranate juice; in a 110 g serving, can be found 85.9 g of water, 0.15 g of protein, 0.29 of FAs, 13.13 g of carbohydrates, 0.1 g of dietary fiber and 12.65 g of total sugars [6]. The edible portion of pomegranate is constituted by 52% of the fruit, 78% of juice, and 22% seeds [8]. Seeds are rich in lipids with 27.2%, 13.2% of protein, 35.3% of crude fiber, 2% of ash, in turn containing 6% of pectin and 4.7 of total sugars [6].

16.6 FUNCTIONAL PROPERTIES AND USE IN NATUROPATHIC MEDICINE

Recently, there has been increased consumption of food products with the capacity to improve the immune system, as well as, to prevent different diseases. Jurenka et al. [9] reported the use of pomegranate (husk rind, juice, seeds, seeds oil, leaves, flowers, roots, and bark) for treatment of diseases, because of its antioxidant, anticancer, and anti-inflammatory activities. Pomegranate roots and bark are traditionally used for intestinal parasite treatment; also, the dried fruit and pulp are used against diarrhea. Roots are used to treat mainly *Taenia solium* and are considered as astringent and anthelmintic.

Pomegranate husks in ancient Egypt were used to treat various diseases such as diarrhea, intestinal worms, colds, and infertility [10]. In India, husks are sun-dried and traditionally used to cure prophylactic malaria [11], which is attributed to its biocompounds such as ellagitannins (ellagic acid and its glycosides). In addition, husk extracts with different solvents (Methanol, ethanol, water, acetone, etc.) have been related to anticancer, antioxidant, anti-inflammatory [12], and antimicrobial activities since they were tested against different microorganisms: *E. coli, Klebsiella pneumoniae, Staphylococcus aureus, Rhizopusoryzae*, etc. These extracts were obtained using as solvents, chloroform, water, ethanol, and acetone, among others [13–15]. Seeds are used to gargle, as well as for treatment of continuous diarrhea, pomegranate seeds are commonly used in traditional medicine in India. In the same way, its flowers are used on the treatment of diarrhea [16].

Pomegranate juice extract has a significant anti-inflammatory [17], antiproliferative [18, 19], anti-tumor properties [17, 20] and potential photo chemopreventive [17]. Fruit extracts are used against cancer. Acetonic extracts from fruits were found to be an inhibitory compound for the ability to induce tumors TPA and help chemopreventive activity in a tumoriferous model of wild and wild-type cells [17]. Because of the different chemical compounds of the fruit, it has been associated with potential therapeutic proprieties, and authors such as Holland and Bar-ya [21] related this fruit with prevention of diverse types of cancers (skin, lung, prostate, and colon), treatment of diabetes, diseases cardiovascular, spills, diarrhea, and dysentery cough and fever.

16.7 CHEMICAL COMPOUNDS OF INTEREST

Most plants produce different chemical compounds (phytochemicals) that offer beneficial effects for their health, although they are not essential nutrients. These compounds protect to plant against external hazards, such as exposure to ultraviolet light, or pathogenic microorganisms [22].

The daily consumption of different phytochemicals is associated with a decrease of risk to different diseases such as different types of cancer, inflammatory, cardiovascular, and neurodegenerative problems. These compounds are present in various parts of the plant (seeds, leaves, husk, etc.) [23]. It has been observed that the daily intake of phytochemicals by humans has a chemopreventive function, as well as inhibiting the conversion of normal cells to cancer cells.

The pomegranate has phenolic compounds, anthocyanins, and vitamin C, which provide a therapeutic effect [24]. It also has a high nutritional value, because it has acids, sugars, proteins, vitamins, polysaccharides, phenols, and minerals [25], remarking its content of proteins, sugars, and minerals [26].

The phytochemical composition of pomegranate consists of at least 124 different compounds [27], among which stand out: procyanidins, anthocyanins, flavanols, organic acids, terpenes, alkaloid, and different acids [5, 9, 12, 22, 28]. Some of these bioactive compounds (BCs) have a high market value such as those included in the family of ellagitannins, punicalagins, anthocyanins, and flavanols [27]. For the extraction of phytochemicals, different solvents have been used such as ethanol, methanol, and acetone-water, chloroform, etc. [29–31].

The most remarkable biological activity of pomegranate phytochemicals is their great antioxidant activity, because of the presence of a high content of polyphenols in fruit, leaves, and bark [12, 22]. Ellagitannins are polyphenols with biological activity; the whole fruit is a source of different ellagitannins [32]. It is known that juice has been used for centuries for different medicinal purposes [33]. Heber [34] said that scientific advances attributed to ellagitannins anti-cancer, anti-inflammatory, angiogenesis, and cell proliferation properties.

While, flavonoids (Figure 16.1) have a structure like the estrogens of mammals. These molecules, also called phytoestrogens, have been associated with low rates of hormone-dependent cancers, such as breast or prostate cancers.

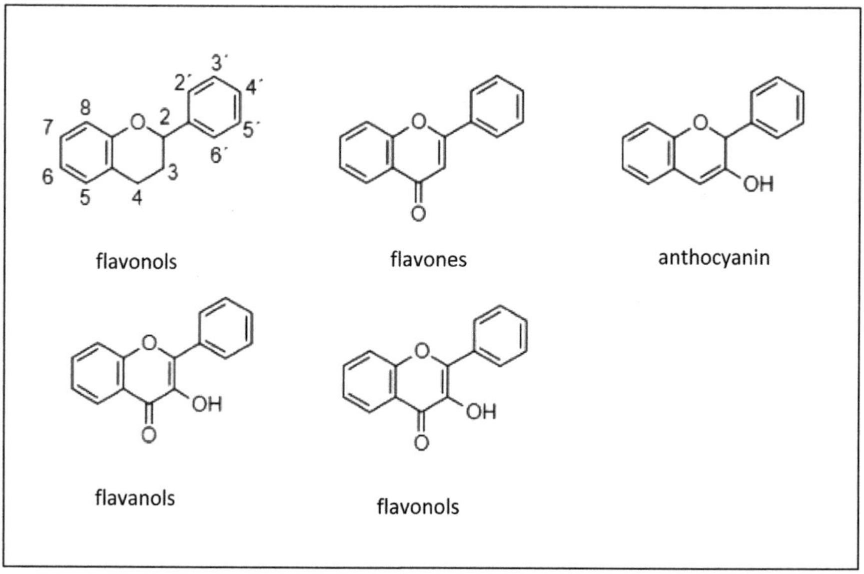

FIGURE 16.1   Structure and type of flavonoids.

16.8   PREVENTIVE USES

Pomegranate has been associated with therapeutic use against different diseases over time by various cultures and civilizations. Use for cancer treatment and prevention [35, 36], antiviral [37], antifungal *in vitro* and *in vivo* [38, 39], and antimicrobial [40, 41]. It is also used against Alzheimer's [42], cardiac treatments [43], diabetes [44], arthritis [45], malaria [46] and protective fibroblasts in skin irradiated with ultraviolet [47].

Researchers reported that polyphenols have a beneficial effect on health and preventive against diseases. It is shown that free radicals stimulate or execrate diseases such as arthritis, cancer, Alzheimer's diabetic complications and complication of Parkinson's [48].

16.9   ANTIOXIDANT PROPERTIES

Humans have developed many applications for this fruit, ranging for fresh consumption, treatment of different diseases; it also has been used during different religious traditions, throughout history. That is why scientists have been interested in this fruit in recent years.

Bassiri et al. [39] mentioned that pomegranate husk contains a significant number of phytochemicals with high antioxidant power; this is due to the activity of ellagic acid, which is the main polyphenol of this fruit [49]. Recent research with extracts of this fruit or its purified compounds reports antioxidant, antiproliferative, anti-invasive, anti-angiogenic, and apoptotic *in vitro* effects [35, 50]. The methanol extracts of this fruit contained 44% of polyphenolic compounds with the presence of gallic acid and quercetin as the major compounds. Singh et al. [51] reported the first antioxidant property, and that the molecules within the husk contain ten times greater activity than that found in the fruit pulp [29]. Extracts of pomegranate husk have the activity of cleaning excessive amounts of hydroxyl radicals and superoxides [52].

Studies in murine animals with liver damage induced by CCl_4 showed that pomegranate husk extracts decreased by 54%, lipid peroxidation values compared to the control group [53]. Another study in murine showed that husk consumption offers a protective activity against tetrachloride carbon. Like pulp extracts, it inhibits the activity of metalloproteinase expression, as does IL-1 beta, which induces tissue destruction [9]

16.10   ANTI-CANCER ACTIVITY

After the discovery of the antioxidant and curative activity of pomegranate, other applications were studied (Table 16.1), identifying the compounds that could help for the prevention of cancer. Cancer is a destructive disease with the ability to create metastases in other body organs [54]. Heber [34] carried out studies on cells and animals and was able to determine that metabolites of the ellagitannins family within pomegranate juice are those that have *in vitro* activity to inhibit the growth of prostate cancer cells and in mice inoculated with human prostate cancer cells. Similarly, purification of extracts has a moderate anti-inflammatory effect and can generate cell apoptosis [17, 55].

TABLE 16.1   Studies with Pomegranate Tissues on Diverse Types of Cancer

Cell Type	Fruit Tissue	References
Colon	Seed oil, juice, and fruit	[19, 56–58]
Lungs	Fruit and leaves	[19, 20, 59]
Breast	Fruit, purified extracts, seed oil	[60–62]
Skin	Fruit, pulp seed oil	[63–67]

Diverse groups of researchers have shown that pomegranate extracts have the antitumor capacity on several types of cells (oral, colorectal, and breast cancer lines), as well as the activity of repairing oxidative damage in rats [18, 33, 35, 40, 49, 68, 69]. In the same way, using pomegranate extracts, obtained cell apoptotic and inhibitory results on cancerous lines of leukemia, and processes associated with the cell cycle [70]. The same compounds of pomegranate extracts have an *in vitro* preventive activity against growth of breast, prostate, colon, and lung cancer; as well as inhibiting the growth of these cancers in preclinical studies using animals [70]. Other studies with pomegranate seed oil have shown prevention of proliferation of several types of tumor cells [65, 66] and reduction of murine skin and breast tumors [61, 67].

16.10.1   SKIN CANCER

The human skin is the first line of defense against heat, sunlight, and different infections [71]. The ellagic acid, as well as, the punicalagins is considered an efficient natural anti-carcinogenic so efficient that start apoptosis to suppress the growth of cancer cells. Pomegranate extracts contain among their molecules ellagic acid, which in mice exposed to UV light and with consumption of pomegranate extracts was observed that prevent edema in the skin peroxidation and helps to maintain stable DNA [72]. Studies with melanomas using pomegranate peel extract, showed a chemoprotective effect for keratocyte cells [54].

It has been observed that pomegranate pulp promotes an antitumor effect against tetradecanephorbol 13-acetate (TPA) which is a tumor promoter [54, 67]. Also, in tests using murine animals inoculated which diverse types of skin cancer treated with pomegranate oils, it was observed that the carcinogenic capacity in those murine was diminished [28, 65]. It is already known, that apoptosis is a protective mechanism against cancer [73], *in vivo* investigations showed that 5% of pomegranate seed oils considerably reduce expression of TPA, which is a molecule that promotes spread of skin cancer [67], demonstrating that various pomegranate tissues have chemopreventive properties against skin cancer and different adverse effects.

16.10.2 PROSTATE CANCER

Prostate cancer is the second leading cause of death in men [74]. The anticancer capacity of pomegranate extracts has been tested in cell culture studies and animal models [18, 65, 75]. *In vitro* studies have shown that pomegranate extracts and ellagitannins regulate expression of the pro-apoptotic genes called Bax and Bak in prostate tumor cells. In addition, these compounds decrease expression of cyclin D1, D2, and E [68, 76], as well as, increase expression of caspases; such as caspase 3, which promotes apoptosis of cancer cells [49]. Studies on animals indicated that oral administration of pomegranate extracts inhibit tumor grows better than antigen specifics to cancer [35]. The first assay in patients with prostate cancer was performed with pomegranate extract treatment, which can also increase expansion of certain specific prostate antigens [59].

Experiments with prostate cancer cells reported inhibition of cancer cell progression, because of low expression levels of IL-6, IL-12p40, IL-1 beta, etc. [77]. In another study, Hong et al. [78], found that pomegranate polyphenols delay expression of genes liked to the production of androgens, as well as, the receptors, which were found in prostate tumor cells. Other researchers have looked for proteins that are activated when the extract is placed, finding the presence of different proteins, which regulate protozoal, anti-apoptosis or angiogenesis activity [74].

16.10.3 BREAST CANCER

Breast cancer is associated with the second cause of death in woman, in all regions of the world [79], only in 2012 about 1.6 million new cases were diagnosticated [80]. Several studies found that in pomegranate juice there were components such as luteolin, ellagitannins, and punic acid, which decreased tumors, and migration, without affecting healthy cells [81], these compounds also have an activity of apoptosis [60, 82–84], cytotoxicity [85] and proliferation [28, 62].

It is also known that pomegranate ellagitannins can inhibit the proliferation of the anti-aromatase pathway in MCF-7 cells, which is responsible for the production of new tumors in the human body. In addition, there is evidence of cytotoxicity of pomegranate extracts, which suppresses *in vitro* the cell proliferation and production of caspase-3 [86]. Another study showed those division steps where extracts attack cells are during the G0/G1 stage in the MCF-7 and MDAMB-231 lines. Also, expression of growth factors and proinflammatory cytokines is increased, which reduces the dose dependence [87]. *In vivo* studies with murine model's tumors of breast cancer, showed a marked anti-proliferative and pro-apoptotic activity [88], which is an indication that pomegranate compounds can confer anticancer effects against diverse types of breast cancer [65].

16.10.4   LUNG CANCER

Lung cancer is leading cancer in the classification of tumors, representing a severe public health problem [33]. According to statistics, 185,000 deaths are due to lung and bronchial tumors in the United States [89]. Cancer is divided into two types of non-small cells (with a higher incidence in advanced cases) [10], including adenocarcinomas, squamous cell carcinomas, large cell carcinomas and small cell cancer [12].

New therapies against cancer are directed as inhibitors or with antibodies [90], but most of these therapies are usually very limited, or resistance to them is generated, which leads to painful side effects. Various studies have found more biological activities of pomegranate; as inhibitory properties for *in vitro* proliferation of A549 cells [82, 91], in treatments directed towards non-small cell cancer, these cells were treated with pomegranate extracts (reducing expression of metalloproteinases, decreasing radicals oxygen-free and potential of the mitochondria) [19, 20, 92].

Several authors reported that pomegranate leaves extracts in contact with lung carcinoma cells alter their growth cell cycle, inducing apoptosis and impede migration to other parts [20, 92], helping as chemopreventive and chemotherapeutic against lung cancer [93].

16.11   FUTURE PERSPECTIVES AND CONCLUSIONS

Currently, cancer continues to be a big issue around the word because, still, the number of victims is very high, which is a concern for all people and promote the searching for the newest solutions. There is evidence of the beneficial properties that pomegranate can confer to human health; this fruit may be a healthy alternative against this disease. Also, the *in vitro* studies revealed that pomegranate juice compounds resist the process of human digestion without changing their structure. Even though, more studies are needed to determine the importance of these molecules in human health. As well as standardizing and optimizing the process to extract the pomegranate compounds safely for the environment, cytotoxic analyzes to all the pomegranate compounds, in order to choose the best compounds to regulate the possible metabolic pathways involved in the processes of cell apoptosis, as well as establishing the route of administration, using studies of pharmacokinetics, pharmacovigilance, and among others.

ACKNOWLEDGMENTS

GMGG thanks the National Council of Science and Technology of Mexico (CONACYT) for the financial support provided during his MSc studies under the scholarship agreement number 861000 Financial support was received from SAGARPA-CONACYT through the project: "Obtención, purificación y escalado de compuestos de extractos bioactivos con valor industrial, obtenidos usando tecnología avanzadas de extracción y a partir de cultivos, subproductos y recursos naturales poco valorados" SAGARPA-CONACYT 2015-4-266936.

CONFLICT OF INTEREST

The authors express no conflict of interest.

KEYWORDS

- **cancer**
- **husk**
- **phenols**
- **pomegranate**
- **tetradecanephorbol 13-acetate**

REFERENCES

1. Shilikina, I. A. On the xylem anatomy of the genus. *Punica L. Bot. Z.,* **1973**, *58*, 1628–1630.
2. Lim, T. K. *Edible Medicinal and Non-Medicinal Plant, 4. Fruits.* Dordrecht, The Netherlands. Springer, **2012**.https://doi.org/10.1007/978-94-007-4053-2 (Accessed on 7 January 2020).
3. CBI (Centre for the Promotion of Imports from Developing Countries). Product Fact Sheet: fresh pomegranates in the European market. **2014** http://www.cbi.eu/market-information/fresh-fruit-vegetables/pomegranates/Europe (Accessed on 7 January 2020).
4. Cocuzza, G. E. M. &, Mazzeo, G. Pomegranate arthropod pests and their management in the Mediterranean area. *Phytoparasitica,* **2016**, *44*(3), 393–409.
5. Teixeira da Silva, J. A., Ranac, T. S., Narzary, D., Verma, N., Meshramf, D. T., Shirish, A., & Ranade, S. A. Pomegranate biology and biotechnology: A review. *Scientia Horticulturae,* **2013**, *160*, 85–107.
6. U.S. Department of Agriculture, Agricultural Research Service (USDA, 2012) USDA National Nutrient Database for Standard Reference, Release 25. Nutrient Data Laboratory Home Page, http://www.ars.usda. gov/ba/bhnrc/ndl (Accessed on 7 January 2020).
7. Pathak, P. D., Mandavgane, S. A., & Kulkarni, B. D. Valorization of pomegranate peels: A biorefinery approach. *Waste and Biomass Valorization,* **2017**, *8*(4), 1127–1137.
8. El-Nemr, S. E., Ismail, I. A., & Ragab, M. Chemical com- position of juice and seeds of pomegranate fruit. *Nahrung,* **1990**, *34*(7), 601–606.
9. Jurenka, M. J. Therapeutic applications of pomegranate (*Punica granatum* L.): A review. *Altern. Med. Rev.,* **2008**, *13*(2), 128–144.
10. Ismail, T., Sestili, P., & Akhtar, S. Pomegranate peel and fruit extracts: A review of potential anti-inflammatory and anti-infective effects. *J. Ethnopharmacol.,* **2012**, *143*,397–405.
11. Bhattacharya, D. *Punica granatum* as a human use, wide-spectrum prophylactic against malaria and viral diseases in India. *Am. Soc. Trop. Med. Hyg.,* **2004**, *71*, 288.
12. Lansky, E. P., & Newman, R. A. *Punica granatum* (pomegranate) and its potential for prevention and treatment of inflammation and cancer. *J Ethnopharmacol,* **2007**, *109*(2), 177–206.
13. Roy, S., & Lingampeta, P. Solid wastes of fruits peels as source of low cost broad spectrum natural antimicrobial compounds— furanone, furfural and benezenetriol. *Int. J. Res. Eng. Technol.,* **2014**, *3*, 273–279.
14. Dahham, S. S., Ali, M. N.,Tabassum, H., & Khan, M. Studies on antibacterial and antifungal activity of pomegranate (*Punica granatum* L). *Am. Eur. J. Agric. Environ. Sci.,* **2010**, *9*, 273–281.
15. Nuamsetti, T., Dechayuenyong, P., & Tantipaibulvut, S. Antibacterial activity of pomegranate fruit peels and arils. *Science Asia,* **2012**, *38*, 319–322.
16. Singh, S., Pandey, P., & Kumar, S. *Traditional Knowledge on the Medicinal Plants of Ayurveda.* CIMAP, Lucknow, India, **2000**.
17. Afaq, F.,Saleem, M., Krueger, C. G., Reed, J. D., & Mukhtar, H. Anthocyanin- and hydrolysable tannin-rich pomegranate fruit extract modulates MAPK and NF-κB pathways and inhibits skin tumorigenesis in CD-1 mice. *Int. J. Cancer,* **2005**, *113*, 423–433.
18. Malik, A., Afaq, F., Sarfaraz, S., Adhami, V. M., Syed, D. N., & Mukhtar, H. Pomegranate fruit juice for chemoprevention and chemotherapy of prostate cancer. *Proc Natl Acad. Sci.,* **2005**, *102*, 14813–14818.

19. Khan, N.,Hadi, N.,Afaq, F., Syed, D. N.,Kweon, M. H., & Mukhtar, H. Pomegranate fruit extract inhibits prosurvival pathways in human A549 lung carcinoma cells and tumor growth in athymic nude mice. *Carcinogenesis,* **2007**, *28,* 163–173.
20. Khan, N., Afaq, F., Kweon, M. H., Kim, K., & Mukhtar, H. Oral consumption of pomegranate fruit extract inhibits growth and progression of primary lung tumors in mice. *Cancer Res.,* **2007**, *67*(7), 3475–3482.
21. Holland, D., & Bar-Ya'akov, I. Pomegranate: Aspects concerning dynamics of health beneficial phytochemicals and therapeutic properties with respect to the tree cultivar and the environment. In: Yaniv Z., & Dudai N. (eds) Medicinal and aromatic plants of the Middle East. *Medicinal and Aromatic Plants of the World,* (vol. 2, pp. 225–239). Springer, Dordrecht. **2014**.
22. Seeram, N. P., Zhang, Y., & Reed, J. D. *Pomegranate Phytochemicals.* In: Seeram, N. P., Schulman, R. N., & Heber, D., (eds.), *Pomegranates: Ancient Roots to Modern Medicine* (pp. 3–29). Taylor & Francis. New York, **2006**.
23. Cerda, B., Ceron, J. L., Tomas-Barberan, F. A., & Espin, J. C. Repeated oral administration of high doses of pomegranate ellagi tannin punicalagin to rats for 37 days is not toxic. *J. Agric. Food. Chem.* **2003**, *51,* 3493–3501.
24. Pearez-Vicente, A. P., Gil-Izquierdo, A., & Garcia-Viguera, C. *In vitro* gastrointestinal digestion study of pomegranate juice phenolic compounds, anthocyanins, and vitamin C. *J Agric. Food Chem.*, **2002**, *50*(8), 2308–2312.
25. Al-Maiman, S. A., & Ahmad, D. Changes in physical and chemical properties during pomegranate (*Punica granatum* L.) fruit maturation. *Food Chem.,* **2002**, *76,* 437–441.
26. Elfalleh, W., Hannachi, H., Guetat, A., Tlili, N., Guasmi, F., Ferchichi, A., & Ying, M. Storage protein and amino acid contents of Tunisian and Chinese pomegranate (*Punica granatum*L.) cultivars. *Genetic. Resour. Crop. Evol.,* **2012**, *59,* 999–1014.
27. Gundala, S. R., Robinson, M. H., & Aneja R. Phytocomplexity: The key to rational chemoprevention. In: Ullah M., & Ahmad A., (eds.), *Critical Dietary Factors in Cancer Chemoprevention* (pp. 39–87). Springer, Cham, **2016**.
28. Viuda-Martos, M., Fernández-López, J., & Pérez-Álvarez, J. A. Pomegranate and its many functional components as related to human health: A review. *Compr. Rev. Food Sci. Food Saf.* **2010**, *9,* 635–654.
29. Li, Y.,Guo, C., Yang, J., Wei, J., Xu, J., &Cheng, S. Evaluation of antioxidant properties of pomegranate peel extract in comparison with pomegranate pulp extract. *Food Chem.,* **2006**, *96*(2), 254–260.
30. Bhandary, S. K., Kumari, S. N., Bhat, V. S., Sharmila, K. P., & Bekal, M. P. Preliminary phytochemical screening of various extracts of *Punica granatum* peel, whole fruit and seeds. *Nitte Univ. J. Health Sci.,* **2012**, *2,* 34–38.
31. Wang, Z., Pan, Z., Ma, H., & Atungulu, G. G. Extract of phenolics from pomegranate peels. *Open Food Sci. J.,* **2011**, *5,* 17–25.
32. Clifford, M. N., & Scalbert, A. Ellagitannins-nature, occurrence and dietary burden. *J. Sci. Food Agric.,* **2000**, *80,* 1118–1125.
33. Longtin, R. The pomegranate: Nature's power fruit? *J. Natl Cancer Inst.,* **2003**, *95,* 346–348.
34. Heber, D. Pomegranate. In: Milner, J., & Romagnolo, D., (eds.) *Bioactive Compounds and Cancer. Nutrition and Health* (pp. 725–734). Humana Press, Totowa, NJ, **2010**.
35. Adhami, V. M., Khan, N., & Mukhtar, H. Cancer chemoprevention by pomegranate: Laboratory and clinical evidence. *Nutr Cancer,* **2009**, *61,* 811–815.

36. Lall, R., Syed, D., Adhami, V., Khan, M., & Mukhtar, H. Dietary polyphenols in prevention and treatment of prostate cancer. *Int. J. Mol. Sci,* **2015**, *16*(2), 3350–3376.
37. Reddy, B. U., Mullick, R., Kumar, A., Sudha, G., Srinivasan, N., & Das, S. Small molecule inhibitors of HCV replication from pomegranate. *Sci. Rep.,* **2014**, *4*, 5411.
38. Anibal, P. C., Peixoto, I. T., Foglio, M. A., & Höfling, J. F. Antifungal activity of the ethanolic extracts of *Punica granatum* L. and evaluation of the morphological and structural modifications of its compounds upon the cells of *Candida* spp. *Braz J. Microbiol.* **2013**, *44*, 839–848.
39. Bassiri-Jahromi, S., Pourshafie, M. S., Ardakani, E. M., Ehsani, A. H., Doostkam, A., Katirae, F., Mostafavi, E. *In vivo* comparative evaluation of pomegranate (*Punica granatum*) peel extract as alternative agents for nystatin against oral candidiasis. *Iran J. Med. Sci.,* **2018**, *43*(3), 296–304.
40. Menezes, S. M., Cordeiro, L. N., & Viana, G. S. *Punica granatum* (pomegranate) extract is active against dental plaque. *J. Herb Pharmacother,* **2006**, *6*, 79–92.
41. Abdollahzadeh, S. H., Mashouf, R. Y., Mortazavi, H., Moghaddam, M. H., Roozbahani, N., & Vahedi, M. Antibacterial and antifungal activities of *Punica granatum* peel extracts against oral pathogens. *J. Dent* (Tehran), **2011**, *8*, 1–6.
42. Subash, S., Essa, M. M., Al-Asmi, A., Al-Adawi, S., Vaishnav, R., Braidy, N., Manivasagam, T., & Guillemin, G. J. Pomegranate from Oman alleviates the brain oxidative damage in transgenic mouse model of Alzheimer's disease. *J. Trad Complement Med.,* **2014**, *4*, 232–238.
43. Mohan, M., Patankar, P., Ghadi, P., & Kasture, S. Cardioprotective potential of *Punica granatum* extract in isoproterenol-induced myocardial infarction in Wistar rats. *J. Pharmacol Pharmacother,* **2010**, *1*, 32–37.
44. Middha, S. K., Bhattacharjee, B., Saini, D., Baliga, M. S., Nagaveni, M. B., & Usha, T. Protective role of *Trigonella foenum graceum* extract against oxidative stress in hyperglycemic rats. *Eur. Rev. Med. Pharmacol Sci.,* **2011**, *15*, 427–435.
45. Rasheed, Z., Akhtar, N., & Haqqi, T. M. Pomegranate extract inhibits the interleukin-1β-induced activation of MKK-3, p38α-MAPK and transcription factor RUNX-2 in human osteoarthritis chondrocytes. *Arthritis Res. Ther.,* **2010**, *12*, R195.
46. Dell'Agli, M., Galli, G. V., Corbett, Y.,Taramelli, D., Lucan-toni, L., Habluetzel, A., Maschi, O., Caruso, D., Giavarini, F., Romeo, S., Bhattacharya, D., & Bosisio, E. Antiplasmodial activity of *Punica granatum* L. fruit rind. *J. Ethnopharmacol.,* **2009**, *125*, 279–285.
47. Pacheco-Palencia, L. A.,Noratto, G.,Hingorani, L., Talcott, S. T., & Mertens-Talcott, S. U., Protective effects of standardized pomegranate (*Punica granatum* L.) polyphenolic extract in ultraviolet-irradiated human skin fibroblasts. *J Agric Food Chem.,* **2008**, *56*, 8434–8441.
48. Dell'Agli, M., Galli, G. V., Bulgari, M., Basilico, N., Romeo, S., Bhattacharya, D., Taramelli, D., & Bosisio, E. Ellagitannins of the fruit rind of pomegranate (*Punica granatum*) antagonize in vitro the host inflammatory response mechanisms involved in the onset of malaria. *Malar. J.,* **2010**, *9*, 208.
49. Syed, D. N., Chamcheu, J. C., Adhami, V. M., & Mukhtar, H. Pomegranate extracts and cancer prevention: Molecular and cellular activities. *Anticancer Agents Med Chem.,* **2013**, *13*, 1149–1161.
50. Faria, A., & Calhau, C. The bioactivity of pomegranate: Impact on health and disease. *Crit. Rev. Food Sci. Nutr.,* **2011**, *51*, 626–634.

51. Singh, R., Singh, M. K., Chandra, L. R., Bhat, D., Singh Arora, M., Nailwal, T., & Pande, V. *In vitro* antioxidant and free radical scavenging activity of *Macrotyloma uniflorum* (Gahat dal) from Kumauni region. *Int. J. Fundam. Appl. Sci.*, **2012**, *1*, 9–11.
52. Kaneyuki, T., Noda, Y., Traber, M. G., Mori, A., & Packer, L. Superoxide anion and hydroxyl radical scavenging activities of vegetable extracts measured using electron spin resonance. *IUBMB Life*, **1999**, *47*(6), 979–989.
53. Moneim, A. E. A. Antioxidant activities of *Punica granatum* (pomegranate) peel extract on brain of rats. *J. Med. Plant Res.*, **2012**, *6*, 195–199.
54. Syed, D. N., Afaq, F., & Mukhtar, H. Pomegranate derived products for cancer chemoprevention. *Seminars in Cancer Biology*, **2007**, *17*, 377–385
55. Adams, L. S., Seeram, N. P., Aggarwal, B. B., Takada, Y., Sand, D., & Heber, D. Pomegranate juice, total pomegranate tannins and punicalagin suppress inflammatory cell signaling in colon cancer cells. *J. Agric. Food Chem.*, **2006**, *54*, 980–985.
56. Kohno, H., Suzuki, R.,Yasui, Y., Hosokawa, M., Miyashita, K., & Tanaka, T. Pomegranate seed oil rich in conjugated linolenic acid suppresses chemically induced colon carcinogenesis in rats. *Cancer Sci.*, **2004**, *95*, 481–486.
57. Dana, N., Haghjooy, J. S., & Rafiee, L. Role of peroxisome proliferator-activated receptor alpha and gamma in antiangiogenic effect of pomegranate peel extract. *Ran J Basic Med Sci.*, **2016**, *19*, 106–110.
58. Larrosa, M., González-Sarrías, A., Yáñez-Gascón, M. J., Selma, M. V., Azorín-Ortuño, M., Toti, S., Tomás-Barberán, F., Dolara, P., & Espín, J. C. Anti-inflammatory properties of a pomegranate extract and its metabolite urolithin: A in a colitis rat model and the effect of colon inflammation on phenolic metabolism. *J. Nutr. Biochem.*, **2010**, *21*, 717–725.
59. Pantuck, A. J., Pettaway, C. A.,Dreicer, R.,Corman, J., Katz, A., Ho, A., Aronson, W., Clark, W., Simmons, G., & Heber, D. A randomized, double-blind, placebo-controlled study of the effects of pomegranate extract on rising PSA levels in men following primary therapy for prostate cancer. *Prostate Cancer Prostatic Dis.*, **2015**, *18*, 242–248.
60. Jeune, M. A., Kumi-Diaka, J., & Brown, J. Anticancer activities of pomegranate extracts and genistein in human breast cancer cells. *J. Med. Food*, **2005**, *8*, 469–475.
61. Mehta, R., & Lansky, E. P. Breast cancer chemopreventive properties of pomegranate (*Punica granatum*) fruit extracts in a mouse mammary organ culture. *Eur. J. Cancer Prev.*, **2004**, *134*, 345–348.
62. Shirode, A. B., Bharali, D. J., Nallanthighal, S., Coon, J. K., Mousa, S. A., & Reliene, R. Nanoencapsulation of pomegranate bioactive compounds for breast cancer chemoprevention. *Int J. Medicine.*, **2015**, *10*, 475–484.
63. Li, H., Wang, Z., & Liu, Y. Review in the studies on tannins activity of cancer prevention and anticancer. *Journal of Chinese Medicinal Materials*, **2003**, *26*(6), 444–448.
64. Miyamoto, K. I., Nomura, M., Sasakura, M., Matsui, E., Koshiura, R., Murayama, T., Furukawa, T., Hatano, T., Yoshida, T., & Okuda, T. Antitumor activity of oenothein B, a unique macrocyclic ellagitannin. *Jpn. J. Cancer. Res.*, **1993**, *84*, 99–103.
65. Kim, N. D., Mehta, R., Yu, W., Neeman, I.,Livney, T., Amichay, A., Poirier, D., Nicholls, P., Kirby, A., Jiang, W., & Mansel, R. Chemopreventive and adjuvant therapeutic potential of pomegranate (*Punica granatum*) for human breast cancer. *Breast Cancer Res Treat*, **2002**, *71*, 203–217.
66. Lansky, E. P., Jiang, W., Mo, H., Bravo, L., Froom, P., Yu, W., Harris, N. M., Neeman, I., & Campbell, M. J. Possible synergistic prostate cancer suppression by anatomically discrete pomegranate fractions. *Invest. New Drugs*, **2005**, *23*, 11–20.

67. Hora, J. J., Maydew, E. R., Lansky, E. P., & Dwivedi, C. Chemopreventive effects of pomegranate seed oil on skin tumor development in CD1 mice. *J. Med. Food,* **2003**, *6,* 157–161.

68. Malik, A., & Mukhtar, H. Prostate cancer prevention through pomegranate fruit. *Cell Cycle,* **2006**, *5*, 371–373.

69. Yang, C. S., Landau, J. M., Huang, M. T., Newmark, H. L. Inhibition of carcinogenesis by dietary polyphenolic compounds. *Ann. Rev. Nutr.,* **2001**, *21,* 381–406.

70. Dahlawi, H., Jordan-Mahy, N., Clench, M., McDougall, G. J., & Le Maitre, C. L. Polyphenols are responsible for the proapoptotic properties of pomegranate juice on leukemia cell lines. *Food Sci. Nutr.,* **2013**, *1,* 196–208.

71. Tonk, M., Vilcinskas, A., & Rahnamaeian, M. Insect antimicrobial peptides: potential tools for the prevention of skin cancer, *Appl. Microbiol. Biotechnol.,* **2016**, *100*(17), 7397–7405.

72. Afaq, F., Hafeez, B., Syed, D. N., Kweon, M. H., & Mukhtar, H. Oral feeding of pomegranate fruit extract inhibits early biomarkers of UVB radiation induced carcinogenesis in SKH-1 hairless mouse epidermis. *Photochem. Photobiol.,* **2010**, *86*(6), 1318–1326.

73. Lowe, S. W., & Lin, A.W. Apoptosis in cancer. *Carcinogenesis,* **2000**, *21,* 485–495.

74. Lee, S. T., Wu, Y. L., Chien, L. H., Chen, S. T., Tzeng, Y. K., & Wu, T. F. Proteomic exploration of the impacts of pomegranate fruit juice on the global gene expression of prostate cancer cell. *Proteomics,* **2012**, *12*(21), 3251–3262.

75. Albrecht, M., Jiang, W.,Kumi-Diaka, J., Lansky, E. P.,Gommersall, L. M., Patel, A.,Mansel, R. E.,Neeman, I., Geldof, A. A., & Campbell, M. J. Pomegranate extracts potently suppress proliferation, xenograft growth, and invasion of human prostate cancer cells. *J. Med. Food,* **2004**, *7*, 274–283.

76. Sadik, N. A., & Shaker O. G. Inhibitory effect of a standardized pomegranate fruit extract on Wnt signaling in 1, 2-dimethylhydrazine induced rat colon carcinogenesis, *Digest. Dis. Sci.* **2013**, *58* (9), 2507–2517.

77. Wang, L., Alcon, A., Yuan, H., Ho, J., Li Q. J., & Martins-Green M. Cellular and molecular mechanisms of pomegranate juice-induced anti-metastatic effect on prostate cancer cells. *Integr. Biol.,* **2011**, *3*(7), 742–754.

78. Hong, M. Y., Seeram, N. P., & Heber, D. Pomegranate polyphenols down-regulate expression of androgen-synthesizing genes in human prostate cancer cells overexpressing the androgen receptor, *J. Nutr. Biochem.,* **2008**, *19*(12), 848–855

79. Chen, H. S., Bai, M. H., Zhang T., Li, G. D., & Liu M. Ellagic acid induce cell cycle arrest and apoptosis through TGF-beta/Smad3 signaling pathway in human breast cancer MCF-7 cells. *Int. J. Oncol.* **2015**, *46*(4), 1730–1738.

80. Ferlay, J., Soerjomataram, I., Dikshit, R., Eser, S., Mathers, C., Rebelo, M., Parkin, D. M., Forman, D., & Bray, F. Cancer incidence and mortality worldwide: Sources, methods and major patterns in GLOBOCAN 2012. *Int. J. Cancer,* **2015**, *136*(5), E359–E386.

81. Rocha, A., Wang, L., Penichet, M., & Martins-Green, M. Pomegranate juice and specific components inhibit cell and molecular processes critical for metastasis of breast cancer, *Breast Cancer Res. Treat,* **2012**, *136*(3), 647–658.

82. Modaeinama, S., Abasi, M., Abbasi, M. M., & Jahanban-Esfahlan, R. Anti tumoral properties of *Punica granatum* (Pomegranate) peel extract on different human cancer cells. *Asian Pac. J. Cancer Prev.,* **2015**, *16*(14), 5697–5701.

83. Joseph, M. M., Aravind, S. R., Varghese, S., Mini, S., & Sreelekha, T. T. Evaluation of antioxidant, antitumor and immunomodulatory properties of polysaccharide isolated from fruit rind of *Punica granatum. Mol. Med. Rep.,* **2012**, *52*, 489–496.

84. Dikmen, M., Ozturk, N., & Ozturk, Y. The antioxidant potency of *Punica granatum* L. Fruit peel reduces cell proliferation and induces apoptosis on breast cancer. *J. Med. Food,* **2011,** *14*(12), 1638–1646.

85. Toi, M., Bando, H., Ramachandran, C., Melnick, S. J., Imai, A., Fife, R. S., Carr, R. E., Oikawa, T., & Lansky E. P. Preliminary studies on the anti-angiogenic potential of pomegranate fractions *in vitro* and *in vivo, Angiogenesis,* **2003,** *6*(2), 121–128.

86. Dai, Z., Nair, V., Khan, M., &Ciolino, H. P. Pomegranate extract inhibits the proliferation and viability of MMTV-Wnt-1 mouse mammary cancer stem cells *in vitro, Oncol. Rep.,* **2010,** *24*(4), 1087–1091.

87. Costantini, S., Rusolo, F., De Vito, V., Moccia, S., Picariello, G., Capone, F., Guerriero, E., Castello, G., & Volpe M. G. Potential anti-inflammatory effects of the hydrophilic fraction of pomegranate (*Punica granatum* L.) seed oil on breast cancer cell lines. *Molecules,* **2014,** *19*(6), 8644–8660.

88. Mandal, A., & Bishayee, A. Mechanism of breast cancer preventive action of pomegranate: disruption of estrogen receptor and Wnt/beta-catenin signaling pathways. *Molecules,* **2015,** *20*(12), 22315–22328.

89. Howell, A. B., & D'Souza, D. H. The pomegranate: Effects on bacteria and viruses that influence human health. *Evid Based Complement Alternat Med.,* **2013,** Article ID 606212.

90. Turrini, E., Ferruzzi, L., & Fimognari, C. Potential effects of pomegranate polyphenols in cancer prevention and therapy, *Oxid. Med. Cell. Longev.,* **2015,** Article ID 938475.

91. Seidi, K., Jahanban-Esfahlan, R., Abasi, M., & Abbasi, M. M. Anti tumoral properties of *Punica granatum* (Pomegranate) seed extract in different human cancer cells. *Asian Pac. J. Cancer Prev.,* **2016,** *17*(3), 1119–1122.

92. Li, Y., Yang, F., Zheng, W., Hu, M., Wang, J., Ma, S., Deng, Y., Luo, Y., Ye, T., & Yin, W. *Punica granatum* (pomegranate) leaves extract induces apoptosis through mitochondrial intrinsic pathway and inhibits migration and invasion in non-small cell lung cancer in vitro. *Biomed Pharmacother,* **2016,** *80*, 227–235.

93. Khan, N., Afaq, F., & Mukhtar, H. Cancer chemoprevention through dietary antioxidants: progress and promise. *Antioxid Redox Signal,* **2008,** *10*, 475–510.

CHAPTER 17

Biomass Fractionation to Bio-Based Products in Terms of Biorefinery Concept

MARCELA SOFÍA PINO,[1] LORENA PEDRAZA SEGURA,[2]
ROLANDO ACOSTA,[3] MARÍA EVANGELINA VALLEJOS,[4] ELISA ZANUSO,[5]
ROSA M. RODRÍGUEZ-JASSO,[1,6] HÉCTOR TORIBIO CUAYA,[2]
JAVIER LARRAGOITI KURI,[2] MARÍA CRISTINA AREA,[4]
DEBORA NABARLATZ,[3] and HÉCTOR A. RUIZ[1,6]

[1]*Biorefinery Group, Food Research Department, Faculty of Chemistry Sciences, Autonomous University of Coahuila, 25280, Saltillo, Coahuila, Mexico, Tel.: (+52) 844-416-12-38, E-mail: hector_ruiz_leza@uadec.edu.mx (H. A. Ruiz)*

[2]*Ibero-American University, 01219 CDMX, Mexico*

[3]*Interfase, School of Chemical Engineering, Industrial University of Santander, Bucaramanga, AA678, Santander, Colombia*

[4]*Institute of Materials of Misiones (IMAM), 3300, Posadas, Misiones, Argentina*

[5]*CEB-Centre of Biological Engineering, University of Minho, Campus Gualtar, 4710057 Braga, Portugal*

[6]*Cluster of Bioalcoholes, Mexican Centre for Innovation in Bioenergy (Cemie-Bio), Mexico*

ABSTRACT

Lignocellulosic biomass, fruit, and vegetable waste are nowadays being a potential feedstock for biofuels production and compounds with high added-value with applications in food, energy, and materials industries in terms of biorefinery, impacting in the circular bioeconomy. Therefore, in this chapter, an overview of the status of biorefineries development, fractionation of biomass using hydrothermal and acid pretreatment, purification of ligno-cellulosic residual biomass hydrolysates after pretreatment and high added-value compounds as bioplastics, xylooligosaccharides (XOS), xylitol, and enzymes is presented.

17.1 INTRODUCTION

In recent years the concept of biorefinery has attracted attention, this due to the versatility in the production of high added-value compounds and biofuels, this concept, or philosophy comes from the typical petroleum refinery [1–4]. According to our research group (www.biorefinerygroup.com), the biorefinery concept integrates processes for biomass conversion in terms of sustainability and bioeconomics. Moreover, biorefineries from biomass have evolved from the 1st "1G" (from crops), 2nd "2G" (from agroindustrial residues-lignocellulosic biomass) and 3rd "3G" generation (from aquatic biomass as algae) [5, 6].

Renewable fuels association reported that in the USA, 200 ethanol biorefineries plants are operating in 2016, producing 57.72 billion liters from the following sources: corn starch 95%, sorghum/barley/wheat starch 3%, cellulosic biomass 1%, and food/beverage 1%, these values indicate that corn starch still dominate in the bioethanol biorefineries and that cellulosic biomass (lignocellulosic materials) represents only a small proportion of the total production [7]. In this context, and in terms of a bioethanol-biorefinery 1G, the Ethanol Industry Outlook 2017 (http://www.ethanolrfa.org) reported that 1 bushel (25.40 kg) of corn-processed produces: 10.78 liters of denatured ethanol, 7.48 kg of nutrient co-products as distillers grains for animal feed (rich in protein and crude fiber), 0.29 kg of corn distillers oil (co-product used as a feed ingredient or feedstock for biodiesel production) and 7.71 kg of biogenic carbon dioxide (promising raw material to produce chemicals via hydrogenation of CO_2) [8, 9].

To our knowledge, eleven large-scale commercial biorefineries 2G are operating around the world. For example, (1) Beta renewables in Crescentino, Italy. This plant is based entirely on the PROESA™ technology and Biochemtex, the bioethanol production is from rice straw, wheat straw and Arundo Donax, also this biorefinery plant is totally self-sufficient, producing electricity from lignin; (2) GranBio is the first commercial-scale plant for bioethanol 2G from sugarcane straw and bagasse in Alagoas State in Brazil; (3) Raízen is a company than in partnership with the Iogen Corporation produces cellulosic ethanol (bioethanol 2G) from the by-products of sugarcane (bagasse and straw), this plant is located in Piracicaba – São Paulo State, Brazil; (4) St1 Biofuels located in Kajaani, Finland is the first facility to produce bioethanol 2G using sawdust as a feedstock; (5) POET-DSM Advanced Biofuels (Project LIBERTY) is a commercial-scale cellulosic biorefinery plant in the production of bioethanol 2G using corn waste (corn cobs, leaves, husk, some stalk) as a feedstock in Emmetsburg, Iowa, USA; (6) Synata Bio Inc. formerly Abengoa Bioenergy is a biorefinery facility located in Hugoton, Kansas, USA for the production of ethanol 2G production and electric cogeneration from cellulosic biomass; (7) DuPont biorefinery cellulosic ethanol plant is in Nevada, Iowa, USA. This plant converts corn stover (corn cobs, leaves, and stalks) to ethanol; (8) Alliance Bio-Products, Inc., subsidiary of Alliance BioEnergy Plus, Inc., is a cellulosic ethanol plant in Vero Beach, Florida, USA; (9) Pacific Ethanol Inc., is a cellulosic biorefinery facility located in Madera, California, USA; (10) Enerkem is a new ethanol unit at its facility is in Edmonton, Alberta, Canada; (11) Praj advanced cellulose ethanol in India. Also, there are demonstration-plant projects as the Inbicon that is a Danish company that produces cellulosic ethanol. Moreover, Golden Cheese is a company in California, USA that produces ethanol, whey powder, and animal feed supplements [10, 11].

On the other hand, biorefineries can be classified based on the type conversion process applied: thermochemical platform biorefineries and biochemical platform biorefineries. According to Haro et al. [12], the thermochemical biorefineries uses the route syngas to the product, and the process as pyrolysis, gasification, and combustion using biomass, organic fraction of municipal solid waste as a feedstock to produces biofuels, biochemicals, heat, and electricity [13]. The sugar or biochemical platform biorefinery uses the sugars from lignocellulosic biomass applying enzymatic hydrolysis process and then converted by microorganisms in the production of biofuels, high added value compounds and biochemical [3, 14–16]. Moncada et al. [17] reported that in the biomass there are metabolites, primary (macromolecules - polysaccharides) and secondary (waxes, oils, etc.). In general, terms, sugar or biochemical platform biorefinery includes different process stages as milling, pretreatment, enzymatic hydrolysis, fermentation for the production high added value compounds and biochemicals of the downstream process. Figure 17.1 shows the scheme of sugar/biochemical platform biorefinery using lignocellulosic biomass in the production of high added-value compounds. This chapter gives an overview of the recent research in the production of high added-value compounds as bioplastics, XOS, xylitol, and enzymes in terms of the biorefinery concept and bioeconomy. Also, the fractionation processes as hydrothermal and acid pretreatment are reviewed.

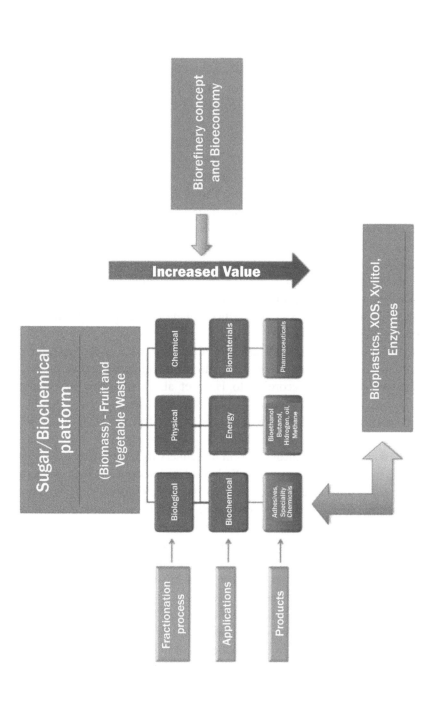

FIGURE 17.1   Scheme of sugar-biochemical platform biorefinery using biomass as a feedstock in the production of high added-value compounds.

Source: Adapted and modified from Ruiz et al. [4].

17.2 FRUIT AND VEGETABLE WASTE IN MEXICAN BIOREFINERY CONTEXT

The global amount of fruit and vegetable waste is estimated at 750 million tons per year [18]. Mexico produces around 76 million tons of agricultural residues each year; besides agricultural and forestry waste, the organic fraction of municipal solid wastes (OFMSW) constitutes an important part of total lignocellulosic biomass [19, 20]. The OFMSW are defined differently by organizations and nations; for example, the European Union (EU) considers food, gardens, and parks waste [21] and Mexico Citys regulation specifies that OFMSW are all biodegradable material [22].

Municipal solid waste is constituted by food waste, garden, and green areas residues and other organic matter, mainly. Its amount and composition vary with country areas, and the urban or rural origin and the organic fraction is estimated at 38%, 70% of OFMSW is from food waste and 24% of garden plants [23]. Nevertheless, other reports consider that organic matter is 52% of MSW and half of this is from food and gardening [20]. In Mexico City solid waste proceeds from homes, markets, stores, and services like restaurants and hotels. Among markets the food supply centers, like the Central de Abasto (CEDA), contribute with huge amounts of vegetal residues. CEDA is the greatest wholesale market; it occupies an area of 327 ha, produces a money flux valuated in 9000 USD per year, from commercial operations, and distributes 35% of fruit and vegetable national production [24].

The CEDA is divided into sections, according to the products offered; agricultural goods are concentrated in two corridors: "Flowers and vegetables" and "Fruits and legumes." Each week 1000 tons of products, including groceries, are received at this supply center, for this reason, produces around 4% of the residues generated by Mexico City (13,000 tons/day), that is, 585 tons/day according to official reports from 2016 [22]. However, the administration of CEDA estimates 10% of total MSW, 1300 tons/day, with 700 tons of vegetables and fruit waste representing a challenge for its management [24]. As well as food waste many fruits and vegetables on good conditions are discarded for producers at the point of sales and are collected for retail or to prepare food [25].

The composition of organic waste from CEDA varies along the year by harvesting time and festivities, so the rate among fruits, vegetables, and flowers changes. Analysis from 2009, performed by the end of the year, shows predominance on lettuce leftovers in the flowers and vegetables corridor meanwhile in fruits and legumes section there was not a predominant product. The data obtained are representative only for the season when samples were collected because heterogeneity and variability of the biomass, frequent analysis must be done.

To explore the potential of organic waste from CEDA at least six samples were collected, in different seasons and years and characterized according to methodology reported [26]. Table 17.1 depicts its average composition, containing more hemicellulose than cellulose; consequently, there will be more pentoses than hexoses if the polysaccharides are hydrolyzed, unlike sugar cane bagasse, a homogenous material, and rich in cellulose. Comparison with other data for MSW from Mexico City (cellulose 21.1%, hemicellulose 5.1%, lignin 13.5%, and protein 15.2%), data for cellulose and hemicellulose content show considerable variation, probably because of the OFMSW from CEDA does not include animal protein and contains variable amounts of stems and leaves from flowers [27]. On the other hand, pectin fraction can be extracted for commercial purposes besides enhancement of enzymatic hydrolysis of lignocellulosic biomass [28].

TABLE 17.1 Average Composition of OFMSW From CEDA [26]

Sample	Sugarcane bagasse	Municipal solid waste
Component	Composition (%) Mean ± SD	
Cellulose	43.89 ± 0.49	19.38 ± 0.15
Hemicellulose	33.38 ± 1.74	35.29 ± 0.53
Acid lignin	8.80 ± 0.81	20.63 ± 0.34
Alkaline lignin	8.86 ± 0.07	7.36 ± 0.38
Ash	4.51 ± 0.13	12.62 ± 0.18
Moisture	49.34 ± 2.14	9.40 ± 0.33

*Standard deviation.

Source: Adapted and modified from Ref. [26].

OFMSW varies in relation with the type and amount of fruit and vegetable present in the waste; Table 17.2 depicts an average composition of some food products and the differences, according to the portion analyzed [29, 30].

OFMWS from food markets as CEDA can be the feedstock for the production of biofuels [31], polyhydroxyalkanoates (PHAs) [32], and lignin derivatives [33] or for a biorefinery concept and biochemical platform, where all the streams can be applied to obtain valuable products like enzymes, bioplastics, nutraceuticals, animal feed, etc. [34].

TABLE 17.2   Fruits and Vegetable Composition in Dry Basis

Sample	Cellulose (%)	Hemicellulose (%)	Lignin (%)
Tomato (dry pomace)	8.60	5.33	2.50
Apple (dry pomace)	8.81	5.44	2.98
Orange peel	11.93	14.46	2.17
Banana peel	11.45	25.52	9.82

Source: Adapted and modified from Refs. [29, 30].

17.3   BIOMASS FRACTIONATION PROCESSES

According to Zanuso et al. [1], the fractionation of biomass for second-generation biorefinery using pretreatment processes is very important since this process allows the separation of the main components of the biomass as cellulose, hemicellulose, and lignin. In the following sections, two types of promising pretreatments are addressed for the concept of biorefinery.

17.3.1   *HYDROTHERMAL PRETREATMENT (AUTOHYDROLYSIS)*

Pretreatment is one of the steps that integrate the biorefinery process. The use of lignocellulosic biomass requires a pretreatment step to disrupt the main components of the material (cellulose, hemicellulose, and lignin) and obtain rich-cellulose solids accessible for the enzymatic hydrolysis as well as high-added products from hemicellulose and lignin. Research on different pretreatments approaches classified them as physical, physic-chemical, chemical, and biological [35]. Among the physic-chemical, hydrothermal pretreatment has been considered a promising technology since the only reagent used during the treatment is water hence reducing overall costs and avoiding equipment corrosion [36]. This pretreatment is also known under different names as auto-hydrolysis, hot compressed water, and liquid hot water [37].

During hydrothermal pretreatment, LCM are treated with water at high temperatures (150–230°C) which entails high pressure into the system leading the water to penetrate into the lignocellulosic biomass. As mention before, the only reagent used is water, and at high temperatures, it acts as autocatalyst by autoionization of hydronium ions (H_3O^+) [5, 38]. As the water penetrates into the material, hemicellulose (xylan, arabinan) is solubilized into the liquid fraction (LF) as monosaccharides and oligosaccharides whereas the solid fraction retains the cellulose and insoluble lignin. Further, the hydrothermal pretreatment can reduce the cellulose crystallinity retained in the solid fraction, which is important for the enzymes as they are more effective in amorphous cellulose [39]. For the case of lignin, it is relocated in the solid fraction during the pretreatment enabling better access for the enzyme to the biomass [5, 40]. Therefore, after hydrothermal pretreatment is obtained a solid rich in cellulose and non-soluble lignin, where a variety of compounds can be found depending on the operational conditions used during the pretreatment. Also, xylooligosaccharides (XOS) are produced in the LF and can be considered as high-added value compounds such as XOS, furfural, 5-hydroxymethylfurfural (HMF), phenolic compounds, acetic acid, and formic acid among others furfural, HMF, phenolic compounds, acetic acid, and formic acid among others in terms of the biorefinery concept. However, these compounds might influence in further enzymatic hydrolysis and fermentation steps as they can act as inhibitors [41]. Hence the importance of hydrothermal pretreatment study to obtain the maximum desired compounds.

Several hydrothermal pretreatment studies have been carried out using agave bagasse as raw material, a residue from the Mexican tequila industry. For instance, Perez-Pimienta et al. [42] evaluated the recalcitrance of the agave bagasse after hydrothermal pretreatment at 180°C for 30 min. They reported a xylose-rich liquid recovered after the pretreatment which can be purified to obtain a high-added value compounds as mention before. Glucan composition increased from 31.2% for untreated agave bagasse to 51.6% (dry basis) after pretreatment. Later, Aguilar et al. [43] compared isothermal and non-isothermal pretreatment using agave bagasse. Results showed the maximum glucan concentration was 56.03% (dry basis) at 180°C for 30 min of residence time, but no significant difference between residence time in the pretreatment process from 20, 30, 40, and 50 min was observed after saccharification.

Hydrothermal pretreatment has demonstrated to be a feasible strategy for fractionation of biomass in terms of biorefinery. The production of cellulose-rich solids leads to efficient enzymatic hydrolysis reducing the enzyme loading thus, the overall process cost in biofuels production, in some cases for the production of ethanol. On the other hand, the LF with the solubilized hemicellulose can be considered to obtain high added value products that contribute to make the process economically viable but also to other paths of ethanol production by fermenting C5 sugars.

17.3.2   *ACID PRETREATMENTS*

Chemical pretreatments break the organic matter by means interaction of polysaccharides and lignin functional group substances with different substances, to hydrolyze the polymers. Alkalis, acids, solvents, and oxidizing agents are the most used substances, besides ionic liquids.

Acid pretreatments can be implemented with concentrated (30–70%) at moderated temperatures (i.e., 40°C) or diluted acids (i.e., 0.1%) at high temperatures (160–200°C) and employing H_2SO_4, HCl, HNO_3, H_3PO_4, acetic, and maleic acids; among this sulfuric acid is commonly used [44]. The principal disadvantage of sulfuric acid pretreatment lies in its corrosive nature. Consequently, process equipment must be resistant to corrosion and it's necessary to neutralize effluent streams. Also, sugars, and lignin react in acid media yielding by-products as furfural, HMF, phenolic compounds, and organic acids that can inhibit equally subsequent enzymatic and fermentative processes [45–47].

17.3.2.1 DISADVANTAGES AND ADVANTAGES OF CONCENTRATED AND DILUTED ACID PRETREATMENTS

As previously mentioned there are two acid pretreatment conditions: concentrated (30–70%) and diluted (<10%); the difference is the energy required to hydrolyze the biomass and inhibitor compounds production. Furthermore, using concentrated acid solutions implies the need for corrosion-resistant equipment, a recovery step, and strict safety rules for the staff. Cellulose hydrolysis by mean concentrated acid pretreatment is an efficient process, but in the case of whole lignocellulosic biomass inhibitors as aldehydes and organic acids are produced. By contrast, diluted acid pretreatment is widely used, since it produces less corrosion of process equipment and acid solutions are less harmful, too. The most significant differences between both methods are the temperature, higher in the diluted acid pretreatment and the solids load [48]. This factor is a crucial process variable, given that technical and economical feasibility is related to the same. Generally, for research purposes, solids loading in lignocellulosic biomass suspensions vary among 1–1% (w/v). Under this condition, it is possible to evaluate acid hydrolysis performance based on reducing sugars release or in subsequent enzymatic hydrolysis and fermentation yields.

Acid pretreatments can be operated under batch, semi-batch, and continuous regimes; then reactor design must be adapted to operating conditions. Time residence plays an important role in deciding on the regime, and it is influenced by factors such as temperature, pressure, acid concentration, solid load, biomass composition, and its reactivity.

Chemical reactions are heterogeneous and are conducted in a two-phase system: a solids suspension of lignocellulosic biomass and an aqueous phase catalyzed by a proton donor, which starts material hydrolysis, increasing its solubility in aqueous media. This phenomenon was corroborated experimentally, concluding that hemicellulose solubilizes in acid media following pseudo-first order kinetics with respect to remaining hemicellulose in solid [49–51]. As a result, different sets of consecutive and irreversible reactions demonstrate that hemicellulose possesses one more reactive area, depending on its structure, rate of monomeric units and the position in the plant cell wall. Lignin reacts in acid media releasing soluble lignin, and amorphous cellulose can be partially depolymerized to glucose.

Sugar (pentoses and hexoses) dehydration reactions lead to furfural and 5-HMF, respectively. Figure 17.2 depicts a diagram with the possible reactions that take place during diluted acid pretreatment of corncob, where final degradation products are not specified.

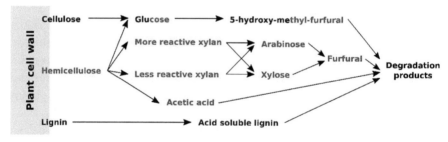

FIGURE 17.2   Possible reactions occurring in diluted acid pretreatment of corncob.

According to specialized literature and in our own experience, if the content of hemicellulose and acid-soluble lignin in lignocellulosic biomass are around 25–32% and 8–12%, respectively, the recommended first step is diluted acid pretreatment, for example, to obtain xylose syrups with low concentrations of inhibitors. On the contrary, obtaining of furfural and 5-HMF requires concentrated acid.

Among lignocellulosic biomass pretreated in acid conditions are agroindustrial, municipal, and forestry waste; operation conditions and differences between untreated and pretreated material composition are shown in Table 17.3.

Frequently, operation temperatures fluctuate from 100 to 160°C with diluted acid, contrary to the process at high concentration, with a maximum value of 70°C. Often acid and alkaline pretreatments are sequential, to obtain materials with high cellulose content that allow for glucose-rich syrups for

TABLE 17.3 Acid Pretreatment of Different Lignocellulosic Biomass

Acid Pretreatment	Material	Acid	Acid Concentration	Solid Load	Temperature and Time	Particle Size	Untreated Material	Pretreated Material	References
							Composition		
Diluted	Jerusalem Artichoke Stalks	H_2SO_4, HCl, and HNO_3	2% and 5% w/v	10% w/v	121°C/ 60 min	Between 0.2–0.5 cm	Cellulose: 12.69%, Hemicellulose: 12.69%, Lignin: 20.24%	Cellulose: 42.16–77.27%, Hemicellulose: 1.31–5.78%, Lignin: 20.70–27.41%	[52]
Diluted	Alamo Switchgrass	H_2SO_4	0.5% w	10% w/v	160°C/ 10, 20, 25 and 30 min	< 1 mm	Glucan: 38.18%, Xylan: 26.96%, Arabinan: 2.97%, Klason Lignin: 20.8%	Glucan: 61.99–62.33%, Xylan: 1.75–4.80%, Klason Lignin: 33.20–35.92%	[53]
Diluted	Sugarcane bagasse and straw, 1:1 w/w mixture of both	H_2SO_4	1% w/v	5% w/v	100°C/30 min	< 0.3–0.85 mm	Bagasse + Straw, Holocellulose: 72.24%, α-cellulose: 41.99%, Hemicellulose: 30.25%, Lignin: 23.74%, Ashes: 3.54%	Holocellulose: 68.50%, α-cellulose: 39.67%, Hemicellulose: 28.83%, Lignin: 23.02%, Ashes: 3.44%	[54]
Diluted	Elephant grass	H_2SO_4	20% w/w	5% w/v	121°C/30 min	< 0.841 mm	Glucan: 33.85%, Hemicellulose: 23.93%, Lignin: 14.15%	Glucan: 43.13%, Hemicellulose: 10.72%, Lignin: 23.32%	[55]

TABLE 17.3 *(Continued)*

Acid Pretreatment	Material	Acid	Acid Concentration	Solid Load	Temperature and Time	Particle Size	Composition		References
							Untreated Material	Pretreated Material	
Diluted	*Agave salmiana* leaves	H₂SO₄	4% v/v	10% w/v	121°C/ 30, 60 and 90 min	Between 0.25 and 0.425 mm	Cellulose: 20.67%, Hemicellulose: 3.74%, Lignin: 26.05%	51.25–61.14% Cellulose: 55.31–58.90%, Hemicellulose: 0.07–0.85%, Lignin: 27.77–35.95%	[56]
Diluted	Poplar wood chips	H₂SO₄	1.0% w/w	10% w/v	120 and 150°C/ 20, 30, 40, 60 and 80 min	20 mm (H) x 20 mm (W) x 3 mm (D)	Glucan: 42.25%, Xylan: 14.80%, Arabinan: 0.29%, Acetyl groups: 2.59%, Acid insoluble lignin: 21.75%, Acid soluble lignin: 4.32%	Glucan: 42.50–55.06%, Xylan: 8.95–14.99%, Acid insoluble lignin: 19.23–22.33%, Acid soluble lignin: 2.96–4.89%	[57]
Diluted	Olive tree biomass	H₂SO₄	1–3% w/v	20% w/v	130°C/60–12 min	≤4 mm	Cellulose: 31.3%, Hemicellulose: 23.6%, Lignin: 23.2%, Extractives: 8.9%	Cellulose: 40.2%, Hemicellulose: 3.3%, Acid insoluble lignin: 42.8%, Ash: 4.5%	[58]
Diluted	banana fruit	H₂SO₄	1.5% v/v	20% w/v	95°C/600 min	none	Cellulose: 4.0%, Hemicellulose: 4.4%, Starch: 49.2%, Lignin: 4.2%, Sugars: 3.0%	Syrup rich in glucose, 95% conversion of Starch	[59]

bioethanol fermentation. Also, furfural, levulinic acid, and other organic acids can be obtained by means acid pretreatment under a biorefinery scheme, on the condition that the process can be scaled up [60–62].

17.3.2.2   PRETREATMENT SCALE-UP PRETREATMENT

To scale up, the acid pretreatment is necessary to take some elements into account, in addition to temperature, pH, solid load, etc. These can be not much evident, as the increase in the volume of biomass once it has contact with an acid solution and absorbs water; to get a homogeneous reaction, mixing is necessary, trough mechanical agitation commonly. The challenge comes from the fact that, for an economically feasible process, solid load must be from 30–45% (w/v); mixing of high solids suspensions implies special impellers and high energy input.

Thus, the main factors for scale-up are:

1. To acquire knowledge of the biomass composition (untreated biomass).
2. To define the products expected from the whole process, for example, biofuels, organic acids, xylitol, etc., and the volume of production.
3. Selection of operation regime, based on the reaction kinetics and the mass and heat transfer phenomena.
4. Operation conditions: minimally consume of catalyst (acid) and water, energy consumption and generation of hazardous or contaminant effluents.
5. Pretreatment processes must be versatile in its application over a wide range of lignocellulosic materials, without considerable modifications of the equipment, only adjustment of operating conditions; this is the first step to biorefining.

17.3.3   PURIFICATION OF LIGNOCELLULOSIC RESIDUAL BIOMASS HYDROLYSATES AFTER PRETREATMENT FRACTIONATION

The hydrolysis in the pretreatment stage of LCM's allows obtaining a solid fraction rich in fermentable sugars (cellulose) and lignin, as well as a LF containing mainly hemicellulose (xylan) derived oligosaccharides, as was previously mentioned. However, thermochemical pretreatment (hydrothermal process) provides, in addition to XOS, low molecular weight compounds, and other impurities. Within these impurities different compounds derived from sugars can be found, such as furfural, HMF, acetic acid, formic acid, levulinic acid and acetyl furan; also organic compounds derived from lignin (CDL) with significant presence of gallic acid and vanillin, and nitrogen compounds such as amines and fatty acids (FAs) [63–66]. The concentration of these impurities in the hydrolysates depends on the conditions of the hydrolysis process and the composition of the lignocellulosic material used. The presence of these impurities in the crude hydrolysates avoids having a product with chemical, structural, and homogeneous characteristics that can be useful for the pharmaceutical and food industry; as well as these compounds can inhibit subsequent fermentation processes within the concept of biorefinery.

Different chemical, physical, and biological methods can be used to remove these minor components from the hydrolysate. Most refining processes start with vacuum evaporation, which not only increases the product concentration but also removes volatile compounds. Besides this, solvent precipitation is a useful technique for the removal of non-saccharide compounds. Vasquez et al. [67] performed this technique using three different solvents (ethanol, 2-propanol and acetone) at the same conditions (up to a solvent to liquor volume ratio of 10:1), obtaining the XOs precipitation yield in the following order 2-propanol > acetone > ethanol (93.8, 89.0 and 52.1 wt%, respectively). However, whereas the desired features of an ideal precipitation are a high percentage of non-volatiles dissolved, higher hemicellulose-derived content recovery and lower monosaccharides content, the authors show that ethanol is highly selective towards the precipitation of XOs, acetyl groups, and uronic acids. Moreover, authors such as Swennen et al. [68] have analyzed the effect of increasing the ethanol concentration during solvent precipitation (0–90% v/v), finding that the degree of polymerization of the oligosaccharide fractions recovered after ethanol precipitation decrease with the increase in the ethanol concentration, allowing controlling the selectivity towards the desired fractions by controlling this parameter. However, the recovery and purification of the XOs fractions depend on the nature of the raw material and the type of solvent used for purification.

Additionally, the adsorption process using activated carbon as adsorbent has been used for the purification of hydrolysates from LCM. The adsorption process was used alone or in combination with other treatments for the removal of undesired products from lignin fraction (phenol compounds) or for selective separation of the oligosaccharides with different degrees of polymerization (DP). According to Montane et al. [69], after adsorption onto activated carbon in a continuous process in column, the analysis of the recovered product showed that monosaccharides and xylooligomers were partially adsorbed on the carbon (22.5 wt.%), while lignin-related products were removed from the solution in a higher proportion (63.9 wt.%). Moreover, authors such as Tan et al. [70] and Wang et al. [71] used adsorption onto activated carbon to purify hydrolysates after enzymatic treatment, recovering 90.5 wt% of the oligomers present in the sample, obtaining in addition carbohydrates with a DP between X2 and X5. The use of activated carbons has also been reported by Chen et al. [72], achieving a total adsorption of XOs from hydrolysis, followed by a fractionation by dilution with ethanol at different concentrations; however, the yield of XOs recovered is very low compared to the other authors mentioned (47.9 wt% of initial XOs).

The ion exchange resins have been used for the purification of hemicellulose hydrolysates, allowing the selectivity towards the removal of certain undesired compounds. Chen et al. [64] performed purification processes using resins Amberlite IR120Na (a strong cation exchange resin), Amberlite IR96 (a weak anion exchange resin) and Amberlite FPA90Cl (a strong anion exchange resin) sequentially, finding that there is no significant difference between the use of strong anion and weak anion resins. After the ion exchange purification sequence, the XOs recovery from treatment was very high (>90 wt.%). It is noteworthy that the strong cation exchange resins are frequently used for water softening by exchange with metal ions, while the strong anion exchange resins are generally used to remove organic acids or color agents, as the phenolic compounds present in biomass hydrolysates derived from lignin. However, authors as Buruiana et al. [73] have reported higher XOs recovery yield using a strongly basic anion-exchange resin (Amberlite IRA-400) compared to a weakly basic anion-exchange resin (Amberlite IRA-96), 84.1 and 71.3%, respectively. The loss of oligomers after contact with the resins can be attributed to that some of the oligomers present in the hydrolysate are anchored to phenolic fragments (lignin) that can be retained by the resin. Otherwise, Wang et al. [74] used a mixed bed of strongly acidic and basic resins, allowing eliminating small residues of lignin-derived compounds due to its simultaneous ion exchange and adsorption. At the same time, cation exchange resin in the mixed bed could capture calcium ions. For these reasons, the use of mixed bed assures a stable pH environment and minimizes saccharides degradation [74].

Finally, the use of ultrafiltration (UF) applies to oligomers separation and purification. UF is a separation process based in the use of a membrane to separate and concentrate substances without undergoing phase changes, using a pressure gradient as driving force (trans-membrane pressure - TMP) to separate molecules based on their molecular size. Membrane techniques are frequently used in the XOs processing for different purposes, including production by enzymatic reactions [75], removal of XOs with an undesired degree of polymerization range [76], refining, and concentration. Membrane technology (UF and nanofiltration) is perceived as the most promising purification strategy for the production of high purity and concentrated oligosaccharides [77]. According to Swennen et al. [68] the molecular weight distribution of the permeate fraction obtained using a membrane of 5 kDa MWCO mainly showed oligosaccharides having molecular weight between 180 and 700 Da, allowing obtaining an oligosaccharides mixture with low polymerization degree, and a retained fraction containing oligosaccharides of both low and high polymerization degree.

Additionally, when using membranes of higher MWCO (10 and 30 kDa), the levels of low and medium molecular weight compounds in the permeate fractions increased, while those in the retained fractions decreased, making the separation less specific. The reduction of permeate fluxes can cause the presence of lower molecular weight components than the MWCO of membrane in the retained fraction as a result of the presence of higher molecular weight compounds accumulated on the surface, acting as an additional filtration layer. For example, Akpinar et al. [78] reported that the monosaccharides could not be separated efficiently from the oligosaccharides even using a membrane of a 1kDa MWCO, due to most of the oligosaccharides were recovered in the retained together with monosaccharides. In these cases, the implementation of consecutive concentration and diafiltration methodologies can improve the selectivity of the process.

Gullón et al. [79] studied the effect of diafiltration process using 1 kDa membrane, concluding that continuous diafiltration allows: (a) an almost complete removal of monosaccharides; (b) a marked drop of the other non-volatile solutes mass fraction and (c) an increase in the mass fraction of total oligosaccharides (including XOS, GOS, and oligosaccharide substituents). Similar results have been reported by Krawczyk et al. [80] for the purification of hydrolysates from barley husks, finding that both the selection of the operation mode (diafiltration or concentration), and the cutoff of the membrane play important roles in the processing of hydrolysates. Besides this, performing consecutive diafiltration and concentration steps with different membranes (10, 5, 3, 1 and 0.3 kDa) resulted in streams containing polysaccharides derived from hemicellulose of different molar mass distribution, having a low content of monosaccharides and other nonvolatile compounds [81].

17.4 HIGH ADDED-VALUE PRODUCTS IN TERMS OF THE BIOREFINERY CONCEPT

In the following sections, some of the products with high added-value as bioplastics, xylitol, and enzymes that in these days can be important in terms of biorefinery concept and bioeconomy are presented.

17.4.1 BIOPLASTICS FROM LIGNOCELLULOSIC MATERIALS

In recent years, the biorefinery studies were mainly focused on biofuels production. However, the importance of the chemical molecules from biomass, from a chemical point of view, it has become relevant in the development of new biomaterials and bioproducts. For this reason, the scientific and industrial interest has been redirected to the development of routes for the conversion of the biomass into bio-based chemical building blocks for high value-added products [82–84].

Bioplastics are entirely (or partly) produced from biomass (renewable agricultural or forestry materials, wastes, others). Bioplastics have gained importance in recent years because, unlike traditional plastics, they have characteristics such as renewable raw materials and biodegradability. These materials have a significant potential to replace to the fossil raw materials, mainly due to the environmental impact of the conversion processes and derivative products. Biorefineries may be able to produce the same kinds of petroleum-based products but from chemical structures derived from biomass. Several building blocks can be obtained from sugars (cellulose and hemicelluloses) and the phenolic compounds (lignin and tannins), whose chemical structures have an enormous potential for the conversion to high valued products. The building blocks with considerable potential have been selected according to their chemical composition, properties, the technical complexity of production and market potential [85–87]. Among the main chemical structures derived from sugars are ethanol, furfural, HMF, lactic acid, succinic acid, levulinic acid, sorbitol, and xylitol. For lignin, some of the most important chemical structures are vanillin, syringaldehyde [88, 89], various low molecular weight phenolic compounds, dicarboxylic acids (such as malonic, succinic, muconic, maleic acids), and quinone [90, 91]. Currently, the research is focused on the optimizing of the conversion technologies, yields, technological costs, and the development of new technologies (catalysts and bioreactors). Metabolic or synthetic catalysis engineering allows obtaining very specific catalysts for the transformation of biomass into a variety of intermediate and final chemical products. This leads to controlled, safe, and efficient production of chemicals and polymers [92].

The production of bio-based polymer (BBP) (cellulose, xylans, glucuronoxylan, arabinoxylan, glucomannan, and xyloglucan or lignin) are based on the pulp manufacturing processes developed by the paper industry (non-degradative processes) and consist of dissolution treatments using water or organic solvents (organosolv pulp) and additives (catalysts and various reagents). Currently, BBP is used in the production of rayon, cellophane, cellulose nanocrystals, nanocellulose, microcrystalline cellulose, hemicellulosic film, and others [93].

The hexoses and pentoses can be converted mainly to organic acids (levulinic acid, lactic acid, succinic acids, among others), furans (HMF and furfural) and polyols (sorbitol, xylitol, mannitol, others). Lactic acid and its derivatives are widely used in the food industry, chemical, pharmaceutical, plastic, textile, agriculture, and animal feed among others, being polylactic acid one of the applications with greater potential in medicine, electronics, and bioplastics [94, 95]. PLA is a biodegradable polymer that has applications, mainly, in food and beverage containers [96] (Figure 17.3)

PHAs are biodegradable biopolyesters produced by microorganisms for energy storage when they are under low nutrient concentration and high carbon concentration. The identification of cost-effective feedstocks and also suitable strain are keys in the PHA production [97]. PHAs are being considered for applications in packaging, (films, bags, and containers), biodegradable carrier for medicines and disposable items [98].

Bio-polyolefins (BioPO) are produced from biomass but they are non-biodegradable bio-resins like: bio-polyethylene (BioPE), bio-poly(vinyl chloride) (BioPVC) and bio-polypropylene (BioPP). The dehydration of bioethanol to ethylene for the production of BioPE and BioPVC is an alternative that gives greater value to bioethanol than its use as a fuel [99]. These polymers are conventionally obtained from petroleum but can also be obtained from renewable sources [100]. Ethylene can be produced from biomass from the ethanol obtained from the fermentation of the hexose fraction followed by a dehydration process. It can then be polymerized to BioPE by conventional industrial processes based on the technology and processes used to produce PE from petroleum. That is, it does not require new developments or investments in production and processing equipment, which is an important advantage from the industrial point of view since it can continue to use the existing infrastructure. The requirements are the use of low-cost raw materials, the optimization of the processes involved to achieve a competitive production of bioethylene and an attractive economic return for the BioPE. Bioethanol production from bioethanol based on the biorefinery concept could be more sustainable and economical compared to the petroleum-based process [101, 102].

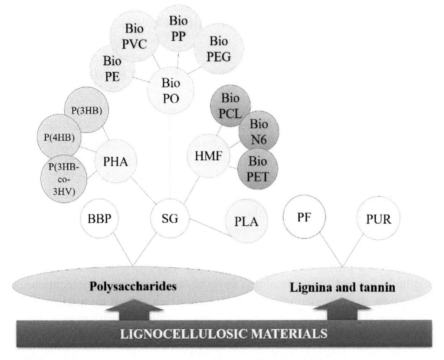

FIGURE 17.3 Family of bioplastics from lignocellulosic materials.

BBP: bio-based polymer; SG: sugar-based polymer; PHA: polyhydroxyalkanoate; P(3HB): poly(3-hydroxybutyrate); P(4HB): poly(4-hydroxybutyric acid); P(3HB-co-3HV): poly (3-hydroxybutyric acid-co-3-hydroxyvaleric acid); BioPO: bio-polyolefinas; BioPE: bio-polyethylene; BioPVC: bio-poly(vinyl chloride); BioPP: bio-polypropylene; HMF: HMF-based polymer; BioPCL: poly-e-caprolactone; BioN6: bio-Nylon 6; BioPET: polyethylene terephthalate; PLA: poly(lactic acid); PF: phenol formaldehyde resin; PUR: polyurethane resin.

HMF is a promising building block because it has a chemical structure (furan ring) and functional groups accessible (hydroxyl group and a formyl group) for further reactions. HMF can be used to synthesize polymers such as poly-e-caprolactone (BioPCL), bio-Nylon 6 (BioN6) and polyethylene terephthalate (BioPET) [103].

The resins urea-formaldehyde, melanin-formaldehyde, and phenol-formaldehyde have traditionally been used as adhesives in the wood industry for the production of wood panels, medium-density fiberboard (MDF), among others. Adhesives play an important role in the forest products industry and are a key factor in the efficient use of wood and other lignocellulosic resources. In recent decades, there has been considerable interest in the development of adhesives that seek to partially or completely replace form-aldehyde (FA)-based mainly on the urea-low content of FA, phenol-lignin, urea-furfural, phenol-HMF, and urea-HMF [104, 105].

17.4.2   XYLITOL FROM FRUIT AND VEGETABLE RESIDUES

Xylitol ($C_5H_{10}O_5$) is a pentose sugar-alcohol that has an equivalent sweetening power to sucrose, is capable of preventing otitis and possesses anti-cariogenic properties (Ur-Rehman, 2013). Therefore, it is mainly used in the food (for dietary and chewing gum), pharmaceutic, and odontological industries. Currently, the global market for xylitol is expected to reach 1 bn USD by 2022 with global production of 266 kton [106].

Fruits like strawberries, plums, and raspberries have low amounts of xylitol, and therefore, extraction from such sources is not cost effective. Hence, xylitol is mostly produced through chemical reduction of xylose, which is a costly and energy-intensive process and has led to research alternative methods through biotechnology. In both, chemical, and biotechnological processes, the first step is to create a xylose-rich stream by treating diverse agricultural residues through an enzymatic or thermochemical pretreatment. Then, the stream is purified (or detoxified) and converted to xylitol through hydrogenation (chemical process) or fermentation with whole cells (typically yeast of the genre *Candida*). Purification of xylitol includes methods like ion exchange resins, activated carbon and chromatography. In the case of the biotechnological process, purification of xylitol from fermented broth is more challenging due to its diluted concentration and the presence of remaining nutrients, polypeptides, pigments, and inorganic salts [107].

The selection of the raw material is mainly based on the availability of the agricultural residue and its composition, which has to be rich in hemicel-lulose and with high xylose content. Corncob is commonly used as a raw material because it is a lignocellulosic residue that fulfills both requirements. Currently, xylitol is also industrially produced from birch wood and wood waste [108].

Several studies have been made for producing xylitol from agricultural wastes such as corncob, sugar cane bagasse, and brewers spent grain and rice straw. Fewer studies have been done for transforming fruit residues as they usually are low in hemicellulose and possess pectin, which makes the pretreatment and purification steps harder, as mentioned above. As an example, while corncob's composition is 45% cellulose, 35% hemicellulose, and 15% lignin, banana peel has a composition of 7.6–9.6% cellulose, 6.4–9.4% hemicellulose, and 6–12% lignin [109]. Table 17.4 shows a summary of the yields of conversion of xylose to xylitol from diverse agricultural wastes.

TABLE 17.4 Biotechnological Production of Xylitol from Agricultural Wastes

Yeast	Substrate	Yield g/g	References
Candida tropicalis	Corncob	0.58	[110]
Pachysolen tannophilus	Olive stones	0.44	[111]
Kluyveromyces sp.	Sugarcane bagasse	0.61	[112]
Candida tropicalis	Banana peel	0.49	[113]

Currently, technological advancements allow for low yields of conversion of agricultural wastes into fermentable sugars (i.e., xylose), and the consequent biotransformation of such sugars into xylitol. Hence, biotechnological production of xylitol is not practiced industrially [114]. The challenge relies on developing a better pretreatment process to convert fruit wastes into an economically feasible raw material and scale-up the bioprocesses from a lab-scale to an industrial scale. Until such work is developed, and as is happens today, chemical conversion process for xylitol production will continue to meet the demand of its global market.

17.4.3 ENZYME

Throughout human's history, enzymes have played an essential role in daily activities, such as brewing, baking, and cheese production, among others. Nowadays, several applications in industrial processes comprise the use of enzymes, including textile, pharmaceutical, food, pulp, and paper, fuel, and chemicals in terms of biorefinery and bioeconomy. Currently, cellulases, hemicellulases, and pectinases have been considered some of the most important biocatalysts in biotechnological applications, representing approximately 20% of the industrial enzymes sales, worldwide [115].

The depolymerization of lignocellulosic materials requires the application of cellulases and accessory enzymes, such as xylanases, that enhance the conversion efficiency. Cellulases and xylanases are crucial in lignocellulosic biomass enzymatic hydrolysis process. They are specific biocatalysts that promote the degradation of cellulose and hemicellulose polymers into soluble monomeric sugars for subsequent use in the production of commercially important metabolites [116]. The degradation of cellulose into glucose in enzymatic hydrolysis of lignocellulosic biomasses comprises the utilization of three specific enzymes that act synergistically in the breakdown of β-1,4 glycosidic bonds present in cellulose backbone, namely, endoglucanases, cellobiohydrolases, and β-glucosidase [117]. On the other hand, xylanases are a complex of glycoside hydrolase enzymes that depolymerize the primary chain of xylan, randomly cleaving β-1,4 glycosidic linkages among D-xylopyranose monosaccharide in xylan structural backbone. Typical xylanases cocktails contain endoxylanases, β-xylosidase, α-glucuronidase, α-arabinofuranosidase, and acetyl xylan esterase [118, 119].

17.4.3.1 ENZYMES PRODUCTION

Enzymes production is considered one of the most important improvement opportunities regarding the bioeconomy feasibility of many biotechnological processes in terms of biorefinery concept. Enzymes production represents high manufacturing costs and is one of the main barriers that have hindered the commercialization of second-generation biomass biorefineries [120, 121]. It has been stated that cellulase enzymes cost represents up to 20% of the overall bioethanol production costs from lignocellulosic materials [117]. Enzymes production encompasses different scientific disciplines, including microbiology, biochemistry, genetics, and protein engineering, which aim to look for lower production investment and guarantee high catalytic efficiency and desirable enzymes stability.

In the enzymes production process, the prime factor to be considered in the selection of the enzymes source. Through the years, cellulases, and xylanases enzymes have been synthesized from different microbial sources, including fungal, bacterial, and protozoal microorganisms [121]; which represent good alternatives for biocatalysts production as they allow to enlarge the cultures quantities in short fermentation time. Industrially, the strains are genetically modified to promote greater production efficiencies. However, recently, agroindustrial, and forestry residues, for instance, sugar cane bagasse, wheat straw, corn stover, and other lignocellulosic biomasses, have been used as growth substrate to produce cellulolytic and xylanolytic enzymes by solid-state fermentation (SSF) induced by microbial sources [120, 122]. This latter approach aims to reduce enzymes production costs through the replacement of costly prime carbon source by inexpensive agricultural wastes and to exploit renewable residues that contain high quantities of the desired enzymes [123].

Cellulases and xylanases are inducible enzymes that can be synthesized either by solid-state or submerged fermentation (SF); nevertheless submerged fed-batch strategy is preferred by commercial biocatalysts producers due to lower manufacturing costs and higher enzymes titer [119, 121]. Commercially, cellulases are mainly produced from fungi and bacteria microorganisms, however, fungal cellulolytic enzymes are more frequently found, due to higher enzyme synthesis concentration compared to the achieved from bacterial cellulases. Additionally, cellulases from fungi present good thermal stability and acid tolerance, which make them attractive for lignocellulosic biomass degradation. Trichoderma and Aspergillus genus are the most commonly used for cellulases production [116, 121]. Likewise, xylanases enzymes are regularly produced from bacteria and filamentous fungi, including *Aspergillus, Trichoderma, Streptomyces, Phanerochaetes, Chytridiomycetes, Ruminococcus, Fibrobacteres, Clostridia,* and *Bacillus.* However, the *Aspergillus* genus has been recognized as the most important xylanolytic enzymes synthesizers [118, 119, 122]. Table 17.5 summarizes some typical microbial sources employed for cellulolytic and xylanolytic enzymes production.

The saccharification of the cellulose and hemicellulose (the two most abundant polysaccharides in nature) contained in biomass is a promising alternative to bioconvert lignocellulosic biomasses into essential industrial products for human's daily activities. However, further research is required to improve enzymes production efficiency and reduce manufacturing costs to develop a cost-effective enzymatic hydrolysis process that allows substituting non-renewable fossil products and chemicals under biorefinery concept.

17.5 CONCLUSION AND FINAL REMARKS

The biomass from lignocellulosic materials, fruit, and vegetable waste, are essential in the production of high added-value compounds as bioplastics, XOS, xylitol, and enzymes in terms of biorefinery concept and bioeconomy. Moreover, pretreatment plays an important and critical stage in the fractionation of biomass. In general terms, this chapter was focused on reviewing the strategies of fractionation biomass as hydrothermal and acid pretreatment, also the purification of lignocellulosic residual biomass hydrolysates after pretreatment fractionation as reviewed and the high added-value compounds. Then, the strategies in the operational conditions of hydrothermal and acid pretreatment are necessary to maximize fractionation of biomass, production of high added-value compounds and reduce operating costs. Therefore, it is necessary to develop and improve the process at the pilot and industrial scale in the production of bioplastics, XOS, xylitol, and enzymes.

TABLE 17.5 Review of Microorganisms for Cellulases and Xylanases Enzymes Production

Enzyme	Microorganism		References
	Genus	**Species**	
Cellulases	*Trichoderma*	*reesei*	[124]
	Aspergillus	*niger*	[125]
	Enhydrobacter	*ACCA2*	[126]
	Pseudomonas	*fluorescens*	[127]
	Bacillus	*subtills*	
	Escherichia	*coli*	
	Serratia	*marcescens*	
	Aspergillus	*fumigatus*	[128]
	Trichoderma	*viride*	[129]
Xylanases	*Aspergillus*	*niger*	[120]
		niveus	
		ochraceus	
	Aspergillus	*fumigatus RP04*	[122]
		niveus RP05	
	Aspergillus	*terricola*	[130]
		ochraceus	
	Bacillus	*subtilis*	[131]
	Penicillium	*chrysogenum*	[123]
	Aspergillus	*niger*	[132]
	Trichoderma	*viride*	[133]
	Aspergillus	*foetidus*	[134]

ACKNOWLEDGMENTS

This project was funded by the Secretary of Public Education of Mexico-Mexican Science and Technology Council (SEP-CONACYT) with the Basic Science Project-2015–01 (Ref. 254808). Marcela Sofía Pino (grant number: 611312/452636) and Elisa also thank the National Council for

Science and Technology (CONACYT, Mexico) for the Master and PhD Fellowship support.

KEYWORDS

- **autohydrolysis**
- **bio-based products**
- **biomass**
- **bioplastics**
- **lignocellulosic material**
- **pretreatment**

REFERENCES

1. Zanuso, E., Lara-Flores, A. A., Aguilar, D. L., Velazquez-Lucio, J., Aguilar, C. N., Rodríguez-Jasso, R. M., & Ruiz, H. A. Kinetic modeling, operational conditions, and biorefinery products from hemicellulose: Depolymerization and solubilization during hydrothermal processing. In: Ruiz, H. A., Thomsen, M. H., & Trajano, H. L., (eds.), *Hydrothermal Processing in Biorefineries* (pp. 141–160). Springer International Publishing, Cham, Switzerland, **2017**.
2. Aguilar-Reynosa, A., Romaní, A., Rodríguez-Jasso, R. M., Aguilar, C. N., Garrote, G., & Ruiz, H. A. Microwave heating processing as alternative of pretreatment in second-generation biorefinery: An overview. *Energy. Convers. Manage.,* **2017**, *136*, 50–65.
3. Ruiz, H. A., Thomsen, M. H., & Trajano, H. L. *Hydrothermal Processing in Biorefineries: Production of Bioethanol and High Added-Value Compounds of Second and Third Generation Biomass.* Springer International Publishing, Cham, Switzerland, **2017**.
4. Ruiz, H. A., Rodríguez-Jasso, R. M., Aguedo, M., & Kádár, S. Hydrothermal pretreatments of macroalgal biomass for biorefineries. In: Prokop, A., Bajpai, R. K., & Zappi, M. E. (eds.), *Algal Biorefineries, Products and Refinery Design*, (vol. 2, pp. 467–491). Springer International Publishing Switzerland, **2015**.
5. Ruiz, H. A., Rodríguez-Jasso, R. M., Fernandes, B. D., Vicente, A. A., & Teixeira, J. A. Hydrothermal processing, as an alternative for upgrading agriculture residues and marine biomass according to the biorefinery concept: A review. *Renew. Sustain. Energy. Rev.,* **2013**, *21*, 35–51.
6. Moncada, J., Tamayo, J. A., & Cardona, C. A. Integrating first, second, and third generation biorefineries: Incorporating microalgae into the sugarcane biorefinery. *Chem. Eng. Sci.,* **2014**, *118*, 126–140.
7. Renewable Fuels Association (RFA). http://www.ethanolrfa.org (accessed on 7 January 2020).

8. Liu, K. Chemical composition of distillers grains, a review. *J. Agric. Food. Chem.*, **2011**, *59*, 1508–1526.

9. Ericsson, K. *Biogenic Carbon Dioxide as Feedstock for Production of Chemicals and Fuels.* Department of Technology and Society, Lund University, Sweden. Available via (http://lup.lub.lu.se/search/ws/files/31711760/Biogenic_carbon_dioxide_as_feedstock_for_production_of_chemicals_and_fuels_IMES_report_103.pdf). **2017.** (accessed on 7 January 2020).

10. Aguilar, D. L., Rodríguez-Jasso, R. M., Zanuso, E., Lara-Flores, A. A., Aguilar, C. N., Sanchez, A., & Ruiz. H. A. Operational strategies for enzymatic hydrolysis in a biorefinery. In Kumar, S., & Rajesh, K. S., (eds.), *Biorefining of Biomass to Biofuels – Opportunities and Perception.* (pp. 141–160), Springer International Publishing, Cham, Switzerland, **2018.**

11. Ethanol producer magazine. http://www.ethanolproducer.com/articles/14479/inside-the-cellulosic-industry. (Accessed on 7 January 2020).

12. Haro, P., Perales, A. L. V., Arjona, R., & Ollero, P. Thermochemical biorefineries with multiproduction using a platform chemical. *Biofuels. Bioprod. Biore.*, **2014**, *8*, 155–170.

13. Haro, P., Ollero, P., Perales, A. L. V., & Vidal-Barrero, F. Potential routes for thermochemical biorefineries. *Biofuels. Bioprod. Biore.*, **2013**, *7*, 551–572.

14. Brethauer, S., & Studer, M. H. Biochemical conversion processes of lignocellulosic biomass to fuels and chemicals – A review. *Chimia.*, **2015**, *69*, 572–581.

15. Ruiz, H. A., Cerqueira, M. A., Silva, H. D., Rodríguez-Jasso, R. M., Vicente, A. A., & Teixeira, J, A. Biorefinery valorization of autohydrolysis wheat straw hemicelulose to be applied in a polymer-blend film. *Carbohydr. Polym.*, **2013**, *92*, 2154–2162.

16. Ruiz, H. A., Martínez, A., & Vermerris, W. Bioenergy potential, energy crops, and biofuel production in Mexico. *Bioenerg. Res.*, **2016**, *9*, 981–984.

17. Moncada, J. B., Aristizábal, M.V., & Cardona, C. A. A. Design strategies for sustainable biorefineries. *Biochem. Eng. J.*, **2016**, *116*, 122–134.

18. FAO. Food wastage footprint: Impacts on natural resources. FAO. Available via http://www.fao.org/docrep/018/i3347e/i3347e.pdf **2013**. (accessed on 7 January 2020).

19. Bustamante, C. A. G., & Cerutti, O. M. *State of the Art of Bioenergy in Mexico.* Publication of the Bioenergy Thematic Network *(RTB) of Conacyt.* Mexico, **2016**.

20. Rodriguez, A., Castrejon-Godinez, M. L., Ortiz Hernandez, M. L., & Sanchez-Salinas, E. Management of municipal solid waste in Mexico. In: *15th International Waste Management and Landfill Symposium.* (pp. 1–7). Cagliari, Italy, **2015**.

21. Campuzano, R., & González-Martínez, S. Characteristics of the organic fraction of municipal solid waste and methane production: A review. *Waste Manage.*, **2016**, *54*, 3–12.

22. Secretariat of the Environment. *Inventory of Solid Waste.* Mexico City, **2016**. https://www.sedema.cdmx.gob.mx/storage/app/media/IRS-2016.pdf (accessed on 7 January 2020).

23. Programa ENRES. Sources of financial resources for projects for the use of Urban Solid Waste (USW) and Special Management Waste (RME) in Mexico. SENER, SEMARNAT, GIZ. **2019.** https://www.giz.de/en/downloads/EnRes_Fuentes_financieras_25.04.2019%20.pdf (accessed on 7 January 2020).

24. Wholesale Market From Mexico City. **2018.** http://ficeda.com.mx/index.php?id=ceda (accessed on 7 January 2020)

25. CEDA. (2010). *Analysis of Organic Waste from the Central de Abasto.* Unpublished data.

26. Toribio-Cuaya, H., Pedraza-Segura, L., Macías-Bravo, S., Gonzalez-García, I., Vasquez-Medrano, R., & Favela-Torres, E. Characterization of lignocellulosic biomass using five simple steps. *J. Chem. Biol. Physic. Sci.*, **2014**, *4*, 28–47.

27. Campuzano, R., & González-Martínez, S. Extraction of soluble substances from organic solid municipal waste to increase methane production. *Bioresour. Technol.,* **2015**, *178*, 247–253.

28. Xiao, C., & Anderson, C. T. Roles of pectin in biomass yield and processing for biofuels. *Front. Plant Sci.,* **2013**, *4*, 67.

29. Szymańska-Chargot, M., Chylińska, M., Gdula, K., Kozioł, A., & Zdunek, A. Isolation and characterization of cellulose from different fruit and vegetable pomaces. *Polymers,* **2017**, *9*, 495.

30. Orozco, R. S., Hernández, P. B., Morales, G. R., Núñez, F. U., Villafuerte, J. O., Lugo, V. L., & Vázquez, P. C. Characterization of lignocellulosic fruit waste as an alternative feedstock for bioethanol production. *Bio Resources.,* **2014**, *9*, 1873–1885.

31. Ravindran, R., & Jaiswal, A. K. Exploitation of food industry waste for high-value products. *Trends. Biotechnol.,* **2016**, *34*, 58–69.

32. Kulkarni, S. O., Kanekar, P. P., Jog, J. P., Sarnaik, S. S., & Nilegaonkar, S. S. Production of copolymer, poly (hydroxybutyrate-co-hydroxyvalerate) by *Halomonas campisalis* MCM B-1027 using agro-wastes. *Int. J. Biol. Macromol.,* **2015**, *72*, 784–789.

33. Morone, P., Papendiek, F., & Tartiu, V. *Food Waste Reduction and Valorisation: Sustainability Assessment and Policy Analysis.* Springer International Publishing, Cham, Switzerland. **2017**.

34. Dahiya, S., Kumar, A. N., Sravan, J. S., Chatterjee, S., Sarkar, O., & Mohan, S. V. Food waste biorefinery: sustainable strategy for circular bioeconomy. *Bioresour. Technol.,* **2018**, *248*, 2–12.

35. Pandiyan, K., Singh, A., Singh, S., Saxena, A. K., & Nain, L. Technological interventions for utilization of crop residues and weedy biomass for second generation bio-ethanol production. *Renew. Energy,* **2019**, *132*,723–741.

36. Fan, S., Zhang, P., Li, F., Jin, S., Wang, S., & Zhou, S. A Review of lignocellulose change during hydrothermal pretreatment for bioenergy production. *Curr. Org. Chem.,* **2016**, *20*, 2799–2809.

37. Yang, B., Tao, L., & Wyman, C. E. Strengths, challenges, and opportunities for hydrothermal pretreatment in lignocellulosic biorefineries. *Biofuels. Bioprod. Biorefining,* **2018**, *12*, 125–138.

38. Guillón, P., Romaní, A., Vila, C., Garrote, G., & Parajó, J. C. Potential of hydrothermal treatments in lignocellulose biorefineries. *Biofuels. Bioprod. Biorefining,* **2012**, *6*, 219–232.

39. Akhtar, N., Gupta, K., Goyal, D., & Goyal, A. Recent advances in pretreatment technologies for efficient hydrolysis of lignocellulosic biomass. *Environ. Prog. Sustain. Energy,* **2016**, *35*, 498–511.

40. Trajano, H. L., Engle, N. L., Foston, M., Ragauskas, A. J., Tschaplinski, T. J., & Wyman, C. E. The fate of lignin during hydrothermal pretreatment. *Biotechnol. Biofuels.,* **2013**, *6*, 1–16.

41. Zhuang, X., Wang, W., Yu, Q., Qi, W., Wang, Q., Tan, X., Zhou, G., & Yuan, Z. Liquid hot water pretreatment of lignocellulosic biomass for bioethanol production accompanying with high valuable products. *Bioresour. Technol.,* **2016**, *199*, 68–75.

42. Perez-Pimienta, J. A., Flores-Gómez, C. A., Ruiz, H. A., Sathitsuksanoh, N., Balan, V., da Costa, Sousa. L., Dale, B. E., Singh, S., & Simmons, B. A. Evaluation of agave bagasse recalcitrance using AFEX™, autohydrolysis, and ionic liquid pretreatments. *Bioresour. Technol.,* **2016**, *211*, 216–223.

43. Aguilar, D. L., Rodríguez-Jasso, R. M., Zanuso, E., de Rodríguez, D. J., Amaya-Delgado, L., Sanchez, A., & Ruiz, H. A. Scale-up and evaluation of hydrothermal pretreatment in isothermal and non-isothermal regimen for bioethanol production using agave bagasse. *Bioresour. Technol.,* **2018**, *263,* 112–119.

44. Den, W., Sharma, V. K., Lee, M., Nadadur, G., & Varma, R. S. Lignocellulosic biomass transformations via greener oxidative pretreatment processes: Access to Energy and Value-Added Chemicals. *Front. Chem.,* **2018**, *6,* 141.

45. Chheda, J. N., Román-Leshkov, Y., & Dumesic, J. A. Production of 5-hydroxymethylfurfural and furfural by dehydration of biomass-derived mono-and poly-saccharides. *Green Chem.,* **2007**, *9,* 342–350.

46. Montané, D., Salvado, J., Torras, C., & Farriol, X. High-temperature dilute-acid hydrolysis of olive stones for furfural production. *Biomass Bioenergy.,* **2002**, *22,* 295–304.

47. Hayes, D. J., Fitzpatrick, S., Hayes, M. H., & Ross, J. R. The biofine process – production of levulinic acid, furfural, and formic acid from lignocellulosic feedstocks. In: Kamm, B., Gruber, P. R., & Kamm, M., (eds.), *Biorefineries-Industrial Processes And Products: Status Uuo And Future Directions* (pp. 139–164), WILEY-VCH Verlag GmbH & Co. KGaA, Weinheim. **2006**.

48. Singh, A., Das, K., & Sharma, D. K. Production of xylose, furfural, fermentable sugars and ethanol from agricultural residues. *J. Chem. Technol. Biotechnol.,* **1984**, *34,* 51–61.

49. Conner, A. H., & Lorenz, L. F. Kinetic modeling of hardwood prehydrolysis. Part III. Water and dilute acetic acid prehydrolysis of southern red oak. *Wood Fiber Sci.,* **1986**, *18,* 248–263.

50. Belkacemi, K., Abatzoglou, N., Overend, R. P., &Chornet, E. Phenomenological kinetics of complex systems: mechanistic considerations in the solubilization of hemicelluloses following aqueous/steam treatments. *Ind. Eng. Chem. Res.,* *30,* **1991**, 2416–2425.

51. Nabarlatz, D., Farriol, X., & Montane, D. Kinetic modeling of the autohydrolysis of lignocellulosic biomass for the production of hemicellulose-derived oligosaccharides. *Ind. Eng. Chem. Res.,* **2004**, *43,* 4124–4131.

52. Dziekońska-Kubczak, U., Berłowska, J., Dziugan, P., Patelski, P., Pielech-Przybylska, K., &Balcerek, M. Nitric acid pretreatment of *Jerusalem Artichoke* stalks for enzymatic saccharification and bioethanol production. *Energies,* **2018**, *11,* 2153.

53. Kothari, N., Holwerda, E. K., Cai, C. M., Kumar, R., & Wyman, C. E. Biomass augmentation through thermochemical pretreatments greatly enhances digestion of switch grass by *Clostridium thermocellum. Biotechnol. Biofuels.,* **2018**, *11,* 219.

54. Ávila, P. F., Forte, M. B., & Goldbeck, R. Evaluation of the chemical composition of a mixture of sugarcane bagasse and straw after different pretreatments and their effects on commercial enzyme combinations for the production of fermentable sugars. *Biomass Bioenergy,* **2018**, *116,* 180–188.

55. Santos, C. C., de Souza, W., Sant'Anna, C., & Brienzo, M. Elephant grass leaves have lower recalcitrance to acid pretreatment than stems, with higher potential for ethanol production. *Ind. Crops Prod.,* **2018**, *111,*193–200.

56. Láinez, M., Ruiz, H. A., Castro-Luna, A. A., & Martínez-Hernández, S. Release of simple sugars from lignocellulosic biomass of *Agave salmiana* leaves subject to sequential pretreatment and enzymatic saccharification. *Biomass Bioenergy.,* **2018**, *118,*133–140.

57. Liu, W., Chen, W., Hou, Q., Wang, S., & Liu, F. Effects of combined pretreatment of dilute acid pre-extraction and chemical-assisted mechanical refining on enzymatic hydrolysis of lignocellulosic biomass. *RSC Adv.,* **2018**, *8,* 10207–10214.

58. Martínez-Patiño, J. C., Ruiz, E., Romero, I., Cara, C., López-Linares, J. C., & Castro, E. Combined acid/alkaline-peroxide pretreatment of olive tree biomass for bioethanol production. *Bioresour. Technol.*, **2017**, *239*, 326–335.

59. Velásquez-Arredondo, H. I., & Ruiz-Colorado, A. A. Ethanol production process from banana fruit and its lignocellulosic residues: Energy analysis. *Energy*, **2010**, *35*, 3081–3087.

60. Siripong, P., Duangporn, P., Takata, E., & Tsutsumi, Y. Phosphoric acid pretreatment of *Achyranthes aspera* and *Sidaacuta* weed biomass to improve enzymatic hydrolysis. *Bioresour. Technol.*, **2016**, *203*, 303–308.

61. Mamman, A. S., Lee, J., Kim, Y., Hwang, I. T., Park, N., Hwang, Y. K., Chang, J. & Hwang, J. Furfural: Hemicellulose/xylose derived biochemical. *Biofuels, Bioprod. Biorefin.*, **2008**, *2*, 438–454.

62. Girisuta, B., Danon, B., Manurung, R., Janssen, L. P. B. M., & Heeres, H. J. Experimental and kinetic modelling studies on the acid-catalysed hydrolysis of the water hyacinth plant to levulinic acid. *Bioresour. Technol.*, **2008**, *99*, 8367–8375.

63. Carvalheiro, F., Garrote, G., Parajo, J. C., Pereira, H., & G, F. M. Kinetic modeling of brewery's spent grain autohydrolysis. *Biotechnol. Prog.*, **2005**, *21*, 233–243.

64. Chen, M. H., Bowman, M. J., Cotta, M. A., Dien, B. S., Iten, L. B., Whitehead, T. R., Rausch, K. D., Tumbleson, M. E., & Singh, V. Miscanthus × giganteus xylooligosaccharides: Purification and fermentation. *Carbohydr. Polym.*, **2016**, *140*, 96–103.

65. Garrote, G., Falqué, E., Domínguez, H., & Parajó, J. C. Autohydrolysis of agricultural residues: Study of reaction byproducts. *Bioresour. Technol.*, **2007**, *98*, 1951–1957.

66. Kabel, M. A., Bos, G., Zeevalking, J., Voragen, A. G. J., & Schols, H. A. Effect of pretreatment severity on xylan solubility and enzymatic breakdown of the remaining cellulose from wheat straw. *Bioresour. Technol.*, **2007**, *98*, 2034–2042.

67. Vázquez, M. J., Garrote, G., Alonso, J. L., Domínguez, H., & Parajó, J. C. Refining of autohydrolysis liquors for manufacturing xylooligosaccharides: Evaluation of operational strategies. *Bioresour. Technol.*, **2005**, *96*, 889–896.

68. Swennen, K., Courtin, C. M., Van Der Bruggen, B., Vandecasteele, C. & Delcour, J. A. Ultrafiltration and ethanol precipitation for isolation of arabino xylooligosaccharides with different structures. *Carbohydr. Polym.*, **2005**, *62*, 283–292.

69. Montane, D., Nabarlatz, D., Martorell, A., Torne, V. & Fierro, V. Removal of lignin and associated impurities from xylo-oligosaccharides by activated carbon adsorption. *Ind. Eng. Chem. Res.*, **2006**, *45*, 2294–2302.

70. Tan, S. S., Li, D. Y., Jiang, Z. Q., Zhu, Y. P., Shi, B., & Li, L. T. Production of xylobiose from the autohydrolysis explosion liquor of corncob using *Thermoto gamaritima* xylanase B (XynB) immobilized on nickel-chelated Eupergit C. *Bioresour. Technol.*, **2008**, *99*, 200–204.

71. Wang, T. H., & Lu, S. Production of xylooligosaccharide from wheat bran by microwave assisted enzymatic hydrolysis. *Food Chem.*, **2013**, *138*, 1531–1535.

72. Chen, M. H., Bowman, M. J., Dien, B. S., Rausch, K. D., Tumleson, M. E., & Singh, V. Autohydrolysis of *Miscanthus x giganteus* for the production of xylooligosaccharides (XOS): Kinetics, characterization and recovery. *Bioresour. Technol.*, **2014**, *155*, 359–365.

73. Buruiana, C. T., Gómez, B., Vizireanu, C., & Garrote, G. Manufacture and evaluation of xylooligosaccharides from corn stover as emerging prebiotic candidates for human health. *LWT – Food Sci. Technol.*, **2017**, *77*, 449–459.

74. Wang, X., Zhuang, J., Jiang, J., Fu, Y., Qin, M., & Wang, Z. Separation and purification of hemicellulose-derived saccharides from wood hydrolysate by combined process. *Bioresour. Technol.,* **2015**, *196*, 426–430.

75. González-Muñoz, M. J., Domínguez, H., & Parajó, J. C. Depolymerization of xylan-derived products in an enzymatic membrane reactor. *J. Memb. Sci.,* **2008**, *320*, 224–231.

76. Pinelo, M., Jonsson, G., & Meyer, A. S. Membrane technology for purification of enzymatically produced oligosaccharides: Molecular and operational features affecting performance. *Sep. Purif. Technol.,* **2009**, 70, 1–11.

77. Qing, Q., Li, H., Kumar, R., & Wyman, C. E. Xylooligosaccharides production, quantification, and characterization in context of lignocellulosic biomass pretreatment. In: Wyman, C. E., (ed.), *Aqueous Pretreatment of Plant Biomass for Biological and Chemical Conversion to Fuels and Chemicals* (pp. 397–422). John Wiley & Sons, Ltd., United Kingdom. **2013**.

78. Akpinar, O., Erdogan, K., Bakir, U., & Yilmaz, L. Comparison of acid and enzymatic hydrolysis of tobacco stalk xylan for preparation of xylooligosaccharides. *LWT – Food Sci. Technol.,* **2010**, *43*, 119–125.

79. Gullón, P., González-Muñoz, M. J., & Parajó, J. C. Manufacture and prebiotic potential of oligosaccharides derived from industrial solid wastes. *Bioresour. Technol.,* **2011**, *102*, 6112–6119.

80. Krawczyk, H., Persson, T., Andersson, A., & Jönsson, A. S. Isolation of hemicelluloses from barley husks. *Food Bioprod. Process.,* **2008**, *86*, 31–36.

81. Gonzalez-Muñoz, M. J., Rivas, S., Santos, V., & Parajó, J. C. Fractionation of extracted hemicellulosic saccharides from *Pinus pinaster* wood by multistep membrane processing. *J. Memb. Sci.,* **2013**, *428*, 281–289.

82. Dammer, L., Carus, M., Raschka, A., & Scholz, L. *Market Developments Of and Opportunities for Biobased Products and Chemicals.* **2013**. https://www.eumonitor.nl/9353000/1/j4nvgs5kjg27kof_j9vvik7m1c3gyxp/vjken6y2ivvo/f=/blg338557.pdf. (Accessed on 7 January 2020).

83. Golden, J, S., & Handfield, R. B. Why biobased? *Opportunities in the Emerging Bioeconomy.* **2014**. https://www.biopreferred.gov/files/WhyBiobased.pdf. (Accessed on 7 January 2020).

84. Snyder, S. W. An introduction to commercializing biobased products: opportunities, challenges, benefits, and risks. In: Snyder, S. W., (ed.), *Commercializing Biobased Products: Opportunities, Challenges, Benefits, and Risks* (pp. 1–7), Royal Society of Chemistry, Cambridge, UK, **2015**.

85. Holladay, J. E., White, J. F., Bozell, J. J., & Johnson, D. Top value-added chemicals from biomass. Volume II—Results of screening for potential candidates from biorefinery Lignin. **2007**. https://www.pnnl.gov/main/publications/external/technical_reports/PNNL-16983.pdf (Accessed on 7 January 2020).

86. Vandermeulen, V., Van der Steen, M., Stevens, C. V, & Van Huylenbroeck, G. Industry expectations regarding the transition toward a biobased economy. *Biofuels, Bioprod. Biorefining,* **2012**, *6*, 453–464.

87. Werpy, T. and Petersen, G. Top value added chemicals from biomass. Volume I—Results of screening for potential candidates from sugars and synthesis gas. **2004**. https://www.nrel.gov/docs/fy04osti/35523.pdf. (accessed on 7 January 2020).

88. Maity, S. K. Opportunities, recent trends and challenges of integrated biorefinery: Part I. *Renew. Sustain. Energy Rev.,* **2015**, *43*, 1427–1445.

89. Maity, S. K. Opportunities, recent trends and challenges of integrated biorefinery: Part II. *Renew. Sustain. Energy Rev.,* **2015**, *43*, 1446–1466.

90. Ma, R., Guo, M., & Zhang, X. Selective conversion of biorefinery lignin into dicarboxylic acids. *Chem. Sus. Chem.,* **2014**, *7*, 412–415.

91. Zhang, X., Tu, M., & Paice, M. G. Routes to potential bioproducts from lignocellulosic biomass lignin and hemicelluloses. *Bio Energy Res.,* **2011**, *4*, 246–257.

92. De Corato, U., De Bari, I., Viola, E., & Pugliese, M. Assessing the main opportunities of integrated biorefining from agro-bioenergy co/by-products and agroindustrial residues into high-value added products associated to some emerging markets: A review. *Renew. Sustain. Energy Rev.,* **2018**, *88*, 326–346.

93. Area, M. C., Vallejos, M. E. Bio-products and bio-materials from biorefinery waste agro and forestoindustriales. In: Area, M. C., & Park, S. W., (eds.), *Panoramaof The Cellulose And Paper Industry and Lignocellulosic Materials* (pp. 120–151), Universidad de Misiones. **2016**.

94. Jantasee, S., Kienberger, M., Mungma, N., & Siebenhofer, M. Potential and assessment of lactic acid production and isolation – a review. *J. Chem. Technol. Biotechnol.,* **2017**, *92*, 2885–2893.

95. Pang, X., Zhuang, X., Tang, Z., & Chen, X. Polylactic acid (PLA): Research, development and industrialization. *Biotechnol. J.,* **2010**, *5*, 1125–1136.

96. Castro-Aguirre, E., Iñiguez-Franco, F., Samsudin, H., Fang, X., & Auras, R. Poly (lactic acid)—mass production, processing, industrial applications, and end of life. *Adv. Drug Deliv.,* **2016**, *Rev., 107*, 333–366.

97. Sathya, A. B., Sivasubramanian, V., Santhiagu, A., Sebastian, C., & Sivashankar, R. Production of polyhydroxyalkanoates from renewable sources using bacteria. *J. Polym. Environ.,* **2018**, *26*, 3995–4012.

98. Kumar, P., & Kim, B. S. Valorization of polyhydroxyalkanoates production process by co-synthesis of value-added products. *Bioresour. Technol.,* **2018**, *269*, 544–556.

99. Le Van Mao, R., Nguyen, T. M., & McLaughlin, G. P. The bioethanol-to-ethylene (B.E.T.E.) process. *Appl. Catal.,* **1989**, *48*, 265–277.

100. Mohsenzadeh, A., Zamani, A., & Taherzadeh, M. J. Bioethylene production from ethanol: A review and techno-economical evaluation. *Chem. Bio. Eng. Reviews*, **2017**, *4*(2), 75–91.

101. Becerra, J., Figueredo, M., &Cobo, M. Thermodynamic and economic assessment of the production of light olefins from bioethanol. *J. Environ. Chem. Eng.,* **2017**, *5*, 1554–1564.

102. Chieregato, A., Ochoa, J. V., & Cavani, F. Olefins from biomass. In: Cavani, F., Albonetti, S., Basile, F., Gandini, A., (eds.), *Chemicals and Fuels from Bio-Based Building Blocks* (pp. 1–32), Wiley-VCH Verlag GmbH & Co. Weinheim, Germany, **2016**.

103. Zhang, D., & Dumont, M. J. Advances in polymer precursors and bio-based polymers synthesized from 5-hydroxymethylfurfural. *J. Polym. Sci. Part A Polym. Chem.,* **2017**, *55*, 1478–1492.

104. Esmaeili, N., Zohuriaan-Mehr, M. J., Mohajeri, S., Kabiri, K., & Bouhendi, H. Hydroxymethyl furfural-modified urea formaldehyde resin: synthesis and properties. *Eur. J. Wood Wood Prod.,* **2017**, *75*, 71–80.

105. Ghafari, R., DoostHosseini, K., Abdulkhani, A., & Mirshokraie, S. A. Replacing formaldehyde by furfural in urea formaldehyde resin: Effect on formaldehyde emission and physical–mechanical properties of particleboards. *Eur. J. Wood Wood Prod.,* **2016**, *74*, 609–616.

106. Global Xylitol Market Research Report. *Xylitol – A Global Market Overview*, **2017**. http://industry-experts.com/verticals/food-and-beverage/xylitol-a-global-market-overview. (Accessed on 7 January 2020).

107. Gurgel, P. V., Mancilha, I. M., Pecanha, R. P., & Siqueira, J. F. M. Xylitol recovery from fermented sugarcane bagasse hydrolyzate. *Bioresource Technol.,* **1995**, *52*, 219–223.

108. Du Pont. Nutrition & Health. http://www.dupontnutritionandhealth.com/products/xivia.html (Accessed on 7 January 2020).

109. Howard, R. L., Abotsi, E., Rensburg, E. L. J. V., & Howard, S. Lignocellulose Biotechnology: Issues of Bioconversion andEnzyme Production. *Afr. J. Biotech.,* **2003**, *2*, 602–619.

110. Ling, H., Cheng, K., Ge, J., & Ping, W. Statistical optimization of xylitol production from corncob hemicellulose hydrolysate by *Candida tropicalis* HDY-02. *New Biotechnol.,* **2011**, *28*, 673–678.

111. Saleh, M., Cuevas, M., Juan, F., & Sanchez, G. S. Valorization of olive stones for xylitol and ethanol production from dilute acid pretreatment via enzymatic hydrolysis and fermentation by *Pachysole ntannophilus. Biochem. Eng., J.,* **2014**, *15*, 286–293.

112. Kumar, S., Dheeran, P., Singh, S. P., Mishra, I. M., & Adhikari, D. K. Bioprocessing of bagasse hydrolysate for ethanol and xylitol production using thermotolerant yeast. *Bioprocess Biosyst Eng.,* **2015**, *38*, 39–47.

113. Rehman, S., Nadeem, M., Ahmad, F., & Mushtaq, Z. Biotechnological production of xylitol from banana peel and its impact on physicochemical properties of rusks. *J. Agr. Sci. Tech.,* **2013**, *15*, 747–756.

114. Dasgupta, D., Bandhu, S., Adhikari, D. K., & Ghosh, D. Challenges and prospects of xylitol production with whole cell bio-catalysis: A review. *Microbiol. Res.,* **2017**, *197*, 9–21.

115. Howard, R. L., Abotsi, E., Jansen, van R. E. L., & Howard, S. Lignocellulose biotechnology: Issues of bioconversion and enzyme production. *African J. Biotechnol.,* **2003**, *2*, 602–619.

116. Michelin, M., Ruiz, H. A., Silva, D. P., Ruzene, D. S., Teixeira, J. A., & M, M. L. T. Cellulose from Lignocellulosic Waste. In: Ramawat, K. G., & Mérillon, J., *Polysaccharides Bioactivity and Biotechnology*, (pp. 475–511). Springer International Publishing, Switzerland, **2015**.

117. Pino, M. S., Rodríguez-Jasso, R. M., Michelin, M., Flores-Gallegos, A. C., Morales-Rodriguez, R., Teixeira, J. A., & Ruiz, H. A. Bioreactor design for enzymatic hydrolysis of biomass under the biorefinery concept. *Chem. Eng. J.,* **2018**, *347*, 119–136.

118. Moreira, L. R. S., & Filho, E. X. F. Insights into the mechanism of enzymatic hydrolysis of xylan. *Appl. Microbiol. Biotechnol.,* **2016**, *100*, 5205–5214.

119. Motta, F. L., Andrade, C. C. P., & Santana, M. H. A. A Review of xylanase production by the fermentation of xylan: Classification, characterization and applications. In: Chandel, A., & Da Silva, S. S., (eds.), *Sustainable Degradation of Lignocellulosic Biomass – Techniques, Applications and Commercialization* (pp. 251–275). Intech Open, **2013**.

120. Betini, J. H. A., Michelin, M., Peixoto-Nogueira, S. C., Jorge, J. A., Terenzi, H. F., & Polizeli, M. L. T. M. Xylanases from *Aspergillus Niger, Aspergillus niveus* and *Aspergillus ochraceus* produced under solid-state fermentation and their application in cellulose pulp bleaching. *Bioprocess Biosyst. Eng.,* **2009**. *32*, 819–824.

121. Zhang, X., & Zhang, Y. P. Cellulases: Characteristics, sources, production, and applications. In: Yan, S., El-Enshasy H., & Thongchul, N. (eds.), *Bioprocessing Technologies in Biorefinery For Sustainable Production of Fuels, Chemicals, and Polymers*. (pp. 131–146). John Wiley & Sons. **2013**.

122. De Carvalho Peixoto-Nogueira, S., Michelin, M., Betini, J. H. A., Jorge, J. A., Terenzi, H. F., & Polizeli, M. D. L. T. Production of xylanase by *Aspergilli* using alternative carbon sources: application of the crude extract on cellulose pulp biobleaching. *J. Ind. Microbiol. Biotechnol.,* **2009**, *36*, 149–155.

123. Ho, H. L. Xylanase production by *Bacillus subtilis* using carbon source of inexpensive agricultural wastes in two different approaches of submerged fermentation (SmF) and solid state fermentation (SsF). *J. Food Process. Technol.,* **2015**, *6*, 437.

124. Ellilä, S., Fonseca, L., Uchima, C., Cota, J., Goldman, G. H., Saloheimo, M., Sacon, V., & Siika-Aho, M. Development of a low-cost cellulase production process using *Trichoderma reesei* for Brazilian biorefineries. *Biotechnol. Biofuels,* **2017**, *10*, 1–17.

125. Alriksson, B., Rose, S. H., van Zyl, W. H., Sjode, A., Nilvebrant, N. O., & Jonsson, L. J. Cellulase production from spent lignocellulose hydrolysates by recombinant *Aspergillus Niger. Appl. Environ. Microbiol.,* **2009**, *75*, 2366–2374.

126. Premalatha, N., Gopal, N. O., Jose, P. A., Anandham, R., & Kwon, S. W. Optimization of cellulase production by *Enhydrobacter* sp. ACCA2 and its application in biomass saccharification. *Front. Microbiol.,* **2015**, *6*. 1046.

127. Sethi, S., Datta, A., Gupta, B. L., & Gupta, S. Optimización of cellulase production from Bacteria isolated from soil. *ISRN Biotechnol.,* **2013**, *7*.

128. Cherian, E., Dharmendira, K. M., & Baskar, G. Production and optimization of cellulase from agricultural waste and its application in bioethanol production by simultaneous saccharification and fermentation. *Manag. Environ. Qual. An Int. J.,* **2016**, *27*, 22–35.

129. El Baz, A. F., Shetaia, Y. M. H., Shams Eldin, H. A., & El Mekawy, A. Optimization of cellulase production by *Trichoderma viride* using response surface methodology. *Curr. Biotechnol.,* **2018**, *7*, 19–25.

130. Michelin, M., Polizeli, M. de L. T. M., Ruzene, D. S., Silva, D. P., Ruiz, H. A., Vicente, A. A., Jorge, J. A., Terenzi, H. F., & Teixeira, J. A. Production of xylanase and β-xylosidase from autohydrolysis liquor of corncob using two fungal strains. *Bioprocess Biosyst. Eng.,* **2012**, *35*, 1185–1192.

131. Terrone, C. C., Freitas, C. de, Terrasan, C. R. F., Almeida, A. F. de, & Carmona, E. C. Agroindustrial biomass for xylanase production by *Penicillium chrysogenum*: Purification, biochemical properties and hydrolysis of hemicelluloses. *Electron. J. Biotechnol.,* **2018**, *33*, 39–45.

132. Park, Y., Kang, S., Lee, J., Hong, S., & Kim, S. Xylanase production in solid state fermentation by *Aspergillus Niger* mutant using statistical experimental designs. *Appl. Microbiol. Biotechnol.,* **2002**, *58*, 761–766.

133. Goyal, M., Kalra, K. L., Sareen, V. K., & Soni, G. Xylanase production with xylan rich lignocellulosic wastes by a local soil isolate of *Trichoderma viride. Brazilian J. Microbiol.,* **2008**, *39*, 535–541.

134. Cunha, L., Martarello, R., De Souza, P. M., De Freitas, M. M., Barros, K. V. G., & Filho, E. X. F., Homem-De-Mello, M., & Magalhães, P. O. Optimization of xylanase production from *Aspergillus foetidus* in soybean residue. *Enzyme Res.,* **2018**, *7*.

Index

β

β-D-fructofuranosidase, 215, 216
β-fructosidase, 215
β-fructosylinvertase, 215
β-glucosidase, 301, 304, 305, 307, 416

A

A. acidoterrestris, 23, 24, 27–29
Abioticmicrobial contamination, 64
Acetaldehyde, 155
Acetate esters, 233
Acetic acid, 24, 25, 140, 144, 156, 172, 402, 408
Acetone, 155, 156, 258, 380, 381, 408
Acidemia, 332
Acidification, 11, 140, 171, 172, 356
Acidilactici, 181, 302
Acidolysis, 221, 328, 334
Acidoterrestris, 23, 24, 27–29
Acquired virulence factors, 49
Actinomycetaceae, 142
Actinomycetes, 221
Adenocarcinomas, 386
Adhesion, 72, 115, 173, 210, 211, 213, 215, 262
Aeromonas hydrophila, 65, 74
Agglutination, 66, 143
Aglycones, 305
Agrochemicals, 55, 65, 105
Agro-food wastes, 290–299
    toxic compounds, 290
        alkaloids, 297
        cyanide, 295
        cyanogenic glycosides, 295
        glucosinolates, 296
        gossypol, 294
        oxalates, 296
        phorbol, 294
        phytic acid, 293
        protease inhibitors, 298
        saponin, 291
        tannins, 292

Agro-industrial wastes, 306, 309, 310
Alcoholic
    beverages, 109, 171, 181, 233
    drinks, 233
    fermentation, 110, 111
Alcoholysis, 221, 328, 334
Aldolase, 142, 172
Alimentarius, 168
A-linolenic acid (ALA), 323, 329, 333, 344
Alkaline
    soils, 379
    tolerance, 115
Alkaloids, 129, 288–292, 297, 298, 362
Allium macranthum, 360
Amentoflavone, 267
Amino
    acids, 112, 154, 171–173, 178–180, 220, 235, 239, 241, 262, 290, 293, 295–297, 299, 303–305, 321, 356, 358, 368
    groups, 215, 236
Aminolysis, 221
Amphipathic
    membrane, 211
    molecules, 115, 320
    structures, 262
Anaerobes, 171
Angiotensin-converting enzyme (ACE), 179, 303
Anilides, 258
Animal excreta, 73
Antagonist activity, 168
Anthelmintic, 257, 380
Anthocyanins, 263, 265, 270–272, 381
Antibacterial
    activity, 74, 253
    effects, 261
Antibiotic-resistant microbial strains, 75
Antibiotics, 87, 146, 148, 150, 168, 169, 175, 176
Anticancer, 288, 291, 304, 325, 377, 380, 385, 386

activity, 383
  breast cancer, 385
  lung cancer, 386
  prostate cancer, 385
  skin cancer, 384
Anticarcinogenic activity, 303, 304
Antigens, 48, 66, 385
Anti-inflammatory effects, 332
Antimicrobial, 4, 6, 11, 12, 18, 19, 22, 25,
    32, 64, 87, 103, 108, 134, 249, 251, 263,
    288, 292
  activity, 12, 93, 113, 116, 168, 169, 180,
    249, 251–254, 256–259, 262, 275–277,
    291, 297, 380
  agents, 94, 129, 131, 249, 251, 253, 255,
    257, 258, 262, 263, 276
  capacity, 3, 4, 104, 251
  compounds, 129, 253, 258, 276
  effect, 12, 31
  effectiveness, 106, 115
  peptides, 66
  potential, 129, 134, 258
  properties, 109, 131, 251, 252, 260, 275,
    277
  ratio, 11
  spectrum, 109, 115
  substances, 150, 171, 180
Antinutritional compounds, 290, 292, 302,
    303, 308
Antinutritive factors, 174
Antioxidant, 66, 129, 131–134, 163, 164,
    168, 173, 179, 249, 250, 253, 254, 256,
    257, 259, 260, 262, 271, 275, 277, 288,
    289, 292, 297, 300, 302–306, 309, 325,
    326, 332, 377, 380, 382, 383
  activity, 113, 114, 256, 268, 304, 305
  compounds, 131, 133
  polyphenols, 173
  properties, 383
Anti-parasitic properties, 377
Antiproliferative, 250, 383
Aqueous ozone spray, 29
Arachidonic acid (AA), 323, 333, 334, 345
Aromatic
  components, 173
  compounds, 214, 232, 304, 308, 309
  profile, 233
Artificial feeding, 67
Ascospores, 14, 15, 17

Aspartic proteinases, 367
Aspergillus
  niger, 215, 217, 218, 224, 236, 301, 357
  oryzae, 149, 224
  versicolor, 307, 308
Atmospheric
  composition, 126
  plasma, 55
Atole sour, 166
Autocatalyst, 401
Autohydrolysis, 419
Autoionization, 401
Auto-oxidation products, 262
Ayran, 24, 25

B

Bacillus
  amyloliquefaciens, 302
  cereus, 87, 105, 106, 252
  coagulans, 17, 152
  prodigiosus, 218
  pumilus, 76
  subtilis, 106, 259, 303
Bacteremia, 54, 72
Bacterial
  cells, 23
  hydrophobicity, 23
  membranes, 12
  pathogens, 54, 69
Bactericide treatment, 27
Bacteriocin, 12, 16, 18, 19, 113–116, 129,
    130, 146, 147, 150, 171, 174, 179–181
Bacteriocinogenic potential, 113
Bagasse, 250, 396, 400, 402, 405, 414–416
Barotolerance mechanisms, 107
Beta-conglycin, 303
Betapol, 344
Bifidobacterium, 142, 146, 148, 149,
    152–154, 167, 168, 170, 175, 176, 178, 303
  bifidum, 148
  longum, 152
  species, 142
Bioactive
  antioxidant compounds, 133
  compounds (BCs), 74, 123, 133–135,
    249–251, 253–256, 259, 260, 262–266,
    269–271, 273–277, 290, 299–301, 306,
    319, 325, 333, 338, 342, 343, 381
  peptides, 178, 179, 302, 303, 305, 371

Bioactivity, 215, 253, 276, 304
Bio-based
  polymer (BBP), 412, 413
  products, 419
Biocomposites, 133, 134, 302
Biocompounds, 288, 305, 306, 309, 310, 380
Bioeconomics, 396
Bioeconomy, 395, 397, 411, 415–417
Bioenrichment, 174
Bioethanol, 396, 397, 407, 412, 413, 416
Biofilm, 56, 109, 115, 210, 211, 262
Biogenic amines, 111
Biological activity, 131, 178, 305, 382
Biologically active molecules, 163, 164
Biomass, 109, 110, 180, 210, 211, 218, 242,
  275, 301, 305, 306, 395–404, 406, 407,
  409, 411, 412, 415–419
  acidification, 165
  fractionation processes, 401
    acid pretreatments, 402
    hydrothermal pretreatment, 401
Biomolecules, 134, 255, 256, 260, 263, 277,
  305
Bioplastics, 396, 397, 400, 411–413, 417–419
Bio-poly(vinyl chloride) (BioPVC), 412, 413
Biopolyesters, 412
Bio-polyethylene (BioPE), 412, 413
Biopolymer, 23, 344
Bio-polyolefins (BioPO), 412
Bio-polypropylene (BioPP), 412, 413
Biopreservation, 12, 129, 180
Bioproduction, 243
Bioproducts, 274, 411
Biorefinery, 395–397, 399–402, 407, 408,
  411, 413, 415–417
Biosensors, 75, 221
Biostructures, 212
Biosurfactants, 115
Biosynthetic capability, 111
Biotechnological processes, 133, 215, 221,
  236, 288, 309, 414, 416
Biotin, 171
Biotransformation, 232, 241, 274, 300, 301,
  304, 415
Bivalve mollusks, 70, 72, 74
Blastobotrys adeninivorans, 304
Blood coagulation, 331
Bone fragments, 89

Botulism, 70
Brevibacterium linens, 236
Brochothrix thermosphacta, 9
Butylated hydroxyl
  anisole (BHA), 256
  toluene (BHT), 256
Byssochlamys nivea, 15, 17

C

Caffeic acids, 257, 273, 303
Campylobacter jejuni, 105
Cancer, 132, 150, 173, 176, 177, 256, 289,
  294, 297, 298, 300, 302, 303, 344, 364,
  381–387
Candida
  albicans, 257, 258
  rugosa, 222
  utilis, 217, 306
Caprenin, 345
Capsicum annuum L., 32
Capsid, 108
Captex, 344
Carassius auratus auratus, 65
Carbohydrates, 110, 139, 142, 143, 147,
  153, 154, 167, 178, 180, 272, 287, 288,
  290, 292, 293, 308, 321, 325, 331, 340,
  366, 380, 409
Carbon, 110, 127, 143, 171, 172, 174, 182,
  210, 214, 217, 220, 237, 242, 272, 292,
  294, 296, 301, 305, 321, 325, 330, 356,
  383, 396, 408, 409, 412, 414, 416
  dioxide (CO_2), 110, 127, 143, 144, 167,
    171, 172, 182, 272, 301, 396
Carboxylic
  acid, 219, 220
  bonds, 218
  functional group, 320
Cardiovascular diseases, 132, 336, 345
Cardosins, 367
Carotenoids, 173, 254, 256, 257, 269, 273,
  275, 276, 306, 320, 325, 342
Carvacrol nanoemulsions, 276
Caseinolytic activity, 172
Catabolism enzymes, 172
Catalytic
  activity, 220, 301, 358, 368
  mechanism, 219, 358
Cavitation, 22, 23, 269

Cell
  membranes, 12–14, 21, 26, 115, 239, 242,
    243, 255, 261, 262, 271, 319, 324, 332
  surface, 6, 11, 66
  wall transition phase, 21
Cellobiohydrolases, 416
Cellular membrane, 115, 116
Cellulase, 215, 224, 273, 308, 415–417
Cellulolytic enzymes, 306, 417
Cellulose, 147, 272, 306, 397, 400–402,
    404–407, 411, 412, 415–417
Cellulosic materials, 212
Centers for Disease Control and Prevention
    (CDC), 75, 105
Central de Abasto (CEDA), 399, 400
Ceramides, 324
Cereals, 112, 132, 137, 138, 146, 152–156,
    171, 181, 288, 293, 298
C-glycosides, 272
Chaetomium crispatum, 215
Chemical
  compounds, 381
  hazards, 87
  methods, 106, 126
Chemo-organotrophs, 171
Chemopreventive, 381, 385, 386
Chemoprotective effect, 384
Chemotaxis, 49
Chicken manure (CM), 67, 77, 401
Chitin oligomers, 64
Chitosan, 9, 13, 130, 214, 215
Chlorine dioxide (ClO$_2$), 24, 25, 55, 109, 118
Chloroform, 258, 259, 319, 381, 382
Chlorogenic acid, 263, 303, 365
Cholecystokinin (CCK), 299, 310
Cholestasis, 332
Chorismate, 235
Chromatography, 74, 214, 250, 414
Chymosin, 355–358, 367–371
Cinnamaldehyde, 25, 27, 28, 32, 56, 130
Cirrhosis, 72, 298
Citrobacter
  braakii, 66
  freundii, 65, 66
Clarias gariepinus, 67
Clostridium
  botulinum, 70, 86, 87, 93, 105, 107, 253
  perfringens, 87, 105

Coccoid, 167
Cocoa butter (CB), 335–338, 345
  equivalents (CBE), 335–338
  extenders (CBEX), 335
  improvers (CBI), 335, 338
  replacers (CBR), 335, 336
  substitutes (CBS), 236, 239, 240, 242,
    335, 336
Codex alimentarius, 31, 196
Cohobation, 250, 264
Cold
  atmospheric plasma, 55
  chain, 85, 95
    distribution, 95
    identification and traceability, 95
    preservation, 95
    storage, 95
Coliforms, 6–10, 24, 25, 30, 31, 67, 141
Coliphage, 64
Colletotrichum gloeosporioides, 267
Colonization, 147, 148, 173, 211
Commercialization, 31, 105, 116, 344, 416
Compounds derived from lignin (CDL), 408
Control
  CDC (control and prevention), 75, 105
  points (CP), 85, 88, 99, 100
Conventional hurdle technologies, 5, 7
Corilagin, 273
Coriolopsis rigida, 308
Cotyledons, 53
Coupling extraction techniques, 239
Coyonoxtle trunks, 212
Critical
  control points (CCP), 69, 85, 87, 88, 91,
    98–100
  strength, 19
Cross-contamination, 89, 90, 92, 94, 105,
    195, 197, 198, 200, 201
Crustacean growth, 66
Cryptosporidium, 63, 75, 105
Culture medium, 221, 241, 243, 300
Cupuassu, 9, 12
Cyclospora cayetanensis, 105
Cymene, 8, 12, 130, 253
Cynara cardunculus, 359, 361, 368
Cysteine, 180, 211, 356, 358
  proteases (CPs), 356
Cytochrome catalase, 171

Cytoplasmic, 11, 21, 180
Cytotoxic effects, 109
Cytotoxicity, 72, 386

D

Daidzein, 302, 305
Daidzin, 305
Dalton (Da), 292, 410
De *novo* synthesis, 235, 237
Debaryomyces hansenii, 304
Deboning, 89
Decarboxylation, 112
Degrees of polymerization (DP), 408, 409
Dehydration, 4, 9, 87, 126, 236, 404, 412
Dehydroellagitannin, 273
Dehydrotomatine, 362
Denaturation, 11, 12, 330
Depolymerization, 415
Dermatologists, 343
Dermatophytic fungi, 258
Destabilization, 115, 342
Deteriorative microorganism, 4, 11
Detoxification, 173, 174, 290, 306, 308, 309
Detoxifiers, 112
Dextrose, 331
Diacetyl metabolites, 155
Diacylglyceride, 219
Diacylglycerols (DAG), 323, 324, 340, 342
Diafiltration, 410
Diarrhea, 49, 53, 63, 73, 150, 151, 166,
    168–170, 174–176, 380, 381
Dicentrarchus labrax, 65
Dienes, 129
Dietary fiber, 153, 321, 326, 380
Diffusely adherent (DAEC), 53, 74
Diglycerides, 218–220, 224, 255
Dimethylallyl diphosphate, 325
Diphosphates, 325
Dipole-dipole particles, 214
Divinyl benzene, 213, 214
Docosahexaenoic acid (DHA), 323, 329,
    332–334
Domestic
    discharges, 62
    fauna, 55
Dominant flora, 31
Drugenriched shell, 342
Dry weight (DW), 264, 265, 293, 364

Duchenne muscular dystrophy (DMD), 298,
    310

E

Efflux, 6, 29
Eicosanoids, 323
Electric fields, 13, 20
Electrical
    breakdown, 19
    conductivity, 19
Electrochemical biosensors, 75
Electromagnetic
    radiation, 270
    waves, 13, 31
Electron carriers, 260, 320
Electrophoretogram, 367
Electroporation, 19, 271
Ellagic acid, 292, 300, 301, 380, 383, 384
Ellagitannins, 273, 292, 300, 380–382,
    384–386
Emulsification, 86, 87, 339, 341, 343
Encapsulation, 66, 133, 146, 151, 153, 253,
    338
Endocarditis, 54
Endogenous enzymes, 3
Endophytes, 112
Endospores, 20, 21
Enfleurage, 264
Enteral nutrition (EN), 331
Enteroaggregative *E. coli*, 74
Enterobacter agglomerans, 47
Enterobacteriaceae, 24, 25, 48, 49, 65, 68,
    201
Enterocin, 8, 12, 16, 18, 130
Enterococcus, 115, 149, 152, 168, 171, 172,
    181, 259
    faecium, 152
Enterohemorrhagic *E. coli* (EHEC), 53, 56, 74
Enteroinvasive *E. coli* (EIEC), 53, 73
Enteropathogenic bacteria, 75, 105
Enterotoxigenic *E. coli* (ETEC), 53, 73
Enterotoxin, 63, 74, 173
    toxicity, 63
Environment Protection Agency (EPA), 62,
    323, 332, 333
Enzymatic
    assisted extraction (EAE), 267, 272–275
    deacylation, 323

hydrolysates, 179
hydrolysis, 178, 397, 400–402, 415–417
Enzyme load, 330
Epidemiology, 69
Epithelium interstitial edema, 65
Equilibrium, 155, 219
Equimolar concentrations, 172
Ergosterol, 324
Erwiniaherbicola, 47
Erythrose-4-phosphate, 235
Escherichia coli (E. coli), 7–16, 18, 20–25,
    27–32, 48–51, 53, 55, 56, 58, 67, 69, 71,
    73, 74, 85, 86, 95, 105–107, 113, 114,
    168, 173, 176, 201, 253, 257–259, 276,
    380
Essential oils (EOLs), 12, 108, 118, 129,
    130, 233, 249–251, 253, 255–258, 260,
    262–264, 266, 267, 269, 275–277, 340
Esterases, 173
Esterification, 328
Ethanoic acid, 321
Ethanol, 5, 111, 143, 144, 155, 167, 171,
    172, 258, 265, 269, 271–273, 380–382,
    396, 397, 402, 408, 409, 411, 412
Ethanolic extracts, 258
Ethyl
    acetate, 156, 270
    esters, 233
Eugenol, 10, 12, 258, 276
Eukaryotes, 324
European Union (EU), 46, 54, 58, 71, 72, 399
Exopolysaccharides, 174
Extracellular invertase, 217, 218
Extraction, 108, 123, 133, 134, 181, 232,
    236, 239–243, 249–251, 253, 255,
    257–259, 263–265, 269–274, 277, 289,
    297, 300, 302, 309, 337, 381, 414
Extractive methodologies, 250, 251, 257,
    264, 265
Extrinsic factors, 3, 4, 125, 201

F

Facultative heterofermentative, 142
Fatty acids (FAs), 67, 147, 173, 218, 222,
    252, 255, 277, 302, 320, 322, 329, 345,
    365, 380, 408
Febrile gastroenteritis, 54
Fenneropenaeus chinensis, 66

Fermentation, 67, 103, 104, 109–112,
    115–118, 137–144, 155, 156, 163, 164,
    166–168, 171–174, 178, 180, 181, 210,
    215, 217, 236, 237, 239, 243, 288–290,
    300–306, 308, 309, 356, 397, 402, 403,
    407, 408, 412, 414, 416, 417
Fermented
    chicken manure (FCM), 67
    dairy products, 141, 166, 173, 176, 182
        healthy benefits, 175
    milk products
        functional properties, 173
        origin, 164
Filamentous fungi, 149, 210, 215, 221, 300,
    301, 306, 308, 417
Fillets, 29
Fish
    contamination, 68
    oil (FO), 328, 332, 344
Flagellar, 48, 49
Flavanols, 381
Flavanones, 263
Flavones, 129, 263
Flavonoid, 130, 253, 254, 257–260, 263,
    272, 292, 342, 364, 365, 382
    glycosides, 260
Flavonols, 129
Floroglucinols, 130
Fluoroquinolones, 64
Folic acid, 171
Food and Agriculture Organization (FAO),
    31, 46, 69, 153
Food and Drug Administration (FDA), 55,
    62, 75, 77, 105, 232, 345
Food
    applications, 341, 345
    contamination, 118
    matrix, 17, 72, 104, 106–108, 117, 253,
        276
    microbial-contamination, 104
    preservation, 3–5, 32, 104, 106, 107, 112,
        115–117, 124, 126, 127, 131, 133, 134,
        138, 139, 171, 174, 193, 277
        conventional and emerging technolo-
            gies, 126
        fermentation, 139
    fermented products culture, 139
    lactic acid bacteria (LAB) role, 139

plant-derived compounds use, 128
strategies, 106
chemical methods, 108
physical methods, 106
technology-future perspectives, 117
spoilage microorganisms, 3
Food safety, 4–6, 31, 45, 48, 55, 56, 63, 69,
75, 85, 86, 91, 92, 95, 96, 105, 128, 129,
133, 134, 140, 195–203
implications, 132
modernization act (FSMA), 55, 203
quality control programs, 96
good manufacturing practices (GMP), 97
hazard analysis and critical control
point (HACCP), 98
sanitation standard operating procedures (SSOP), 97
Foodborne
diseases, 3, 46, 63, 197, 203
illness, 46–48, 53, 62, 195, 200
infections, 4, 198
pathogens, 3, 12, 56, 69, 72, 77, 97,
106–108, 116, 128, 202
Formaldehyde (FA), 68, 77, 147, 218,
222–224, 252, 255, 320, 321, 323, 324,
326, 328–334, 336–338, 342, 344, 413, 414
Formate, 144, 167, 171
Fortification, 151
Fragile, 6
Free fatty acids (FFAs), 173, 255, 260–262
Fresh weight (FW), 263, 268, 270
Fructans, 154
Fructo-oligosaccharides (FOS), 175, 216,
218, 225
Fruit
nutrients, 379
pericarp, 31
Functional
additives, 112, 135
benefits, 118
foods, 137, 138, 145, 151, 153, 154, 165,
173, 174, 182, 319, 320, 325, 326
Fusarium oxysporum, 267, 307

G

Galactose, 172, 174
Gallic acid, 8, 258, 259, 263, 264, 301, 383,
408
equivalents (GAE), 264, 265, 272

Gamma rays, 31, 117
Ganoderma applantaum, 308
Gastroenteritis, 49, 63, 70, 73
Gastrointestinal
disorders, 146, 150, 167, 177
tract (GIT), 139, 142, 146–148, 150, 156,
173, 176–179, 323, 331, 338
Gelation, 342
Gene transfer applications, 343
Generally
modified microorganisms (GMOs), 357
recognized as safe (GRAS), 12, 140, 149,
215, 225, 232, 243, 251, 265, 345
Genistein, 302, 305
Genomes, 57, 357
Geotrichum penicillium, 236
Geraniin, 273
Germination, 211, 357, 358
Gliocladium viride, 215
Global
food safety initiatives (GFSI), 203
value of billions of dollars (USD), 111,
399, 414
Glucanases, 273
Glucose, 115, 172, 174, 215, 216, 236, 237,
292, 295, 297, 404, 406, 416
Glucosinolates, 289, 296, 297
Glutamate, 235, 236
Glyceraldehyde-3-phosphate (GAP), 143
Glycerides, 173
Glycerol
backbone, 222, 331, 333
molecule, 326
Glycerolipids, 320, 323, 328
Glycerophospholipids, 320, 324
Glycitein, 302, 305
Glycitin, 305
Glycoalkaloids, 359, 362, 364
Glycolysis, 235
Glycon, 219
Glycosides, 129, 259, 263, 269, 272, 290,
292, 295, 319, 380
Glycosphingolipids, 324
Good
hygienic practices (GHP), 69, 73, 77, 203
manufacturing practices (GMP), 69, 89,
92, 97, 98, 100, 200–203
Gracilaria verrucosa, 74

Gram
  negative
    bacteria, 14, 19, 26, 31, 56, 66, 74,
      115, 252, 253, 258, 259
    enterobacterium, 73
    organisms, 47
    strains, 257, 258
  positive
    bacteria, 19, 66, 141, 181, 252, 258, 259
    organisms, 47
Growth
  inhibition, 7, 8, 15, 20, 24
  reduction, 9

H

Haematoxylon brassiletto, 56
Halotolerant bacteria, 70
Harbor microorganisms, 47
Hazard
  analysis and critical control point
    (HACCP), 69, 73, 98–100
  pollution, 62
Heat shock proteins (HSPs), 106, 107
Helicobacter pylori, 168, 170, 177, 297
Helminths, 105
Hemicellulases, 273, 415
Hemicellulose, 272, 400–402, 404–408,
  410, 414–417
Hemin group, 171
Hemolymph, 66, 74
Hemolytic uremic syndrome (HUS), 53
Hemorrhagic colitis, 53
Hepatitis, 72
  A, 63, 69, 105
Heterofermentative, 142, 143, 171, 172
  bacteria, 172
  sugar metabolism, 142
Hexaphosphoric ester, 293
Hexose, 144, 171, 172, 400, 404, 412
  isomerase, 172
  monophosphate, 172
High
  hydrostatic pressure (HHP), 13–19, 107,
    117
  pressure sterilization, 14
  temperature PEF, 21
  voltage electron beams, 31
Histamine, 68

Homeostasis, 5, 11, 330
Homofermentative, 142, 171–173
Homogeneity, 210
Homogeneous dispersion, 212, 335
Homogenization, 22, 269, 343
Human milk fat (HMF), 333–335, 402–404,
  408, 411–414
Humidity, 67, 91, 97, 125, 133, 134, 181
Hurdle technology, 3–6, 12, 14, 21, 23, 32,
  87, 127
Husk, 380, 381, 383, 387, 397
Hybridization, 57
Hydrocarbons, 76, 253, 255
Hydrogel particles, 343, 344
Hydrogen
  cyanide (HCN), 295, 310
  peroxide (H2O2), 22, 55, 171, 174
Hydrolysates, 179, 359, 368, 396, 408–410,
  417
Hydrolysis, 173, 181, 216, 221, 223, 295,
  303, 305, 328, 334, 356, 358, 370, 397,
  401, 403, 404, 407–409
Hydrolytic enzymes, 273, 309
Hydrolyze triglycerides, 218
Hydronium ions, 401
Hydrophilic-hydrophobic interface, 211
Hydrophobic
  anchors, 320
  compounds, 6, 264
Hydrophobicity, 219
Hydrophobins, 211, 215
Hydroquinones, 325
Hydrostatic pressure, 15
Hydrothermal process, 407
Hydroxyl group, 252, 262, 272, 293, 294, 413
Hydroxymethylfurfural (HMF), 402
Hygienic and manufacturing procedures
  (GHP/GMP), 69
Hygienic practices, 69, 73, 74, 77, 98, 105,
  112, 198, 203
Hypersensitivity, 89
Hypocholesterolemia, 173
Hypothetical transmission networks, 57

I

Immobilization process, 134
Immune-modulating effects, 332
Immunoassays, 57

Immunocompetent individuals, 54
Immunostimulation, 174
Immunosuppressed patients, 72
In situ product removal techniques (ISPR), 232, 243, 236, 239
Individualized freezing (IQF), 29
Industrial fermentation, 118
Industrialization, 109, 140, 302
Inflammatory
  activity, 169
  bowel disease (IBDs), 150, 177, 328
  properties, 177, 254, 289, 332
Infrared, 26, 108
Inhabitants, 45, 49, 57, 173
Inhibitory effect, 9, 15, 31, 239, 252, 257
Intensification, 239, 243
Intensity, 17, 20, 22, 26, 107, 181, 262
Interesterification, 218, 221, 323, 334, 336, 337, 344
Internal buffer capacity, 11
International Lipid Classification and Nomenclature Committee (ILCNC), 320
Intoxication symptoms, 74
Intracellular
  invertase, 217
  material, 25, 29, 108
Intrahepatic biliary cholestasis, 331
Intramolecular disulfide bonds, 211
Intraperitoneally injection, 65
Intrinsic properties, 4
Invertase, 209, 215–218
  action mechanism, 216
  microorganisms producers, 217
  production, 217
Invertebrates, 66
Ionizing radiation, 56, 127
Irradiation, 4, 29–32, 56, 117
Ischnoderma benzoinum, 236
Isoamyl acetate, 233
Isoeugenol, 130
Isoflavones, 304, 305
  aglycones, 304
Isoforms, 217
Isopentenyl diphosphate, 325
Isoprenoids, 325
Isorhamnetin, 259
Isothiocyanate, 130
Istituto Zooprofilattico Sperimentale of Piemonte, Liguria, and Valle d'Aosta (IZSPLV), 201

K

Keratocyte cells, 384
Ketone, 233
Klebsiella pneumoniae, 258, 259, 380
Kluyveromyces
  lactis, 304
  marxianus, 168, 232, 236, 239, 304
Knowledge-attitude-behavior (KAB), 197, 203
Komagataella pastoris, 304
Kombucha, 110

L

Lactate dehydrogenase (LDH), 144
Lactic acid, 7, 9, 16, 18, 19, 24, 47, 103, 110, 115, 118, 137, 138, 140–144, 156, 163–167, 171–174, 182, 302, 305, 411–413
Lactic acid bacteria (LAB), 7, 9, 16, 18, 24, 25, 47, 103, 110, 112, 113, 115, 116, 118, 137–145, 148, 154–156, 165, 167, 171–176, 178–180, 182, 302
  application, 140
  bacteriocins, 116
  biosurfactants, 115
  classification, 141
  fermentation metabolism, 143
    pyruvate and lactate catabolism, 144
    sugars catabolism, 143
  metabolism, 140
  physiology, 140
  use by humans, 141
Lactobacillus, 9, 14, 15, 47, 113, 115, 116, 138, 141, 142, 146, 148, 152, 154, 156, 167–169, 171, 172, 175, 176, 178, 181, 302, 303, 305
  acidophilus, 152
  casei, 148, 152, 303
  plantarum, 14, 15, 116, 148, 152, 154, 169, 181, 302, 303
Lactococcus, 113, 115, 149, 167, 171, 172, 180, 181
  lactis, 113, 180
Lactoferrin, 16, 18, 129, 130
Lactoflavin, 171
Lactones, 129
Lactoperoxidase, 16, 18, 130
Lactose, 146, 151, 171–174, 176, 237, 238

Lauric acid (LaA), 252, 321, 329, 344, 380
Lectins, 66, 290, 307, 308
Lettuce, 25, 31, 49, 53, 54, 56, 114, 399
Leuconostoc, 47, 115, 142, 168, 169, 171, 302
Levulinic acid, 407, 408, 411, 412
Light-absorbing pigments, 320
Lignocellulosic, 302, 306, 396, 397, 399–401, 403, 404, 407, 408, 413–417, 419
   biomass, 397, 403
   material, 408, 411, 419
   residual biomass, 407, 417
Linamarin, 116, 117, 290
Linear alcohols, 325
Linoleic acid (LA), 61, 255, 263, 267, 268, 321, 323, 329, 331, 333, 335, 344, 365
Lipases, 172, 173, 209, 215, 218–224, 291, 306, 308, 327, 333
   action mechanism, 219
   industrial applications, 221
      animal feed food supplements, 223
      detergent industry, 222
      food industry, 222
      organic synthesis, 223
      paper industry, 222
      racemic acids and alcohols, 223
   interfacial activation, 219
   microorganisms produced, 221
Lipid, 23, 139, 180, 216, 222, 251, 252, 255, 256, 306, 319, 320, 324, 325, 328, 330–333, 335, 342, 343, 345, 380
   crystals, 342
   emulsions (LE), 331, 332
   fraction, 255
Lipids biochemistry, 320
   fatty acids (FAs), 255, 262, 263, 320, 343, 345, 365, 380, 408
      glycerolipids, 323
      glycerophospholipids, 324
      long-chain fatty acids (LCFA), 321
      medium chain fatty acids (MCFA), 321–323, 331, 333
      prenol lipids, 325
      short-chain fatty acids (SCFA), 321–323, 345
      sphingolipids, 324
      sterols, 324
Lipolytic properties, 221

Lipophilic compounds, 264
Lipopolysaccharides, 6, 174
Lippiacitriodora, 9, 13
Liquid
   chromatography, 214
   crystalline phase, 21
   culture (LC), 74, 210, 221, 255, 274, 342
   fermentation, 210
   fraction (LF), 401, 402, 407
   matrix, 269
   nutrition, 331
Listeria monocytogenes, 48, 58, 69, 72, 85–87, 94, 105, 106, 169, 181, 253, 257
Listeriosis, 54, 72
Lithospermum erythrorhizon, 9, 13
Local breakdown, 19
Log cycles, 14, 17, 18, 21–25, 28, 29, 31, 32
Logaritmic reduction, 7, 8
Long-chain
   fatty acids (LCFA), 255, 321, 322, 331, 344, 345
   triacylglycerols, 331
Lycopene, 254, 267, 269, 270, 273, 306
Lyophilization, 126
Lysophospholipids, 328
Lysozyme, 15, 18, 129, 130

M

Maceration, 250, 259, 264, 270, 300, 358
Macromolecules, 23, 292, 397
Macro-morphology, 115
Magnetic
   fields, 127
   nanoparticles (MNP), 212, 214, 215, 225
Magnetism, 134
Magnetite, 214, 225
Malabsorption, 177, 224, 323
Maleic acids, 403, 411
Maltodextrins, 224
Mango seed
   almond fat (MAF), 337
   fat (MFK), 336, 337
Mano-thermosonication, 23
Meat
   products unit operations, 90
      cooking, 93
      curing, 92
      injection, 92

massaging, 92
post-packing thermal treatment, 94
raw meat thawing, 91
slicing and packing, 94
trimming (size reduction), 92
tumbling, 92
raw materials, 87, 89, 91
Medium
chain triacylglycerols (MCTs), 331–333
density fiberboard (MDF), 413
Membrane
cell potential, 19
fluidity, 13, 325
potential, 116, 260, 261
proteins, 13, 260
technology, 410
Meningitis, 54
Mentha piperita, 15, 17, 20
Mesophiles, 6–10, 15, 16, 20, 24, 30, 31, 172
Mesophilic bacteria, 25
Metabiotic effect, 48
Metabolic
activity, 104, 110, 146, 166
depletion, 5, 32
exhaustion, 5
syndrome, 150
Metagenomic studies, 117
Metagenomics, 111
Metallic nanoparticles, 214
Metalloproteinase, 383, 386
Methyl group, 320
Methylated purine alkaloid, 298
Methylxanthine, 298
Mexican biorefinery context, 399
Microbial
cell, 116, 128, 260
community, 110
contamination, 55, 64, 89, 104, 105, 112, 133
count, 12, 17, 25
deterioration, 32
DNA, 26
growth, 3, 4, 14, 18, 21, 22, 32, 91, 97, 115, 126, 129, 166, 175, 214, 242, 274, 301
inactivation, 14, 19, 22, 29, 32, 106, 108, 271
inhibition, 16

inoculum, 300
lipases, 223
load, 4, 6, 19, 21, 23, 27–29, 93, 105–107, 128, 193, 203
membrane level, 108
population, 112, 117
processes, 265, 274
quality, 46, 85
reactivation, 26, 91
reduction, 5, 14, 18, 21–23, 25, 28, 32
safety, 5, 55, 131, 140
strains, 111, 112, 117
Microbiological
agents, 125
analysis, 32, 202
criteria, 56
margins, 133
quality, 55, 56, 85, 86
studies, 165
Microbiota, 47, 48, 57, 70, 87, 112, 126, 129, 146–148, 150, 166, 167, 175–177
Microbubbles, 22
Microcins, 150
Microencapsulated lipases application, 223
Microencapsulation, 133, 240, 276
Microflora, 29, 146–148, 150, 216
Micronutrients, 153, 302
Microorganism, 5, 14, 56, 57, 64, 70, 72–74, 107–111, 113–115, 128, 149, 150, 168–171, 200, 210, 213, 215, 217, 221, 232, 238, 241, 288, 290, 300, 302, 305, 418
Microphotography, 212
Microstreaming, 22
Microwave (MW), 31, 32
apparatus, 270
assisted extraction (MAE), 251, 265, 267, 270
treatment, 31
Milk
clotting
activity to proteolytic activity (MCA/PA), 356, 367, 370
enzymes, 357, 359, 365, 371
fat globules (MFG), 334
membrane (MFGM), 334
Molds, 7–9, 14–16, 19–21, 23–26, 29–31, 47, 56, 104, 111, 115, 125, 252

Molecular weight, 232, 261, 291, 292, 367–370, 407, 410, 411
Mollusk predators, 69
Monascusruber, 7
Monocarboxylic acids, 321
Monocytes, 53
Monocytogenes, 9, 12, 15, 16, 18, 19, 23, 24, 27, 28, 48, 50–54, 58, 69, 72, 73, 85–87, 94, 95, 105–107, 109, 113, 114, 181, 202, 253, 257
Monogastric
    animals, 223, 293
    diets, 224
    nutrition, 223
Monoglycerides, 218, 224, 252, 255
Monomeric
    sugars, 416
    units, 345, 404
Monosaccharide, 401, 408–410
    xylose, 308
Monounsaturated fatty acid (MUFA), 255, 321, 322, 329, 332, 333
Motility, 49, 72, 294
Mucor griseocyanus, 215
Multiflora thiouchin, 300
Multiresistant strains, 75
Mycelial fungi, 211
Mycelium, 218
Mycotoxigenic fungi, 67
Mycotoxins, 67

N

Nanocomposites, 134
Nanoemulsions, 276, 338, 340–343
Nanoencapsulation, 133–135
Nanomaterials, 134, 212
Nanoparticles, 134, 214
Nanoseconds, 271
Nanostructured lipid carriers (NLC), 342, 343
Nanotechnology, 133
Narrow activity spectrum, 13
Nasogastric tube, 331
Natamycin, 129
National Oceanic and Atmospheric Administration (NOAA), 62
Native microorganisms, 14, 111
    protective roles, 111

animal-derived products, 112
    vegetable-derived products, 112
Natural antimicrobials, 11, 12, 18, 19, 25, 134, 251, 256
    action mechanism, 260
        cell lysis, 261
        electron transport chain disruption, 260
        enzyme activity inhibition, 261
        oxidative phosphorylation uncoupling, 261
    bioactivities, 256
        antimicrobial activity, 257
        antioxidant activity, 256
        gallic acid and flavonoids, 258
    stability, 275
    vegetable by-products, 262
Naturopathic medicine, 380
Necrotizing, 150, 177
    enterocolitis (NEC), 151, 177
Negative redox potential, 70
Neosartorya fischeri, 15, 17
Neurological diseases, 132
Neuroprotection, 325
Neurospora crassa, 308
Neutralization, 256
Neutrophils, 53
Nicotinamide-adenine-dinucleotide (NAD), 144, 172
Nicotinic acid, 171
Nisin, 9, 12, 15, 18, 20, 21, 116, 129, 130, 180, 181
Nonconventional dairy substrates, 155
Non-cultivable state, 57
Non-dissociated acids, 166
Non-enzymatic methods, 273
Non-invasive syndromes, 54
Non-pathogenic strains, 64
Non-thermal technologies, 55, 126, 271
Non-volatile components, 155
Norovirus, 64, 69
Novel
    hurdle technologies, 13
        high hydrostatic pressure (HHP), 13
        other novel technologies, 29
        pulsed electric field (PEF), 19
        thermosonication, 22
        ultrasonication, 22
    probiotic foods, 155

technologies, 13, 14, 19, 26, 28
Nucleic acid hybridization, 57
Nucleophile, 219, 220, 356

O

Obligate
  heterolactic species, 172
  homofermentative, 142
Ogataea polymorpha, 304
Oleic acid (OA), 240, 242, 255, 263, 321,
    329, 331, 335, 337, 344
Oleuropein, 258
Oligomerization, 12
Oligomers, 292, 409
Olive oil (OO), 181, 221, 233, 332
Onopordum turcicum, 360
Opuntia ficus indica, 212
Oral cavity, 142, 232
Oreochromis niloticus, 65, 67
Organic
  acids, 12, 13, 55, 107, 110, 115, 129, 134,
    140, 144, 171, 221, 223, 290, 292, 304,
    305, 381, 403, 407, 409, 412
  compounds, 68, 232, 296, 319, 408
    derived from lignin (CDL), 408
  fraction of municipal solid wastes
    (OFMSW), 399, 400
  materials, 250, 263
  solvents, 250, 264, 265, 319, 342, 412
Organoleptic
  characteristics, 141
  properties, 137, 138, 155
Oscillatory pulse, 19
Osmotic concentration, 4
Oxidative stress, 331, 332
Oxide-reduction potential, 4, 166
Oxyanion cavity, 220
Oysters, 64, 72
Ozone treatments, 29

P

P. expansum, 24, 31, 221
Paecilomyces variotii, 308
Palm oil (PO), 259, 336
Palmitic, 263, 307, 321, 335–338, 380
  acid (PA), 263, 267, 307, 323, 329,
    333–335, 344, 356, 367–370, 380
  oleic palmitic (POP), 335, 336, 338

oleic stearic (POS), 335–338
Panax ginseng, 10, 13
Pantothenic acid, 171, 380
Parenteral nutrition (PN), 331, 332
Pasteurization, 6, 14, 19, 24, 25, 31, 94, 95,
    106, 124, 271
Pathogen, 19, 45, 53–55, 63, 65, 69, 70, 73, 75
  contamination, 48
  detection, 45
Pathogenic
  bacteria, 45, 48, 49, 53, 63, 69, 70, 94,
    106, 109, 110, 140, 146
  growth, 11
  microbiota, 47
  microorganisms, 14, 47, 85, 128
  organisms, 147, 171
Pathogenicity, 63
Pathogens, 4, 5, 11, 13, 18, 31, 48, 53–56,
    61, 62, 66, 67, 69, 74, 86, 95, 97, 105,
    112, 115, 117, 124, 146, 147, 150, 168,
    173, 174, 253, 290, 294, 296–298, 358
Pathotypes, 49, 56
Pectic substances, 308
Pectinases, 224, 273, 415
Pediocin, 129, 130, 181
Pediococcus, 115, 142, 167, 171, 172, 181,
    302, 303
  acidilactici, 18
Penicillium
  canesense, 222
  corylophilum, 8
  expansum, 30, 31
  pinophilum, 218
  simplicissmum, 308
  strains, 218
Pentose, 172, 400, 404, 412
  phosphate, 235
Peptidases, 172, 173
Peptides, 12, 18, 21, 25, 66, 74, 116, 173,
    178–181, 214, 293, 303, 304, 356, 358,
    359, 368, 369
Periplasmic enzymes, 6
Permeability, 6, 19, 181, 239, 260, 262, 271,
    291
Permeabilization, 12, 271
Peroxidase, 307
Peroxidation, 331, 332, 383, 384
  values, 383

Pesticide-degradation, 76
PH
  homeostasis, 11
  value, 11, 165
Phagocytosis, 66
Pharmacovigilance, 387
Phenolic, 253, 263, 272–274
  acids, 254, 263, 272
  compounds, 22, 129, 253–255, 257, 258,
    260, 261, 263, 269, 272–274, 276, 290,
    292, 303, 305, 381, 402, 403, 409, 411
Phenols, 130, 251, 252, 303, 304, 381, 387
Phenylalanine, 235–237, 241, 243, 296
Phenylethanol, 233, 235
Phenylpyruvate, 235–237
  decarboxylase, 235
Phosphatases, 224
Phosphoenolpyruvate, 235
Phosphoketolase, 142, 172
Phospholipids (PL), 21, 255, 262, 323, 324,
  328
Phosphosphingolipids, 324
Photobacterium phosphoreum, 9
Photochemical action, 26
Photoinhibition, 357
Photolyase, 26
Photomorphogenesis, 357
Photons, 26
Photosensitive compounds, 125
P-hydroxybenzoic acids, 302
P-hydroxyphenyacetic acids, 305
Physicochemical characterization, 155
Physiological characteristics, 141, 210
Phytic acid, 289, 293, 305
Phytoalexins, 134, 135
Phytochemicals, 254, 381–383
Phytoestrogens, 382
Pichia pastoris, 306
Plant
  cells permeabilization, 271
  extracts, 12, 129
  proteases, 357–359
Plantaricin, 181
Plant-derived
  coagulants, 371
  compounds
    biological activity mechanisms, 129
  preservatives, 135

Plasma membrane, 13, 29, 115, 181, 270
Pleurotus
  ostreatus, 308
  pulmonaris, 308
Pluricellular organisms, 129
PO mid fraction (PMF), 336, 337
Polar molecules, 270
Polyamino saccharide, 214
Polygalacturonase, 307
Polyhydroxy, 292, 324
Polyhydroxyalkanoates (PHAs), 400, 412
Polyketides, 320
Polymeric materials, 108
Polymerization, 178, 213, 408, 410
Polymers, 112, 134, 212, 214, 292, 344,
  403, 412, 413, 416
Polypeptide chain, 219
Polyphenolic compounds, 292, 299, 300, 383
Polyphenols, 74, 133, 254, 256, 257, 263,
  264, 269–274, 276, 290, 293, 300, 301,
  308, 382, 383, 385
Polypropylene, 32, 240
Polysaccharides, 130, 154, 250, 272, 340,
  381, 397, 400, 402, 410, 417
Polystyrene (PS), 213, 214
Polyterpenes, 325
Polyunsaturated fatty acids (PUFA), 67,
  255, 277, 321–323, 329–334
Polyurethane, 212, 218, 241, 413
Pomegranate, 269, 300, 377–387
  adaptation, 378
  origin, 378
Poria subvermispora, 308
Porosity, 213, 270
Potassium
  metabisulfite, 8, 11
  sorbate, 10, 13, 20, 22
Prebiotics, 138, 153, 173, 175, 178, 289, 333
Prephenate, 235, 236
  dehydratase enzyme, 236
Preservative fermentation, 116
  economical advantage, 116
Pretreatment, 22, 396, 397, 401–404, 407,
  414, 415, 417–419
  scale-up pretreatment, 407
Probiotics, 66, 137, 138, 145, 146, 148–152,
  156, 164, 171, 173–175, 177, 182
  bacteria, 147, 154

characteristics, 137, 153, 155
foods, 138, 145, 146, 153, 155, 176
microbial and functional aspects, 145
  desirable probiotic characteristics, 149
  improve health effects, 150
  microorganisms, 148
  probiotics action mechanisms, 146, 148
microorganisms
  anti-inflammatory effect, 177
products, 137, 138, 145, 148, 151, 154,
  155, 176
  cereal-based functional products, 153,
    154
  cereals use, 154
  non-dairy and cereal, 151
Procarcinogens, 147
Procyanidins, 381
Produce
  production, 45, 46, 55
    global situation, 46
  quality, 56
Proinflammatory cytokines, 177, 386
Propionic acids, 115, 321
Protease, 173, 179, 215, 222, 224, 291, 298,
  308, 355–359, 365–368, 370, 371
  cocktail, 273
  inhibitors, 290, 291
Protected designation of origin (PDO), 357,
  359
Protein
  adducts, 324
  kinase C (PKC), 294, 310
Proteolytic
  activity, 356–358, 366–368
  bacteria, 68
  capacity, 369
  degradation, 224
  enzymes, 179, 379
  system, 178
Proton motive force, 260, 261
Protozoa, 105
Pseudogymnoascus roseus, 222
Pseudomonas
  aeruginosa, 74, 253, 259
  flourescens, 47
  putida, 74
Psychrophilic aerobic counts, 203
Psychrotrophic bacteria, 19

Pulse
  electric field technology, 271
  light, 107, 108
Pulsed
  electric field (PEF), 16, 19–21, 108, 127,
    265, 270, 271
  light treatments, 26
  x-ray, 127
Punicalagins, 381, 384
Pyknotic nuclei, 65
Pyrimidine
  dimmers, 26
  nucleotides, 108

Q

Qualified presumption of safety, 180
Quality control programs, 97, 98, 100
Quercetin, 257, 259, 273, 383
Quinic acid, 292, 365
Quinolones, 64
Quinones, 325

R

Racemic mixtures, 223
Radiofrequency (RF), 31, 32
Radioisotopes, 31
Radula symbol, 117
Raffinose, 216
Raw materials, 85–87, 89, 91, 95, 96, 99,
  100, 109, 111, 138, 139, 153–155, 164,
  166, 223, 231, 260, 275, 338, 411, 412
  additive raw materials, 89
  meat raw materials, 87
Ready to eat (RTE), 86, 88, 90–92, 94, 95,
  123
Reflux method, 259
Renin, 173
Reuterin, 16, 18, 129, 130
Rheological evaluations, 155
Rheum palmatum, 9, 13
Rhizopus
  arrhizus, 224
  oligosporus, 305, 307
Rhizosphere, 112
Ribosomes, 14, 180
Ricin, 308
Rosmarinus officinalis, 13
Rotavirus, 69, 169, 170, 175, 176

S

Saccharification, 402, 417
Saccharolipids, 320
Saccharomyces
 boulardii, 149, 175
 cerevisiae, 111, 217, 232, 236, 237, 240, 241, 304
 pastorianus, 111
Saccharose, 217
Salmonella, 6, 7, 10, 13, 16, 23–25, 27, 48–51, 55, 56, 58, 63–65, 67, 69, 71, 73, 75, 86, 87, 95, 105, 113, 114, 168, 169, 175, 176, 201, 253, 257, 259, 276
 enterica, 13, 49, 65
 serovars, 65
 species, 63
 tiphimorium, 170
Salmonellosis, 49, 63, 73
Salvia lavandulifolia, 13
Sanitation standard operating procedures (SSOP), 98, 100
Saponins, 288–291, 307, 308, 365
Saprophytic microbiota, 47
Saturated fatty acid (SFA), 255, 321, 329, 330, 333
Scanning electron microscope (SEM), 212, 213
Scheffersomyces stipites, 304
Schizosaccharomyces pombe, 236, 304
Schwanniomyces occidentalis, 304
Sclerotinia sclerotiorum, 267
Seafood, 11, 61–63, 67–70, 72–75, 77, 132
 aquaculture, 66
 safety, 68
Sepsis, 331, 344
Septicemia, 54, 63–65, 70, 72
Serine proteases (SPs), 298, 356, 358, 368, 371
Serotype, 48, 53, 72
Shewanella
 baltica, 9
 putrefaciens, 65
Shiga toxin producer *E. coli* (STEC), 53
Shikimate pathway, 235
Short-chain fatty acids, 147
Silverleaf, 362, 371
Sodium
 benzoate, 20, 22

 chloride, 18
 dodecylsulfate (SDS), 321, 368–370
 hypochlorite, 7, 25, 109
 tripolyphosphate (STP), 8, 12
Solamargine, 362, 364
Solanum, 291, 357, 359–362, 364–366, 370, 371
 elaeagnifolium, 357, 361, 362, 364, 365, 370, 371
Solid
 bioprocess, 301
 lipid
  nanoparticles (SLN), 338, 342, 343
  particle (SLP), 123, 342
 state, 225
  culture (SSC), 267, 274, 275
  fermentation (SSF), 210, 212, 213, 215, 218, 300–302, 304, 305, 308, 416
 filamentous fungi, 215
 substrates, 211
Sonochemical products, 22
Soxhlet extraction, 264
Soybean oil (SO), 61, 302, 328, 331, 332, 403, 405, 406
Spectrophotometric techniques, 301
Sphingolipids, 320, 324
Spoilage
 microorganisms, 3, 12
 organisms, 5
Spore reduction, 15
Spray drying, 343
Squamous cell, 386
Stachyose, 216
Staphylococcal enterotoxin, 71, 74
Staphylococcus aureus, 74, 87, 105, 113, 168, 201, 252, 253, 258, 259, 380
Steam distillation, 250, 264
Stearic, oleic, stearic (SOS), 335–338
Sterols, 173, 255, 291, 324, 365
Strain pathogenicity, 63
Streptococcus, 115, 142, 148, 149, 167, 171, 172, 175, 252, 305
Stress-regulation genes, 108
Structured lipids (SL), 320, 323, 326–331, 334, 344, 345
 functional foods applications, 330
  cocoa butter alternatives (CBA), 335, 336, 338

commercial products, 344
conventional emulsions and nanoemulsions, 339
enteral and parental feed, 330
filled hydrogel particles, 343
human milk analogs, 333
lipid base delivery systems, 338–343
solid lipid particles (SLPS), 342
preparation, 325
enzyme-catalyzed synthesis, 326
functional and nutraceutical foods, 326
Styrene, 213, 231, 233
Sublethal damage, 271
Submerged fermentation (SF), 209–211, 217, 218, 225, 417
Substrates molar ratio, 330
Succinic acid, 411
Sugarcane, 6, 11, 396
Sulfamethoxazole, 65
Supercritical fluid extraction (SFE), 251, 265, 267, 272
Superparamagnetism, 214
Surface
adhesion fermentation (SAF), 210
tension, 269, 272, 339
Symbiotic microorganisms
anti-inflammatory effect, 177
Synergistic effect, 6, 11, 12, 18, 22, 25, 104, 107, 305, 309
Synthesize polymers such as poly-e-caprolactone (BioPCL), 413

T

Taenia solium, 380
Tail necrosis, 66
Tannases, 308
Taxonomic groups, 115, 358
Tepache, 166
Termitomyces clypeatus, 217
Terpene units, 325
Terpenoids, 319
Tetradecanephorbol 13-acetate (TPA), 381, 384, 385, 387
Tetragenococcus, 115
Therapeutic applications, 74
Thermal
inactivation kinetics, 11
treatment, 6, 14, 17, 19, 21, 23, 25, 26, 28

Thermochemical pretreatment, 407, 414
Thermodynamics, 134
Thermolabile molecules, 265
Thermophiles, 7, 172, 305
Thermophilic
fermentation, 67
growth, 115
Thermoplastic polymer, 213
Thermosonication, 23–25
Thermostable direct hemolysin, 70
Thiamine, 171, 380
Thymol, 12, 130, 253, 262
Thymus
mastichina, 13
vulgaris, 9, 13
Tocopherols, 132, 256
Toxic
compounds, 68, 174, 287–290, 302
peroxidation, 262
Toxicity, 62, 109, 129, 212, 214, 239–241, 256, 265, 272, 275, 294, 297, 362
Training process, 194
characteristics, 197
trainee characteristics, 195
training, 194, 195
Trametes versicolor, 306, 307
Trans-cynnamaldehyde, 10, 12
Transesterification, 218, 223, 345
Transmembrane
potential, 271
pressure (TMP), 410
Triacylglycerols (TAG), 323, 326, 328, 330, 331, 333–338, 340, 342, 345
Trichomes, 53
Triglycerides, 173, 219, 220, 222, 223, 255
Trimethoprim, 65
Trisaccharide, 362
Trompillo, 371
Trypsin inhibitors, 289, 303
Typhimurium, 24–29, 49, 257

U

Ubiquinones, 325
Ulcerative colitis (UC), 177
Ultrafiltration (UF), 409, 410
Ultrasonic waves, 107, 269
Ultrasonication, 22–25
Ultrasound, 22–25, 107, 265, 269

assisted extraction (UAE), 265, 269
  waves, 269
Ultraviolet (UV), 26, 381, 383
  light, 55, 56, 384
United States of Agriculture's Food Safety
  and Inspection Service (USDA-FSIS), 75
Urban emissions, 105
Uronic acids, 408

V

Vaccinium angustifolium, 29
Vegetable
  by-products, 249–251, 255, 256, 259,
    260, 263, 265, 269, 274, 275, 277
  mesocarp, 256
Vegetarian probiotic products, 151
Vegetative cells, 6, 31
Verotoxic *E. coli* (VTEC), 74
Vibrio
  cholerae, 71, 72
  harvey, 74
  parahaemolyticus, 63, 70, 71, 76, 105
  vulnificus, 72
Vibriosis, 70

Viruses, 31, 64, 66, 69, 105, 107, 115, 146,
  257, 261
Viscosity, 269, 272, 330, 340
Volatile
  compounds, 155, 233, 408
  flavor compounds, 155

W

Water
  activity (Aw), 4, 7–9, 27, 70, 87, 99, 135,
    264, 330
  distillation, 250, 264

X

Xanthophyllomyces dendrorhous, 304
Xylanase, 215, 224, 273
  enzymes, 416, 417
Xylitol, 396, 397, 407, 411, 412, 414, 415,
  417, 418
Xylooligomers, 409
Xylooligosaccharides (XOS), 396, 397, 402,
  407, 410, 417, 418
Xylopia aetiopica, 7, 11

*For Product Safety Concerns and Information please contact
our EU representative GPSR@taylorandfrancis.com Taylor & Francis
Verlag GmbH, Kaufingerstraße 24, 80331 München, Germany*

T - #0147 - 230425 - C478 - 234/156/21 - PB - 9781774634875 - Gloss Lamination